EMBANKMENT DAMS

James L. Sherard Contributions

Edited by Sukhmander Singh

Geotechnical
Special Publication No. 32

Sherard Memorial Volume

Published by the
American Society of Civil Engineers
345 East 47th Street
New York, New York 10017-2398

ABSTRACT

This volume is dedicated to the memory of Dr. James L. Sherard, who made many contributions to embankment dam engineering. Twenty-one of his papers related to this topic were selected for this memorial volume. A large portion of these are considered seminal papers of great value to the profession. The papers are preceded by an introduction prepared by either a co-author or someone closely associated with the work. Papers of lasting value as references on subjects such as seismic aspects, dispersive clays, hydraulic fracturing and filters have been included. His last two papers on the concrete face rockfill dam are the last two papers in this memorial volume.

Library of Congress Cataloging-in-Publication Data

Sherard, James L.
Embankment dams: James L. Sherard contributions/edited by Sukhmander Singh and Edward B. Perry.
p. cm.—(Geotechnical special publication; no. 32)
Prepared for release at the ASCE Stability and Performance of Slopes and Embankments II Conference held June 1992 in Berkeley.
"Sherard memorial volume."
Includes bibliographical references and indexes.
ISBN 0-87262-897-3
1. Earth dams. I. Singh, Sukhmander. II. Perry, Edward B. (Edward Belk) III. Title. IV. Series.
TC543.S515 1992
627'.83—dc20 92-15661
CIP

Library of Congress Catalog Card No: 92-15661
ISBN 0-87262-897-3
Manufactured in the United States of America.

Preface

The Embankment Dams and Slopes Committee of the Geotechnical Engineering Division of ASCE conceived, at a meeting held at Northwestern University in June 1989, to prepare a memorial volume in honor of Dr. James L. Sherard. In June, 1991, at its meeting in Denver, it was decided to have the volume ready for release at the time of the ASCE Stability and Performance of Slopes and Embankments - II Conference at Berkeley, California, in June 1992. The following members constitute the sub-committee on the Sherard memorial volume:

Prof. Sukhmander Singh, Santa Clara University; Chairman
Prof. Iraj Noorang, San Diego State University
Prof. Edward B. Perry, U.S. Army Engineer Waterways Experiment Station
Mr. Richard L. Volpe, Earth Sciences Associates

A substantial portion of Dr. Sherard's professional career was devoted to advancing the state-of-the-art of embankment dams. Twenty-one of his papers related to this topic were selected for this memorial volume. A large portion of these are considered seminal papers of great value to the profession. The papers are preceded by an introduction prepared by either a co-author or someone closely associated with the work.

The late Edward C. Pritchett had just enthusiastically taken up the task of contacting some of the co-authors when he passed away. Thanks to the efforts of several individuals who responded promptly to our requests for introductions and other memorabilia, we were able to meet the target date. Mrs. Elizabeth Sherard and Dr. Sherard's daughter Claire were extremely helpful and provided photographs and other details. Especially, J. Barry Cooke's keen interest, guidance and active participation provided inspiration to finish the work on time.

Shiela Menaker of ASCE provided remarkable management of the publication of the book. We also acknowledge permission to use copyrighted material, given by ASTM, The Asphalt Institute, John Wiley & Sons, Institution of Civil Engineers (Geotechnique), International Congress on Large Dams, and Water Power and Dam Construction for articles contained in their publications.

Sukhmander Singh, M. ASCE
Editor
Edward B. Perry, M. ASCE

James L. Sherard

1925 - 1987

Biography

James L. Sherard was born in San Francisco, California on September 20, 1925. His engineering education began at the University of California at Berkeley, where he graduated in 1946 with a B.S. in Civil Engineering. At Harvard, two years later, he received his Masters Degree majoring in Soil Mechanics and Foundation Engineering.

Under the guidance of Karl Terzaghi, Jim received his doctorate from Harvard in 1952. His thesis on earth dam performance was the result of four years of academic and practical study with organizations related to earthfill dams. Consequently, his career covered the broad field of embankment dam engineering.

In 1953, Jim approached Woodward-Clyde with a proposal to open an office in Denver, Colorado, having known Woodward-Clyde members while a student at U.C. Berkeley. He opened the office and several more, and the organization became Woodward-Clyde-Sherard. However, running offices was not for Jim. He wanted to write a book on embankment dams, and was assigned full time to that task. For five years he researched literature, visited dams and talked with organizations and dam engineers. Earth and Earth-Rock Dams" was published in 1963 by Wiley and Sons, written by Sherard and co-authored by members of the firm. The book is a comprehensive and valuable reference on earth, earth core rockfill and impervious face rockfill dams. Although Jim did some consulting on dams while with Woodward-Clyde-Sherard, he decided to leave the technical and management duties to the firm, and he became an individual consultant on dams in 1967.

Jim's dam consulting practice developed rapidly, and soon his assignments included projects across the U.S. and in many foreign locations. Individual assignments came from owners, engineering firms and contractors in countries such as Brazil, Colombia, Greece and Venezuela. For instance, in Venezuela he consulted on about 30 dams during his 20 years of consulting practice.

Jim was a member of the Owner's Board of Consultants on major hydro projects such as Guri-Venezuela, Xingo-Brazil, Guavio-Colombia and Lubuge-China. His intensity and perspective were invaluable not only to the embankment dam aspects, but to the other civil engineering features of a hydro-electric project as well.

His busy consulting practice did not keep him from writing investigating papers into particular elements of embankment dams of all types. His doctorate thesis was on earth dams, other papers covered a wide variety of topics such as asphalt impervious face rockfill dams, seismic aspects, dispersive clay, hydraulic fracturing and filters. His last two papers were on the concrete face rockfill dam. His curiosity and intense investigative interest resulted in papers of permanent value as recognized by his peers in ASCE.

Papers were published primarily in ASCE-GT Journals. ASCE's recognition of Jim Sherard's work, together with that of his co-authors, began with the award of the Arthur M. Wellington Prize in 1973 for the paper "Hydraulic Fracturing in Low Dams of Dispersive Clay." He was awarded the Norman medal twice: in 1985 for "Filters for Silts and Clays" and in 1986 for "Hydraulic Fracturing in Embankment Dams." In 1991 he was awarded, posthumously, the Middlebrooks Award for the paper "Critical Filters for impervious Soil." He was to give the 1988 Terzaghi Lecture.

Jim received the special Service Award of the ASTM in 1977 for his work as co-chairman and co-editor of ASTM SPT 623 "Dispersive Clays, Related Piping, and Erosion in Geotechnical Projects." Jim was Co-Chairman of the 1985 ASCE Symposium on the Concrete Face Rockfill Dam.

Just preceding and during his terminal illness he was working on two papers which were to be in a GT Journal together with the many discussion of the Symposium Papers. The GT Journal became a Special Memorial Issue: In Memorium, James L. Sherard, Fellow, ASCE, 1925-1987: a tribute by ASCE.

A later special recognition came from U.C. Berkeley Geotechnical Society in 1991. In accordance with the tradition of a " Distinguished Lecture Series" on the day of the graduation banquet, "The James L. Sherard Memorial Lectures" was initiated in 1991. The first lecture was given by Barry Cooke, his friend and fellow UCB Alumnus.

As an international dam consultant Jim spent evenings at job sites and became well acquainted with fellow consultants and clients. In many parts of the world Jim is fondly remembered, not only for his value to the project, but for his bridge, tennis and his warm and enthusiastic presence. For those who did not have the pleasure of knowing him personally, Jim will be long remembered by reference to his technical papers.

J. Barry Cooke
April 5, 1992

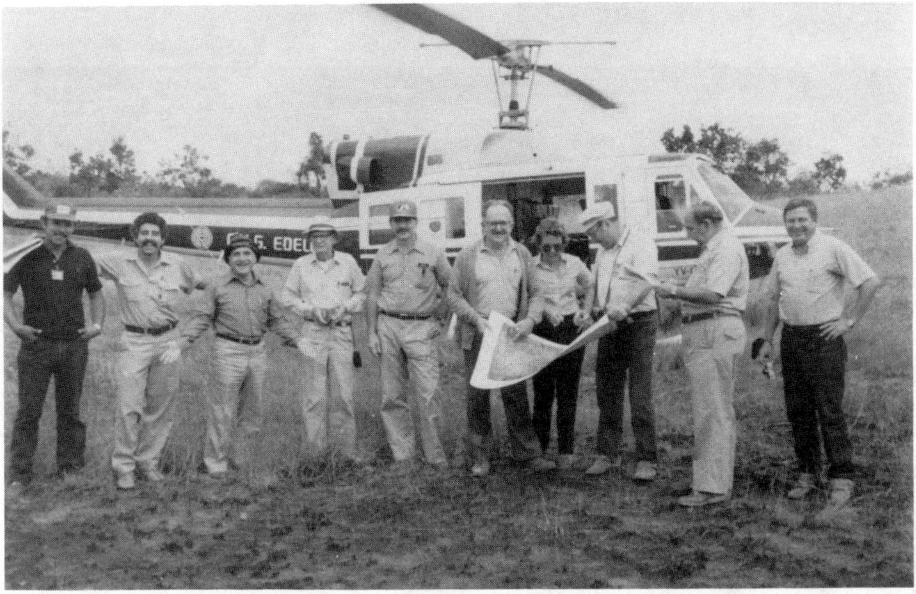

CONTENTS

GEOTECHNICAL SPECIAL PUBLICATIONS

1) TERZAGHI LECTURES
2) GEOTECHNICAL ASPECTS OF STIFF AND HARD CLAYS
3) LANDSLIDE DAMS: PROCESSES RISK, AND MITIGATION
4) TIEBACKS FOR BULKHEADS
5) SETTLEMENT OF SHALLOW FOUNDATION AND COHESIONLESS SOILS: DESIGN AND PERFORMANCE
6) USE OF IN SITU TESTS IN GEOTECHNICAL ENGINEERING
7) TIMBER BULKHEADS
8) FOUNDATIONS FOR TRANSMISSION LINE TOWERS
9) FOUNDATIONS AND EXCAVATIONS IN DECOMPOSED ROCK OF THE PIEDMONT PROVINCE
10) ENGINEERING ASPECTS OF SOIL EROSION DISPERSIVE CLAYS AND LOESS
11) DYNAMIC RESPONSE OF PILE FOUNDATIONS—EXPERIMENT, ANALYSIS AND OBSERVATION
12) SOIL IMPROVEMENT—A TEN YEAR UPDATE
13) GEOTECHNICAL PRACTICE FOR SOLID WASTE DISPOSAL '87
14) GEOTECHNICAL ASPECTS OF KARST TERRAINS
15) MEASURED PERFORMANCE SHALLOW FOUNDATIONS
16) SPECIAL TOPICS IN FOUNDATIONS
17) SOIL PROPERTIES EVALUATION FROM CENTRIFUGAL MODELS
18) GEOSYNTHETICS FOR SOIL IMPROVEMENT
19) MINE INDUCED SUBSIDENCE: EFFECTS ON ENGINEERED STRUCTURES
20) EARTHQUAKE ENGINEERING & SOIL DYNAMICS (II)
21) HYDRAULIC FILL STRUCTURES
22) FOUNDATION ENGINEERING: CURRENT PRINCIPLES AND PRACTICES
23) PREDICTED AND OBSERVED AXIAL BEHAVIOR OF PILES
24) RESILIENT MODULI OF SOILS: LABORATORY CONDITIONS
25) DESIGN AND PERFORMANCE OF EARTH RETAINING STRUCTURES
26) WASTE CONTAINMENT SYSTEMS: CONSTRUCTION, REGULATION, AND PERFORMANCE
27) GEOTECHNICAL ENGINEERING CONGRESS
28) DETECTION OF AND CONSTRUCTION AT THE SOIL/ROCK INTERFACE
29) RECENT ADVANCES IN INSTRUMENTATION, DATA ACQUISITION AND TESTING IN SOIL DYNAMICS
30) GROUTING, SOIL IMPROVEMENT AND GEOSYNTHETICS
31) STABILITY AND PERFORMANCE OF SLOPES AND EMBANKMENTS II (A 25-YEAR PERSPECTIVE)
32) EMBANKMENT DAMS - JAMES L. SHERARD CONTRIBUTIONS

Introduction to Paper 1

Rockfill Dams: Steel-Faced Dam

by James L. Sherard
Transactions, American Society of Civil Engineers,
Vol 125, Part 2, 1960, pp. 465-472

Prepared by J. Barry Cooke

Though the steel-faced dam is no longer an economic type, the paper describes advantages of the upstream membrane type of dam and the cost disadvantages of the steel-faced dam.

The paper illustrates his early interest not only in earth dams, but in rockfill dams. The paper was in the only Special ASCE Transaction volume on a specific subject, "Symposium on Rockfill Dams", 711 pages. I was Chairman of the Symposium Committee and could write the introduction to the paper. Jim visited PG&E (JBC) in connection with his thesis on cracking in earth dams. He visited PG&E's 1850-80 gold mining dams.

AMERICAN SOCIETY OF CIVIL ENGINEERS

Founded November 5, 1852

TRANSACTIONS

Paper No. 3079

ROCKFILL DAMS: STEEL-FACED DAM

By James L. Sherard,[1] F. ASCE

SYNOPSIS

Speed of construction and structural integrity were the chief advantages gained by the use of a steel-plate facing on a rockfill dam built in Venezuela in 1958. At present, steel-plate facing is seldom considered for use on earth and rockfill dams, although a number of such dams have been built in the past and have seen many years of satisfactory service. The principal reasons for not considering steel facing the dams have been concern about possible corrosion of the steel and cost. However, the advantages offered by steel-plate impervious membranes for dams are such as to justify the consideration of this type of construction under some site conditions. Experience with structures in the Western United States gives weight to this statement.

RIO LAGARTIJO LOW DAM

In recent years the demand for domestic water in Caracas, Venezuela, has grown considerably faster than anticipated, and a severe shortage developed in 1957. During 1957 and 1958, water had to be severely rationed and a supplemental supply had to be brought into the city in tank trucks from sources outside the system. The cost of the supplemental water and the inconveniences

Note.—Published, essentially as published here, in October, 1959, in Civil Engineering, Vol. 29, No. 10, pp. 698-701. Positions and titles given are those in effect when the paper was approved for publication in Transactions.

1. Partner, Woodward, Clyde, Sherard & Associates, New York, N. Y.

caused by the shortage led the Instituto Nacional de Obras Sanitarias (INOS) to accélerate the rate of construction of new storage facilities.

Previously planned for construction in the early 1960's was a large reservoir to be formed by an earthfill dam 164 ft high on the Rio Lagartijo, about 40 miles southeast of Caracas. The size of this dam and the incomplete state of the design studies made it impracticable to attempt its completion in time to alleviate the water shortage anticipated during the 1959 dry season. Consequently it was decided to proceed immediately with the construction of a low dam on the Rio Largartijo some 3,300 ft upstream from the site chosen for the main dam. The low dam was expected to have a useful life of only two or three years. Because of the temporary nature of this structure, it was desirable to construct it as economically as possible, consistent with complete safety. But the designers had to keep in mind the fact that although the dam was planned to have a life of only a few years, it might be used for a much longer period.

In April 1958, the Venezuelan firm of OTEHA, C.A., was commissioned to prepare plans and supervise construction of the "low dam," to have a height of 79 ft. Construction started in May, filling of the reservoir started in September, and the dam was completed in December—all in 1958. The water stored during the rainy season of 1958 provided a sufficient additional supply for Caracas to prevent serious shortage in 1959.

The brief time available for design and construction, and the imminence of the annual rainy season, prohibited consideration of a conventional earthfill dam.

Because of the very limited time available and the fact that no borrowpit explorations or laboratory studies had been made, the engineers concentrated immediately on a rockfill design. Alternate plans were prepared for rockfill dams having upstream impervious membranes of reinforced concrete and plate steel. The contractor made cost estimates for each and concluded that the plate steel was the cheaper.

A section through the dam, as constructed, is shown in Fig. 1. The foundation layer is made up of a dense stratum of river gravel. Underlying the

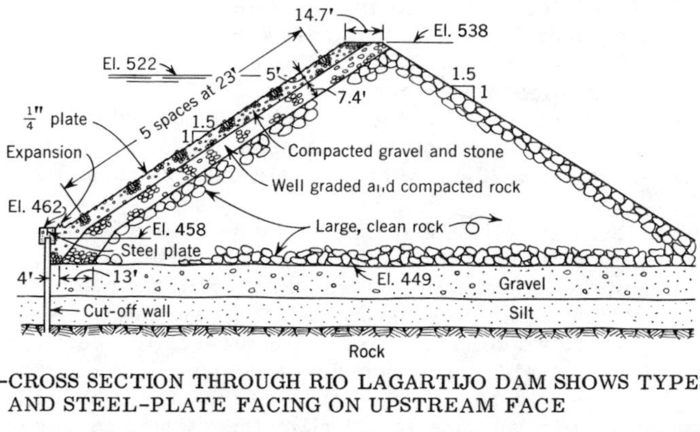

FIG. 1.—CROSS SECTION THROUGH RIO LAGARTIJO DAM SHOWS TYPES OF FILL AND STEEL-PLATE FACING ON UPSTREAM FACE

gravel is a layer of sandy silt. Time was not available to remove the alluvium down to bedrock. Time was not even available for a careful field and laboratory exploration of the characteristics of the silt layer. A few small-diameter Shelby-tube samples were obtained from various points in the silt layer, and some classification tests and unconfined compression tests were performed.

Stability calculations, based on the results of the unconfined compression tests, indicated that the embankment could be constructed with the slopes and height indicated without danger of foundation shear failure. It was expected that considerable foundation settlement would take place during and after construction. However, since the dam was sufficiently flexible to withstand large differential settlements without rupture, the designers were not particularly concerned about this.

A cutoff wall or seepage barrier, to prevent excessive percolation through the foundation, was provided by constructing a 20-in.-wide concrete wall at the upstream toe. This cutoff wall was constructed according to the patented ICOS process. The trench was excavated to bedrock by drilling holes and excavating between them with a clamshell bucket. During the excavation the trench was kept filled with a bentonite slurry, which prevented the walls from caving and eliminated the need for sheathing. After the trench was excavated to bedrock, tremie methods were used to place the concrete, which displaced the bentonite slurry mixture. Vertical pipes were placed in the trench for subsequent grouting of the bottom of the trench, and to permit drilling into the bedrock for further foundation grouting.

Rock used for the embankment was a hard, sound, bulky limestone, with a maximum size of 10 tons. Only clean, large, hard rock was placed in the embankment, which was constructed in layers ranging from 10 ft to 13 ft. It was sluiced continuously with nozzles of the largest size that could be handled by hand. The sluicing was used to consolidate the embankment and to wash the fines into the interstices of the rock embankment. The 7-ft-thick layer of well-graded fine rock on the upstream slope was obtained as quarry-run fines. Gravel for the 5-ft-thick layer came from deposits in the stream bed. Both of these sloping layers were placed after the main embankment was completed. They formed a cushion to prevent high concentration of stress between the upstream steel face and the large individual rocks in the embankment.

Details of the steel-plate connections are shown in Fig. 2. To provide flexibility in the steel membrane, which would allow it to deform without rupturing or developing extensive tensile stresses, transverse expansion joints were provided vertically up the face on 33-ft centers. An expansion joint consists of a simple fold in a steel plate, which permits considerable movement of the steel at the joint. All joints were continuously welded.

The steel facing is anchored into the underlying gravel layer by bolts embedded in concrete, also shown in Fig. 2. These are spaced 33 ft apart horizontally and 23 ft parallel to the slope.

At the upstream toe of the dam, the steel plate is connected to the top of the concrete cutoff wall, using a double-expansion joint. At the abutments the steel plate is sealed into a 20-in.-wide concrete wall extending 3 ft to 5 ft into the rock.

The steel plates, connections and expansion joints were fabricated in Caracas. Large sections of the upstream face were connected with field bolts before completion of the continuous fillet welds united all the plates and connections into a watertight unit. Oddshaped plates to fit the natural contours of the abutment were cut and welded in the field. The steel plates were placed directly on top of the gravel layer, which had been carefully compacted to a plane sloping surface on grade. In the middle of the day the plates became so hot that they were difficult to handle. Work on them was done only during the morning and evening hours, between 5 and 10 a.m. and 4 and 10 p.m. Connecting the plates when they were hot and expanded was thus avoided.

This construction procedure differs from that on several dams in the United States, where steel plate was laid on gravel. During construction of the United States dams, the steel plates were placed on carefully leveled lightweight steel angles. After the plates were fastened to the angles, the space between the plates and the embankment was backfilled with gravel. The angle frames were left embedded in the backfill, the legs acting as anchors for the plates. This method resulted in a slightly smoother and more attractive face than was obtained by placing the plates directly on gravel. About eight weeks were required for completing the construction of the steel face for the Rio Lagartijo Dam (Fig. 3).

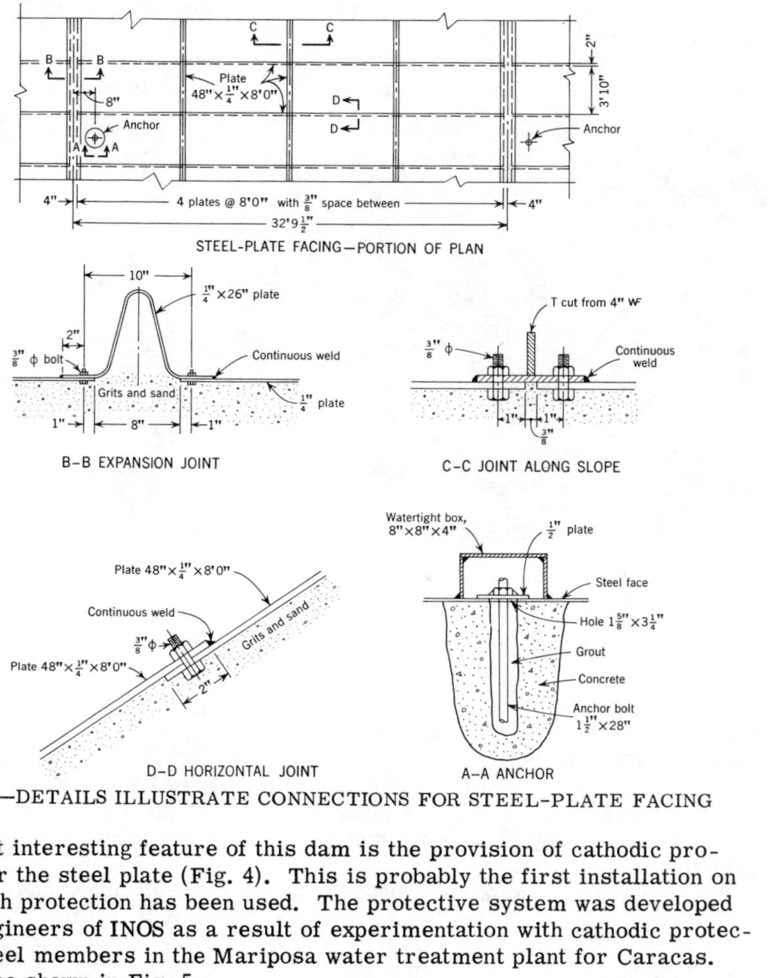

FIG. 2.—DETAILS ILLUSTRATE CONNECTIONS FOR STEEL-PLATE FACING

A most interesting feature of this dam is the provision of cathodic protection for the steel plate (Fig. 4). This is probably the first installation on which such protection has been used. The protective system was developed by the engineers of INOS as a result of experimentation with cathodic protection of steel members in the Mariposa water treatment plant for Caracas. Details are shown in Fig. 5.

The cathodic protection system is operated from 110-v alternating current that is fed into a rectifier. The rectifier is grounded directly on the steel plate with a copper cable near the crest of the dam. The positive lead is through shunts to used steel rails that run vertically down the slope of the dam. The steel rails are separated from the steel face by 2-in. rubber insulators. The shunts provide each steel rail with a potential of about 1 v.

Using this system, the upstream steel face of the dam becomes the cathode and the expendable steel rails the anode. Computations indicate that the steel rails will corrode at a rate of about 100 kg per year.

Experiments made by the corrosion engineers of INOS indicate that more current will flow in the system during the first few months after the cathodic protection system is installed than later. Carbonates will deposit on the steel, providing a thin protective covering and a deterrent to the flow of electricity. Consequently the power necessary for cathodic protection will diminish with time.

Performance Record.--The reservoir was almost completely filled by the end of the 1958 rainy season. The level receded as water was drawn off for the Caracas water supply during the 1959 dry season. The reservoir is again filling during the 1959 wet season.

The dam is in excellent shape. No measurements of embankment settlement have been taken, but settlement is not so great as to be noticeable on the

FIG. 3.—STEEL PLATE TERMINATES AT CREST, NOTE EXPANSION JOINT AND RIDGE MADE BY 4-IN. WIDE FLANGE SPLIT INTO A T

FIG. 4.—STEEL-PLATE FACING FROM CREST, CATHODIC PROTECTION BEING INSTALLED

crest of the upstream face. Even though the steel-plate facing was not painted, and the cathodic protection was not installed until July 1959, rusting of the steel plates during the first year was negligible (Fig. 6). Considerable leakage appeared downstream from the dam. Through the use of dyes and observation by a diver, it was established conclusively that the leakage was through the upper part of the foundation cutoff wall. The steel face was intact and watertight.

The Rio Lagartijo Low Dam, described above, was designed and the construction supervised by the Caracas engineering firm of OTEHA, C.A. Work was carried out under the direction of Dr. Diego Ferrer-Fernandez, M. ASCE. Construction was done by the firm of Technica Constructora C. A. of Caracas.

STEEL PLATE ON U.S. DAMS

While there have been very few dams constructed in the United States with steel-plate upstream faces, experience with these dams has generally been good. Corrosion of the steel plate has not resulted in severe maintenance problems.

Rockfill Skagway Dam was built about 1900 in the Rocky Mountains near Victor, Colo. It has a height of 75 ft. The steel plate on the upstream face has a thickness of 1/2 in. at the bottom, 3/8 in. at mid-height, and 1/2 in. at the top. The facing was "riveted and calked in the same thorough manner as in boiler practice."

In the past 30 years the steel face has been given two complete treatments of cleaning, chipping, buffing and painting. The last of these was in 1940. Concerning its present condition, the owner states:

> "Through the years the steel face has become pitted, but not to the extent to become alarming . . . . We are not able to examine the fill side of the steel, but have examined the portion extending above the fill on the cap of the dam at ground level and find that it has suffered less damage than the water side. . . . The steel face has not received the maintenance it should have but . . . it has stood the years very well."

When it is considered that this steel facing is about 60 years old and that the quality of the steel is probably much less uniform and desirable than that in use today, this is a very good recommendation for the durability of steel plate on the upstream face of a dam.

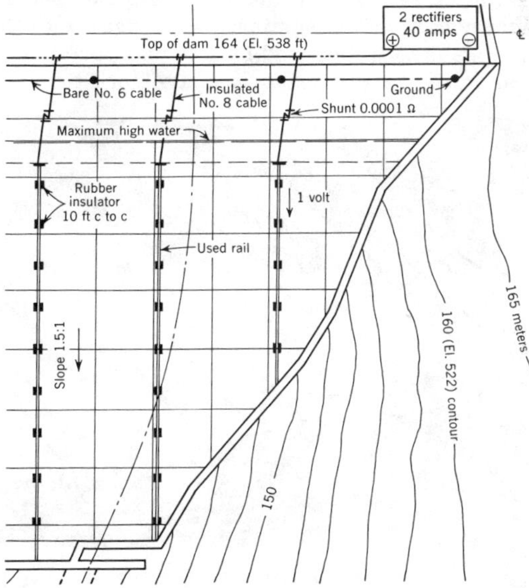

FIG. 5.—CATHODIC-PROTECTION FEATURES ARE INDICATED ON END OF DAM, SEEN IN PLAN. USED RAILS ACT AS ANODES; THE STEEL-PLATE FACING ITSELF SERVES AS CATHODE

Gravel-filled El Vado Dam, with 1/4-in. steel plate on the upstream face, was constructed in New Mexico in 1934. The dam has a maximum height of 175 ft and a 1.5 to 1 upstream slope. The maintenance forces report that "the steel plate is in excellent condition at the present time and with minor inexpensive repair work from time to time should have a life as long as the dam."

Two other earth or gravel dams 80 to 100 ft in height, constructed in the Rocky Mountains in Colorado in the early 1930's, have 1/4-in. steel plate on the upstream face. Detailed information on the maintenance of these dams is lacking but the writer has inspected their steel faces several times in the past 10 years. They appear to be in excellent shape and there are no signs of undue corrosion. From experience with these structures, it can be concluded that corrosion has not been a serious problem on steel-faced dams and that, with reasonable maintenance, the steel face can be considered permanent.

ADVANTAGES OF STEEL-PLATE FACING

Flexibility, watertightness and best distribution of reservoir water pressure are advantages attained by steel-plate facing. The steel plate itself is very flexible and can withstand a great deal of embankment movement without danger of rupture. Through the use of simple and economical expansion joints, of the type used at Rio Lagartijo Dam, the steel-plate facing can be made sufficiently flexible to withstand whatever settlements an embankment or foundation is likely to undergo.

For all practical purposes, steel-plate facing is completely impermeable. When supported on a uniform layer of compacted gravel, the steel plate is not required to resist bending or tensile stresses and acts only as a water barrier or impervious membrane. Used in this fashion, a thin steel plate is capable of acting as an impervious membrane for a reservoir of any depth.

The further upstream the water pressure is resisted by the dam, the greater the stability of an earth or rockfill dam when the reservoir is full. The steel upstream membrane resists pressure on the plane of the upstream face, which is the optimum condition. Assuming proper drainage, there is no problem of decreased slope stability caused by internal pore pressures, following "rapid drawdown" or in the "steady state seepage condition." The dam is not

FIG. 6.—UPSTREAM FACE OF RIO LAGARTIJO DAM AFTER ONE SEASON OF USE

subject to the weakening influence of water seeping through the embankment. Consequently earth dams with steel faces are very different structures from ordinary earth dams and can be constructed on steeper slopes.

Steel-plate facing provides complete protection against the erosive forces of waves.

DRAWBACKS

There are three principal disadvantages in the use of steel plate: (1) the cost, (2) the possibility of damage due to corrosion, and (3) the necessity for drawing down the reservoir level occasionally for inspection and possible repairs. The cost is a major disadvantage, particularly for low dams. At the present time in the United States, it should be possible to put a 1/4-in. steel plate in place on the upstream face of a dam for $20 to $25 per sq yd. For a low dam the cost of the steel plate would be an appreciable part of the total cost of the structure. For high dams, which are thick at the base, the cost of the steel plate in relation to the cost of the embankment becomes much less. Consequently, the higher the dam, the better the competitive position of a steel facing.

Fear of corrosion has probably been the main deterrent to the use of steel plate on dams. However, experience has shown, as discussed above, that the few steel-faced dams in the United States have been remarkably resistant to corrosion. Advances in corrosion engineering should make it possible to reduce this danger to zero, although many engineers would consider this unnecessary.

It would probably be wise to use a steel face on the upstream slope of a dam only where it would be possible, without excessive cost and inconvenience, to completely drain the reservoir at intervals to inspect the condition of the face and repair the steel plate if necessary. On the other hand, any earth or rockfill dam may develop trouble of sufficient magnitude to make it extremely desirable to drain the reservoir completely. Also, some repair work can be done on the steel face by divers, which is an advantage not enjoyed by earth dams with internal clay cores that may develop cracks or leaks. Therefore the necessity for drawing down the water level constitutes only a slight disadvantage for a steel-plate face.

In summary, the writer believes that the data available indicate that steel-plate facing should receive more consideration from dam designers than it has in the past. Some site conditions that make it especially desirable to consider a steel upstream face will now be discussed.

CONDITIONS FAVORING STEEL FACING

No Impervious Soil.--Where impervious soil for an earth core is not available within an economical haul distance, an upstream diaphragm of steel plate may serve the purpose very well.

Possibility of Embankment Cracking.--Recently some earth dams have cracked as a result of differential settlement. This is particularly alarming since there are no reliable criteria by which the possibility of cracking can be analyzed and steps taken to avoid it. The problem is becoming increasingly severe as the height of dams becomes greater and greater. In some dams, intricate and expensive drains and graded filters have been provided to minimize the danger of seepage through possible embankment cracks. The use of an impervious membrane of steel plate would eliminate all danger of failure caused by excessive seepage through embankment cracks.

Economy with Hard Rock.--Where the abutments and foundation for a dam consist of strong rock, and where an ample supply of strong granular embankment material is available, the steel-plate upstream membrane may result in the most economical design, even if clay is available for an impervious core. If a clay core is used, the embankment will probably be constructed with a slope of 1 on 2 1/2 upstream and perhaps 1 on 2 downstream. If the embankment is constructed using only the strong granular material available and an upstream steel face, there will be no danger of embankment sliding, with water held completely outside the structure. Consequently the embankment slopes can be made as steep as practicable from the construction standpoint, which may be 1 on 1.2 or steeper. Obviously the steel-face design would have a much smaller embankment volume and might be considerably cheaper.

Severe Wave Action.--Where severe wave action is expected in the reservoir, and rock for a riprap blanket involves an exceptionally long haul, the cost of any type of suitable wave protection may approximate that of steel plate.

A CURRENT REVIEW OF EXPERIENCE WITH ASPHALTIC CONCRETE IMPREVIOUS MEMBRANES ON THE UPSTREAM SLOPE OF EARTH AND ROCKFILL DAMS

By James L. Sherard*

At some sites where an earth or earth-rock dam is being considered, studies may indicate that the most economical and satisfactory dam would be obtained with a pervious embankment of soil or rock or an embankment of doubtful permeability and an impervious membrane of manufactured material on the upstream slope as the water barrier. Membranes from manufactured material have been constructed satisfactorily with reinforced concrete, asphaltic concrete and plate steel. Reinforced concrete (of Portland cement) has been used most frequently, especially with rockfill embankments. Asphaltic concrete and plate steel have been used with approximately equal frequency -- both have been used on less than a dozen major dams; however, asphaltic concrete has been used more frequently in the last few years. While no dams of any appreciable size have been constructed with the newly developed plastic materials, these appear to offer a fourth possibility which may, possibly in combination with concrete or asphaltic concrete, be used in future years.

When considering the use of any type of membrane for an embankment dam, the designer naturally looks to past experience with similar structures as a starting point and a guide for his efforts. Consequently it is desirable at frequent intervals to summarize the knowledge available concerning the performance history of dams with asphaltic concrete membranes, and to benefit as well as possible from the experience gained by those of our fellow engineers who have had the opportunity (and the courage) to use this relatively new and untried though promising type of dam. This paper summarizes the results of the author's attempts, as part of his practice, to learn as much as possible about this interesting subject.

Before continuing to a discussion of asphaltic concrete water barriers specifically, it is of interest to summarize in general terms some of the principal advantages tc be obtained from the use of upstream impervious membranes on embankment dams. First, by placing the membrane on the upstream slope and providing a suitable drain behind it, we pervent seepage water from the reservoir from entering the embankment. This eliminates the reduction in the stability associated with the development of seepage pore pressures within the embankment and consequently increases the margin of safety against shear failure of both the upstream and downstream slopes. Secondly, the further upstream that the water pressure is resisted by the dam, the greater the mass of the embankment which acts to resist the water pressure. Thirdly, by resisting the water pressure on the plane of the upstream slope, the resultant total force exerted by the reservoir on the dam is directed downward into the foundation with as great an inclination as possible, which gives the optimum condition of stability from the standpoint of the foundation. These actions combine to give the embankment dam with an upstream impervious membrane a greater margin of safety against shear failure than any other type of earth and rockfill dam. Consequently, for a given safety factor, the embankment slopes can be made steeper and the embankment volume smaller than for dams using rolled impervious earth for the core (or for dams in which a vertical impervious membrane of manufactured material is used).

In addition, the upstream impervious membrane of manufactured material has several other advantages. Some of the more important of these are:

*Woodward-Clyde-Sherard and Associates, Consulting Engineers

1. The impervious membrane is exposed for inspection and repairs. If the water can be pulled down, repairs can be made in the dry and, if necessary, work can be done with divers below the reservoir surface. For dams with rolled earth cores, through which concentrated leaks develop, we normally have to content ourselves with the safe control of the leakage and in extreme cases where the leakage is too great to be tolerated, we have to resort to grouting or other means which are usually cumbersome and expensive.

2. The membrane can serve a secondary function as wave protection whereas for earth dams under certain circumstances, wave protection in the form of rock riprap can be a relatively expensive item approaching the cost of an impervious membrane.

3. The membrane of manufactured material is not subject to progressive erosion under the action of water which breaks through a concentrated leak as is a rolled earth core ("piping" (Further, with suitable design details, manufactured membranes can be made to be self-sealing).

4. By bulkheading the impervious membrane and underlying drainage layer into sections, the locations of leaks which do develop can be determined which may be very difficult in dams with earth cores.

A dam with an impervious membrane on the upstream slope is quite a different animal from the earthfill dam or the rockfill dam with a rolled earth core. In the author's opinion, the advantages which can be obtained with this kind of dam have not received the recognition which they deserve. This is due primarily to the natural reluctance in the profession to use relatively "untried" methods on a structure where the failure is capable of such a disaster at a time ir the development in the art of dam building when a rolled earth dam can be made essentially "fool-proof". There is also the ever-present and very real problem of deterioration of the membrane of manufactured material; whereas the basic constituents comprising a rolled earth core can be considered as the final product of nature's process of deterioration and consequently will not be modified during the lifetime of a reservoir project. Nevertheless, there ar sufficient advantages offered by the use of manufactured membranes such that dams using this design will be constructed with more frequency in future years than in the immediate past period and for major dams as well as smaller structures.

Some of the circumstances which may favor the use of an impervious membrane of manufactured material are as follows:

1. No Impervious Soil Available. Where there is no impervious soil for an earth core availab within an economical haul distance, obviously a manufactured impervious membrane shoul be considered. This situation frequently exists in the high mountain country where the litt soil which is available is usually of a basically pervious nature.

2. Economy With Hard Rock Foundation. Where the abutments and the foundations for a dam consist of strong rock and where there is an ample supply of strong granular embankment material available, the upstream impervious membrane of manufactured material may result in the most economical design even if fine-grained soil is available for an impervious core. If a rolled earth core is used, the embankment will probably have to be constructed with a slope of 2.5:1 upstream and perhaps 2:1 downstream. If the embankment is constructed using only the strong granular soil or rock and an upstream membrane of impervious manufactured material, there is no danger of shear failure or excessive deflection of the crest and the slopes can be made as steep as practicable from the construction standpoint. The higher the dam, the greater the advantage of the upstream membrane.

3. Where Wave Protection is Costly. Where severe wave action is expected in the reservoir and rock for a riprap blanket involves an exceptionally long haul, the cost of any type of suitable wave protection may approximate the cost of an impervious upstream membrane which could also serve as wave protection.

4. Danger of Embankment Cracking. Recently, some large dams have cracked as the result of differential settlement. This is particularly alarming to the earth dam designer since there are no reliable criteria by which the possibility of cracking can be analyzed and steps taken to avoid it. The problem is becoming increasingly severe as the height of dams becomes greater and greater, and these dams are usually located in narrow rockwalled canyons. In some dams, in which the designers were concerned about possible cracking, intricate and expensive drains and graded filters have been provided to minimize the danger of seepage through the embankment cracks. The use of a flexible impervious membrane on the upstream slope essentially eliminates danger of failure caused by excessive seepage through embankment cracks.

5. Limited Construction Time in Bad Weather. Using conventional methods of moisture-density control, it is not possible to compact most fine-grained soils while it is raining or to get much work accomplished in wet weather. Therefore, in circumstances where long rainy seasons exist and/or where the construction period is limited, a dam constructed of pervious soil or rock which can be placed in wet weather with a manufactured impervious membrane may be the most economical and satisfactory solution.

6. Extensive Grouting Required. If the foundation requires an extensive and/or time-consuming grouting program, the upstream impervious membrane may offer a considerable saving in construction time since the grouting can be carried out at the same time that the embankment is being constructed, whereas under ordinary conditions, a dam with a conventional rolled earth core cannot be started until the grouting is essentially completed.

7. Expensive Appurtenant Structures. Where the outlet conduit or other appurtenant structures are imbedded within or abutted against the embankment, the length and the cost of these structures is often directly related to the width of the base of the dam. Where these appurtenant structures have appreciable cost with respect to the embankment, the most economical design for the total structure may be obtained with a dam with an upstream impervious membrane which has a minimum base width.

While asphalt has been used to surface the upstream slope of quite a few low dams for the purpose of erosion control, the total number of earth and rockfill dams constructed with asphaltic concrete as the impervious membrane is small. A large percentage of these have been constructed in the last decade and the experiences have been so universally satisfactory that there can be little doubt (1) that asphaltic concrete is perfectly satisfactory material for the purpose -- even for high dams, (2) that it offers certain advantages over alternate construction material and (3) that it will be more widely used in future.

Though it is clear that asphaltic concrete can be safely used, until more structures are built and the performance records evaluated, the design of the membrane (i.e., the thickness, mix, drainage provisions, surface protection, etc.,) still must be considered in the nature of a developmental process for high dams. This situation provides a fascinating challenge to the designing engineer. There is little doubt that considerably better and more economical design details for asphaltic concrete membranes for high dams will be developed and used in the next few years.

The following sections of the paper contain a description of the design and construction and of the performance history of 8 dams with asphaltic concrete membranes -- 4 in Algeria, 3 in Germany and 1 in Colorado. The records of these 8 dams constitute a major portion of all the available experience with the use of asphaltic concrete membranes on the upstream slope of high dams. The attached appendix contains a summary of some of the more important dimensions of the dams and the details of the asphaltic concrete membranes.

GHRIB DAM (ALGERIA)

Ghrib Dam in Algeria is the first of the major dams in modern history to be constructed with an impervious membrane of asphaltic concrete. At the time of its conception and design in the early Nineteen Thirties, the dam was a complete and bold break with past precedent. It is still one of the highest dams, (235 ft.), retains the largest reservoir (230,000 acre-feet), and has the steepest upstream slope (steeper than 45°) of any structure with an asphaltic concrete membrane. The successful performance of this dam over more than 20 years of reservoir operation, under extreme conditions of temperature, is one of the strongest supporting elements for the conclusion that asphaltic concrete is eminently suitable for the purpose.

The Ghrib Dam was designed and constructed by French engineers. Tests for the design of the asphaltic concrete mix were carried out in the Shell Laboratories in Holland. The design and construction are described in great detail by Thevenin in Ref. 1. For design purposes it was estimated that the temperature of the outer surface of a slab in the normal summer sun would go as high as 70°C (about 160°F) and it was necessary to provide a membrane which would withstand such temperatures on the steep slopes without dangerous softening and downward creep.

The embankment of the dam was constructed of large, bulky limestone rocks fitted together carefully with cranes in the form of massive dry masonry in such a way that the large rocks were in contact and the small rocks filled the interstices. The final percentage of voids in the embankment was approximately 26%. The total embankment volume was 700,000 cubic meters. Because of the carefully constructed, tightly packed rockfill embankment, the post construction settlement of the dam crest in the first 20 years of operation was only about 12 in. The horizontal component of the movement of the crest during this time was also roughly 12 in. in the downstream direction.

As shown in Table 1, the asphalt concrete membrane was only 12 centimeters (about 4.8 in.) in thickness and this was sandwiched between a 12 centimeter thick lower drainage layer of porous concrete, and a 10 centimeter thick outer protective layer of porous concrete which was reinforced with galvanized steel wire mesh. The steepness of the upstream slope varied from approximately 1:1 at the upstream toe to 0.7:1 (or steeper than 45°) at the crest of the dam. The area of the asphaltic concrete membrane is approximately 13,000 square meters.

Before placing the lower layer of porous concrete, the voids in the upper surface of the rockfill embankment were filled with mortar in order to find the outer rocks together and to provide a smooth surface on which to place the overlying membrane. This mortar was placed by hand labor and rammed into the spaces between the rocks. The porous concrete drainage layer was placed for the purpose of catching and controlling any water which found its way through leaks in the asphaltic concrete membrane. It contained interior open drains which ran vertically down the slope at intervals, discharging into a control gallery at the upstream toe of the dam. The porous concrete was constructed of uniform aggregate and had an average content of air voids of 25%.

The control gallery is a reinforced concrete conduit, founded on bedrock, which runs along the upstream toe of the dam. The interior dimensions are large enough to provide entrance and working space for men. Leakage water which gets past the asphaltic concrete membrane and is caught in the underlying porous concrete layer is discharged into the gallery through pipes. The locations at which the water enters gives some indication concerning the position of the leaks in the membrane. All the dams with asphaltic membranes discussed in this paper have similar control galleries at the upstream toe except the Montgomery Dam in Colorado.

The aggregate for the asphaltic concrete was obtained from crushed hard limestone and dune sand. The finally prepared aggregate with a well-graded material varying from a maximum of about one inch downward with approximately 10% finer than the No. 200 sieve. To this was added a quantity of filler equal to 10% by weight of the aggregate. The content of asphalt binder was 8% by weight.

The asphaltic concrete was prepared hot by methods similar to those used at present. It was transported to the dam in heated cars, dumped on the slope from a movable steel scaffolding and spread by hand labor with rakes. The 12cm thick slab was placed in two layers of equal thickness, and compacted with a smooth steel roller which operated horizontally on the slope. The weight of the roller was adjusted as a function of the slope to give a contact pressure of approximately one ton per linear foot. Each layer was given 20 to 30 passes with the roller.

The only bond between the asphaltic concrete and the underlying porous concrete was a layer of asphalt paint sprayed on the surface. The two layers of asphaltic concrete were fused together in the same manner: hot, pure asphalt was sprayed on the surface of the first layer before the second was placed. This gave a perfect bond between the layers which could only be distinguished later by a slight enrichment of the content of asphalt binder. Samples subjected to bending tests indicated that there was no weak plane at the surface of juncture between the two layers.

Construction joints were treated in the same manner as the joint between the two layers; i.e., they were reheated, painted with hot asphalt and specially compacted. The joints were checked with water pressure tests during construction. Approximately 100 square meters of surface were completed per day on the average.

The asphaltic concrete membrane was bonded to the concrete structures with a flashing consisting of laminated cotton cloth under the asphaltic concrete slab saturated with Flintkote Emulsion.

The outer layer of porous concrete was provided as thermal and physical protection for the asphaltic concrete slab. It was nominally reinforced with steel wire mesh as shown on the sketch (see appendix) and it was also anchored at the top with a reinforced concrete longitudinal beam sunk into the rockfill embankment. This anchorage was provided in an effort to minimize the tendency for the whole upstream membrane to slip down the slope.

The reservoir capacity was so large in comparison to the stream flow and the water use requirements that several years were required for the water surface to reach the maximum level for the first time. In 1939, when the reservoir reached a new high, some rupturing of the asphaltic concrete evidently took place due to differential settlement. This rupturing was manifested by concentrated leakage appearing in the control gallery. No repairs were made and, even though the reservoir level remained at the same elevation, the leakage gradually decreased and, at the end of several months, had practically stopped, indicating a remarkable facility for self-repair in the asphaltic concrete. After 1939, no further leakage developed and at the present time the membrane is completely water tight.

Trouble developed with the outer layer of porous concrete soon after construction: the reinforcing mesh corroded and the layer deteriorated and gradually peeled off. In 1953, the outer layer of porous concrete was completely removed from the surface of the dam. Parenthetically, it is of interest to note that this is only one in a number of experiences which are available to indicate that porous concrete is not a suitable material for use on the exposed upstream face of a dam.

In 1954-5 -- approximately 20 years after the asphaltic concrete had been installed -- and after the outer layer of porous concrete had been completely removed to expose the whole asphaltic concrete surface, an extensive series of tests were performed for the following purposes:

1. To determine the then existing condition of the asphaltic concrete membrane.

2. To obtain general information concerning the aging influence of asphaltic concrete when use in this fashion.

3. To determine what type of protection should be applied to the outer surface of the asphaltic concrete to replace the porous concrete protective layer which had been removed.

Samples were cut from the membrane at various locations and extensive tests were performed in three laboratories independently. In addition, tests were performed in place on the dam slope to determine the velocity of shock waves and the modulus of elasticity of the membrane and to determine the insulating properties of several prospective materials for a new outer protective layer.

From the laboratory tests, the following conclusions were obtained:

1. Aging of the asphaltic concrete was characterized by a moderate loss of plasticity and a concomitant increase in the compressive and shearing strength.

2. The influence of aging was much less where the asphaltic concrete had been below the water level for long periods of time. It was most pronounced in sections which had been above the high water level and insignificant in parts that had been continuously immersed.

3. Aging was greater where the asphalt had been less well compacted and was more porous.

4. The permeability was not influenced by aging.

The shock wave tests did not give any striking results; however, it was established that the velocity of wave transmission varied from approximately 3,000 meters per second at the top of the membrane to 3,800 meters per second at the bottom, corresponding to a range in moduli of elasticity from 85,000 kg/cm^2 at the top, to 103,000 kg/cm^2 at the bottom.

From laboratory tests, it was decided that the asphaltic concrete membrane could be left bare if the temperature did not rise above the creep point, which was established at approximately 50° centigrade. Test sections of the exposed asphaltic concrete were covered with 3 types of materials:

1. Materials operating by reason of their coefficient of thermal insulation (aerated concrete, etc.).

2. Materials operating by reason of the fact that they reflect heat (light-colored paints).

3. Materials which provided both actions (synthetic resins of light color).

Temperature measuring probes were installed at the surface of the membrane and at different depths in the asphalt below the various test sections. Measurements were made of the actual temperatures which developed under the action of the hot August sun (1954).

From these tests, the following conclusions were made:

1. The asphaltic concrete in itself had a considerable thermal inertia. On a hot day when the surface of the unprotected asphaltic concrete reached 60° centigrade, the temperature -- at the middle of the 12-centimeter thick slab -- reached a maximum of only about 40°C.

2. The addition of white paint kept the temperature at the surface of the asphaltic concrete at a minimum of 16°C below the value which occurred when no protection was used, and well below the temperature at which creep might be expected.

As a result of this experimentation, the surface of the asphalt was painted with a coat of white paint. Up to the present time, the performance of this thermal protection for the slab has been excellent.

It is difficult to leave the subject of the Ghrib Dam without an expression of gratitude to those French engineers who had the insight, the engineering skill, and the courage to execute this very interesting project. There can be no question that the experience of the Ghrib Dam is a milestone in the development in the art of dam building and there have been few comparable single advances. The use of a 5-inch thick asphaltic concrete slab so successfully on a dam with such a height and steep slopes at a site with high temperatures and a large reservoir is an amazing and unprecedented piece of civil engineering.

While there is little doubt that the development of the use of asphaltic concrete for membrane for high dams has been unduly delayed because of the natural inertia in the profession resisting change, there can also be no doubt that the acceptance of this method of construction would have not reached its present status if it had not been for the precedent of the Ghrib Dam.

BOU-HANIFIA DAM (ALGERIA)

The Bou-Hanifia Dam which was constructed several years later, also in Algeria, was patterned closely after the design of the Ghrib Dam. The embankment slopes and construction were approximately the same, though the height was somewhat less.

Taking advantage of the experience gained with the unsatisfactory performance of the outer protective layer of porous concrete on the Ghrib Dam, this was replaced at the Bou-Hanifia Dam with a 12-centimeter thick monolithic slab of ordinary reinforced Portland cement concrete. Also, in an attempt to provide a little more coherence to the asphaltic concrete membrane, the lower of the two compacted layers was reinforced with a welded steel mesh. Otherwise, the details of the membrane are the same as at Ghrib -- as shown on the sketch in the appendix.

The performance of the Bou-Hanifia Dam has been successful and uneventful with negligible leakage passing through the asphaltic concrete membrane.

IRIL-EMDA DAM (ALGERIA)

The Iril-Emda Dam, constructed in 1952-4, was the 3rd of the 4 dams with asphaltic concrete membranes in Algeria. With a maximum height of approximately 245 feet and a membrane area of more than 66,000 square meters, it is the largest dam of this type which has been constructed to date.

The embankment was constructed from broken limestone in the form of a well-graded mixture of cobbles, gravels and fines and can be considered as a compacted gravel-filled dam. It was constructed in layers with vibratory compactors. In the upstream third of the dam, under the membrane, the embankment was compacted in 12 in. thick layers. The layer thickness was increased to a maximum of about 3 ft. in the downstream third of the embankment.

The steepness of the upstream slope was made approximately 1.6:1 which was the inclination at which it was estimated that the material could be held without raveling during construction while the membrane was being placed.

The membrane is essentially the same as that used at the Ghrib and Bou-Hanifia Dams. The outside protective layer was constructed of monolithic reinforced concrete, varying in thickness from eight centimeters at the crest to 15 centimeters at the toe and reinforced with 0.6% steel in each direction.

The construction of the asphaltic concrete was handled with standard road-building equipment. The layers were compacted with 15 passes of a 5-ton smooth steel roller running up and down the slope.

For the design of the membrane at the Iril-Emda Dam, in addition to reviewing the performanc of the two previous dams, extensive laboratory tests were performed on the asphaltic concrete (see Ref. 7). These included among the other more or less standard tests, an attempt to study the self-repairing properties of the asphaltic concrete. For these special tests, specimens were made in the laboratory which were essentially complete cross-sections of the whole mem brane, including the upper and lower concrete layers. These specimens were placed in a steel cylindrical permeability testing apparatus of the type usually used for earth specimens. For the test, a hole was made in the asphaltic concrete layer and the rate of seepage through the specimen as a function of time was measured under various hydraulic gradiants. In all cases, the leakage decreased with time, evidently due to the plastic flow of the asphaltic concrete under the pressure caused by the water gradient which had a tendency to close the leakage channel by squeezing. At a water pressure equivalent to a head of 160 ft. or more, nearly com plete self-sealing occurred on all samples.

Also as a part of this testing program, tension tests were performed on full-size reinforced concrete slabs to determine the amount of steel which would be necessary in the outside protective layer to force the slab to crack in many hair line fissures rather than a few larger cracks which might rupture the asphaltic concrete membrane. The results of these tests led to the use of a quantity of steel equal to 0.6% of the area of the concrete as shown in the sketch Parenthetically, this is a little more steel than would normally be considered necessary for this purpose.

The performance of the dam has been completely satisfactory. The maximum post-construction settlement of the crest has been only 2.5 in. indicating the efficacy of the vibratory compaction.

The leakage measured in the control gallery when the reservoir reached maximum height for the first time was about 0.3 cfs (coming through the entire 66,000 square meters of surface). With time, this total amount of leakage decreased to approximately half.

SARNO DAM (ALGERIA)

The Sarno Dam was the last of the four Algerian dams constructed with asphaltic concrete membranes. It is an earth dam compacted in layers with a sheepsfoot roller from a coarse gravelly cobble mixture with a relatively small percentage of fine-grained sizes.

Because the dam was considerably lower (and the upstream slope much flatter) than the other three structures, it was decided to use a single asphaltic concrete layer with a thickness of only 8 centimeters as the water barrier, and to provide no outer protective layer of concrete (see Appendix). Another innovation in this dam was the use of an underlying porous drain of asphaltic concrete instead of Portland cement concrete. The purpose was to make the porous drain less brittle and more likely to follow embankment settlements without cracking.

Some trouble was experienced during construction making the membrane impervious and more filler and asphaltic binder had to be added as a change from the original mix design in order to make the membrane impervious. During construction, every square meter of the membrane surface was subjected to a field permeability test and a number of sections were torn up and replaced when they did not indicate the specified degree of impermeability.

When the reservoir was full for the first time, no leakage was observed to enter the control gallery. The performance has been in every way satisfactory.

GENKEL DAM (GERMANY)

Genkel Dam is the first of a series of 5 German dams which have been built in the last decade using essentially the same details for the impervious membrane. As shown in the sketch, the membrane design was a departure from that used by the French engineers in Algeria in that the asphaltic concrete impervious layer is exposed directly on the surface of the dam with no protective covering. The membrane is also more complex with a gravel drain under the upper slab and a second layer of asphaltic concrete under the gravel drain.

The embankment was constructed of quarried rock which broke into relatively small pieces and most of the material was in the range between one and fifteen inches. The rock was hauled to the embankment surface with rubber-tired hauling equipment and spread in layers approximately 18 in. thick. The hauling equipment was routed over the construction surface in order to obtain as much compaction as possible. The surface of the layers was subsequently thoroughly compacted with 2 1/2 ton vibrating tampers suspended from cranes and to a lesser extent with smooth steel rollers. This type of compaction with heavy tamping and vibrating plates has been used for a considerable number of rock fill dams in Germany since World War II. The exterior surface of the rockfill embankment on the upstream slope directly below the asphaltic membrane was also compacted with a 5,000 lb. weight raised and dropped 10 to 15 ft. in a tamping action. The embankment consisted of approximately 400,000 cubic meters of rock.

The 2 1/4:1 exterior slope was chosen even though it was believed that a steeper slope might have been feasible because of the novelty of the asphaltic membrane.

The outer impervious layer of asphaltic concrete was placed in 3 thin (about 1.2 in.) layers. These were compacted with light vibratory plates which were believed to give as good compaction as heavy static rollers and which did not require heavy equipment on the crest of the dam to act as anchors.

The upper surface of the final asphaltic concrete membrane was covered with two layers of mastic placed with a special spreading machine to a total thickness of about 0.5 centimeters to "seal the pores".

The drain immediately under the outer asphaltic concrete impervious layer consisted of gravels graded roughly between 1 1/2 and 3 in. with enough asphalt to bind the stones together but not enough to reduce the permeability. The drainage layer was bulkheaded off with strips of asphaltic concrete both horizontally and vertically into sections so that any leaks through the membrane could be located.

At the contact of the asphaltic membrane with the abutments, the surface of the membrane was covered with a layer of rock riprap for a distance of approximately 7 ft. in order to provide a protection against rock rolling down the abutment and striking the dam.

The surface of the membrane was given a slight camber primarily for esthetic reasons since experience indicates that a perfectly plain surface gives the impression of a "shrunken chest". Another beneficial feature of the camber is that it should prevent the development of any tensile stresses developing in the membrane due to embankment settlement.

The maximum post-construction settlement at the crest of the embankment was less than 1/2 in. Such low settlements have been consistently obtained on the rockfilled dams in Germany which are compacted with the vibrating tampers.

Concerning the performance of the dam, the Owner reports, "The membrane has been satisfactory in all respects." No maintenance or repairs of any kind have been required. The reservoir has been in operation since the Spring of 1953. The total leakage through the membrane (11,000 square meters) which discharges into the gallery consists only of drops of water probably due primarily to condensation of moist air in the drainage layer.

Ice on the reservoir reaches a maximum thickness of about 10 in. but does not adhere to the membrane, probably because warm air rising from the gallery in the drainage layer keeps the membrane sufficiently warm to prevent a bond from forming. During most of the Winter, the ice is observed to stop short of the face of the dam and there is a space of several inches of water between the ice sheet and the membrane indicating that a considerable quantity of heat is supplied from the interior of the dam.

A small amount of creep of the thin surface layer of mastic has occurred on some of the hotter days. (The temperature of the asphaltic concrete surface reaches 110°F or higher). This has had negligible influence on the appearance of the membrane and no influence on its primary characteristics. Because of this slight creep, the designers in future will reduce the thickness of the mastic seal coat to 4 millimeters or less and perhaps apply it by spraying instead of spreading.

WAHNBACH DAM (GERMANY)

Wahnbach Dam is the last of the series of 5 dams which have been completed in Germany with essentially the same asphaltic concrete upstream membranes. Both the embankment and the membrane were constructed in essentially the same manner as the Genkel Dam. The upstream slope of the embankment was steepened to 1.6:1 which the designer considered the maximum on which the work would be carried up in an economical and safe fashion.

The reservoir has been in operation since 1957 and has required no repairs or maintenance of any kind.

HENNE DAM (GERMANY)

The embankment and the membrane at the Henne Dam were designed and constructed in essentially the same manner as for the Genkel and Wahnbach Dams. The upstream slope was made flatter than would have been otherwise necessary because the dam is located immediately above a densely populated community, in order to prevent any possible criticism concerning the safety of the structure. In addition, a second "line of defense" was built into the dam in the form of a sloping internal core zone (bremzone). This zone consisted of a tight compacted mixture of quarried rock with a sufficient quantity of a rich mortar of asphalt and sand to fill all the voids and make the zone impervious.

Because of the disaster which could possibly result from the failure of this dam, it was felt desirable to design a structure which would, in principle and in fact, be safe against any possibility of failure, even if the upstream impervious membrane of asphaltic concrete were completely and suddenly removed by some activity.

The reservoir has been in operation since 1955 and the dam has been completely successful with no measurable leakage getting past the membrane. Ice with a maximum thickness of about 17 in. developed on the reservoir during the Winter of 1957 but this adhered only very lightly to the membrane and no damage of any type occurred when the ice layer fractured under the action of the fluctuating reservoir level.

MONTGOMERY DAM (U.S.A.)

Montgomery Dam, completed in Colorado in 1957, is a dumped rockfill structure with a maximum height of 113 ft. and a provision for raising to an eventual height of 150 ft. It is located at an elevation of approximately 11,000 ft. in the Colorado Rockies. With its 12-in. thick, asphaltic concrete face, it was an extreme departure from any previous practice in the United States but not radical when compared with the thinner facings used on the higher dams in Algeria and in Germany.

Extensive tests were carried out in the asphalt laboratories of the U. S. Bureau of Reclamation in order to determine an asphaltic concrete mix which would be suitable for the extreme range of temperatures expected as well as the possible influence of the thick ice sheet on the asphaltic concrete membrane. The tests also included an attempt at an evaluation of the influence of the relative action of waves beating on the upstream slope. The test program was probably the most thorough of its type ever carried out (Ref. 15).

The embankment was constructed from large hard quarried granite dumped and sluiced in 30-ft. layers, following the American practice for construction of dumped rockfill dams. The principal problems of design and construction are described in great detail in Fefs. 13 and 14.

Montgomery Dam is unprecedented in a number of respects. It is the only dam in which a single membrane of asphaltic concrete has been used without a protective layer or underlying drainage layer. It is the only dam with an asphaltic concrete face which has been constructed as a dumped rockfill. It does not have a drainage gallery in the upstream toe which can be entered to collect and evaluate the seepage which gets through the membrane. In addition to these differences, the dam is constructed at a site where the asphaltic concrete membrane will be exposed to much more severe cold than has been experienced at any of the other similar dams. In fact, the winter temperatures to which the Montgomery Dam will be exposed probably will be as severe as those to which any dam will ever be exposed.

The asphaltic concrete was mixed and spread with equipment commonly used for highway construction. The membrane was constructed in three layers 3 to 4 in. thick with an underlying leveling course. The material was compacted with a 3,000-lb. steel drum vibrating roller which was run up and down the slope. The entire membrane (approximately 28,000 sq. yds.), was completed in a single 2-month construction season.

After two years of reservoir operation, the total crest settlement had been less than 0.1 ft. No noticeable quantity of leakage through the membrane developed. At the end of the first winter, the effect of the ice on the surface of the pavement had been negligible. A few minor surface cracks in the surface of the asphaltic concrete membrane developed above the high water level along the construction joints. The cracks had a maximum depth of about 4 in. and widths varying between 1/4 and 1/2 in. The cracks were undoubtedly due to the fact that the paving machine could not operate as effectively on the extreme upper portion of the asphaltic concrete membrane at the crest of the dam. The cracks were cleaned out and filled with sand and asphalt using the local practice for repair of highway pavements.

SUMMARY OF EXPERIENCE

In summary, it is of interest to recapitulate briefly some of the main lessons which have been learned concerning the basic properties of asphaltic concrete when used for this purpose.

Permeability

Both laboratory tests and field experience indicate that asphaltic concrete can be made completely water-tight for all practical purposes. A well-graded aggregate of fine gravel and sand with 10-20%fines passing the No. 200 sieve and 8-10%of asphalt binder has been used in most cases. Laboratory permeability tests under high gradients on representative samples have shown that essentially no water will get through a well-designed mix. The quantity of water which can be forced through the well designed and constructed asphaltic concrete slab is so small that it is not reasonable to assume that Darcy's Law applies or to assign a coefficient of permeability to the material.

Only in the case of Sarno Dam, where a single layer of asphaltic concrete was used, was there any trouble in the field to obtain a sufficiently impervious membrane and this was remedied by a slight increase in the content of filler and asphalt binder. A very thin membrane will satisfactorily resist several hundred feet of water head. In no case was there any noticeable tendency for the water percolating through the asphaltic concrete to push the asphalt binder out of the pores. Probably the surface coats of mastic used on the German dams provide an extra degree of impermeability since all the surface pores through which water could enter are sealed.

Strength and Maximum Slope Steepness

In none of the dams was there any damage caused by creep of the asphaltic concrete down the slope under its own weight. Much laboratory research and all the field experience indicate that satisfactory asphaltic concrete mixes can be designed to act as exposed membranes on a 1.5:1 slope except perhaps in the very hottest climates. Asphaltic concrete has high thermal inertia. Even when the outer surface exposed to the sun becomes quite hot, the temperature within the layer a few inches below the surface remains considerably lower during the whole day. The experience at Ghrib Dam indicates that slopes of 1:1 or steeper can be used in very hot climates if a reflecting coat of light-colored paint is provided.

Cracking, Flexibility, and Resistance to Erosion by Leaks in Cracks

None of the dams studied developed large post-construction settlements; therefore, we have no reliable fund of experience to indicate the maximum differential settlements which can be tolerated. Almost no records of cracking due to differential settlement are available.

At both Ghrib and Iril-Emda Dams some concentrated leakage developed in the first few years of reservoir operation, and the quantity of leakage decreased to practically zero with time, although no repairs were made. This "self-sealing" action of asphaltic concrete due to squeezing under water pressure was verified by laboratory tests (Ref. 7). From these results it can be concluded that at least for small leaks under high heads, there is no danger of progressive erosion or "piping" of asphaltic concrete membranes but, on the contrary, the leaks sealed themselves especially under high water pressures.

Resistance to Wave Action

The majority of the eight dams retained relatively small reservoirs which would not generate large waves. Because of its 230,000 acre-feet of storage capacity, it is probable that at least moderately severe wave action developed against the Ghrib Dam. In none of the dams was there any indication reported of erosion of the outer surface of the membrane due to wave action.

Only for the Montgomery Dam were any laboratory tests performed for the purpose of estimating the resistance to wave action (Ref. 15). In these tests, small asphalt samples were splashed in and out of a bath of water literally millions of times and the amount of erosion observed. While it is impossible to correlate the erosive action obtained in such a test with the action of the waves beating against a dam, the results of the tests indicated that specimens of good asphaltic concrete are remarkably resistant to the erosion of rapidly splashing water.

Another indication of the erosion resistance of asphaltic concrete slabs on the upstream face of dams was obtained from the performance record of a test section constructed in 1951 by the U. S. Bureau of Reclamation in the Bonny Reservoir in Eastern Colorado. Unfortunately, a large part of the embankment behind the asphaltic portion of the test section was eroded away a few years after it was placed in operation through no fault of the asphaltic concrete facing. However, the outer surface of the asphaltic concrete facing which remains, at least when the author visited the site in 1960, was not appreciably eroded by the splashing waves.

From these comments, it appears that while asphaltic concrete undoubtedly has high resistance to the erosive action of waves, more information in the form of additional performance records of actual dams in service is needed before any general or reliable conclusions can be drawn concerning design criteria. Probably exposed membranes of the type used in the German dams would not be resistant indefinitely to the battering of large waves, especially if the water contained an appreciable content of solid matter or floating logs which would act as battering ram during a storm. On the other hand, the author believes that future experience may prove that exposed slabs of properly designed asphaltic concrete may be more resistant to wave action than would be generally anticipated particularly if the outer surface is sealed with mastic or other coating.

Cold Weather and Ice Sheets

The experience with the German dams and the Montgomery Dam and the laboratory testing carried out for the Montgomery Dam indicates that it is extremely unlikely that ice adhering to the asphaltic concrete membrane during the winter will tear away any of the outer portions of the membrane. In the German dams, the ice did not bond strongly to the membrane because of the existence of warmer air in the internal drainage layer. Probably a similar phenomenon occurs at Montgomery Dam since the interior of the embankment is warmer than the exterior air. Also, the tests performed for the Montgomery Dam design studies indicated that even when the ice is solidly frozen to the membrane by lowering the temperature of both in a refrigerating room in the laboratory, the ice sheet itself will fail when loaded at a cantilever before any asphaltic concrete will be torn from the membrane.

There is no indication that freezing of the asphaltic concrete itself, even to very low temperatures, has any appreciable detrimental effect on its properties; i.e., no ice segregation or reduction of the density of the asphaltic concrete occurred.

Also, there has been no record of cracking due to expansion or contraction of membranes caused by temperature changes even though the range of temperature to which the slabs have been exposed is great and no expansion joints are provided in the membranes: evidently, any tendency for the development of compressive or tensile stresses in the membrane because of temperature changes is dissipated by plastic flow.

Time Change of Properties

Of the eight dams described, only the Ghrib Dam has been in operation long enough to allow an evaluation of the influence of time on the properties and the functioning of asphaltic concrete membranes. Consequently, there is not yet a great fund of information on the influence of aging on asphaltic concrete when used as the impervious membrane on the upstream slope of a dam.

Nevertheless, the experience at Ghrib Dam plus our general knowledge concerning the long-time performance of asphaltic concrete when used for surfacing roads and airports, surely allows us to draw the conclusion that properly designed and constructed asphaltic concrete can be expected to have at least as long a life span as reinforced Portland cement concrete or plat steel.

CONCLUSIONS

Though the experience available is not extensive, it is sufficient to conclude that properly designed asphaltic concrete, not greatly different in properties than normally used in highway surfacing, is a perfectly satisfactory material for the construction of an impervious upstream membrane on an earth or rockfill dam. On the other hand, there is not enough performance experience as yet to allow the formulation of any specific criteria concerning design details.

At any given site the design of the membrane still leaves considerable room for judgment and ingenuity.

From the details of the 8 dams described, we can isolate the following basic design provisions:

1. An exterior slab of reinforced Portland cement concrete (Ghrib, Bou-Hanifia and Iril-Emda Dams). The purpose of this slab is to provide physical and thermal protection for the softer underlying membrane of asphaltic concrete. It provides protection against the erosive action of waves and especially against the battering of floating logs and pieces of ice. It also provides protection against damage from rocks which might roll down the abutments and strike the face of the dam.

   Secondarily, the outer concrete slab keeps the temperature of the asphaltic concrete below the value to which it would rise if it were exposed directly to the sun's rays and allows the use of a richer asphalt mix on a steeper slope than would otherwise be possible. The protective layer probably also delays oxidation and aging of the asphaltic concrete.

2. An exposed asphaltic concrete slab with no protective covering (Genkel, Wahnbaçh, Henne, Sarno, and Montgomery Dams). The principal advantage of the exposed membrane is the fact that cracks or tears in the membrane can be readily observed and repaired.

3. A membrane consisting of a number of layers with an internal drainage layer which discharges into concrete gallery at the upstream tow (all dams except Montgomery). The primary advantage of the internal drain is that leaks through the outer impervious membrane are trapped and never allowed to enter the embankment and that the quantity and location of the leakage can be determined. For the design in which there is another impervious layer below the drainage layer, the membrane has a second line of defense.

4. Single asphaltic concrete slab (Montgomery Dam only). The principal advantage of the single slab is economy and speed of construction. The main disadvantage is that it is impossible to locate where leakage erupting at the toe of the dam may be coming from. Also, this this type of design can only be used in conjunction with an underlying drain which has sufficient capacity to carry away any conceivable leakage without causing high hydrostatic pressures in the dam.

In future years, improved designs will undoubtedly be developed. Probably, designs other than those used at the 8 dams studied can be devised to take better advantage of the "self-sealing" property of the asphaltic concrete which is one of its most intriguing attributes. At a high dam recently, the author considered seriously the use of a membrane consisting of a thin layer of very stiff asphalt binder (say 1/2 in. thick) sandwiched between two thin slabs of Portland cement concrete which had sufficient steel reinforcing to prevent the development of large cracks. Using this design, it was reasoned that the water pressure forcing the two concrete slabs together would be sufficient to cause the asphalt to squeeze into and seal all cracks which developed in both the upper and lower concrete slabs making what seemed to be a fool-proof and economical design.

LIST OF REFERENCES

1. Jean Thévenin, "The Ghrib Dam", Travaux (Paris), August 1958, p. 141

   (Excellent description of the construction of the Ghrib Dam and its performance. Also includes detailed results of laboratory and field tests made of the asphaltic concrete membrane in 1954 after 20 years of reservoir operation. In English.)

2. Jean Thévenin, "Les Barrages Algériens avec Masque Amont en Béton Bitumineux", Travau (Paris), November 1960, p. 690

   (Brief review of design and construction details and performance history in 1960 of the four Algerian dams with asphaltic concrete membranes. In French).

3. J. C. Ott, "La Construction du Barrage de Bou-Hanifia", Bulletin Technique de la Suisse Romande, February 1944, p. 25

   (Detailed description of the design and construction details of Bou-Hanifia Dam written soon after construction. In French).

4. J. C. Ott, "Comportment du Barrage de Bou-Hanifia", Bulletin Technique de la Suisse Romand, February 1955, p. 57

   (Thorough discussion of the first 10 years of performance history of Bou-Hanifia Dam. In French).

5. Jean Guilhamon and Henri Castelnan, "Tassement des Materiaux Constutifs et de la Fondation (Barrage de l'Iril-Emda)", Fifth Congress on Large Dams, Paris 1955, Vol, III, p. 389

   (General discussion of the Iril-Emda Dam construction with emphasis on the measurements of embankment and foundation settlement. In French).

6. J. Lordet et al, "Hydro-electric Development of the Oued Agrioun", Fifth International Congress on Large Dams, Paris, Supplement to "Travaux" No. 247 (1955), p. 190

   (Complete treatment of the design and construction of Iril-Emda Dam. In English).

7. R. Vormeringer, "Constitution d'un Masque Souple et Raccordement aux Ouvrages Rigides. (Barrage de l'Iril Emda)", Fifth Congress on Large Dams, Paris, (1955), Vol. IV, Communication 22, p. 951.

   (Very thorough and detailed treatment of the Iril-Emda Dam embankment design and construction with emphasis on the laboratory tests which were made for the analysis of the asphaltic concrete membrane. In French).

8. H. Castenow and Y. Haller, "Barrage d'Iril Emda", La Technique Moderne, Construction (Paris), September 1954.

   (Excellent description of the construction of Iril-Emda Dam. In French).

9. J. Salva, "The Sarno Dam", Fifth International Congress on Large Dams, Paris, Supplement to Travaux No. 247, p. 179

   (General description of design and construction of the Sarno Dam written at the end of the construction period. In English).

10. P. Schmitz, "Die Genkeltalsperre", Bitumen (Hamburg, Germany), 1953, p. 204

(Detailed discussion of the design and construction of Genkel Dam and the first year of performance. In German).

11. H. W. Koenig, "Der Neubau der Hennetalsperre des Ruhrtalsperrenvereins", Publication of the West-deutschen Wasserwertschaftsverbandes E. V., Essen 1957

(General description of the main features of the Henne Dam. In German).

12. H. Press, "Talsperren", (German Book", Wilhelm Ernst & Sohn, Berlin, 1958

(Contains description in sketchy fashion of some details of Genkel, Henne and Wahnbach Dams. In German).

13. G. R. Scott, "Montgomery Dam - A Rockfill Structure", Sixth Congress on Large Dams, New York, Report 92, Question 20, 1958

(Good general treatment of the design and construction of Montgomery Dam. In English).

14. F. W. Scheidenhelm, J. B. Snethlage and A. N. Vanderlip, "Montgomery Dam - Rockfill with Asphalt Concrete Deck", Journal Power Division, American Society of Civil Engineers, February 1958, Paper 1556

(Detailed description of all the principal elements of the design and construction of Montgomery Dam. Discussions contain first three years of performance history. In English).

15. A. N. Vanderlip, F. W. Scheidenhelm and J. B. Snethlage, "Laboratory Investigation of Asphaltic Concrete, Montgomery Dam, Colorado", Proceedings Association of Asphalt Paving Technologists (Montreal) February, 1958, p. 383

(Very thorough discussion of laboratory testing carried out for the design of the asphaltic concrete membrane at Montgomery Dam. In English).

16. H. Haas, "Technische Eigenschatten von Asphaltbeton fur Wasserbauzwecke", Die Bautechnik (Berlin), September 1959, p. 339

(Detailed summary of laboratory tests carried out in connection with the use of asphaltic concrete slabs on the German dams. In German).

SUMMARY OF PERTINENT DETAILS

DAMS WITH ASPHALTIC CONCRETE UPSTREAM MEMBRANES

1) Dam name (location) 2) Date of completion 3) Approximate maximum height 4) Upstream slope 5) Embankment type 6) Owner	Details of Upstream Asphaltic Concrete Membrane	References (See list attached)
1) Ghrib Dam (Algeria) 2) 1937 3) 235 feet 4) 0.7:1 (at crest) 1:1 (at toe) 5) Large bulky quarried limestone fitted together with cranes in form of rough dry masonry embankment with the larger rocks in contact and the smaller rocks filling the voids. 6) Direction de l'Hydraulique et de l'Equipement Rural, 135, Rue Michelet, Alger	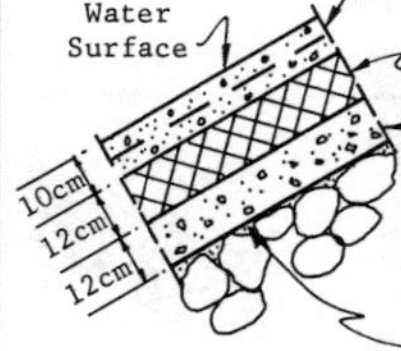 Porous concrete reinforced with 5mm galvanized steel wires in 90mm grid (removed completely in 1953). Asphaltic concrete in two equal layers, A/C = 8-10%. Porous concrete with open drains running to gallery at upstream toe. Voids in rockfill embankment surface filled with sand-cement mortar	1. 2.
1) Bou-Hanifia Dam (Algeria) 2) 1939 3) 180 feet 4) 0.8:1 (at crest) 1:1 (at toe) 5) Same as Ghrib Dam 6) Same as Ghrib Dam	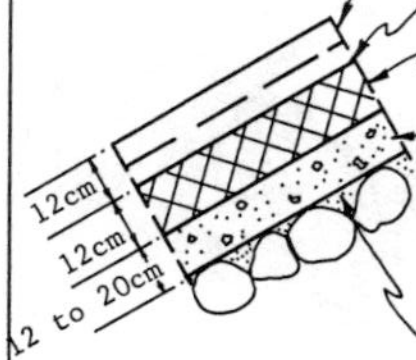 Monolithic reinforced concrete. Layer of paper. Asphaltic concrete placed in 2-6cm layers with grill of steel mesh in the lower layer. Porous concrete with open drains to downstream toe Voids filled with sand-cement mortar	3. 4.

1) Iril-Emda Dam (Algeria)
2) 1954
3) 245 feet
4) 1.6:1 (top)
   1.7:1 (bottom)
5) Broken limestone in the form of sand-gravel-cobble mixture compacted in layers 12 to 40 inches thick with vibrating compactor and wetted with volume of water equal to half the volume of rock.
6) Electricité et Gaz d'Algérie
   2, rue Jean Rameau, Alger

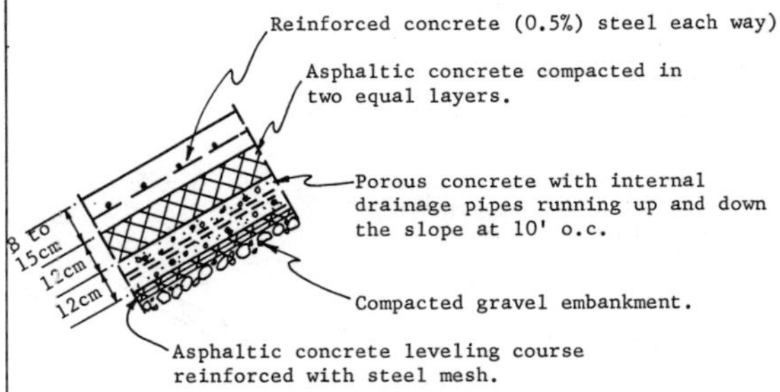

5.
6.
7.
8.

1) Sarno Dam (Algeria)
2) 1955
3) 90 feet
4) 2:1 (top)
   2.5:1 (bottom)
5) Homogeneous embankment of coarse gravelly soil compacted with sheepsfoot roller and water content control.
6) Same as Ghrib Dam

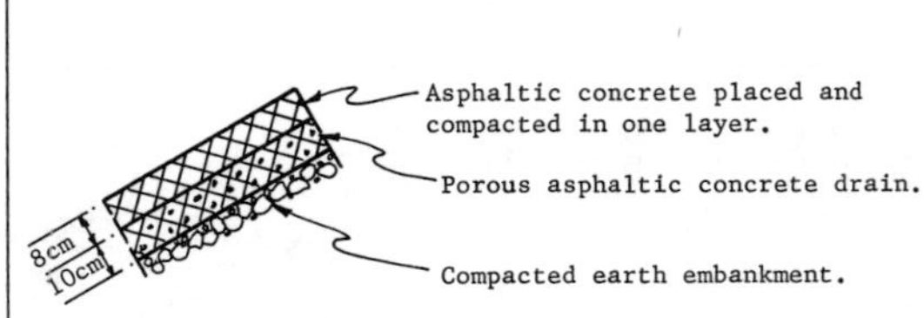

2.
9.

1) Genkel Dam (Germany)
2) 1953
3) 110 feet
4) 2.25:1
5) Quarried rock in small sizes spread in 18 inch layers and compacted with heavy steel tampers.
6) Genkeltalsperre
   Aggerverband
   Niedersessmer, Rhld.
   Bernberger Str.
   Bundesrepublik, Deutschland

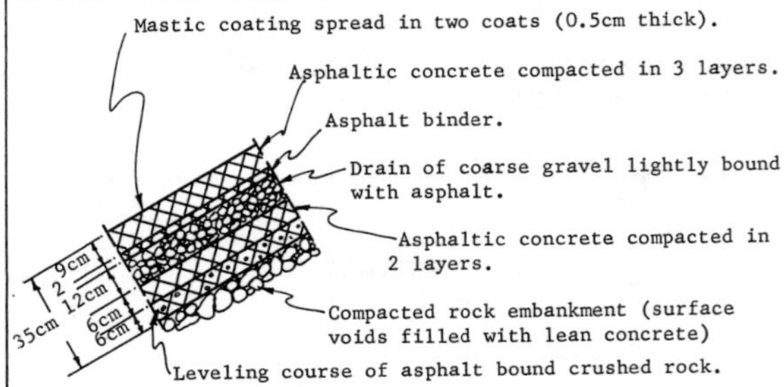

10.
11.

1) <u>Henne Dam</u> (Germany) 2) 1955 3) 190 feet 4) 2.1:1 5) Same as Genkel Dam 6) Ruhrverband und Ruhrtalsperrenverein Essen Kronprinzenstrasse, 37 Bundesrepublik, Deutschland	Membrane similar to Genkel Dam	10. 11.
1) <u>Wahnbach Dam</u> (Germany) 2) 1957 3) 170 feet 4) 1.6:1 5) Same as Genkel Dam 6) Wahnbachtalsperrenverbandes Siegburg Kronprinzenstrasse, 17 Bundesrepublik, Deutschland	Membrane similar to Genkel Dam	12.
1) <u>Montgomery Dam</u> (Colorado, U.S.A.) 2) 1957 3) 113 feet 4) 1.7:1 5) Large quarried granite dumped and sluiced in 30-foot layers with 10-foot thick layer of pervious screened rock under upstream membrane. 6) Department of Utilities City of Colorado Springs, Colorado	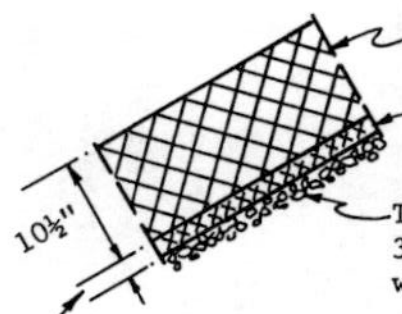	13. 14. 15.

Journal of the
SOIL MECHANICS AND FOUNDATIONS DIVISION
Proceedings of the American Society of Civil Engineers

EARTHQUAKE CONSIDERATIONS IN EARTH DAM DESIGN[a]

By James L. Sherard,[1] M. ASCE

INTRODUCTION

In 1963, the writer was engaged by the State of California Department of Water Resources to make a study of the influence of the hazard of earthquakes on the design of embankment type dams, Sherard, (1966). This paper summarizes some of the main conclusions. All the opinions contained herein are solely those of the writer and in no way have they been endorsed or received any official status or recognition by the Department of Water Resources.

The purpose of the study was simply to scrutinize in a concentrated fashion what information is available to aid the civil engineer in answering the basic question: "How should earth dams be designed differently in regions of earthquake activity?" The writer spent approximately half time on the project over a period of about 18 months. A large part of this time was spent in discussions and conferences with specialists in the various related disciplines of seismology, geology, dam engineering, and soil mechanics.

In the course of the effort, the writer had conferences with selected people in various parts of the United States. He visited and discussed the problems in Mexico and Japan, and observed the results of the great 1964 Alaska earthquake at firsthand. Typically, the study went in stages. A series of discussions provided the subject matter for several days of reflection and led to the formulation of additional questions to put to the experts in the subsequent interviews and conferences.

The paper contains no large or original contributions to the existing knowledge in the various scientific specialties which, in combination, com-

Note.—Discussion open until December 1, 1967. Separate discussions should be submitted for the individual papers in this symposium. To extend the closing date one month, a written request must be filed with the Executive Secretary, ASCE. This paper is part of the copyrighted Journal of the Soil Mechanics and Foundations Division, Proceedings of the American Society of Civil Engineers, Vol. 93, No. SM4, July, 1967. Manuscript was submitted for review for possible publication on January 30, 1967.

[a] Presented at the Aug. 22-26, 1966, ASCE Soil Mechanics and Foundations Division Conference on Stability and Performance of Slopes and Embankments, held at Berkeley, Calif.

[1] Vice-Pres., Woodward, Clyde, Sherard and Assoc., Consulting Engineers and Geologists, Oakland, Calif.

prise practical earthquake engineering for dams. The contribution is rather in the scrutiny, from the viewpoint of the dam engineer, of some of those important questions for which at the present time (1967) there are no wholly satisfactory answers. In this realm of problems peculiar to earthquake engineering where the design engineer has neither the direct experience nor the facts to guide him, but where he must nevertheless make decisions and proceed to build structures, it is necessary for him to extrapolate from indirect and sometimes tenuously related experience and to rely on his intuition and imagination to an extent which he would certainly avoid if he had an alternative course of action. One main purpose of the study was an attempt to sample and summarize what might have been called the "collective intuition" existing among able specialists for the purpose of providing the dam designer a reference framework against which he can compare his own intuition, judgment, and decisions.

LESSONS LEARNED AT HEBGEN DAM

The experience of Hebgen Dam in Montana during the August 17, 1959, earthquake is one of the few instructive and well-documented examples of a moderate-sized earth dam located near the epicenter of a strong earthquake. Through a fortunate combination of circumstances the dam did not fail, and it was put back into service after a minimum of repairs. Almost every dangerous effect of an earthquake was present, however: major faulting, large permanent deformations of the earth's crust, strong earth shaking, landslides, and a major rockfall. Consequently the case justifies the most careful scrutiny by the earth dam designer. Since the experience has been well described in the literature, it is not repeated in detail here; see, for example, Ball, (1960) Hadley, (1964) Ryall, (1962) Sherard et al., (1963) Steinbrugge & Cloud, (1962) Tocher, (1962) U. S. Forest Service, (1959) Wiegel & Camotim, (1962) Witkind, (1962). One does not have to reflect very long on the events which occurred before concluding that an extremely fortunate array of conditions prevented a complete failure—with such a large reservoir, the failure would have been a disaster even in that sparsely populated area.

First, it was lucky that the main Hebgen fault was located up on the valley wall several hundred feet from the dam. If it had run down the center of the stream channel (as faults more commonly do), it would have torn the dam in two. The concrete core wall would probably have been shattered and the embankment ripped open. It is likely that failure by progressive erosion of the dam embankment would have resulted in a period of hours, so that the whole reservoir would have been released in one flood wave.

Second, it was only the most fortunate coincidence that the rock foundation of the dam, which supported the concrete core wall, sank 9.7 ft at the same time that the 20-mile-long, 340,000 acre-ft, reservoir went down an average of more than 10 ft. If the dam had sunk and the reservoir had not (or had risen), it would have been necessary to pass a great deal of water over the dam and through the spillway. Since the bottom of the spillway channel was practically destroyed by the shaking, and since a soil deposit about 100 ft thick lay underneath, it seems almost certain that the spillway would have failed and the water would have burst through a breach in the right abutment.

Third, it was lucky that the block of foundation rock underlying the dam moved in one piece with essentially no tilting. With such a large total movement (9.7 ft), it is easy to imagine that there could have been several feet of tilting, or even of abrupt differential displacement between immediately adjacent points in the foundation causing a large fissure in the concrete core wall and embankment possibly leading to a rapid failure by piping.

Fourth, it was lucky that a large landslide of the type which occurred at several places along the bank of the reservoir did not develop near or at the right abutment of the dam. The topography of these slides and the material involved were not much different from the topography and material of the dam abutment—itself probably an old slide. A landslide at this location could have led to rapid piping and a breach through the right abutment, or it could have demolished the spillway channel.

Fifth, it was lucky that the massive rockslide occurred 7 miles downstream from the dam. If this slide had fallen into the reservoir instead, there might have been a failure like the one at Vaiont Dam. It is worth noting that the river valley offered a good dam site at the location of the rockslide and that construction of a dam there, with a much larger reservoir than Hebgen Dam's had been considered seriously in the past.

Sixth, it was lucky that the spillway did not have to be used before it was repaired. If a flood had occurred which could not have passed through the outlet pipe, or if the outlet pipe had been put out of operation by the earthquake, the water would have had to go over the damaged spillway and it is problematical how long it could have lasted.

Because Hebgen Dam did not fail, its experience has had little or no influence on current design practice. Unfortunately, the average designer knows little about the facts of this case and considers it simply one more demonstration of the intrinsic resistance of earth dams to earthquake damage. In fact, this experience justifies much more attention by the profession than it has received. If only one of the results of the earthquake discussed in the preceding paragraphs had been less fortuitous, it seems apparent that the dam could have failed and released the whole of the large reservoir in one disastrous flood wave. If one speculates only a little further and imagines that this event occurred in a populated region of California instead of a relatively remote part of Montana, then we can be sure that it would have received infinitely more attention by the profession and the public and would already have had a dominant influence on our current dam building activities. Reviewing the 1959 Hebgen Dam experience in this light may be, as some reviewers of this manuscript have remarked, approaching the realm of excessively pessimistic speculation; however, in the writer's opinion the facts justify the conclusion that we were very fortunate that the dam did not fail and that to view the experience otherwise would be unrealistic.

SIGNIFICANCE OF THE EMPIRICAL NATURE OF EARTH DAM DESIGN

Several features of the earth dam, which make it different from most other civil engineering structures, have fundamental importance in connection with the problem of building safe structures in regions of potentially strong earthquake shaking:

1. Earth dam design is nearly wholly empirical, in the sense that the

majority of the important decisions governing the safety of the structure (such as the selection of the embankment materials, the zoning, the core thickness, the freeboard, the thickness and details at the top of the dam) are made arbitrarily using judgment guided by precedent and past experience.

2. The value of the property that might be damaged in a failure can be much greater than the original cost of the dam or of the cost of repairing the dam after failure.

3. Under certain conditions, a failure can result in a greater loss of human life than is conceivable as the result of a failure of most other man-made structures.

Present (1967) standards of practice for the design and construction of earth dams, using which the engineer is confident that he can build completely safe structures (at least when strong earthquakes are not considered), have developed in a step-by-step process over the past 50 yr or 60 yr. In the years prior to 1910, the highest earth dams had a height of about 100 ft. By 1925, the maximum had reached about 200 ft, and by 1945, about 400 ft. At the present, a number of dams with heights of 600 ft and more are being considered and built. At each stage, the confidence to build higher dams has been based on the satisfactory performance of the previous structures. In addition, in the last 20 yr to 25 yr, movements and pore water pressures which developed inside many dams have been measured. The fact that the values measured were not much different from those which could have been predicted gave engineers more confidence in constructing dams of unprecedented heights. The development of soil mechanics as a practical engineering specialty during the same period also greatly strengthened the confidence of the earth dam designer; however, soil mechanics has served basically as a complementary tool and could not have taken the place of the performance experience.

In 1967, therefore, our confidence in our ability to build safe dams is derived primarily from the satisfactory performance of literally hundreds of large dams and thousands of small ones. In direct contrast to this great body of available experience with the performance of dams under ordinary conditions, very few dams have been shaken severely by earthquakes. Most of those which have been shaken were old, relatively small dams, built with less exacting standards of design and construction than current practice requires. Some of these dams were undamaged by the shaking, some failed completely, others were seriously endangered in one way or another. For no dam which has been badly shaken by an earthquake have instrumental measurements been made of the severity of the shaking. Hence, in order to guide and support design procedures for dams under ordinary conditions of performance, there is available literally thousands of "dam-years" of experience with dams of all kinds under all conceivable circumstances all over the world and, in addition, a large volume of measurements with instruments. The equivalent experience available concerning the probable behavior of dams shaken by earthquakes is relatively so much smaller as to be insignificant.

The writer believes that these ideas are fundamental and that it should be clearly recognized that the dam designer is in the rather less-than-comfortable position where: (1) present design practice (and our confidence in it) has come largely from experience with the performance of structures in the past; (2) experience with the performance of dams during earthquakes is much too limited to provide any appreciable guidance or any general confidence. Nevertheless, it is generally agreed that safe dams can be built at most sites in seismic regions and that it is reasonable and prudent to do so. The writer certainly agrees with this; however, he also believes that there is a tendency in the profession to gloss over the above facts with too little consideration of the influence which they should have on the engineer's decisions.

Confidence that earth dams can be built which will be safe against earthquakes, and that these will not have to be greatly more expensive or conservative than dams in nonseismic areas, comes from a number of sources. In the writer's opinion, a large part of this confidence stems from the basic knowledge that embankments of compacted earth and rock generally do not lose their strength when strained beyond a failure strain or when subjected to a number of rapidly alternating pulses of strain. Therefore, while some deformation and damage may occur in the embankment during the short period of time that it is being shaken, after the earthquake is over it is reasonable to assume that the dam should be as stable as before. The more doubt which may exist in any given case that the foundation or a portion of the embankment may lose an appreciable part of its strength or experience large deformations during shaking, the less confidence that is justified and the greater the need for arbitrary conservatism.

CONSERVATISM JUSTIFIED AT A GIVEN SITE

Influence of the Consequences of a Failure.—Under the philosophy commonly governing design practice, the degree of conservatism built into a dam is not generally varied greatly as a function of the amount of damage which would be caused by a failure. Every dam is made "completely safe" and, therefore, it is not necessary for the designer to examine in a conscious fashion the likelihood or consequences of a failure.

When the earthquake hazard is not considered, there can be little argument with this approach. As stated previously, however, in a region which may be shaken by a strong earthquake, the engineer does not have the degree of confidence in his ability to build a completely safe dam that he would like to have for such an important structure. For this basic reason, it seems apparent to the writer that a different philosophical approach is needed for the design of dams in earthquake areas. A dam whose failure would cause a great loss of life should be built considerable more conservatively than one whose failure is potentially less hazardous. While no specific rules can be given regarding the extra degree of conservatism needed in any given case (or the amount of money that should be spent to achieve it), these should depend to a large degree on the risk or likelihood that a very strong earthquake could create conditions which are more severe than the worst conditions anticipated by the sesigner, and hence that the dam could fail.

The writer has found no discussion in the literature of the desirability of increasing the conservatism built into dams in earthquake areas as a function of the increasing consequences of failure. There have been a few interesting examples, however, where the fear of bombing or sabotage has led to similar thinking. One of these was the 190-ft-high rockfill Henne Dam built in Germany in 1953 by the government agency which develops and controls the Ruhr River and its tributaries, Press, (1958); Koenig, (1957), Schuster, (1957). The dam is located in a place where its failure could send a wall of water into a large city only a few miles downstream, and in a region where the people still remember well the great disaster caused by the failure of a dam by bombing during World War II. Because of this, the engineers provided special details in the dam crest to strengthen it against bombing or sabotage. This extra safety more than doubled the cost of the dam. Other similar dams in the Ruhr basin, whose failures would have had less catastrophic results were built in the same period by the same agency without this additional conservatism.

Another example is the High Aswan Dam, currently under construction on the Nile, whose failure could result in a human catastrophe of unparalleled proportions. The width of the top of this dam is being made three to four times the width which would probably be chosen under normal circumstances for an important dam in the United States. For both the above dams, the designers were thinking primarily of providing extra safety at the top against the hazard of damage during war (bombing from the air or sabotage). During his study of the subject, however, the writer has concluded that there is a great deal of similarity between the hazard of sabotage or bombing and that from earthquakes: (1) Both occur at rare intervals on occasions and with a likelihood that the designer is essentially powerless to predict; (2) both are likely to damage the dam primarily in the upper part; (3) both are likely to cause final failure as the result of progressive erosion of an initial concentrated leak which gets out of control.

Possibility of Failure.—The problem of evaluating the possibility of a dam's failure is greatly more complex and difficult than the study of the consequences. Our information on the factors involved comes primarily from what we know and what we can speculate about the types and likelihood of damage or failure to be expected from earthquake action. From this body of information, it can be concluded generally that the probability of failure for a given dam from a strong earthquake increases with the following main factors:

1. Decreasing foundation strength, especially under shaking conditions;
2. decreasing ability of the foundation to withstand concentrated leaks safely;
3. decreasing ability of the dam itself to withstand concentrated leaks safely, particularly at the top;
4. increasing probability or magnitude of fault movement in the foundation;
5. increasing abutment steepness and irregularity;
6. increasing likelihood of large landslides into the reservoir;
7. increasing likelihood of slides on the abutments;
8. decreasing speed with which the reservoir level could be lowered by discharging water through the outlet conduits;
9. decreasing height of the dam above the normal pool level;
10. increasing vulnerability of the spillway and outlet works to earthquake damage;
11. decreasing resistance of the dam crest to erosion from overtopping.

Influence of Dam Height.—Study of the above list leads to the general conclusion that the risk from most of the main sources of potential trouble increases with increasing height of dam and depth of water in the reservoir. Other things being equal, the higher the dam and the deeper the water, the greater the hazard of foundation shear failure, abutment slides, large slides into the reservoir, embankment cracking and concentrated leaks, and crest settlement and overtopping. Because the water depth in the reservoir is an important factor in the majority of these modes of failure, it seems reasonable to deduce that the risk of complete failure may increase rapidly with the dam height, probably at least with the square. This fact, combined with the fact that the consequences of a failure increase with increasing depth of the reservoir, leads to the conclusion that, other things being equal, the degree of conservatism justified should be considered to depend to a great extent on the dam height.

Stability of Soil Foundations.—A second major influence on the risk of failure is the strength of the foundation. One of the few points during the course of this study on which the writer found almost universal agreement among engineers consulted was serious concern over the possibility that strong earthquake shaking could cause shear failure (or large shear strains) in many soil foundations which might otherwise be quite satisfactory. There was also almost universal concern over our relatively poor ability to evaluate the problem. Except for deposits of stiff, insensitive clays and very dense cohesionless soils, it is not possible to be completely confident that any given soil foundation might not fail, or at least strain an intolerable amount, during strong earthquake shaking. We are only just starting to learn something about the analysis and behavior of soils under repeated dynamic stresses of the type that might be induced during a strong earthquake.

The present (1967) practice of providing a safety factor of about 1.1 under a static horizontal force of 10% to 15% of gravity, using conventional laboratory tests for evaluation of the foundation strength, gives no assurance whatsoever that the dam will be safe under strong earthquake shaking. Since the ground shaking motion in a strong earthquake apparently has acceleration pulses ranging between 30% and 50% of gravity, this analytical procedure can give no confidence that large shearing movements could not occur.

Finally, there can be little argument with the basic conclusion that the evaluation of the stability of soil foundations is one of the weakest elements of current earthquake engineering for dams and that improvements in our ability to analyze the problem are desperately needed. Because of this and the fact that a foundation failure might release the whole reservoir in a few minutes, a very conservative attitude should be adopted at the present time (1967) for any dam in an earthquake area on a soil foundation. The combination of a soil foundation of suspect strength and severe consequences of failure surely justifies the extreme of conservatism. At this limit, if the foundation material cannot practicably be excavated to a considerable depth, the designer should consider carefully the prudence of building the reservoir at all. It he decides to go ahead with construction, he cannot depend on the fact that dams on similar foundations have been satisfactory under ordinary conditions to give him any confidence whatever; the conditions existing during an earthquake are completely different.

As a sidelight here, the writer concluded there is a strong possibility that one of the most probable types of earthquake failure may be the movement of the whole dam, Fig. 1. (That is, a shear failure on a more or less horizontal plane in the upper portion of the foundation). There is no space available here for a full exposition of all the evidence, but it includes both theoretical considerations and observations from experience. Many slides in natural slopes or bluffs caused by earthquakes occur on essentially horizontal sliding planes.

Also one of the few well-documented cases of a complete failure of a dam during an earthquake (the Sheffield Dam during the 1925 Santa Barbara, California, earthquake) was one in which essentially the whole embankment moved downstream as a mass. This general conclusion includes an important element of speculation because the experience is so limited. Nevertheless, there is no question in the writer's mind that there is sufficient supporting evidence to make it prudent for the designer of a dam on a soil foundation in an earthquake area to assume that the most critical mode of shear failure may be the movement of the whole dam on a more or less horizontal plane in the upper part of the foundation (rather than the more conventionally assumed shear slides of the individual upstream or downstream slopes).

Variation of Conservatism in Present Practice.— Under present (1967) practice, many types of dams are built on foundations of widely varying character-

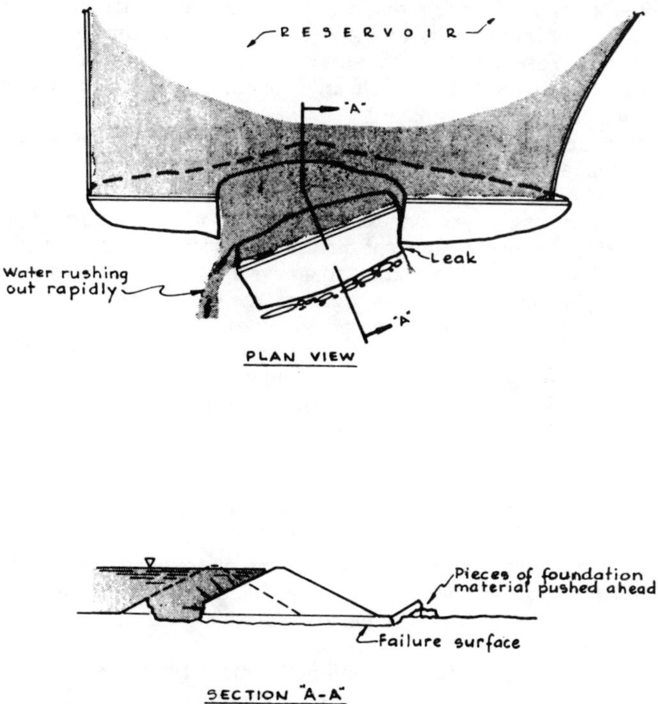

FIG. 1.—POSSIBLE FOUNDATION FAILURE

istics. All are designed and built according to modern standards, and the probability of failure of any one is extremely low. Nevertheless, there are enormous differences in the fundamental safety of these dams. For some, a complete failure of any kind is almost impossible; for others, a failure of one kind or another could very well occur if the conditions were somewhat more hazardous than foreseen.

At one extreme, for example, might be a rockfill dam with a core of material which is not susceptible to piping, good transistion zones, and a foundation of massive hard rock. At the other extreme might be a homogeneous dam which is constructed of a material which does not have high resistance to piping and which is placed on a foundation with low strength and/or low piping resistance. Though both dams may be designed and built by the same competent engineer using the best modern practice for the design details in each case and giving equally conscientious attention to the design and construction, there is no question that the first dam is a much safer structure than the second. It may be 100 or 1,000 times less likely to fail from any possible unexpected conditions which could arise.

In present practice, these differences in inherent safety are not given a great deal of consideration by designers. If, in a hypothetical example, the engineer had two sites for a given reservoir and had the choice of building either of the two dams described in the preceding paragraph (and further if his calculations showed that the safer dam would cost 15% or 20% more than the other), using current design thinking he might have difficulty justifying in his own mind a decision to build the more expensive dam. This is true because both structures would be considered to meet current acceptable standards, and it is difficult to attach an economic value to the much greater intrinsic safety of the first dam.

These differences in the intrinsic safety of different types of dams have the utmost importance from the standpoint of earthquake design. Because we do not have the analytical tools or other means to eliminate all doubt that a given dam might fail during a strong earthquake, we must acknowledge and take into consideration the large inherent differences in the safety of various types of dams and foundations. The extra conservatism which should be built into a given dam to increase its safety against the earthquake hazard should be governed to a large degree by these considerations. Where several different types of dams are technically feasible and could be built to retain an important reservoir, first consideration should be given to the safest type even though it may cost considerably more than the minimum cost structure which might conceivably be used in a nonseismic area.

IMPORTANCE OF GOOD ZONING

Except for the evaluation and treatment of foundations of soil or weak rock, the most important design decisions from the standpoint of increasing the safety against the earthquake hazard are those which concern the zoning of the embankment.

There can be little question that one of the most likely actions of an earthquake will be to cause concentrated leaks through the dam. Leaks can develop through differential settlement cracks caused by embankment slumping under the shaking action, landslides on the abutments which rupture or displace the dam, movements on faults, foundation shifting due to shear strain, and cracks in the outlet conduits. Hence, the better the embankment zoning from the standpoint of providing resistance to the progressive erosion of leaks (or the ability to seal leaks), and from the standpoint of putting the strongest embankment material where it will do the most good in reducing embankment slumping under the shaking action, the greater the intrinsic safety of the dam.

It seems clear to the writer that it should be assumed by the designer that any dam in an earthquake area may develop concentrated leaks. Hence, even low earth dams on sound foundations should have at least a single cohesionless filter ("chimney") extending the full height of the dam for leakage control. A homogeneous dam with a horizontal drain under the downstream slope or a toe drain of moderate dimensions should be considered inadequate. The higher the dam and the greater the uncertainties which may exist concerning the possibility of leaks caused from any source of earthquake action, the greater the need for good zoning and several lines of defense against concentrated leakage in the form of cohesionless filters and transitions.

Transitions and Filters.—One of the best lines of defense available is a relatively wide transition or filter zone consisting of a sand-gravel mixture well graded between a maximum size of several inches and fine sand sizes. This material is not very pervious. Depending on the gradation and the fraction passing the No. 200 sieve, the coefficient of permeability will typically range between 50 ft per yr and 5,000 ft per yr. Thus, even if the core of the dam is badly shattered, as long as the transition remains intact it will prevent the velocity and quantity of leakage from exceeding moderate values. Because of its gradation, this material has ideal filter and self-healing properties. The coarseness of such a zone gives it an inherent stability, in the sense that it cannot penetrate the voids of even very coarse shell material and that it will probably be caught in and tend to seal and control leakage in the largest open cracks which might develop in the foundations and abutments.

Transitions between earth cores and coarse cobble or rockfill zones are sometimes made of two or three adjacent bands of sand and gravel with gradually increasing coarseness from the core outward. Theoretically, as long as there are no great embankment deformations, such a design provides ideal leakage control and filter protection. For earthquake design, however, a single wider transition of a well graded sand-gravel mixture should be considered superior. If the core of the dam were ruptured or offset as the result of any of the possible actions of an earthquake, the wider transition of well-graded material would withstand more severe conditions before losing its effectiveness.

On some dams for which banded transitions are used, the finest band directly adjacent to the core is a medium sand or even finer. In earthquake regions, it seems apparent that it is undesirable to have an individual thick zone of fine sand located downstream from the core. It is conceivable that water breaking through a leak in the core could find an exit through the dam or foundation which bypasses the coarser filter zones. Large quantities of the fine transition could then be carried out of the dam. Therefore, it is desirable that the finest transition material should have an appreciable percentage of gravel size particles. Preferably, it should be well graded and have a D_{85} size over 1 in.

Thin zones of fine filter have usually been provided just downstream from a fine clay core because the coarser well-graded transition will not meet the theoretical requirements for filter action needed by the fine-grained core. In fact, however, these standard filter criteria have not been derived from tests on clays. In general, it can be shown by laboratory tests that a coarse transition material consisting of a well graded sand-gravel mixture will provide adequate filter action for almost any clayey core, Kassiff, et al., (1965). Hence, any theoretical advantage to be obtained from a zone of fine sand filter just downstream from the core is likely to much less important than the potential disadvantage of this material under the extreme conditions of earthquake damage.

Desirability of Processing Embankment Materials.—In addition to utilizing the available construction materials in their natural state to obtain good zoning, the designer can frequently obtain a much safer dam by processing (screening and blending) materials for the various zones. By screening, he can make the shells stronger and more pervious and, as a result, minimize the probable embankment slumping due to earthquake shaking. By processing the filters and transistions, he can provide the most reliable and effective zones for control of leakage. Through blending, he can enormously improve the leak resistance in the core material. Because of the importance of good zoning, all the forces which tend to make the designer consider processing his materials under ordinary circumstances should be considered stronger for dams in earthquake areas. This was certainly a major conclusion from this study which deserves emphasis. For dams in earthquake areas, the designer should always give serious consideration to improving the properties of the available embankment materials by processing them.

It may be desirable to process any one or all of the zones of the dam. Frequently, many zones can be derived from one material—in the case where there is a large nearby deposit of talus, a very coarse alluvium, or other soil consisting of a mixture of large cobbles or rock, gravels and finer soils. In its natural state, material of this kind is not very desirable for use in any zone of an earthquake resistant dam. However, when it is screened on screens of various sizes, it can frequently be separated into sources of material for a well-graded core, graded transitions, and strong free-draining shells which can be used in zones to construct the best type of earthquake-resistant dam.

Strong arguments can be made that most important dams in earthquake areas should, as a minimum, have at least one transition (leakage control) zone of processed materials because of the difficulty in assuring that natural soil deposits will have the desired uniformity of properties. The major exception is in that circumstance where there is an adequate supply of well-graded sand and gravel available from which a major zone of the embankment can be constructed so that it can be assured that a large portion of the dam will have high resistance to concentrated leaks.

TOP OF DAM DESIGN DETAILS

Certainly it is no great exaggeration to state that the top of the earth dam is the most vulnerable single location from the standpoint of earthquake hazard. It is the part where the core is most likely to be damaged by cracks or shear displacements, where wide cracks in the core can open and stay open, and where leaks at the abutment contacts are most likely to occur. At the same time, because it is the thinnest part and because the internal pressures are not high, it has the least inherent safety against concentrated leaks. Also, if the dam is overtopped, the details of the crest will determine to a large degree the rate at which the top of the dam will be eroded.

While this conclusion is generally held in the profession, in current practice the details of the crest zoning are not made much, if any, more conservative in an earthquake region than they would be in a nonseismic area, and the freeboard may or may not be increased by a moderate amount. This discrepancy between opinion and practice is probably best explained by the fact that there is no clear evidence from experience indicating a need to change present practice; that is, no well-publicized failure has been clearly attributed to inadequacies in the crest details. Nevertheless, it is not difficult to be concerned that these details are not being made sufficiently conservative, or at least that the conservatism at the crest is not consistent with the safety inherent in the other main elements of the dam. Also, it is apparent that a relatively small amount spent to increase the safety of the top of the dam may have a large influence on its ultimate safety in an earthquake.

Sometimes the design of the upper part, particularly the part above the normal pool level, is considered less important because it has to withstand less water pressure. The zoning is frequently governed to a large degree by the effort to make the dam easier to build at the top, where the surface is smaller and the construction becomes increasingly awkward. Occasionally, the quality of the construction is even relaxed to accelerate the work.

Insofar as current practice does not take into consideration the special vulnerability of the crest to earthquake damage, it seems apparent that this practice should be changed. As a design criterion, it should be assumed that after an earthquake the reservoir water level may be at the top of the dam and that cracks running completely through the upper part of the dam may have opened to the maximum depth which could be sustained by the strength of the core material (50 ft or more for a tough clay). Zoning should be provided to control the leaks which could develop through these cracks. And since there is less room in the upper part for thick zones of leak-resistant materials, the quality of the soils comprising the zones and the interrelationship of the zones are more important than they are at lower elevations.

The use of the ideal transition material previously mentioned, (i.e., a sand-gravel mixture well graded from coarse gravel sizes to the No. 200 sieve) has special merit near the top of the dam. This material will not sustain open cracks of any magnitude for any great length of time, it will not let much water pass through the zone, and it provides inherent stability against internal erosion. If a thick zone of this material is backed up by a downstream zone of compacted rockfill, it is difficult to imagine leakage developing which could get out of control under the worst conceivable circumstances.

For the typical high, zoned earth-rock dam, with an upstream shell of quarried rock, there is no advantage to be gained (from the standpoint of the stability of the upstream slope) by extending the main rockfill zone all the way to the top. If, in such a dam, the upper 50 ft or so of the shell is built wholly of a cohesionless sand-gravel mixture, the ability of the crest to control concentrated leaks would be appreciably increased. From this reasoning, it can be suggested that the main consideration guiding the designer in selecting the material and zoning for the upstream shell at the top of a high dam should be safety against concentrated leaks. Thus, a well-graded, cohesionless sand-gravel mixture is preferable to a free-draining zone of larger rock.

In summary, it seems reasonable to conclude that the optimum zoning for the top of an earthquake-resistant dam will have the following main elements:

1. Upstream shell or thick upstream transition of a well graded, cohesionless, sand-gravel mixture;
2. core material with high resistance to concentrated leaks;
3. thick and conservatively designed downstream transition; and

4. downstream shell of compacted rockfill or coarse, clean cobbles.

The greater the justification for conservatism, the greater the desirability of providing zoning at the top containing all of these elements.

LEAKAGE RESISTANCE OF CORE MATERIAL

The filter and transition zones provide the first line of defense against earthquake-induced concentrated leaks through the dam. If thick, adequately graded, cohesionless transitions are provided, a leak can only get out of control in extreme cases of embankment distortion of foundation movement. Because of the uncertainties regarding the worst conditions which might result from a severe earthquake, however, it is desirable to anticipate the contingency that a filter zone might be wholly or partially removed or that it might be reduced in effectiveness. The greater the need for conservatism at a given site (because of the greater probability or consequences of failure as cited earlier), the greater the desirability of providing a core with some inherent resistance to concentrated leakage as a secondary defense line against earthquakes.

At the present time (1967), the relationship between soil properties and resistance to leakage is not well understood and there are no standard means for investigating it directly in the laboratory. But although nothing can be said about their merits in a quantitative sense, the soils with the greatest and least resistance are known with some certainty, and it is possible to guess, in a rough fashion, about the relative position of the intermediate ones. In actual fact, the leakage resistance is not a completely independent property of the core material; it depends on the size of the initial leakage channel and the velocity of flow. Also, if the leak discharges from the core into another zone of coarser soil or rock, the action of the leak in the core will depend on the size of the pores of the coarser soil. It seems clear that the most important soil properties are: (1) The plasticity of the fine fraction; (2) the average or maximum size of the coarse particles; and (3) the gradation of the coarse fraction (as measured by the shape of the grain-size curve).

If the soil is fine-grained, then the leakage resistance is wholly a result of the cohesion of the material. There can be little question that fine-grained and cohesionless silts and very fine, uniform sands have the least resistance of all materials. Even a small crack and a leak with low velocity can carry away these particles and lead to rapid progressive erosion.

At the other extreme of fine-grained soils is the highly plastic, tough clay which can withstand high velocities of flow on its surface without appreciable erosion. Evidence of its erosion resistance is given by the facts that it exists in the beds of some rapidly flowing streams and that many small homogeneous dams built of tough clay have safely withstood large concentrated leaks for years, Sherard et al., (1963) pp. 127-129. Consequently, very tough clay is probably among the better materials. Even under the worst conditions of cracking and leakage, it is unlikely that there would be a great deal of erosion with a core of this soil.

Thus, fine-grained soils extend over the whole range from the poorest to the best materials. Differentiating between the intermediate soils in this range is not easy. Probably all silts (ML and MH in the Unified Soil Classification

System) are relatively poor materials not worthwhile to consider separately. For inorganic clays, the leakage resistance is probably strongly dependent on the Atterberg limits; the higher the liquid limit and the higher the position above the A-line on the plasticity chart (Unified Soil Classification System), the higher the leakage resistance, Sherard, (1959).

For cores consisting of a mixture of coarse-grained soils and fines, the resistance to leakage may be derived from both the cohesion of the fines and the action of the coarse particles in the flowing water. The coarse particles may contribute to the leak resistance in several ways.

First, regardless of the width of the leakage channel, the larger and heavier the individual soil particle, the more rapidly the water must be flowing to carry it along and move it out of the core. Hence, the coarser particles are less likely to move out of the core and the more likely to become wedged in the leakage channel where they start to build up a dam of progressively smaller particles to seal the leak.

Second, soil particles which are larger than the average width of the crack cannot be moved along the crack regardless of the water's velocity. If the core has an appreciable percentage of these larger particles, it is unlikely that many smaller particles could be moved out of the crack before a natural filter was built up to stabilize or seal the leak. Because of this and because the width of cracks formed in the core will probably be generally on the order of several inches, increasing the maximum size of the soil particles by an inch or two may very well have a large influence on the leakage resistance of the material.

Probably the best core material from this standpoint is a well-graded sand-gravel mixture with a maximum size of 4 in. to 6 in. For a crack 4 in. wide or less, it seems apparent that there would not be much danger of progressive erosion with such a material. The larger particles would not move out of place, and the finer particles would move slightly under the action of the leak but would soon become lodged between the large gravels, forming a natural filter which would seal a leak.

There are no quantitative guides for evaluating the merits for this purpose of different mixtures of gravel, sand, and fines. As stated above, probably the size of the coarsest fraction is very important. Other things being equal, the inherent leakage resistance of a gravel-sand-fine mixture may well increase with the square of the D_{85} size or even at a faster rate, especially for the potentially large cracks at the top of the dam.

Finally, there is no quantitative way to compare the relative merits of a clay of high plasticity with a well-graded, coarse sand-gravel mixture. Both are good materials, but their ability to resist concentrated leaks derives from completely different actions.

Table 1 provides a tentative classification system for core materials on the basis of resistance to concentrated leaks put together by the writer using the reasoning of the preceding pages. It represents a compromise between an effort to avoid extending the system too far into the realm of speculation and the desire to give the design engineer a framework which he can use in making practical decisions. The quantitative criteria and even the relative position of the various groups have been chosen arbitrarily; future research will undoubtedly show that the order should be changed. Nevertheless, the kind of errors that Table 1 may contain, because it has been extended somewhat beyond existing knowledge, cannot lead to a dangerous situation. The table should provide a useful design tool until a better one becomes available.

Where there is a choice between several types of materials for the core of a dam which may be subjected to an earthquake, it seems apparent that the resistance to concentrated leakage should be the main factor in the decision. The other properties of the core material (i.e., the strength, the compressibility, and the potential construction difficulties) generally vary within a much smaller range and are less important to the safety of the structure.

TABLE 1.—AN ATTEMPT AT AN APPROXIMATE CLASSIFICATION OF CORE MATERIALS ON THE BASIS OF RESISTANCE TO CONCENTRATED LEAKS

1	Very Good Material
	a. Very well-graded coarse mixtures of sand, gravel, and fines. D_{85} coarser than 2 in.; D_{50} coarser than 1/4 in. If fines are cohesionless, not more than 20% finer than the No. 200 sieve.
2	Good Materials
	a. Well-graded mixture of sand, gravel, and clayey fines. D_{85} is coarser than 1 in. Fines consisting of inorganic clay (CL) with plasticity index greater than 12.
	b. Highly plastic tough clay (CH) with plasticity index greater than 20.
3	Fair Materials
	a. Fairly well-graded, gravelly, medium to coarse sand with cohesionless fines. D_{85} coarser than 3/4 in.; D_{50} between 0.5 mm and 3.0 mm. Not more than 25% finer than the No. 200 sieve.
	b. Clay of medium plasticity (CL) with plasticity index greater than 12.
4	Poor Materials
	a. Clay of low plasticity (CL and CL-ML) with little coarse fraction. Plasticity index between 5 and 8. Liquid limit greater than 25.
	b. Silts of medium to high plasticity (ML or MH) with little coarse fraction. Plasticity index greater than 10.
	c. Medium sand with cohesionless fines.
5	Very Poor Materials
	a. Fine, uniform, cohesionless silty sand. D_{85} finer than 0.3 mm.
	b. Silt from medium plasticity to cohesionless (ML). Plasticity index less than 10.

For very critical dams, it could be argued that conservative design practice would require not only thick and well-graded transitions both upstream and downstream, but also a core constructed from materials which would fall into Groups 1 or 2 of Table 1.

As for the materials in Groups 4 and 5, there is serious question about whether an important dam in a seismic area should ever have a core made from one of these. A conservative view of the problem would maintain that such materials should not be used except where failure of the dam would not cause a disaster, and that if they are used, it should only be in connection with excellent transition zones and in an embankment designed so that the dam should be safe even if large volumes of the core were carried away by piping.

SPILLWAY CONSIDERATIONS

In common current (1967) practice, not much consideration is given to the influence of the spillway design on the safety of the dam during an earthquake. The basic decisions concerning the spillway are so strongly dominated by other factors (i.e., economy, topography, foundation conditions) that the spillway's effect on the earthquake resistance of the dam is not taken into account. (It is common, of course, to give consideration in the design to earthquake damage to the spillway itself from landslides. In earthquake areas, the general preference in the profession for an open-channel spillway over a tunnel is even greater than in nonseismic areas. Wider excavations and flatter slopes have been used for open-channel spillways to decrease the likelihood that the spillway may be blocked by earthquake-induced slides.)

In contrast to this general practice, there can be little doubt that the greater the distance between the normal reservoir water level and the top of the dam, the safer the dam will be against the possibility of failure in an earthquake. In the extreme case where the dam is strongly shaken and the embankment slumps considerably or where the top of the dam is left badly cracked and leaking after the earthquake, even a few feet of extra height can be important; it can quite literally provide the difference between failure and the needed time to carry out remedial action to save the dam. Since the spillway type and details can have a large influence on height of the dam above the normal pool level, it follows that the spillway decisions are important in the design effort to make the dam safe against earthquake action. Sometimes, by leaning one way or another in a decision for the spillway, the dam can be made considerably safer against earthquakes without increasing the total cost of the project.

Another point which should influence the design considerations for the spillway is that it may be necessary to use the spillway immediately following the earthquake. Landslides into the reservoir, an uplift of the reservoir floor with respect to the dam, or a settlement of the dam with respect to the reservoir can cause an almost instantaneous reduction in the reservoir storage volume. Hence, there may be no time to repair structural damage to the spillway before it is necessary to spill. The typical spillway, moreover, is probably more likely to be damaged structurally by an earthquake than by other cause.

Storing Floods.—It is becoming increasingly common to provide a combination of dam height and spillway capacity that allows only a part of the design flood to be taken immediately through the spillway while the rest is stored temporarily in the reservoir by letting the water level rise. The determination of the portion of the flood which will be stored is usually made on the basis of economic studies. In these studies, several tentative designs are made with different combinations of dam height and spillway size and the total cost computed for each combination, Fig. 2. Depending on a number of major factors, such as the topography, the design flood, the size of the reservoir, and the relative cost of the dam and spillway, the minimum total cost for spillway and dam can range in practice anywhere between the two extremes: (1) that which stores none of the flood, and (2) that which stores all of the flood. The larger the cost of the spillway with respect to the cost of the dam and the larger the surface area of the full reservoir with respect to the design flood, the greater the economic advantage to building the dam higher to store the flood. Also, the greater may be the influence which the spillway design will have on the earthquake resistance of the dam.

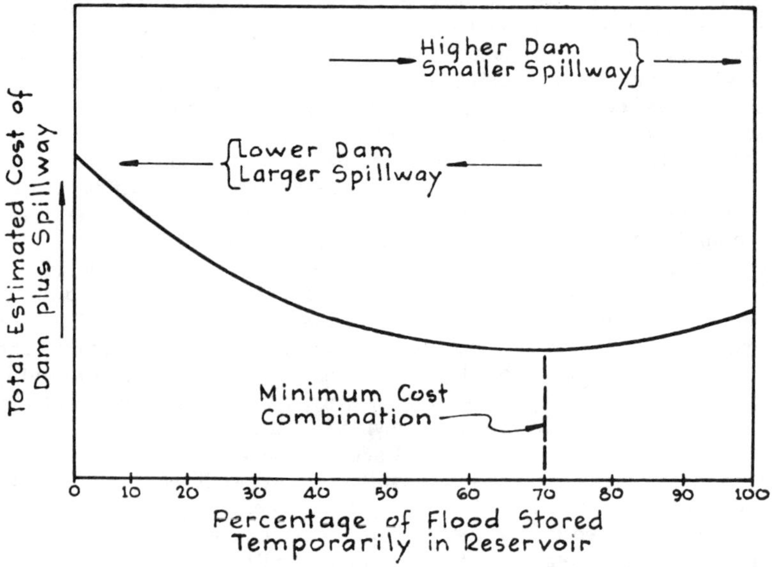

FIG. 2.—TYPICAL PLOT OF RESULTS OF COST STUDIES OF DIFFERENT COMBINATIONS OF SPILLWAY SIZE AND DAM HEIGHT

Near the combination of spillway size and dam height which gives the minimum cost, generally a curve representing the relationship between total cost and spillway size is relatively flat and the designer can go in either direction from the minimum point without increasing the total combined cost of dam and spillway by a great deal. If the decision is left to the spillway designer, his interests would normally cause him to lean toward a larger spillway and lower dam. In this way he could get a larger spillway for about the same cost, and he would have a somewhat larger safety factor against the possibility of a totally unexpected flood. Conversely, from the standpoint of increasing the safety of the dam against earthquakes, there is no doubt that it is preferable to go in the other direction—toward a smaller spillway and greater vertical distance between the normal water level and the top of the dam. For a given case where two alternative possible designs could be used at about the same total cost, one may have 15 ft or more greater freeboard than the other. This extra height, which is gained at no appreciable extra cost, can have an enormous influence on the safety against earthquakes.

Another reason for working in the direction of the smaller spillway and higher dam occurs when the spillway becomes 30 ft or 40 ft wide. At this width, the spillway chute is inherently much stronger, since the sidewalls and the bottom slab can be made to act together as a rigid frame structure. For the wide spillway, the bottom slab has little reinforcement from the sidewalls and is much more vulnerable to displacement and damage.

Gated Spillways.—In an earthquake area there are two important reasons for selecting an ungated spillway. First, it is more likely that the gates may not function properly after the earthquake. Second, the distance between the normal water level and the top of the dam is greater in the case of an ungated spillway.

The damage to Ponteba Dam in Algeria during the 1954 earthquake, Thevenin, (1955); Thevenin and Lemay, (1964); Walters, (1962), where the gates were wedged in their guides and had to be burned out to free the channel, gives clear evidence that the gates themselves can easily be rendered inoperable. In general, the tolerance to differential settlement in the typical gate is so small that large strains are not required to put them completely out of commission.

The only advantage of a gated spillway from the earthquake standpoint is in a situation with the following conditions: (1) Where the concrete sill of the gated spillway is considerably lower with respect to the top of the dam than the crest which would be used for the alternate ungated spillway; or (2) where the reservoir size and the river channel downstream are such that it would be feasible to pull the gates completely free of the water and use the full capacity of the spillway to lower the reservoir in a period of days or perhaps a few weeks. In such a situation, the ability to pull the water level down further might have an important influence on the safety of a dam which is badly shattered at the top by an earthquake.

Open Channels Versus Tunnels.—Other things being equal, it is generally agreed that an open-channel spillway is a basically safer structure than a tunnel spillway. This superiority of the open-channel spillway is even greater for dams in earthquake areas. The main reason is that the tunnel mouth is more likely to be blocked by a smaller landslide than would block an open channel. Also, movements on faults could cause worse trouble for tunnels than for open channels. Perhaps an even more important consideration is that a landslide in an open channel can be removed fairly easily, whereas, if the tunnel is blocked by some action of an earthquake, it may be impossible to open the spillway rapidly. For this reason particularly, it seems apparent that there are strong disadvantages to building tunnel spillways in earthquake regions, and that they should be avoided if at all possible.

Erodible Soil Foundations.—The experience with Hebgen Dam in Montana during the 1959 earthquake clearly indicates one of the ways that a complete failure of a dam can occur from earthquake damage to the spillway. The spillway was founded on a soil abutment and the concrete lining was badly shattered. If it had been necessary to use it in the days following the earthquake, progressive erosion of the material under the broken slab could easily have led to a breach in the abutment and complete failure. If, however, the spillway had been built on the other abutment, which was sound rock, it is probable that the concrete would have been less badly damaged; and in any event, the rock foundation would have been able to withstand the erosion from the flowing spillway water with much less likelihood of complete failure.

It is always a possibility that the typical spillway built on a soil abutment around the end of the dam may be taken out completely by a landslide. Short of this extreme, the hillside may simply slip downward several feet twisting and deforming the spillway. For these reasons, it is surely justified to spend considerably more money to put the spillway on a stable foundation of rock if at all possible.

Expansion Joints in Spillway Chutes.—In common practice, there are two alternate details used for reinforced concrete spillway chutes by different engineers at any given site. One group breaks the spillway into discrete lengths, using expansion joints with shear keys in the joints between them. In this case, the reinforcing steel is not run through the joints. The other group builds the spillway as a monolithic structure with no expansion joints and with the reinforcing steel run through all the construction joints. The argument for the first design is that expansion and contraction of the slab from temperature changes and the creep down the hill can be absorbed in the expansion joints without building up stresses in the structure. The argument for the second is that by tying the whole structure together, there is less likelihood that a single slab could be popped out when the spillway is discharging.

Whatever the arguments for and against these two fundamentally different design details under ordinary circumstances, it seems obvious that if the foundation of the spillway chute is moved several feet in an earthquake (which it is difficult or impossible to assure could not happen for many spillways founded on relatively steep natural hillsides), then the monolithic slab is the better detail. There is a much better chance that the continuously reinforced structure will hang together long enough to pass water safely following an earthquake. If such large strains are imposed on a spillway channel which consists of independent slabs separated by expansion joints, it is easy to imagine that one of the slabs could be raised enough that the flowing water would wash it out. With one slab gone, others could follow in rapid succession.

EVALUATION OF THE SEISMICITY OF THE SITE

As part of the earthquake design problem, it has been common practice for the engineer to make some attempt to estimate the severity of the shaking to which the dam may be subjected as a function of the geographic location of the site or the local geologic conditions. For consideration of most of the principal design features which govern the fundamental safety of the dam against failure by earthquake action (for example, the choice of the embankment materials, zoning, crest details, and foundation treatment), he does not need this estimate in quantitative terms, since these aspects of the design are not amenable to analytical treatment in any event. If calculations of the stability of the structure are to be made, however, it is necessary to have something quantitative.

As a main result of this study, the writer has concluded that it is reasonable to assume that all dams located in any large general area of seismic activity such as the state of California may be equally severely shaken regardless of their location within the area. This conclusion can be expressed alternatively: "The conservatism built into a given dam located anywhere in a general large area of potential strong earthquakes should not depend to any appreciable extent on the results of a study of the 'seismicity' of the specific site which leads to an estimate of the 'maximum probable' shaking which the dam may experience during its lifetime." There is no space available in this paper for a complete exposition of all the arguments which have led to this conclusion; however, they are all based fundamentally on the opinion that there is not yet enough information or knowledge available concerning the prediction of large earthquakes to permit a civil engineer building a dam to adopt a less conservative or more refined position.

It should be emphasized that the conclusion stated above does not imply in any way that the accumulated knowledge in the fields of geology and seismology is not of great value to the dam engineer in connection with the earthquake design problems. On the contrary, the advice of the specialist in these fields is frequently invaluable. He is the best man to evaluate the hazard of fault movements and he can estimate the relative seismicity of various sites; that is, in the circumstance where the dam can be built at several locations, he can give some guidance as to the relative likelihood that each will be strongly shaken. While it is possible to differentiate among various sites within a general seismic region on the basis of the relative likelihood of a severe shock in any given 10-yr period, it is not possible to assure that a major earthquake will not occur anywhere in the region during the lifetime of a dam. Most of the seismologists and geologists consulted in the course of the study agree in principle with this conclusion.

In this regard, it is pertinent and instructive to examine and compare the practice and philosophical approach commonly used for design of buildings and with those justified for dams. The Uniform Building Code which is used for buildings by most of the smaller cities and rural communities in California divides the state into two areas from the standpoint of seismic design forces for buildings. Most of the state is categorized Zone 3, and the central valleys and a part of northern California is Zone 2. Zone 3 requires design for the highest earthquake forces specified by the code. Other things being equal, the earthquake forces specified for buildings in Zone 3 are twice those for Zone 2.

In the course of this study, a number of thoughtful and able structural engineers who work in various parts of the state were asked their opinion regarding the use of different seismic design forces in different geographic areas in California. Almost all replied that they prefer to design all buildings in the state for the Zone 3 criteria. They said that they do not believe enough is known about earthquake prediction to be confident that any one area in the state could not be shaken as severely as others. Hence, it can be considered common practice, at least among a large group of consulting structural engineers, to design buildings in California for the same high seismic forces, regardless of their location in the state. If this approach is justified for buildings, it seems apparent that the justification is much stronger for dams for a number of reasons.

First, because many more buildings than dams have been shaken by strong earthquakes, the structural engineer has a body of experience to draw upon in predicting how a certain intensity of shaking will affect his building. The earth dam engineer does not and never will have comparable information. Second, for a building frame which consists of essentially elastic materials, the maximum stresses that are likely to develop in all the major elements can be calculated with some degree of assurance. Thus, the structural engineer has some confidence that his building will actually withstand the specific earthquake he is designing against. Moreover, because of the safety factors used, the principal structural elements of a typical well-designed building can absorb considerably more earthquake load than the design values without catastrophic results. Where a very strong earthquake shakes a building which had been designed to withstand only a moderate shock, the building will probably not fail completely and kill all its occupants, even though it may be badly shattered.

In earth dam design, however, current knowledge simply does not permit the engineer to judge reliably whether a dam just might be designed for a moderate earthquake would also withstand a more severe shock without failure. The adequacy of only a small part of the earth dam design can be evaluated analytically (analysis against shear failure) and the reliability of these stability analyses is several orders below that for any computations used in the analysis of buildings.

A third reason why it is desirable to adopt a more conservative attitude for dams than for buildings is that the influence of the foundation is quite different for the two structures. Since even a skyscraper is not as heavy as a small dam, the character of the building's foundation is not of the same order of importance as that of the foundation under a dam. And if the dam foundation is suspect, it is the single most difficult aspect to evaluate.

Fourth, the existence of the reservoir behind a dam justifies another whole magnitude of conservatism when comparing the practice for dams and buildings. If a building is still standing after an earthquake, then it has fulfilled the designer's primary requirement of public safety, even though it may be badly cracked. In contrast, deformation and cracking in an earth dam may be only the beginning of far more serious trouble.

Fifth, the average anticipated lifetime of the ordinary reservoir is likely to be much greater than that of the average building; hence, the dam has a much greater chance of being shaken by a strong earthquake.

The writer has no fixed opinion concerning the criteria which should be used to define the extent of the geographic area over which it is desirable to assume that all dam sites may be shaken equally severely; however, he has concluded that in all probability it would be wise to include at least all that portion of the western United States located to the west of the major active north-south Wasatch Fault system in central Utah. Following this line of reasoning, the writer would assume a very strong earthquake could shake any dam site in broad geographic areas surrounding any point in the world where there has been historical evidence of moderate to strong earthquakes or where there is any geologic evidence of recent tectonic activity (recent faulting, etc.).

SEICHES

Bodies of water are frequently set into oscillation (seiches) by earthquake ground shaking, evidently when the dominant period of the shaking motion approximates the natural frequency of the water body. It has been speculated that these seiches could endanger an earth dam and in a few recent cases design engineers have made rough attempts to calculate the maximum height which such waves might reach.

It is well known that lakes, bays, and other ponds of water do develop seiches and that they may occur hundreds of miles away from the epicenter of a strong earthquake, for example, see Richter, (1958), p. 110. However, no records were found of seiches in which the vertical amplitude of the oscillation was more than a few feet. One of the most spectacular cases developed at the Hebgen reservoir in 1959. The oscillating water apparently reached a height of about 10 ft above the normal level. But probably this oscillation was caused mainly by the great subsidence and tilting of the reservoir bottom rather than by a resonance phenomenon associated with ground surface shaking. Even as a result of this extreme excitation, the seiches did not create a serious threat to the dam.

A number of small ponds and tanks relatively far away from the epicenters of earthquakes have developed waves with the heights of several feet. A well-studied example is given by Steinbrugge and Moran (1957) for the 1954 Nevada earthquake. Three small tanks (less than 300 ft long) with water depth of less than 20 ft were set into oscillation in Sacramento, California. In this case, the dominant ground surface motion in Sacramento had periods between 5 sec and 9 sec, which approximate the natural period of oscillation of the small water bodies.

An interesting theory for the computation for the height of wave which might be generated by a given earthquake ground shaking has been developed by G. W. Housner (1957). This theory is probably quite good for application to small tanks with natural periods of 6 sec or less. According to Housner, there is no reason that the theory itself could not be used for large bodies of water; however, to use it for reservoirs more than a few hundred feet long, it would be necessary to know more than is presently known about long-period ground shaking motion. Housner's theory does indicate that the height of the wave which will be generated from this action is not greatly dependent on the depth of the water in the reservoir.

Many of the geologists and seismologists consulted in the course of this study believe it is very unlikely that seiches with a height of more than a few feet will occur on large reservoirs. The basic reason for this conclusion is simply that many lakes have been shaken by strong earthquakes; yet there are no records of seiches with dangerous heights.

Finally, even if a seiche could conceivably reach a height of as much as 30 ft above the still water level in a large reservoir and send a 20-ft wall of water over the top of the dam a number of times, since it will surely require a period of at least several minutes for the wave to travel from one end of the reservoir to the other, the total amount of time during which water will pour over the dam crest cannot be great. From all these observations, the writer has concluded that, while we still cannot absolutely rule out the eventuality that ground surface shaking may someday generate seiches large enough to damage an embankment dam, it seems almost inconceivable that this action could threaten failure. Because of this and because there is no reliable means to do it in any event, the writer believes that it is a fruitless exercise during the design of a dam to endeavor to compute the height of a seiche which might be generated by the ground surface shaking. There can be little question that the hazard of waves generated by landslides, of the type which caused the disaster at Vaiont Dam, is greatly more imminent and likely.

SUMMARY OF CONCLUSIONS

There was a real possibility that the Hebgen Dam could have failed as the result of the 1959 Montana earthquake. If it had, it would probably have had a major influence on present design practice for earth dams. Consequently, it behooves the profession to understand this experience and consider it as being relatively important in the evolution of our ability to build safe dams.

We are confident that embankment dams which would be safe against strong earthquake shaking can be built at most sites with suitable foundations and that it is reasonable and prudent to do so. A large single part of this confidence derives from the general knowledge that compacted soils and rocks do not lose an appreciable portion of their shearing strength when they are strained beyond the normal failure strain or when they are subjected to rapidly alternating pulses of strain such as are generated during an earthquake. Hence, regardless of the momentary conditions which may exist during the earthquake, after the shaking is over, the dam will be as stable as before. The more doubt which we may have in any given case that the strength of the foundation or of a portion of the embankment may be reduced by the earthquake action, the less confidence we have in our ability to build a safe dam and the more arbitrary conservatism is justified.

Under current common design practice, the degree of conservatism built into a dam is not varied as a function either of the geographic location of the dam (with respect to how much damage would be caused by a failure) or as a function of the type of dam or foundation (which governs the likelihood of a failure in the event that an untoward condition develops). Under this current philosophy, every dam is designed in such a fashion as to make it "completely safe" and, therefore, it is not necessary to examine the likelihood or consequences of a failure. For dams in seismic areas, a fundamental change in this philosophy is needed. All other things being equal, a dam whose failure would cause a great loss of life should be built more conservatively than one whose failure is potentially less hazardous. No specific rules can be given regarding the extra degree of conservatism needed in a given case (or the amount of money which should be spent to achieve it); we can, however, list the factors which should influence the decision, among which are: (1) The nature of the foundation and abutments; (2) the height of the dam; (3) the materials available for the construction of the dam.

Another way to view the above conclusion is to scrutinize the fact that there are great differences in the intrinsic safety of different dams at different sites although all may be designed and constructed according to acceptable modern criteria and practice. This is true in the sense that some would safely withstand almost any concentrated leakage which might conceivably develop (or any effort at sabotage) whereas, others, because of the nature of the embankment or foundation, could fail if the conditions became much more severe than those foreseen by the designer. These differences in intrinsic safety—which depend primarily on the conditions existing at the site rather than on the actions and decisions of the designer—are not given great consideration in current practice. For dams in seismic areas, they should be considered of the greatest importance.

The following embankment design provisions have a fundamental and vital influence on the safety of a dam against the earthquake hazard:

1. Zoning to provide high resistance to concentrated leaks and to put the strongest embankment material where it will do the most good;

2. special details at the top of the dam: better zoning; wider dam; higher freeboard;

3. choice of (or improvement of to obtain) a core material with the

highest resistance to progressive erosion of leaks.

Where there is a choice of several materials for the core, the relative safety of the materials under the action of concentrated leaks (i.e. to resist erosion and to seal the leak) should always be considered an important factor in the decision. Generally, the other main properties of the core material (strength, compressibility, etc.) are of a much lesser order of importance to the safety of the dam against the earthquake hazard.

A fundamentally safer dam from an earthquake standpoint can frequently be made by processing (screening and blending) soils and rocks for both the shells and core of the embankment.

There are a number of major design decisions regarding the spillway and the interrelationship of the spillway and the dam that may have an important influence on the safety of the dam against the earthquake hazard.

The designer is not justified in making the effort to compute the magnitude of the oscillations (seiches) of the reservoir body which might be generated by the shaking of the bottom.

ACKNOWLEDGMENTS

The writer wishes to express his deep appreciation to the California Department of Water Resources (William E. Warne, Director; Alfred R. Golzé, Chief Engineer; Robert B. Jansen, Division Engineer, Division of Safety of Dams) for making the study possible and for permission to summarize some of the results in this paper. Thanks are especially due to W. A. Brown, until recently long-time supervisor of the Division of Safety of Dams, who gave more than generously of his time, advice and inspiration in support of the effort. The writer wishes also to express his gratitude to his colleagues on the Department of Water Resources' Consulting Board for Earthquake Analysis (Hugo Benioff, Clarence R. Allen, George W. Housner, H. Bolton Seed and Nathan D. Whitman, Jr.) for all their assistance over the last few years.

APPENDIX.—REFERENCES

Ball, R. M., "A Report: Earthquake, August 17, 1959," Montana Power Company 1960; reprinted in the Eleventh Annual Field Conference, Billings Geological Society.

Hadley, J. B., "Landslides and Related Phenomena Accompanying the Hebgen Lake Earthquake of August 17, 1959," Professional Paper 435, Geological Survey, U. S. Department of the Interior, Washington, D. C., 1964.

Housner, G. W., "Dynamic Pressures on Accelerated Fluid Containers," Bulletin of the Seismological Society of America, Vol. 47, 1957.

Kassiff, G., et al., "Analysis of Filter Requirements for Compacted Clays," Sixth International Conference on Soil Mechanics and Foundation Engineering, Canada, Vol. 2, 1965, p. 495.

Koenig, H. W., "Der Neubau der Hennetalsperre des Ruhrsperrenvereins," Westdeutschen Wasserwerschaftsverbandes, E. V. Essen, 1957.

Press, H., Talsperren, Wilhelm Ernst & Sohn, Berlin, 1958.

Richter, C. F., Elementary Seismology, W. H. Freeman and Company, San Francisco, Calif., 1958.

Ryall, A., "The Hebgen Lake, Montana, Earthquake of August 18, 1959; P-Waves," Bulletin of the Seismological Society of America, Vol. 52, 1962, p. 235.

Schuster, D. R., "Dammbau und Asphalbeton—Dichtung der Hennetolsperre," Bitumen (Hamburg), Nos. 8 and 9, 1957.

Sherard, J. L., discussion of "Tractive Resistance of Cohesive Soils," by I. S. Dunn, Journal of the Soil Mechanics and Foundations Division, ASCE, Dec., 1959.

Sherard, J. L., et al., Earth and Earth-Rock Dams, John Wiley & Sons, New York, 1963.

Sherard, J. L., A Study of the Influence of the Earthquake Hazard on the Design of Embankment Dams, report submitted to the California Department of Water resources, July, 1966.

Steinbrugge, K. V., and Cloud, W. K., "Epicentral Intensities and Damage in the Hebgen Lake, Montana, Earthquake of August 17, 1959," Bulletin of the Seismological Society of America, Vol. 52, 1962.

Steinbrugge, K. V., and Moran, D. G., "Engineering Aspects of the Dixie Valley—Fairview Peak Earthquakes," Bulletin of the Seismological Society of America, Vol. 47, No. 4, 1957, p. 335.

Thevenin, J., "The Effects of the September 9, 1954 Earthquake in the Orleansville Region," Terres et Eaux, 1955, No. 24 (in French).

Thevenin, J., and Le May, Y., "Quelques Constations Particulieres Faites Sur Les Deformations de Barrages sous L'Effect de la Charge d'Eau, de la Temperature, et des Seismes," Eighth Congress on Large Dams, Edinburgh, Scotland, Vol. II, 1964, p. 261.

Tocher, D., "The Hebgen Lake, Montana, Earthquake of August 17, 1959," Bulletin of the Seismological Society of America, Vol. 52, 1963, p. 153.

Walters, R. C. S., Dam Geology, Butterworths, London, 1962.

Wiegel, R. L., and Camotim, D., "Model Study of Oscillations of Hebgen Lake," Bulletin of the Seismological Society of America, Vol. 52, 1962, p. 273.

Witkind, I. J., et al., "Geologic Features of the Earthquake at Hebgen Lake, Montana," Bulletin of the Seismological Society of America, 1962, Vol. 52.

Introduction to Paper 4

Piping in Earth Dams of Dispersive Clay

by J.L. Sherard, R.S. Decker and N.L. Ryker
Proceedings, Specialty Conference on Performance of Earth and Earth-Supported Structures, American Society of Civil Engineers, Vol 1, Part 1, 1972, pp 589-626.

Prepared by Norman L. Ryker[1]

In 1971 Jim Sherard was engaged to investigate the common failure of a group of Soil Conservation Service (SCS) floodwater retarding earthfill dams built between 1957 and 1970, eleven being located in Oklahoma and three in Mississippi. These dams failed upon initial reservoir filling from internal erosion along cracks developed in compacted dispersive clays. The uniqueness of the natural dispersion and erodability of these otherwise intact plastic clays intrigued Dr. Sherard, leading him to further expand this investigative research at his own expense to include other earthfill projects known to him to have had similar problems. He therefore became involved with researching information from projects located in several other states and many foreign countries. In one parallel to the study, Jim unraveled some mystery involving a piping accident in the late 1940's with a large dam in eastern Oklahoma.

While a small number of engineers in the United States and Australia were familiar with dispersive clays, it was not until Dr. Sherard happened upon the scene that the civil engineering profession became aware of the nature and identity of these problem soils. His presentations of the following two papers at the Purdue specialty conference in 1972 were the vehicles that brought this about. The paper "Hydraulic Fracturing in Low Dams of Dispersive Clay" was awarded the Arthur M. Wellington prize by ASCE in 1973.

I first met Jim Sherard at the beginning of his work for the SCS. It was immediately apparent to me that here was an exceptional engineer, a man who devoted his complete self, an outstanding knowledge, energy and zeal, to his work. He met people easily and was a remarkable communicator. By example he readily stimulated others to enthusiastically share their insight, knowledge and experiences with him. These two papers bear out these facts.

[1]Soil Mechanics Engineer, Soil Conservation Service, U.S. Department of Agriculture, Stillwater, OK.

PIPING IN EARTH DAMS OF DISPERSIVE CLAY

By James L. Sherard[1], F.ASCE;
Rey S. Decker[2]; and
Norman L. Ryker[3]

INTRODUCTION - PREVIOUS STUDIES

Australian Research

During the last decade research on the failures of small dams, carried out largely in Australia, has provided new insights into the erodibility of clay soils (1,2,7,12,13,14,15,20,23,34)*. The Australian experience is based primarily on study of failures in low clay dams retaining small farm ponds. The main conclusions from these Australian researches are as follows:

1. There are some clays existing in nature which are highly erodible, even when compared to cohesionless fine sands and silts ("dispersive" clays).

2. These clays erode in the presence of water by a process called "dispersion" or "deflocculation". This occurs when the repulsive forces (electrical surface forces) between individual clay particles exceed the attractive (van der Waals) forces so that when the clay mass is in contact with water, individual clay particles are progressively detached from the surface and go into suspension. If the water is flowing, the dispersed clay particles are carried away.

3. The main property of the clay governing the susceptibility to dispersion piping is the quantity of dissolved sodium cations in the pore water relative to the quantities of other main basic cations (calcium and magnesium). The sodium acts to increase the thickness of the diffused double water layer surrounding individual clay particles and hence to decrease the attractive force between the particles, making it easier for individual particles to be detached from the mass.

4. A second main factor governing the susceptibility of the clay mass to dispersion piping is the total content of dissolved salts in the reservoir water. The lower the content of dissolved salts in the water, the greater the susceptibility of the clay to dispersion piping. Examples are given of dams which did not fail when retaining a reservoir of relatively salty water but failed rapidly when the reservoir was emptied and refilled with pure water.

5. Clays with exchangeable sodium percentages (ESP) of 7 to 10 are moderately dispersive and have been associated with piping failures in clay dams when the reservoir water is relatively pure. Clays with ESP of 15, or more, have serious piping potential (ESP is defined later).

---

1) Consulting Engineer, Berkeley, California; 2) Head, Soil Mechanics Unit, Soil Conservation Service, Lincoln, Nebraska; 3) Soil Mechanics Engineer, Soil Conservation Service, Stillwater, Oklahoma.

*Numbers in parentheses refer to References, Appendix B.

6. When a concentrated leak starts through a clay embankment with high ESP, either of two actions can occur: (1) if the velocity of flow is sufficiently low, the clay surrounding the flow channels swells and progressively seals off the leak, or (2) if the initial velocity is sufficiently rapid, the dispersed clay particles are carried away, enlarging the flow channel at a faster rate than it is closed by swelling, leading to progressive piping failure.

7. There is no good relationship between the ESP and the results of ordinary index tests employed by the soil mechanics engineer. Highly dispersive clays frequently have the same Atterberg Limits, particle size distribution, compaction characteristics, and visual appearance as ordinary erosion resistant clays. Highly dispersive clays associated with piping failure in Australia were generally of low to medium plasticity (liquid limit 30 to 50), but they also give examples of piping failures in dams of clay of high plasticity (7,20).

8. High ESP values and piping potential are generally found to exist in soils in which the clay fraction is composed largely of montmorillonite. Some illites are highly dispersive. High values of ESP and high dispersibility are rare in clays composed largely of kaolin.

Theory of Colloids and Agricultural Soil Science

The Australian conclusions described above are novel and somewhat surprising to the soil mechanics engineer. They are consistent, however, with fundamental laws governing the flocculation and suspension of colloids. Also, the same general relationships have been understood for many years by agricultural soil scientists.

The Schultz-Hardy rule, formulated before 1900 after numerous flocculation experiments with colloids in suspension in test tubes, states that the higher the valence of the cation in solution in the water, the lower the concentration of dissolved salts needed to cause suspended colloidal particles to flocculate. Commonly, if a given colloidal suspension will be flocculated by a certain concentration of divalent calcium salt, then 50 to 100 times the concentration of monovalent sodium salt is needed to cause flocculation (30). Hence, it would be expected that clays with sodium in the pore water would go into and stay in suspension in pure water much easier than clays with pore water containing primarily dissolved calcium salts.

It has been known by agricultural soil scientists for more than 40 years that clays with ESP of 15 or more, and low total salts in the pore water, are highly dispersive (21). Soils of these properties are generally called "nonsaline-alkali" soils, and are sometimes characterized by the general term "black alkali" or as "solonetz" (Russian term) soils. Dispersive clays cause trouble in agriculture because the individual clay particles become suspended in raindrops falling on the ground surface and they are carried downward into the soil with the percolating water where they accumulate and form a dense layer of low permeability, which makes the ground difficult to irrigate and drain.

Soil scientists have also known for many years that agricultural fields consisting of a surface layer of dispersive clay may be damaged, or rendered useless, by the development of sinkholes and erosion tunnels. Where a surface layer of dispersive clay is underlain by pervious sand or gravel, rainwater flows downward through initial leakage channels formed as the result of drying cracks, animal holes, or holes left by decaying roots. The walls of the leakage channel are rapidly eroded forming vertical tunnels and sinkholes. Figure 1 shows a schematic drawing of vertical piping and sinkhole formation caused by downward flow of rainwater in southern Arizona at a location where there was 12 feet of dispersive clays overlying pervious gravel. Similar damage to natural clay deposits by tunnel erosion is described in many papers, mainly in the soil science literature (3,5,9,11,19,21).

Some of the most spectacular early examples of dispersive clay piping were in the southwestern part of the United States, and from these early studies it was believed that the problems were confined largely to soils formed in arid and semi-arid climates. In recent years the same soils and erosion problems have been found to exist in humid climates in various locations. in the United States, and in many other countries (see, for example, 14,19,33).

Soil Conservation Service Experience

The Soil Conservation Service (SCS), which builds many small irrigation, flood control and drainage structures all over the United States, and also deals with farmers from the standpoint of agronomy and soil science, recognized many years ago that certain clayey soils in regions of the western United States are particularly susceptible to erosion and did some pioneer investigations on the subject. For nearly 40 years the Laboratory Dispersion Test, with procedures as described in Appendix A, has been used by the SCS as a measure of "dispersibility" or erodibility of clayey soils (32). As a part of this test the content of 5 micron size particles is measured in the standard hydrometer test, in which a chemical dispersant is used in the water and the soil sample is dispersed in the water with strong mechanical agitation. As seen in Appendix A, a parallel test is then made in which no chemical is used in the water and in which the soil is not dispersed with strong mechanical agitation. The content of 5 micron size which goes into suspension in the water in this latter test is compared with the content of 5 micron sizes measured in the standard hydrometer test. This laboratory dispersion test is now employed as a routine laboratory test by the SCS engineers for structures throughout the United States in areas where there is a history of trouble from dispersion piping. Their experience over many years indicates the following conclusions concerning the applicability of the test:

1. In some areas of the United States where there is a considerable problem with dispersive piping of clays, such as the state of Oklahoma, the dispersion test gives a good measure of the dispersibility of the clay. Over 30% dispersion from the laboratory test is considered to be moderately susceptible to erosion and tests on structures which have had severe erosion usually show some samples with 50-75% dispersion, or more.

2. For tests on ordinary erosion resistant clays in areas of the United States where there is little or no trouble from clay erosion, such as the loessial clays of Iowa and Nebraska, commonly the Laboratory Dispersion Test gives 10-15% dispersion or less.

3. In some areas of the U.S. where structures have been seriously damaged by dispersive clay piping, such as the state of Mississippi, the Laboratory Dispersion Test does not give a consistently good measure of the dispersibility of the clay.

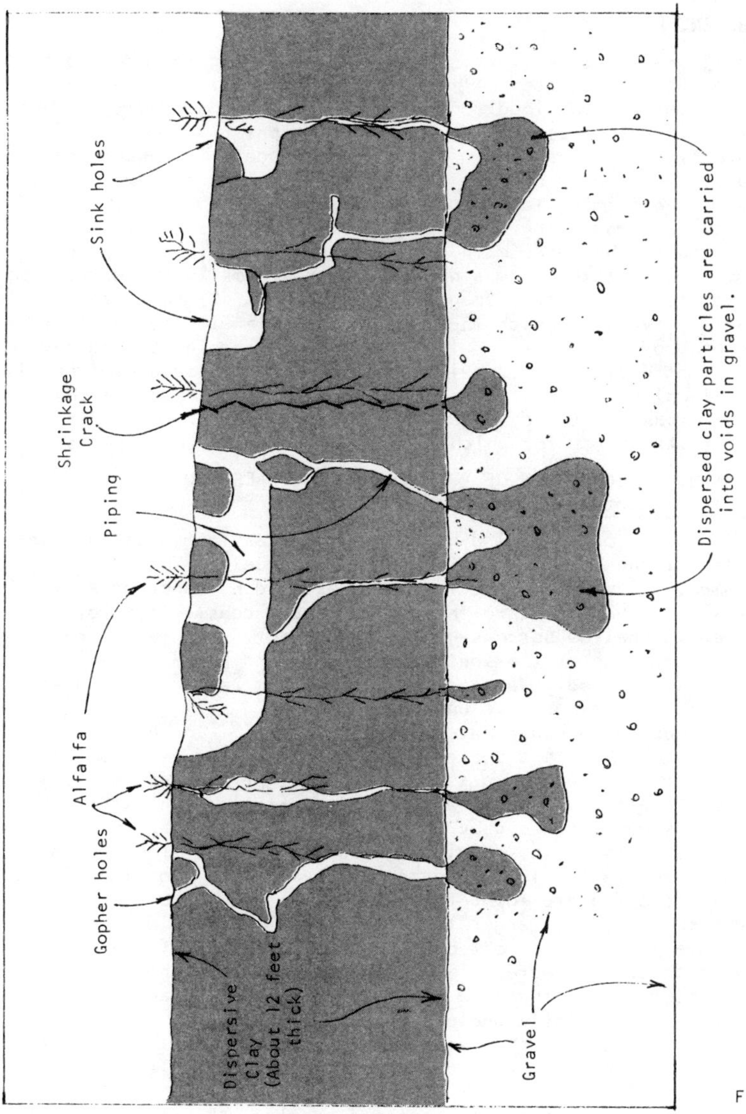

DAMAGE TO AGRICULTURAL FIELDS IN ARIZONA
FROM PIPING AND SINKHOLE FORMATION IN DISPERSIVE
SODIUM CLAYS (After Carroll, 1949)

Figure 1

SCOPE OF PRESENT STUDY

During 1971-72 an investigation was made of piping in three main groups of clay dam structures: (1) a long flood control dike in Venezuela; (2) a number of low flood control dams in Oklahoma; and (3) a number of similar low flood control dams in Mississippi.

The investigation consisted of gathering the details concerning the erosion and piping, taking samples of the typical soils from which the dams were built, performing laboratory tests (both soil mechanics and chemical tests) and analyzing the results.

In addition to the three main groups of structures, tests were made also on soil samples taken from a number of other dams in two categories: (1) dams which failed or were damaged by piping; and (2) dams from clays of various types on which there was no trouble with piping (for the purpose of providing test results on "control" samples of ordinary, erosion resistant clays). In total, more than 150 soil samples from 51 structures (21 which failed or were damaged by piping through the dam from reservoir leaks, 12 which were severely damaged by rainfall erosion and 18 control structures for which no erosion problems developed or were anticipated.

BRIEF PERFORMANCE DESCRIPTION OF PRINCIPAL STRUCTURES STUDIED

Zulia River Flood Control Dike

A continuous flood control dike, about 75 miles long, was built on the right bank of the Zulia River in western Venezuela. This was about 15-20 feet high, 3:1 side slopes, 15 foot crest width, constructed in 6-inch layers with sheepsfoot rollers and moisture-density control from soil excavated in shallow borrows directly adjacent. Soon after completion of construction at the southern end of the dike in 1963-64, certain portions were badly damaged by erosion from rainfall. Two aspects of the erosion damage were unusual.

The first was the unusual nature of the erosion. Instead of eroding gullies in the embankment surface, rainwater entered desiccation cracks on the crest and upper parts of the slope, percolated vertically downward 6 to 10 feet and then flowed out horizontally. The width of the drying cracks was increased by erosion more rapidly than the cracks were closed by swelling. The result was erosion tunnels from rainfall runoff as shown schematically in Figure 2. The vertical portion of the tunnels eroded to a diameter of 6 to 18 inches within two or three seasons and eventually the larger tunnels became 3 feet in diameter, and more (Figure 3). In the most badly damaged sections of the dike the large erosion tunnels were only a few feet apart and the embankment was so badly deteriorated that the crest settled several feet and the dike was rendered essentially useless as a water retaining structure. When exposed by excavation, the walls of the erosion tunnels frequently resemble the appearance of walls of tunnels in high soluble limestone formations, Figure 4.

The second unusual aspect was that the erosion was highly selective. Sections of the dike with lengths of 500 to 1000 meters were very badly eroded and immediately adjacent sections were essentially undamaged by erosion. Studies of the construction records showed no consistent difference to explain the great differences in the performance. Both the eroded and non-eroded sections were constructed with clays of medium to high plasticity with apparently identical construction procedures, at about the same time, by the same people, with similar construction water content, and with no appreciable difference in any of the other

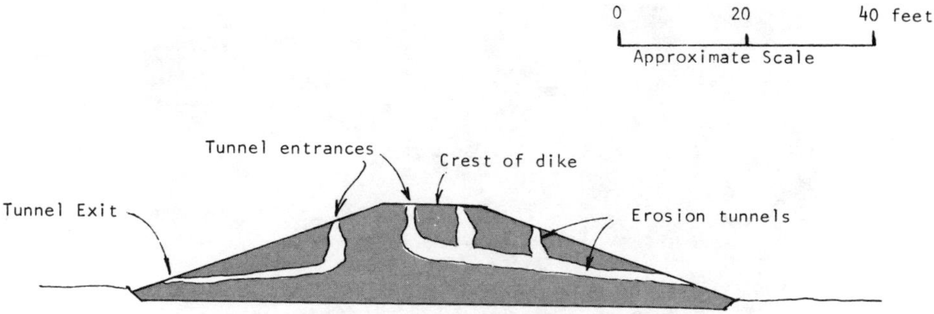

Figure 2. SCHEMATIC DETAILS OF TYPICAL RAINFALL EROSION TUNNELS IN CLAY FLOOD CONTROL DIKE IN BADLY DAMAGED SECTION

Figure 3. TYPICAL APPEARANCE OF VERTICAL EROSION TUNNEL ON CREST IN BADLY DAMAGED AREA OF ZULIA RIVER FLOOD CONTROL DIKE

Tunnels are only a few feet apart and the crest has sagged several feet.

TYPICAL CHUNKS OF CLAY EMBANKMENT TAKEN FROM THE WALLS OF RAINFALL EROSION TUNNELS IN ZULIA RIVER DIKE ("Worm eaten" appearance is a common erosion feature in some dispersive clays.) FIGURE 4

activities which might have been expected to influence the performance of the structure. Later it was shown by tests of the dissolved salts in the pore water of othe clay that there was such a large and consistent difference between the clays in the eroded and non-eroded sections of the dike that there could be no reasonable doubt that it was this property of the clay which was responsible for the trouble.

Oklahoma Flood Control Dams

Starting about 1950, approximately 1500 earthen flood control dams were constructed on tributary streams in Oklahoma under SCS direction. The dams are generally homogeneous dams, built in 6-inch layers with sheepsfoot rollers with conventional practice for moisture-density control. Of these, 11 failed by progressive erosion of an initial leak which developed through or under the dam. The failures usually occurred immediately after the first, rapid reservoir filling. Some of the main facts concerning these failures are given below; descriptions of the failures are given in more detail in a companion paper (24).

1. All the failed dams were homogeneous dams, 22 to 50 feet high, built of clay (CL and CH) derived as a result of weathering of the shale-sandstone bedrocks of the area in the form of in-situ residual clay, slope wash, or alluvium.

2. The embankment clays varied from low to high plasticity, with a typical material having a liquid limit between 30 and 50%, Figure 5.

3. One of the dams consisted wholly of clay of high plasticity with liquid limit varying generally between 60 and 90%.

4. In none of the cases was there any obvious cause of the failure determined. The main common feature among the probable factors listed as failure causes was the fact that some of the clay in each of the failed dams had high dispersion when tested with the SCS dispersion test.

5. Because the failures occurred within a few days (or hours) after the first reservoir filling, it was clear that the initial leak must have been travelling in a concentrated leakage channel, such as a crack (24). The intact compacted clay embankments are much too impervious to allow penetration of water through the dam within a period of several days.

6. Nine of the 11 failed dams are located within a relatively small geographic area (within a circle with a radius of less than 40 miles) in southeastern Oklahoma.

7. In the same geographic area there are several hundred other dams of similar design constructed about the same time. The geologic and soil conditions at all the many dam sites in the general area are very similar. Undoubtedly many dams which did not fail are constructed of clays which are very similar to those of the failed dams.

8. In this same geographic area many of the dams have been damaged by rainfall in the form of erosion tunnels on the crests and slopes, of an almost identical nature as described above for the Zulia River flood control dike and shown in Figure 2. The vertical rainfall erosion tunnels occur in spite of the very best protective growth of bermuda grass, Figure 6. Some of the dams most badly damaged by rainfall erosion are some of those with the best maintained grass covers indicating the strong probability that a heavy growth of good protective grass worsens the problem, because water sucked from the clay by the roots of the grass causes larger drying cracks.

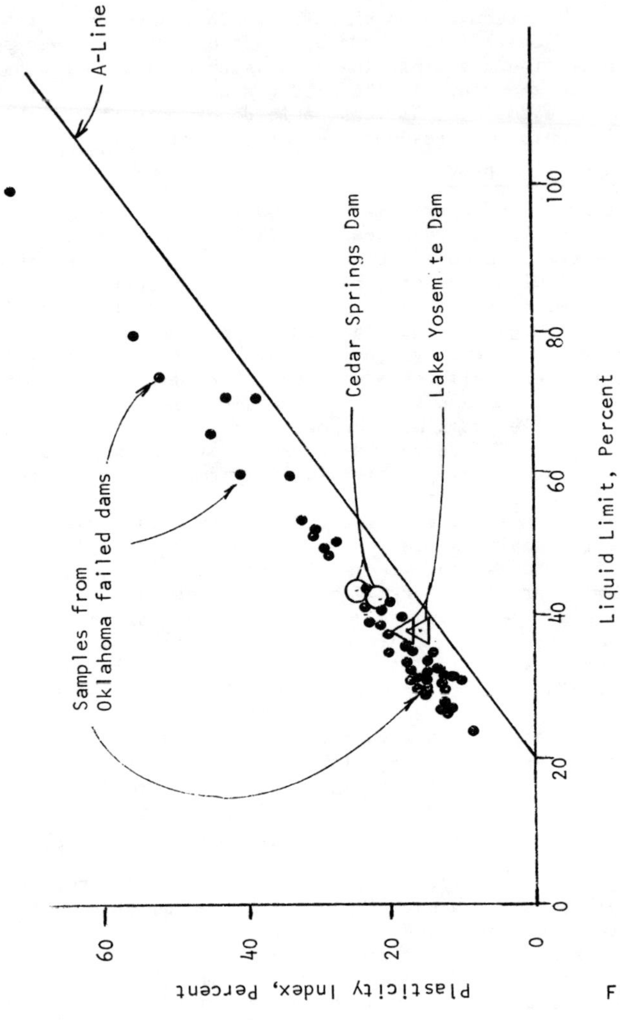

RANGE OF ATTERBERG LIMITS OF CLAY SAMPLES FROM FAILED OKLAHOMA DAMS COMPARED WITH SAMPLES FROM TWO CONTROL DAMS

Figure 5

RAINFALL EROSION TUNNELS ON THE DOWNSTREAM SLOPE OF TYPICAL DAM SHOWING EXCELLENT GROWTH OF PROTECTIVE GRASS AND COMPLETE FREEDOM FROM EROSION ON DAM SLOPING SURFACE
Figure 6

In addition to the 11 failures of relatively low dams between 1957 and 1970, described above, a much larger homogeneous dam in Oklahoma, Wister Dam, was badly damaged by clay piping on the first reservoir filling in 1949 and required extensive repairs before it was finally placed in operation successfully (4,6,23,28). Vertical erosion tunnels from rainfall also developed each year on the downstream slope of Wister Dam in spite of an excellent grass protective cover. Tests on samples of the Wister Dam embankment material, taken in this investigation, indicate that it is a highly dispersive clay with properties similar to the average of those of the clays from the 11 low SCS dams which failed later. Because of this, and because the nature of the unusual piping damage was very similar, it is overwhelmingly probable that the damage to the Wister Dam had the same underlying cause as that of the failures of the 11 low dams (24).

Mississippi Flood Control Dams

In the last 20 years, approximately 400 small, earthen flood control dams have been built in the state of Mississippi, using the same practice described above for the Oklahoma SCS dams. Of these, three dams have failed by breaching and there is a serious problem of tunnel erosion from rainfall on more than 50. The experiences are generally very similar to those described above for the Oklahoma dams with the following special details:

1. The trouble is limited to a geographic area in north-central Mississippi with a length of about 100 miles in the north-south direction and a width of about 30 miles.

2. The sub-soil conditions in the area of damaged dams is very similar and consists of a surface layer, generally about 5 to 30 feet thick, of clay of low to medium plasticity (CL and CL-ML) in the form of a weathered loessial deposit, generally overlying a very soft weathered claystone. The weathered loess is usually the problem material.

3. In the southern part of the state dams built of apparently similar loess have not been damaged by erosion.

4. The general dimensions of the erosion tunnels from rainfall are very similar in Mississippi and Oklahoma, except that the severity of the erosion on the average damaged Mississippi dams is greater. It has been necessary to repair about 20 of the Mississippi dams by excavating and regrading. This work has not been highly successful as the erosion tunnels start again almost immediately.

5. In general the natural ground surface in the area has gentle grass covered slopes giving no easily visible indication of unusual soil conditions; however, erosion of some excavated faces for road construction shows a peculiar pattern of erosion tunnels and deep, narrow gullies, Figures 7 and 8.

6. Many of the dams which have been most badly damaged by rainfall erosion tunnels have excellent grass covers, which has completely prevented any erosion of the surface of the dam slopes in the form of gullies or sheet erosion. On some of the dams the grass is 18 inches or higher and so thick that it is literally dangerous to walk on the slopes because of the hazard of falling into one of the erosion tunnels, which are obscured by the grass.

7. A small geographic area in northeastern Arkansas has very similar soil conditions and identical trouble with a small number of clay dams.

7b. CLOSEUP VIEW OF ERODED FACE OF PHOTO 7a. NOTE INTERNAL TUNNELLING IN CLAY DEPOSIT RESEMBLES LIMESTONE SOLUTION CAVERN (See also Figure 4)

7a. TYPICAL VIEW OF EROSION PATTERN IN EXCAVATED SLOPE IN NATURAL DEPOSIT OF HIGHLY DISPERSIVE CLAY IN MISSISSIPPI

Figure 7

TYPICAL EROSION PATTERNS IN EXCAVATED SLOPE OF NATURAL DEPOSIT IN OKLAHOMA REGION OF HIGHLY DISPERSIVE CLAYS

Figure 8

The climatic conditions existing in the areas of the rainfall erosion tunnels in Venezuela, Oklahoma and Mississippi have important similarities. All three areas are relatively humid, with average annual rainfall between 45 and 55 inches per year. Also in all three areas it is common that heavy concentrated rainstorms occur immediately following periods of hot weather without rain, so that relatively large runoff can be expected to flow into open drying cracks.

ATTERBERG LIMITS

This study confirmed completely the conclusion of the Australian researchers that it is not possible to identify the erodible clay by the results of the index tests commonly employed by the soil mechanics engineer. From the standpoint of Atterberg Limits, particle size distribution, and visual classification (ASTM D2488), there is no discernible, consistent difference between the erosive, dispersive clays taken from the damaged structures and ordinary erosion resistant clays. Clay samples taken from the walls of piping tunnels in the most badly damaged structures would be judged by the civil engineer generally as satisfactory (or good) materials for the core of an embankment dam.

As seen in Figure 5, which shows a plot of the Atterberg Limits for all the samples from the 12 failed Oklahoma dams, the test results are typical of those of ordinary, inorganic clays of medium plasticity. Also shown in Figure 5 are the test results for samples from two of the "control" dams, which test results are indistinguishable from the average of those of the failed Oklahoma dams. The two control dams shown in Figure 5 were chosen for reasons as described briefly below.

Lake Yosemite and Cedar Springs Dams (Controls)

Lake Yosemite Dam is a long, homogeneous dam, about 50 feet high, located near Merced, California, built more than 90 years ago with horsedrawn equipment and no roller. It retains a fairly important reservoir and has been in operation continuously. There has been some seepage appearing on the downstream slope for many years in several places without any erosion or piping.

Cedar Springs Dam is an important storage dam constructed in the last few years in California (29). It is located in a highly seismic area at a site where an active (but minor) fault is believed to exist in the foundation because of the discovery of offset recent alluvial beds directly overlying a bedrock fault. Consequently, the dam was very conservatively designed with rock shells and excellent, thick, well-graded transitions. Also, an effort was made to find a very tough and erosion resistant clay core and material was transported from a lake bed clay deposit at a considerable distance and cost. The clay is very tough and hard at the plastic limit. When rolled near Standard AASHO optimum water content, the feet of heavy sheepsfoot roller commonly penetrated only 3 or 4 inches.

In addition, all the laboratory tests on the clay samples from Lake Yosemite and Cedar Springs Dams indicated that these materials were free of all tendency toward dispersion.

CHEMICAL TESTS

SAR and ESP

Confirming general results of previous studies it was found that there was a very strong correlation between the incidence of piping failure and the results of tests of the clay chemistry. The chemical tests employed in this investigation are routine tests of the soil scientist working in agronomy, but are not generally familiar to the civil engineer at the present time. The tests, which are described in References 21 and 25, are in two main parts.

The first part is a set of tests made on the pore water. For these tests, a sample of the clay at the natural water content is mixed with distilled water until the consistency is approximately that of the Atterberg Liquid Limit, at which point the soil scientist refers to "saturated soil paste". This is allowed to sit for a number of hours until equilibrium takes place between the salts in the pore water and on the cation exchange complex. Subsequently, a small quantity of the pore water is extracted from the saturated soil paste (10 to 25 milliliters) usually using a vacuum and a filter, Figure 16. This extracted pore water is called the "saturation extract". This saturation extract is then tested by water testing procedures employed by the chemist to determine the amounts of the main metallic cations in solution (calcium, magnesium, sodium and potassium), in terms of milliequivalents per liter. For agricultural purposes the results of these tests are commonly given in terms of the "sodium absorption ratio" (SAR), calculated by the formula:

$$SAR = \frac{Na}{\sqrt{\frac{Ca + Mg}{2}}}$$

where Na, Ca, and Mg are dissolved salts in milliequivalents per liter of saturation extract.

In the second type of test the total cation exchange capacity (CEC) of the clay is measured, in terms of milliequivalents per hundred grams of dry soil. Also measured are the relative amounts of the four basic cations on the "exchange complex". The results of these tests are frequently employed in a form of the "exchangeable sodium percentage" (ESP) which is computed by the formula

$$ESP = \frac{\overline{Na}}{CEC} \quad (100)$$

where $\overline{Na}$ is the amount of sodium on the exchange complex, in terms of milliequivalents per 100 grams of dry soil.

Both the ESP and the SAR are measures of the relative amount of sodium in the soil. Theoretically they are related, and it has been shown that for most soils there is a fairly good correlation between them (21). One of the conclusions of the Australian researchers was that for the purposes of evaluating dispersive clays it is sufficient to test the pore water extract, and to estimate the ESP from known correlations.

In this study both the ESP and the SAR were obtained by laboratory tests. The relationship for the two values for all the tests was very good so that it would not have been necessary to make both types of tests, Figure 9. It is concluded that in the future it is sufficient only to test the dissolved cations in the saturation extract and that it is not necessary to perform the more extensive tests needed to determine the CEC and the ESP.

Tests on Saturation Extract

Figure 10 shows the results of tests on the saturation extract of

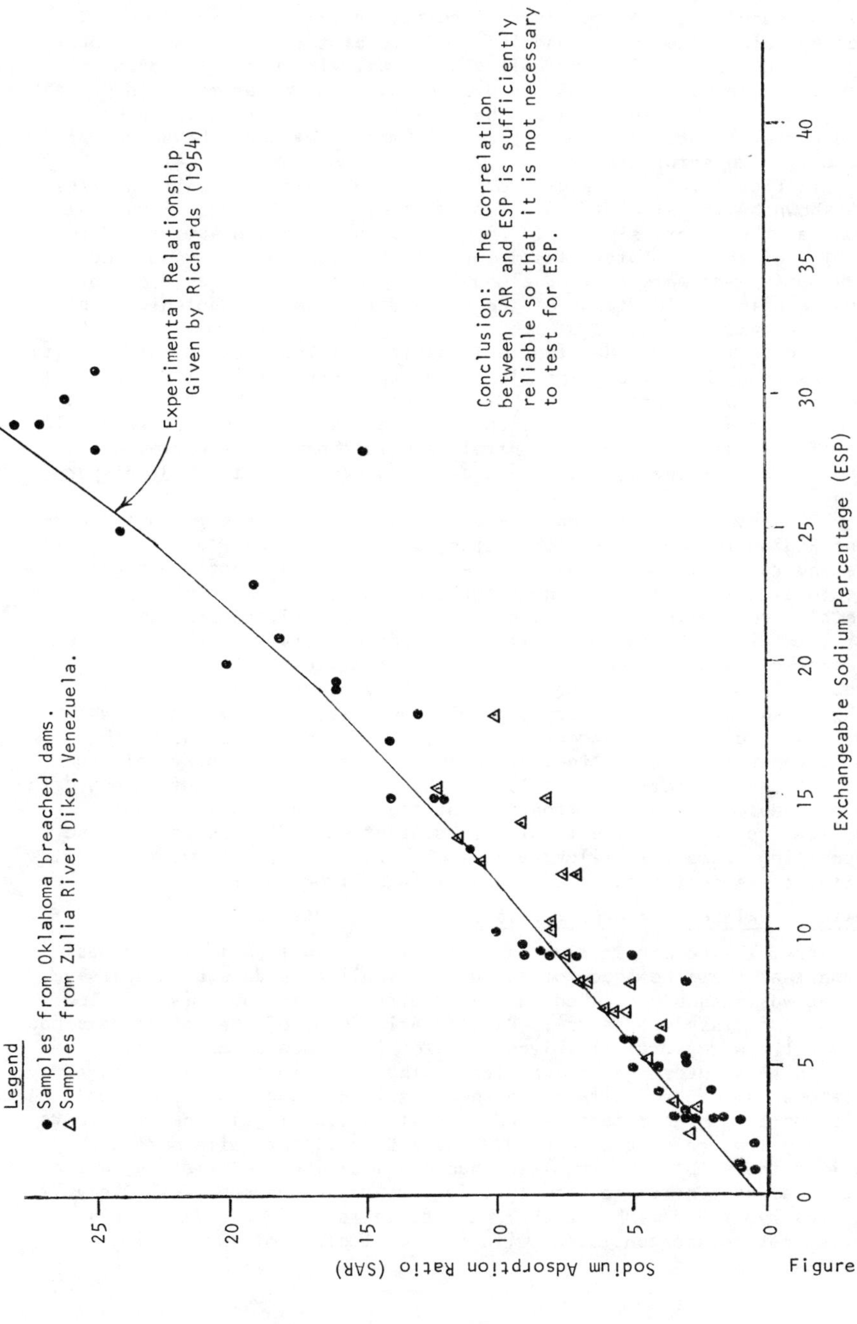

ESP AND SAR VALUES FOR TYPICAL SAMPLES

Figure 9

all the samples taken from dams which failed by breaching or were damaged by rainfall erosion tunnels. In this plot the horizontal coordinate is the Percent Sodium defined as shown. The Percent Sodium is directly related to the SAR and has been used as a parameter to present the results rather than the SAR because it is simpler, gives equally good correlations, and is a direct arithmetic ratio of the amount of sodium in the saturation extract.

In Figure 10, the results of the SCS Laboratory Dispersion Tests are shown by the size of the dot or circle, as indicated. Figure 11 shows another plot, similar to Figure 10, in which are summarized the results of the laboratory tests for the samples from the undamaged clay dams which were carried along in the investigation as "controls", showing also the locations of the dam and the general geologic origin of the clays.

Figure 10 shows that the test results for the great majority of the samples from the failed dams plot above the curved solid line, and all the samples with more than 67% dispersion are above the curved solid line. From Figure 11 it is seen that the test results for essentially all of the samples from the control dams fall below the curved dotted line and all of these, with one exception, have less than 33% dispersion.

These test results indicate a strong correlation between the Percent Sodium in the saturation extract and the incidence of piping failure and rainfall erosion damage. Also, it is clear that the higher the total salt content in the saturation extract, the more resistant is the clay to erosion; that is, clays with total salt content of 1.0 meq per liter in the saturation extract may be dispersive with 50% sodium, whereas clays with 20 or 30 meq. per liter of total dissolved salt probably need 75 to 80% sodium to be highly dispersive.

As shown in the Summary plot of Figure 12, all the samples which were taken directly adjacent to rainfall tunnels plotted above the solid upper transition line, but only in Zone 1, with total salt content of the saturation extract less than 15 meq/liter. Sufficient data is available to suggest strongly that the erosion tunnels do not develop from rain if the total salt content exceeds this value. One supporting example is illustrated in Figure 13, which shows the results of the tests performed on the samples taken from the Zulia River dike.

Tests on Oklahoma Reservoir Waters

From the recent studies in Australia and in Israel (16) it was shown that the likelihood of failure of small clay dams by dispersive piping was much higher when the reservoir water was pure (with a low content of dissolved salts). Particularly, many of the failed dams had reservoirs with total dissolved salts of less than 5 meq/liter.

In this investigation samples of the water in the reservoirs at Oklahoma failed dams were taken in January 1971 and tested for total salt content. These tests show the salt content in all the reservoirs was very low, ranging between 0.3 and 2.5 meq/liter, with more than half of the samples having less than 1.0 meq/liter. Therefore, while there was no control to show that failure may have been less likely if the reservoir waters had higher salt contents, the results of this investigation are consistent with the conclusions of the previous studies.

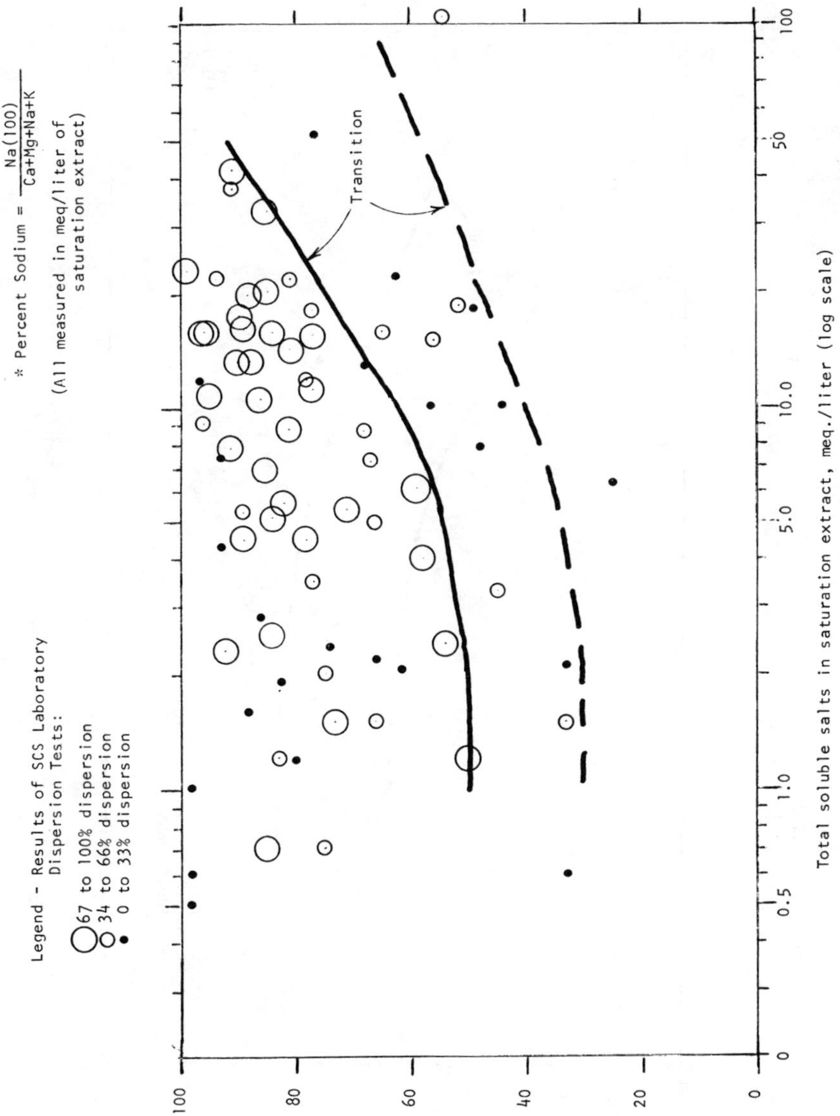

RESULTS OF TESTS ON SAMPLES FROM DAMS WHICH FAILED BY BREACHING OR WERE DAMAGED BY RAINFALL EROSION TUNNELS

Figure 10

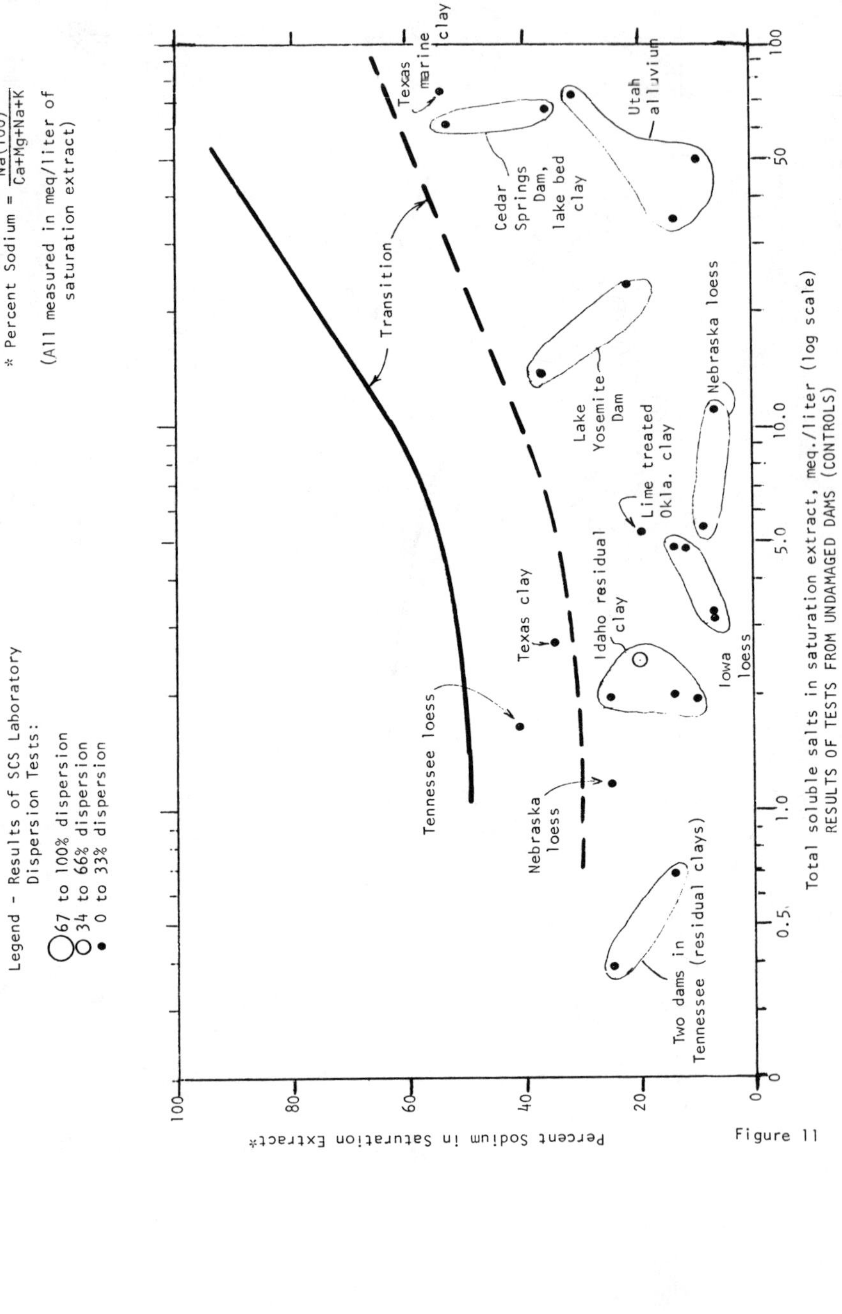
Legend - Results of SCS Laboratory Dispersion Tests:
67 to 100% dispersion
34 to 66% dispersion
0 to 33% dispersion
* Percent Sodium = Na(100) / Ca+Mg+Na+K
(All measured in meq/liter of saturation extract)
Transition
Texas marine clay
Cedar Springs Dam, lake bed clay
Utah alluvium
Lake Yosemite Dam
Nebraska loess
Lime treated Okla. clay
Texas clay
Idaho residual clay
Iowa loess
Tennessee loess
Nebraska loess
Two dams in Tennessee (residual clays)
Percent Sodium in Saturation Extract*
Total soluble salts in saturation extract, meq./liter (log scale)
RESULTS OF TESTS FROM UNDAMAGED DAMS (CONTROLS)

Figure 11

$$* \text{ Percent Sodium} = \frac{Na(100)}{Ca+Mg+Na+K}$$ All in meq/liter of Saturation Extract.

Note: The Zone 4 shown below is the same as the "Transition" shown in Figures 10 & 11 in logarithmic plots.

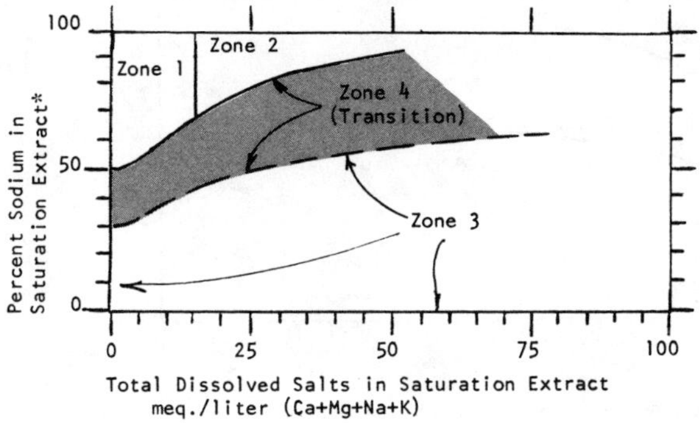

<u>Zones 1 and 2</u> include nearly all of the clay samples from dams which failed by breaching in Oklahoma and Mississippi. Samples generally have high dispersion when tested in the laboratory. Highly erodible clays.

<u>Zone 1</u> includes all samples from 16 clay dams which were damaged by tunnel erosion from rainfall in Venezuela, Oklahoma, Mississippi, Arkansas, Tennessee and Texas.

<u>Zone 3</u> includes the test results for most of the "control" samples. Probable range of ordinary, erosion resistant clays.

<u>Zone 4</u> is the transition zone. Most samples in this zone had low dispersion when tested in the laboratory. The lower boundary of the Zone is not well established by the data.

SUMMARY OF CORRELATION BETWEEN CHEMICAL TEST RESULTS AND DAM PERFORMANCE EXPERIENCE Figure 12

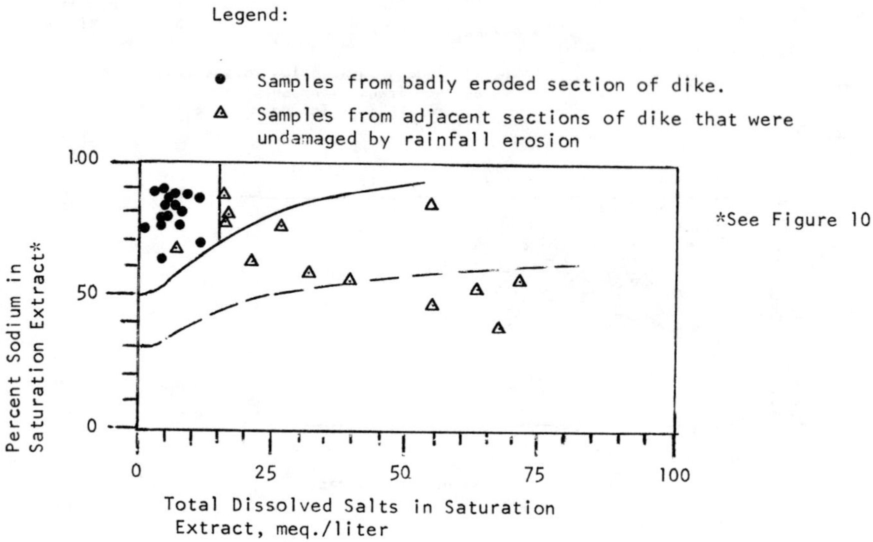

TESTS ON SAMPLES FROM ZULIA RIVER FLOOD CONTROL DIKE

Figure 13

SCS LABORATORY DISPERSION TEST

During the many years that this test (Appendix A) has been employed as a routine design procedure by the SCS engineers, it has been recognized that the test results vary somewhat, depending primarily on the water content of the specimen. Also, different results have been obtained for samples from the same dam when tested by different laboratories, which may have been caused by differences in laboratory techniques or by the fact that samples tested were not identical.

In previous years, fairly extensive investigations have been made of the influence of the water content, by testing identical specimens with different initial water contents. No consistent pattern was obtained from these studies. Sometimes samples with low initial water content showed lower dispersion and vice versa. Usually the quantitative differences in the test results were not great, even when comparing specimens with initial water content ranging between oven dried and fairly wet. In a relatively small percentage of the cases there was a very large difference. The SCS now recommends that the test be performed on samples at the natural water content (Appendix A).

In the current investigation, two laboratory dispersion tests were performed on each specimen (by the procedures of Appendix A) for the purpose of investigating the reproducibility of the test results and the influence of small differences in water purity. Both tests were performed on specimens at the water content at which they were delivered to the laboratory. Distilled water was used for one set of tests and water which had been run through a laboratory demineralizer for the other. The two series of tests were run at different times. With essentially no exceptions, the results of the two tests were very close, with Percent Dispersion slightly higher when using distilled water. From these results it can be concluded without question that, when starting with the same sample in the laboratory at least, and using distilled water, the SCS Laboratory Dispersion Test is a reliably reproducible test.

Oklahoma Failed Dams

In this investigation generally 2 to 5 soil samples were taken from each of the Oklahoma dams which failed by breaching. The samples were taken from shallow pits on the dam surface, or from borrow areas, with the hope of obtaining at least some soil which was representative of the portion of the embankment involved in the piping failure. The results of the tests showed that at least one sample taken from each of the dams had a very high reaction to the dispersion test (60% dispersion or more), and at least one sample for more than half of the dams had 80% dispersion or more. Hence, even with the relatively small number of specimens taken from each dam, chosen by random sampling generally, one or more of the test results showed very high dispersibility.

The results of the tests on the Oklahoma samples gave many examples to support strongly the conclusions that: (1) within borrow pits and embankments, the dispersibility of the clays varied greatly over short distances; and (2) clay samples taken either from the dam or borrow pit at two locations only a few feet apart, which samples have essentially identical visual appearance, Atterberg Limits and particle size distribution, may have radically different Percent Dispersion. Great differences in the erodibility, and in the water chemistry, in similar appearing clay in the same deposit have been also reported by several of the Australian investigators.

Zulia River Flood Control Dike Samples

Two sets of tests were performed on samples taken at different times. One set of tests was performed in the Soil Mechanics and Soil Science laboratories of the Soil Conservation Service in Lincoln, Nebraska, and a second set of tests were performed by parallel laboratories of the Ministry of Public Works in Venezuela, using the same procedures. The average and the range of results from both sets of tests was essentially identical.

For samples taken from the section of the dike which had been badly damaged by rainfall erosion tunnels, 15 test results ranged between 54% and 94% dispersion, with most of the results exceeding 75%.

Mississippi Flood Control Dams

Results of dispersion tests on samples taken from the walls of rainfall erosion tunnels in Mississippi dams were not consistent. Some gave high dispersion and some low. Also, two series of samples were taken from one of the dams which was most badly damaged by rainfall erosion. These were taken by different engineers in the same general location. They were sent to the same laboratory and tested at two different times. In one series the results of the dispersion tests ranged between 13 and 24% and in the other series between 86 and 95%. To investigate this great difference both sets of samples were retested under uniform conditions and the same difference obtained. There is an unknown factor in the sampling-testing procedure which causes large differences in the dispersion test results for the dispersive Mississippi clays and others. Results of chemical tests for both groups of samples were essentially identical, all plotting in Zone 1 of Figure 12. Hence, for the Mississippi loessial clays, the chemical test on the saturation extract is a much better indicator of the behavior than the dispersion test.

Controls

As seen in Figure 11, with only one exception all of the samples from dams which were included in the investigation as "controls" had less than 33% dispersion. Samples from Lake Yosemite and Cedar Springs Dams, described previously, all had less than 10% dispersion.

CLAY MINERALOGY

Mineralogical (X-ray) analyses were performed on 9 typical clay specimens from Venezuela, Oklahoma and Mississippi. The test results were similar for all samples, showing mixed mineral composition generally with an appreciable content of smectite, the active clay group including montmorillonite. Particularly, it was not possible to see any difference in the clay mineralogy between samples of the Venezuelan dike from the area of bad erosion damage and from the non-eroded area.

The Corps of Engineers' Grenada Dam is located within the general geographic area in Mississippi in which many of the low SCS flood control dams have been badly damaged by rainfall erosion tunnels, and it has developed similar tunnels. As a part of a study of this problem, the Vicksburg District, Army Engineers, sent a series of clay samples to Professor R. E. Grim at the University of Illinois for X-ray and differential thermal analysis. Some of the clay samples were taken directly adjacent to rainfall erosion tunnels and others were taken from areas where no erosion occurred. No consistent or important difference was found in the mineral composition of the two groups of samples. The 10 samples analyzed all had about the same percentage of the main minerals as follows:

Montmorillonite	30-50%
Illite	10-25%
Kaolinite	10-25%
Quartz	20-40%

A number of the Australian investigators have the opinion that all dispersive soils probably have an appreciable percentage of montmorillonite. While not much information was obtained in the current study concerning the mineralogy of the clay, the writers have no reason to doubt this conclusion. It seems likely that most of the dispersive soils taken from damaged dams and tested in this investigation have substantial quantities of montmorillonite. It is possible that the main relationships relating dispersibility and sodium, as given in Figure 12, apply only to soils with at least a moderate content of montmorillonite.

EXCEPTIONS TO CORRELATION OF FIGURE 12

In addition to the breaching failures in Oklahoma and Mississippi, samples were taken from two other homogeneous clay dams which failed by piping on the first reservoir filling, the Washington County Dam in Illinois which failed in 1962 (10), and the Stockton Creek Dam in California which failed in 1950 (22, 23). Both dams failed by breaching with failure details very similar to those of the Oklahoma dams. For both dams the chemical test results plotted in the zone of non-dispersive ordinary clays (Zone 3), and the results of the dispersion tests were all low. Hence, from these experiences it must be concluded that either the failure occurred in spite of the fact that the embankment material had the erosion resistance of an ordinary clay or that there may have been localized zones in the embankments of high dispersibility which were not sampled in this investigation. The former is believed to be the more probable.

In addition to the main groups of dams studied, as described earlier, samples were obtained of the core material from four moderately high, zoned dams in various parts of the world which have had trouble with piping in the last few years. As seen in Figure 14, the test results for the Balderhead (31) and the Hyttejuvet (17) Dams fall in the lower part of Zone 3 which is the expected region of ordinary, erosion resistant clays. In spite of this, both have one sample with 50% dispersion. Since both of these dams were damaged by piping of fines from the core, it seems reasonable to conclude that the correlation of Figure 12 does not apply to the materials used for the cores of these two dams, which were both glacial tills. Hence, if the samples tested for the Balderhead and Hyttejuvet Dams are representative of the core material, it would have been impossible to identify them as piping susceptible materials from the results of the chemical tests; however, the results of the SCS Laboratory Dispersion Test would have given an indication that the materials were more susceptible to dispersion than ordinary clays.

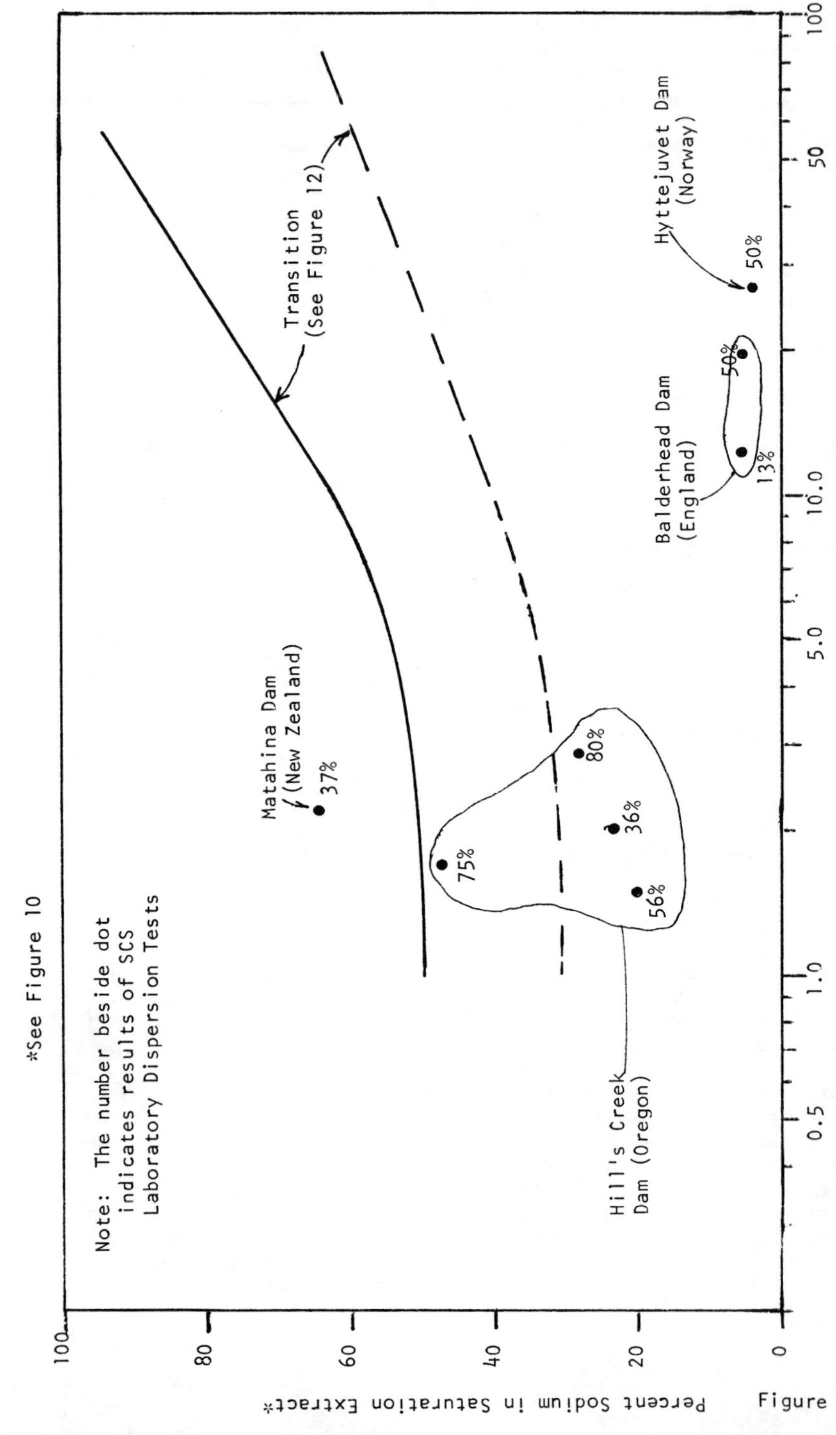

Figure 14

TESTS ON CORE MATERIAL OF ZONED DAMS DAMAGED BY PIPING

Several of the test results for samples of the core of the Hills Creek Dam (described in another paper submitted to this conference) also plot in Zone 3 and have 56 and 80% dispersion, Figure 14. The clay size particles in the core material of the Hills Creek Dam are primarily halloysite. Hence, the samples of the Hills Creek Dam are also exceptions to the correlation of Figure 12, in the sense that the material would not have been easily recognized as a dispersive clay from the chemical tests alone.

The admittedly small amount of experience represented by the test results of the Hills Creek, Balderhead, and Hyttejuvet Dams indicates that certain clay types, which may be relatively susceptible to erosion and piping in the core of a dam, cannot be recognized by the chemical tests but can be identified from the SCS Laboratory Dispersion Test. Possibly these clays are of fundamentally different mineralogical composition than the typical clays of the preponderant number of samples tested from the Oklahoma, Mississippi and Venezuelan structures.

The results of the single test on the core material from the Matahina Dam which was badly damaged by piping (23), a gravelly clay derived by deep in-situ weathering of a greywacke bedrock, are consistent with the main results of this investigation as plotted in Figure 12. As seen in Figure 14, the material has a sufficiently high percentage of sodium to be included in Zone 1.

REPAIR OF RAINFALL EROSION TUNNELS

Lime Treatment in Oklahoma by Soil Conservation Service

When highly dispersive clay is mixed with a small amount of hydrated lime (calcium hydroxide), the properties are changed radically. As shown in Table 1, the plasticity index is generally reduced and the shrinkage limit increased. The reaction to the SCS Laboratory Dispersion Test is reduced to nearly zero by 1 or 2% of hydrated lime. One of the actions of the hydrated lime is to provide calcium cations to exchange with the sodium cations in the clay; however, in addition, the lime causes other chemical reactions (carbonization and pozzolanic reactions) which actually create new minerals and cement individual clay particles together (8, 13,27).

In the last few years four SCS Oklahoma dams have been repaired by plating the crest and slopes with 12 inches of clay mixed with 2 to 3% hydrated lime. Two of these dams were treated during the original construction, because the partially completed embankments were badly damaged by rainstorms. The other two were treated several years after construction because the rainfall erosion tunnels were especially severe. Hence, the lime treated surfaces were placed on dams where it was known from experience that the embankment material was exceptionally susceptible to damage by rainfall.

The treated clay was mixed with dry powdered lime in the borrow area with a pulvermixer, sprinkled to bring the water content near optimum, and allowed to cure at least four days. The lime treated clay was then transported and placed on the dam surface in two 6-inch layers and partially compacted with complete coverage by the tread of a heavy crawler tractor. The cost of the lime treated soil has been of the order of $2.50 to $3.50/c.y. in place on the dam, about half of which represents the cost of the lime and half the cost of processing and placing on the dam.

Although the performance experience with the four dams is still

TABLE 1

INFLUENCE OF HYDRATED LIME ON
PROPERTIES OF HIGHLY DISPERSIVE CLAYS

(Typical Laboratory Test Results

Clay Source	Treatment	Liquid Limit %	Plasticity Index	Shrinkage Limit %	Percent Dispersion % (1)
Oklahoma (CH)	Untreated	64	43	10.4	100
	1% Lime	85	60	13.3	20
	2% Lime	71	43	14.4	7
	3% Lime	64	29	16.1	Trace
	4% Lime	62	25	17.1	"
Oklahoma (CL)	Untreated	30	12	17.1	100
	1% Lime	42	21	19.6	Trace
	2% Lime	38	11	22.0	"
	3% Lime	37	10	21.7	"
	4% Lime	38	9	19.2	"
Mississippi (CL)	Untreated	32	17	15	17
	1% Lime	34	15	16	
	2% Lime	31	10	19	
	4% Lime	33	8	21	
Tennessee (CL)	Untreated	34	14	20	33
	2% Lime	45	20	23	4
	4% Lime	50	31	26	Trace
Texas (CL)	Untreated	37	21	15.1	73
	2% Lime	35	11	18.2	Trace
	4% Lime	35	8	22.0	Trace
Illinois (CL-CH)	Untreated	47	27	18	94
	1% Lime	71	48	24	Trace
	2% Lime	63	35	25	Trace
	4% Lime	56	20	27	Trace

1) SCS Laboratory Dispersion Test (Appendix A).

limited (2 to 4 years), the treatment has been remarkably successful. None of the treated dams has developed any erosion tunnels and there has been essentially no gullying or sheet erosion, Figure 15. The lime treatment raises the pH of the soil, with the consequence that it is difficult to get a growth of grass started rapidly. The pH apparently decreases slowly with time, however, and grass can be made to grow after a few months, especially if the slopes are sprinkled. Figure 11 shows the results of tests on the saturation extract from a sample of the lime treated surface from one of the Oklahoma dams, approximately 2 years after treatment. Very probably test results of this sample before lime treatment would have plotted in Zone 1, Figure 12.

It is speculated that the lime treated surface has been so completely successful in preventing surface erosion because of a number of actions. First, the clay is completely flocculated so that there is very little tendency for individual clay particles to be detached from the mass and eroded. Also, the treated surface apparently has practically no shrinkage and cracking caused by drying in the sun so that little if any water can penetrate through the treated layer. Also, because of the freedom from drying cracks, the treated surface layer protects the underlying embankment from drying, in the same way that the embankment material would be protected by a thick layer of moist sand. Hence, the water content of the underlying dispersive clay remains constant with no tendency for shrinkage cracks.

Zulia River Dike

Consideration was given to repairs by plating the surface with lime treated soil; however, investigations showed that the erosion damage was so severe that it was finally decided to excavate a large part of the total volume of the embankment and rebuild, employing clay which was resistant to rainfall erosion damage, following the criteria of Figure 12.

DISPERSIVE CLAYS ARE NOT RARE IN NATURE

It is well known that clays have a stronger affinity for divalent calcium and magnesium cations than for monovalent sodium. Because of this, dispersive sodium clays of Zones 1 and 2, Figure 12, are less common in nature than the ordinary clays of Zone 3, which have primarily divalent cations.

It is apparent, however, that dispersive sodium clays are far from rare in nature. In many parts of the world they are relatively common, as shown by the literature on the subject (Appendix B). One piece of evidence for this conclusion is given by collections of laboratory tests of the clay chemistry carried out for agricultural purposes in the United States. For example, Ref. 26 is a summary of several hundreds of tests performed on clay samples taken in various parts of the state of Colorado. This shows that there are clays with very high sodium content in many locations on the main Colorado plains, both east and west of the Rocky Mountains. Ref. 11 describes studies of erosion tunnels from rainfall in natural formations in western Colorado.

Other evidence supporting the conclusion that dispersive sodium clays are not rare is given by observations of erosion patterns in natural formations. Once having focused attention on the peculiar patterns of surface erosion in dispersive clays, as illustrated in Figures 4, 7 and 8, the writers are confident of having seen similar

APPEARANCE OF SLOPE OF OKLAHOMA DAM WITH SURFACE LAYER OF LIME TREATED CLAY, TWO YEARS AFTER CONSTRUCTION. DIFFICULT TO START GROWTH OF GRASS BUT NO IMPORTANT EROSION Figure 15

erosion of natural clays in many parts of the world. For example, there can be little doubt that the distinctive erosion pattern characterizing the scenic Badlands of western North Dakota is caused by rainfall erosion in highly dispersive clay. In the North Dakota Badlands the surface topography is dominated by "tunnel erosion, tubes and caves, with sinkholes and irregular pipe orifices" (3).

Twenty years ago one of the writers carried out an investigation of 60 old homogeneous earth dams in the western United States (22). These included a number of dams which failed by piping on the first reservoir filling and some other clay dams which had had leaks through them for years without piping. Looking back at this investigation now, it appears likely that most of the failed dams consisted of dispersive clays and that a much better understanding of the performance of these old structures could have been made at that time if tests of the clay chemistry and dispersion had been performed.

There are apparently several sources for sodium cations in dispersive clays. In the Oklahoma clays, which are derived from marine sediments, the sodium is probably a remnant from the salt water of the sea in which the material was deposited in ancient geologic time. Some of the highly dispersive clays in Illinois are formed as the result of the weathering of a loess deposit, in which chemical disintegration of the feldspar produces an excess of sodium cations (33).

CURRENT STATUS AND RESEARCH NEEDED

Routine Tests for Clay Dispersibility

Undoubtedly future studies will provide large improvements in our understanding of dispersive clay piping, and tests and criteria employed to evaluate clays will be improved.

Present knowledge shows clearly that it is desirable to test clays as a routine procedure during design studies for earth dams and other hydraulic structures in which clay may be subjected to potential erosion and piping.

For the civil engineer the SCS Laboratory Dispersion Test (Appendix A) is the simplest, since the tests can be easily performed in any soil mechanics laboratory. For any deposit of clay being studied a considerable number of tests (such as 10) should be performed on samples taken from different locations. The writers believe that the results of these tests should be interpreted approximately as follows:

1. If many of the results of SCS Laboratory Dispersion Tests in a 10-sample group exceed 40 to 50% dispersion, the material should be considered highly dispersive, regardless of the results of chemical tests on the saturation extract, and it is probably not essential to perform chemical tests.

2. If most of the test results are in the range between 20 and 40% and one or two of the results are above 50%, the clay should be considered at least moderately dispersive.

3. The results for ordinary clays will generally fall in the range between 10 and 40%; however, some clays of high dispersibility and susceptibility to erosion and piping, such as the Mississippi loessial clays, give less than 40% dispersion. Therefore, for clays in this group it is necessary to perform the chemical tests on the saturation extract and employ the criteria of Figure 12.

4. Results for clays which are highly resistant to erosion will generally be in the range between 0 and 15% dispersion. If all the

results of a series of tests fall within this range it is highly probable the materials are resistant to erosion; however, some chemical tests on the saturation extract are probably desirable for confirmation.

The above recommendations and criteria should be considered to apply in the general case where the water retained in the reservoir has a relatively low content of total dissolved salts, such as less than 5 milliequivalents per liter. Undoubtedly, as indicated by colloidal theory and the results of the studies of the Australian researches, the higher the salt content in the reservoir water, the less likelihood that leaks in the dam will cause piping.

At the present time conventional soil mechanics laboratories do not have the specialized equipment or personnel needed to perform the chemical tests on the saturation extract. From a practical standpoint the tests can be performed simply and reliably in two parts. The first part, which consists of removing the saturation extract from the saturated paste can be performed easily in soil mechanics laboratories using the procedure shown schematically in Figure 16. The saturation extract can be then tested in a chemical laboratory of the type equipped to analyze water samples (for such procedures as evaluating water quality for municipal water supply). When performed on a fairly large number of samples, such as several dozen, the unit cost of the chemical tests is of the same order of magnitude as the unit laboratory cost for Atterberg Limits.

Research Needed

The recent developments, initiated primarily by the Australian researchers and extended by the investigations described in this paper, have provided important knowledge for the civil engineer building earth dams and related structures. At the same time it is clear that additional research is needed to improve understanding of the fundamental relationships, and the criteria and tests employed for design purposes. Particularly the following types of investigations would be potentially fruitful.

1. Laboratory and theoretical studies of the SCS Laboratory Dispersion Test, leading to an understanding of the reasons for the variability of the test results for some soils and for the fact that it does not measure the high dispersibility of some clays. Improvement of the procedures of the test or development of a new dispersion test which reliably measures clay dispersion quantitatively.

2. Research of various kinds leading to a strengthening and improvement of the criteria shown in Figure 12. This can be approached by supplementary investigations of the type included in this study; that is, by taking samples from additional clayey dams which were either damaged by piping or display high resistance to erosion. This goal can also be approached in the laboratory by making direct erosion tests on clay samples. One promising means is a technique which has been developed of rotating cylindrical specimens of clay at various velocities in a bath of water. Such tests have been made at the University of Texas (18) and are being currently studied at the University of California at Davis.

3. Studies of the relationship between clay mineralogy and dispersion.

MAIN CONCLUSIONS

1. Certain clays (dispersive clays), which are not rare in nature,

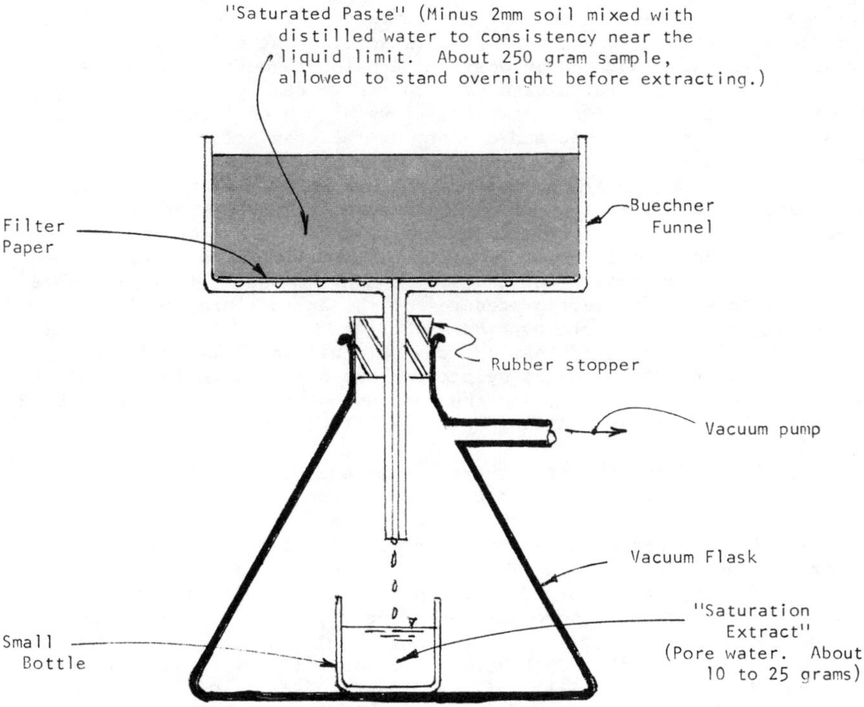

SCHEMATIC DETAILS OF APPARATUS EMPLOYED TO OBTAIN "SATURATION EXTRACT" FOR CHEMICAL TESTS Figure 16

are highly erodible. A considerable number of well constructed, low, homogeneous dams constructed of dispersive clays have failed by piping or have been badly damaged by vertical erosion tunnels from rainfall.

2. Most of the failures have occurred in dams of low to medium plasticity ((CL and CL-CH); however, clays with high liquid limit can also be highly dispersive, and one of the failures occurred in a dam of highly plastic clay (CH).

3. It is not possible to differentiate between highly dispersive clays and ordinary, erosion resistant clays on the basis of visual classification, Atterberg Limits or particle size distribution.

4. In the geographic areas studied, the dispersive clays can be well identified by results of tests of the dissolved salts in the pore water, using standard test procedures of the agricultural soil scientist and the criteria of Figure 12.

5. It is speculated that the clays in all the dams included in this study which were damaged by erosion and piping have appreciable content of montmorillonite, and the conclusions summarized in Figure 12 may not apply to clays of fundamentally different mineralogical composition.

6. For the soils tested there was an adequately reliable correlation between the exchangeable sodium percentage (ESP) and the results of the tests of dissolved salts in the saturation extract, so that no additional value was obtained by performing the tests necessary to determine the ESP.

7. The SCS Laboratory Dispersion Test is a reliably reproducible test. The results vary somewhat depending on the initial water content of specimen.

8. Results of the SCS Laboratory Dispersion Test give a good measure of the dispersibility of the clays studied in Oklahoma and Venezuela, but gave poor results for the highly dispersive Mississippi clays.

9. Laboratory tests to determine clay dispersibility should be performed as a part of the routine classification testing carried out for design of earth dams. Both laboratory dispersion tests (Appendix A) and test of the dissolved cations in the saturation extract are generally desirable.

10. Dams of soils in Zone 1, Figure 12, can be damaged by rainfall erosion tunnels in spite of the very best protective grass cover on the slopes. This is proved beyond doubt by many examples in Oklahoma and Mississippi.

11. The chemistry and dispersibility of clay frequently varies greatly within short distances in apparently uniform deposits.

12. Tunnel erosion due to rain runoff apparently occurs in compacted embankments of dispersive clay only when the total dissolved salts in the saturation extract is less than 15 meq./liter.

13. Additional investigations, both field and laboratory, are desirable to strengthen and extend understanding of the subject.

ACKNOWLEDGEMENTS:

Appreciation is particularly extended to the Soil Conservation Service for permission to publish the results of this study, and especially to Mr. M. M. Culp, Chief, Design Branch, Washington, D.C., for critical review and suggestions. The staff of the SCS Soil Mechanics Laboratory, Lincoln, Nebraska, provided great assistance on all phases of the investigation. Many personnel in the SCS, Oklahoma,

Mississippi, and Arkansas state offices assisted the writers to obtain soil samples and gather information concerning the details of the performance of the various dams. Dr. R. Grossman and his colleagues at the SCS Soil Survey Laboratory, Lincoln, Nebraska, provided great help with regard to the clay chemistry problems. The writers are also grateful to the Venezuelan Ministry of Public Works, Hydraulic Resources Department, Rafael Martinez M., Director General, and Jose R. Velasco G., Director, Design and Construction Division, for permission to publish the material on the Zulia River Dike problem. Dr. Carlos E. Flores C., Chief, Geotechnical Division, provided much valued assistance in Venezuela.

Soil samples from some of the dams outside Oklahoma and Mississippi were supplied by courtesy of a number of colleagues: Balderhead Dam (Dr. P. R. Vaughan, Imperial College, London); Hyttejuvet Dam (Mr. B. Kjaernsli, Norwegian Geotechnical Institute, Oslo); Matahina Dam (Mr. W. M. Duncan, Ministry of Works, Wellington, New Zealand); Hills Creek Dam (Mr. R. J. Pope, Corps of Engineers, Portland, Oregon); Cedar Springs Dam (Mr. D. H. Babbitt, Department of Water Resources, Sacramento, California.) The samples from the Lake Yosemite and Stockton Creek Dams were obtained with the assistance of Messrs. A. M. McClure and C. J. Cortright, Division of Dam Safety, Sacramento, California. Soil samples from the Washington County Dam were obtained through the courtesy and assistance of the Illinois Department of Conservation and Mr. W. E. Hanson, Consulting Engineer, Springfield, Illinois.

. . . . .

APPENDIX A

PROCEDURE FOR SCS LABORATORY DISPERSION TEST

A. PURPOSE

The dispersion test is a measure of the water stable aggregates in a soil. This test is designed to measure the amount of material finer than 0.005mm size in a soil-water suspension that has been subjected to a minimum of mechanical agitation. The fraction finer than 0.005mm in the soil-water suspension divided by the total fraction finer than 0.005mm, determined from a regular hydrometer analysis x 100, is referred to as % dispersion.

B. APPARATUS AND EQUIPMENT

1. Hydrometer jar calibrated at 1000 ml.
2. Calibrated hydrometer with a specific gravity range from 0.995 to 1.030 and marked in 0.001 subdivisions.
3. 500 ml. filtering flask with side tube and one-hole stopper.
4. Vacuum pump.
5. Balance sensitive to 0.01 grams.
6. Thermometer accurate to 0.5°C.
7. Thermometer accurate to 0.5°C.
8. Timer or clock.

(cont'd)

APPENDIX A (cont'd)

C. SAMPLE PREPARATION

The sample is usually shipped to the laboratory in an airtight container to prevent moisture loss, and tested at the natural moisture content. The sample will consist of material passing the No. 10 sieve (2.0mm). The water content of the material finer than the No. 10 sieve is determined.

D. PROCEDURE

1. Place approximately 125 ml. of distilled water in the filtering flask.
2. Weigh the equivalent of 25 grams of dry soil to 0.01 grams and place into the flask with water.
3. Place the rubber stopper into the mount of the flask and connect to vacuum pump.
4. Start pump and apply a full suction.
5. At approximately 3 minutes, 5 minutes and 8 minutes after pumping has started, swirl the flask several times in a rotating manner to assist in removing entrapped air.
6. After a total pumping time of 10 minutes, disconnect from vacuum.
7. Wash the soil-water suspension into the hydrometer jar and add distilled water to the 1000 ml. mark.
8. Shake the cylinder end over end at the rate of approximately 30 times per minute for one minute and record the time at the end of the shaking. This is the start of the sedimentation period. The time interval between Item 2 and Item 8 should not exceed one hour.
9. Determine the percent of material finer than 0.005 mm. in suspension.
10. Compute the percent dispersion as follows:

$$\text{Percent Dispersion} = \frac{\text{\% finer than 0.005mm in soil water suspension}}{\text{\% finer than 0.005mm from hydrometer analysis}} \times 100$$

11. Report as Percent Dispersion.

APPENDIX B - REFERENCES

1. Aitchison, G.D., and Wood, C. C. (1965) "Some interactions of Compaction, Permeability, and Post-construction Deflocculation Affecting the Probability of Piping Failure in Small Earth Dams", Sixth International Conference on Soil Mechanics and Foundation Engineering, Montreal, Vol. 11, p. 442 (Study of piping failures in 60 small dams leads to conclusion that incidence of failure is strongly dependent on the relationship between the amount of exchangeable sodium in the soil and the salt content in the reservoir water.)
2. Aitchison, G. D., Ingles, O. G., and Wood, C. C. (1963) "Post-construction Deflocculation as a Contributory Factor in the Failure of Earth Dams", Fourth Australia-New Zealand Conference on Soil Mechanics and Foundation Engineering, p. 275.
3. Bell, G. L. (1968), "Piping in the Badlands of North Dakota", Sixth Annual Engineering Geology and Soils Engineering Symposium, University of Idaho, College of Engineering, Boise, Idaho.
4. Bertram, G. E. (1967), "Experience with Seepage Control Measures in Earth and Rockfill Dams", Trans. 9th Congress on Large Dams, Istanbul, Vol. 3, p. 91. (Description of serious cracking and piping in Wister Dam of the Army Corps of Engineers.
5. Carroll, P. H. (1949), "Soil Piping in Southeastern Arizona", Regional Bulletin 110, Soil Series 13,)oil Conservation Service, Albuquerque, New Mexico.
6. Casagrande, A. (1950), "Notes on the Design of Earth Dams", Journal Boston Society of Civil Engineers (Describes the leakage and piping at the Wister Dam - "Case A", and gives the opinion that differential settlement cracking was the probable cause.)
7. Cole, D.C.H. and Lewis, J. G. (1960) "Piping Failure of Earth Dams Built of Plastic Materials in Arid Climates", Third Australia-New Zealand Conference on Soil Mechanics and Foundation Engineering, p. 93.
8. Eades, J. L., Nichols, F.P. Jr., and Grim, R. E. (1962), "Formation of New Minerals with Lime Stabilization as proved by field experiments in Virginia", Bulletin 335, Highway Research Board, Washington, D.C.
9. Fletcher, J. and Carroll, P. H. (1948), "Some properties of Soils that are Subject to Piping in Southern Arizona", Proceedings, Soil Science Society of America, Vol. 13, p. 545.
10. Hanson, W. E. (1969), "Deformation Problems in Small Dams", The Military Engineer, March-April, p. 119. (Failure of the Washington County Dam.)
11. Heede, B. H. (1971), "Characteristics and Processes of Soil Piping in Gullies", U. S. Dept. of Agriculture Forest Service Research Paper RM-68.
12. Ingles, O. G., Lang, J. G. and Richards, B. G. (1968), "Pre-equilibrium Observations on the Reconstructed Flagstaff Gully Dam", Symposium on Earth and Rockfill Dams, Talwara (Punjab), India, Nov. 1968, publisher Sarita Prakashan, Meerut, India, Vol. 1, p. 162.

13. Ingles, O. G. (1968), "Soil chemistry Relevant to Engineering Behavior of Soils", Soil Mechanics--Selected Topics, edited by I. K. Lee, American Elsevier Publishing Company, New York.
14. Ingles, O. G., and Aitchison, G. D. (1969), "Soil-water Disequilibrium as a Cause of Subsidence in Natural Soils and Earth Embankments", International Symposium on Land Subsidence (Tokyo), AIHS Publication No. 89, Sept. 1969, Vol. 2, p. 342. (Review and Extension of Research Relating Piping Potential and Soil Chemistry.)
15. James, J. P. and Wickham, R. T. (1970), "Failure and Rectification of the Morwell No. 2 Fire Service Reservoir", Bulletin, Australian National Committee on Large Dams, July 1970, p. 20.
16. Kassiff, G., and Henkin, E. N. (1967), "Engineering and Physico-Chemical Properties Affecting Piping failure of Low Loess Dams in the Neger", Proceedings, Third Asian Regional Conference on Soil Mechanics and Foundation Engineering, Haifa, Vol. 1, p. 13. (Study of 10 low clay dams, including 5 which failed by piping, shows a strong correlation between incidence of failure and the percentage of exchangeable sodium cations in the clay.)
17. Kjaernsli, B., and Torblaa, I. (1968), "Leakage through Horizontal Cracks in the Core of Hyttejuvet Dam", Publication No. 80, Norwegian Geotechnical Institute, Oslo, p. 39.
18. Masch, F. D., Espey, W.H., and Moore, W. L. (1965), "Measurements of the Shear Resistance of Cohesive Sediments", Proceedings, Federal Inter-Agency Sedimentation Conference, Agricultural Research Service Publication No. 970, Washington, D.C.
19. Parker, G.G. and Jenne, E.A. (1967), "Structural failure of Western U.S. Highways Caused by Piping", in Symposium on Subsurface Drainage: Highway Research Record No. 203 (National Research Council-National Academy of Sciences-National Academy of Engineering Publication No. 1534), p. 57-76, 1967. (Damage to embankments and other highway structures has been caused by piping of rain runoff in drying cracks, especially in soils with high content of exchangeable sodium.
20. Rallings, R.A. (1966), "An Investigation Into the Causes of Failure of Farm Dams in the Brigalow Belt of Central Queensland", Water Research Foundation of Australia, Bulletin No. 10, 1966. (An investigation of causes of piping failure of small farm dams. Concludes that strong correlation exists between failure and exchangeable sodium in the soil. Good bibliography of articles relating chemistry of soils to engineering properties for dams.)
21. Richards, L. A. (1954), "Diagnosis and Improvement of Saline and Alkali Soils," U.S. Dept. of Agriculture Handbook No. 60, U. S. Government Printing Office, Washington, D.C.
22. Sherard, J. L. (1953), "Influence of Soil Properties and Construction Methods on the Performance of Homogeneous Earth Dams", Technical Memo 645, U. S. Bureau of Reclamation, Denver, Colorado, January 1953.
23. Sherard, J. L. (1972), "Embankment Dam Cracking", a paper to be included as a part of the book, "Contributions to Embankment Dam Engineering", to be published by John Wiley and Sons, New York in 1972.
24. Sherard, J. L., Decker, R. S. and Ryker, N. L. (1972), "Hydraulic Fracturing in Low Dams of Dispersive Clay", paper submitted for the Conference on Performance of Earth and Earth Supported Structures,

25. Soil Conservation Service (1967), "Soil Survey Laboratory Methods and Procedures for Collecting Soil Samples," Soil Survey Investigations Report No. 1, U. S. Dept. of Agriculture, Washington, D. C.
26. Soil Conservation Service (1967), "Soil Survey Laboratory Data and Descriptions of Some Soils of Colorado", Soil Survey Investigations Report No. 10, U. S. Dept. of Agriculture in cooperation with the Colorado Agricultural Experiment Station, August 1967.
27. Thompson, M. R. (1966), "Lime Reactivity of Illinois Soils", Journal, Soil Mechanics and Foundations Division, ASCE, Sept. 1966.
28. U. S. Army Corps of Engineers (1959), "Review of Soils Design, Construction and Performance Observations, Wister Dam, Oklahoma", Waterways Experiment Station Tech. Report No. 3-505, Vicksburg, Mississippi.
29. USCOLD (1970) General Paper No. 7, Tenth Congress on Large Dams, Vol. IV, p. 806 (Contains brief description and cross-section of Cedar Springs Dam.)
30. van Olphen, H. (1953), "An Introduction to Clay Colloidal Chemistry", John Wiley and Sons, New York.
31. Vaughan, P. R., Kluth, D. J., Leonard, M. W. and Pradoura, H.H.M. (1970), "Cracking and Erosion of Balderhead Dam and the Remedial Works Adopted for Its Repair, 10th Congress on Large Dams, Montreal, Vol. 1, p. 73.
32. Volk, G. M. (1937), "Method of Determination of Degree of Dispersion of the Clay Fraction of Soils", Proceedings, Soil Science Society of America, Vol. 11, p. 561.
33. Wilding, L. P. et al. (1963), "Source and Distribution of Sodium in Solonetzic Soils in Illinois", Soil Science Society of America, Vol. 27, No. 4, p. 432.
34. Wood, C. C., Aitchison, G. D. and Ingles, O. G. (1964), "Physicochemical and engineering aspects of piping failures in small earth dams;" paper presented at the Colloquium on Failure of Small Dams, Nov. 16, 1964, Melbourne, Australia.

Introduction to Paper 5

Hydraulic Fracturing in Low Dams of Dispersive Clay

by J.L. Sherard, R.S. Decker and N.L. Ryker
Proceedings, Specialty Conference on Performance of Earth and Earth-Supported Structures, American Society of Civil Engineers, Vol 1, Part 1, 1972, pp 653-689.

Prepared by Norman L. Ryker

Please see Introduction to Paper 4.

HYDRAULIC FRACTURING IN LOW DAMS OF DISPERSIVE CLAY

By James L. Sherard[1], F.ASCE;
Rey S. Decker[2]; and
Norman L. Ryker[3]

SCOPE OF INVESTIGATION

During the last 20 years, approximately 1500 relatively low, earthen flood control dams were constructed on tributary streams in Oklahoma under direction of the Soil Conservation Service. They were generally homogeneous dams, built in 6-inch layers with sheepsfoot rollers and conventional practice for moisture-density control. As a part of the same program approximately 400 similar dams were constructed in the state of Mississippi. Of these, eleven of the Oklahoma dams and three of the Mississippi dams failed by piping. The failed dams ranged in height between 22 and 50 feet. In general, the post failure studies revealed no conditions which were greatly different from those of the average of many similar dams constructed in the nearby areas. While various conclusions and opinions were formed from study of the individual failures it was not possible to be completely certain that the main cause of failure was known. During the same period of time several thousand similar dams were constructed in other parts of the United States with no piping failures.

During 1971 these failures were restudied with the hope that, by investigating the whole group, patterns would be found that were not apparent during the study of the individual failures. The investigation consisted of gathering and analyzing the details concerning the failures and taking and testing samples of the typical soils from which the dams were built.

As discussed in a companion paper (9)*, a main conclusion from this study is that all the failed dams were constructed of highly erodible, "dispersive" clay. The present paper deals with the causes of the initial leaks which led to the breaching failures by piping.

OKLAHOMA BREACHING FAILURES

Some of the main details concerning the dam failures are tabulated in Appendix A and are summarized below.

Time of Failure

Ten of the eleven dams failed on the first reservoir filling, usually within a short time after completion of construction. In general the reservoirs filled rapidly (within several days or less) following a rainstorm and the initial concentrated leak appeared within a few hours (or a day or two), Figure 1.

---

1) Consulting Engineer, Berkeley, California; 2) Head, Soil Mechanics Unit, Soil Conservation Service, Lincoln, Nebraska; 3) Soil Mechanics Engineer, Soil Conservation Service, Stillwater, Oklahoma.

* Numbers in parentheses refer to References, Appendix B.

MAIN FAILURE TUNNEL AT CANEY COON CREEK SITE 2. Photos were taken approximately 24 hours after the initial small dirty leak appeared at the downstream toe contact. Failure occurred with 15 feet of water in the reservoir. Length of leakage path through the dam is about 300 feet. Approximately 6 hours after these photos were taken, a second independent dirty leak appeared at the downstream toe, about 150 feet away (see Appendix A).

Figure 1

The first two of the eleven failures occurred on May 17, 1957, during the same storm. The other failures occurred at periodic intervals thereafter without any discernible pattern, with the most recent in October 1970.

One of the eleven dams (Leader Middle Clear Boggy Site 15) failed on two separate occasions, in 1965 and 1968. The failures were at locations about 100 feet apart and apparently not related.

Eyewitness Observations

At only three of the failures were eyewitnesses present to observe the events. The failures of the other8 dams were discovered only after the reservoir had been emptied through the failure breach, and sometimes the time of failure was not known within several days.

Geology and Embankment Soil Types

Nine of the eleven failed dams were located in a relatively small area (within a circle with a radius of less than 40 miles) to the south and east of Oklahoma City. The geologic and soil conditions at all the failed dam sites were very similar. The underlying bedrock is a horizontally bedded shale-sandstone, of Pennsylvanian and lower Permian age, which is usually within a few feet of the surface on the dam abutments. In the central part of the valley the dam embankments were usually underlain with 20 to 40 feet of stream alluvium, over bedrock. Total foundation settlements caused by compression of the foundation alluvium were frequently in the range between 6 and 24 inches.

The soils from which the dams are built are generally highly dispersive clays or sandy clays of low to medium plasticity (CL and CL-CH) derived from the bedrock. The only exception is Cherokee Sandy Creek Site 8A, for which the typical embankment material is a clay of high plasticity (CH).

Locations of Piping Tunnels

The failures started with an initial leak which gradually eroded a tunnel. In some cases the entire reservoir volume was discharged through the tunnel. In other cases the erosion of the tunnel proceeded to the point where the roof collapsed, forming a breach.

With the single exception of Cherokee Sandy Creek Site 8A, all failures occurred where the depth to bedrock in the foundation was not great, and was changing relatively rapidly. Hence, with the single exception, the failures occurred at a point where relatively high differential foundation settlement would be expected. While differential settlement cracking was frequently suspected as a cause of failure, in not one of the cases were any differential settlement cracks observed in the embankments, either prior to or after failure.

In two cases, Caney Coon Creek Site 2 and Leader Middle Clear Boggy Creek Site 33, two independent tunnels occurred more or less simultaneously and apparently independently.

Including these dual tunnels, the two failures at Leader Middle Clear Boggy Creek Site 15 and the potential failure at Leader Middle Clear Boggy Creek Site 29 (Appendix A), there were a total of 15 piping tunnels or breachings in the 12 low dams. Of these 15, seven were at or very near the location of the conduit passing through the base of the dam.

In one of the cases where the failure occurred on the plane of the conduit, Little Wewoka Creek Site 17, an eyewitness was present to observe the events from the initial development of the leak. As seen in

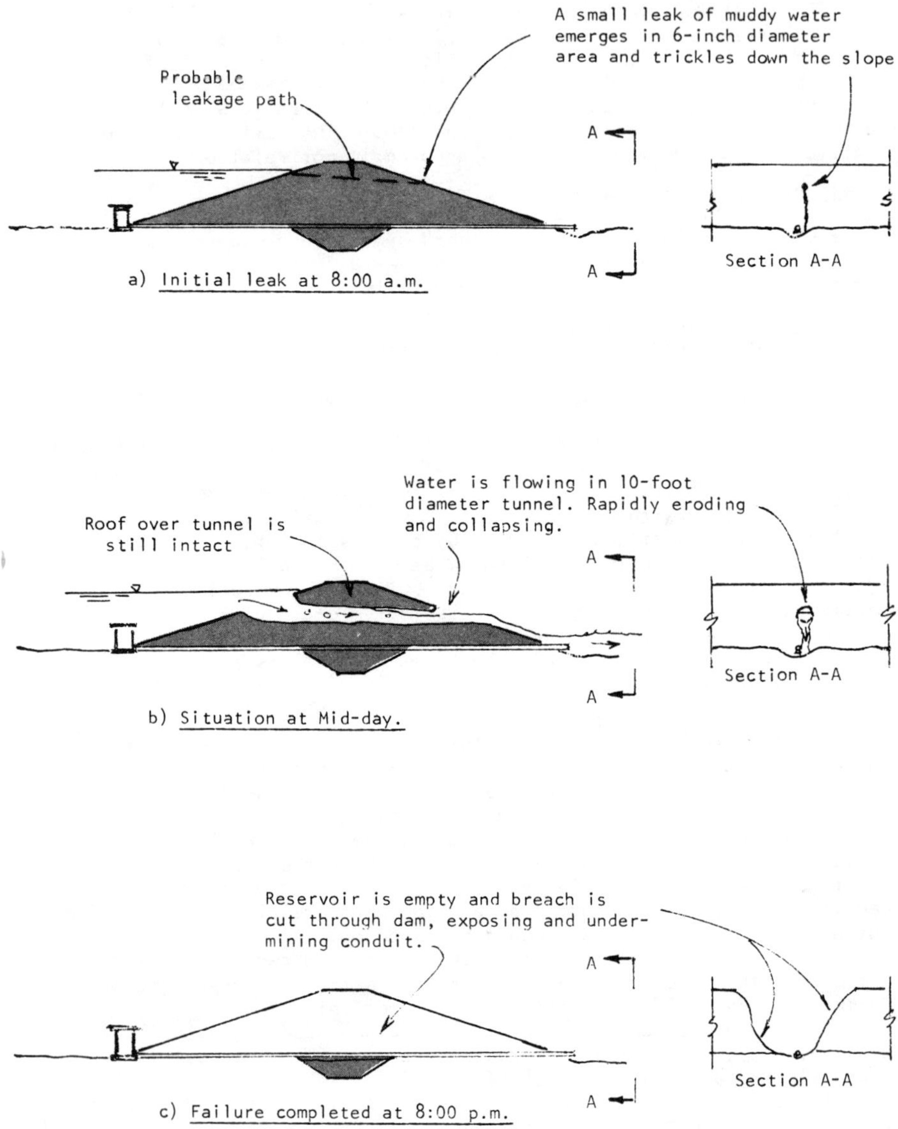

DEVELOPMENT OF PIPING FAILURE AT LITTLE WEWOKA CREEK SITE 17 AS OBSERVED BY EYE WITNESSES (May 21, 1960) Figure 2

Figure 2 the leak started through the embankment at about mid-height and then gradually eroded downward so that it is known that the initial leak did not develop along the conduit, even though the conduit was subsequently exposed. A similar situation apparently nearly occurred at Leader Middle Clear Boggy Site 29 where a muddy leak developed through the embankment at about mid-height, but directly over the conduit: had this leak progressed to failure, very likely the resulting erosion tunnel would have exposed the conduit. In other cases where the conduit was exposed by the failure, the initial leak must have traveled along or near the conduit.

As seen in Appendix A, the piping tunnels frequently emerged in the embankment, just above the foundation, as in the example of Figure 1. For some of these failures, the initial leak may have travelled along the contact between the dam and the foundation. In none of the cases, however, was the failure caused by an initial leak which travelled deep in the foundation, by-passing the dam.

Reservoir Volumes Released During Failure and Damages

A characteristic common to all the failures was the fact that the depth of water in the reservoir was not great, ranging from a minimum of about 5 feet to a maximum of about 20 feet, with the average being in the range between 10 and 15 feet. The reservoir volumes at the time of failure ranged between 30 and 1100 acre-feet. In none of the failures was there any appreciable damage to property downstream, or was loss of life even threatened. In a typical case several hundred acre-feet of water were discharged more or less gradually over a period of time ranging between a half day to several days, the maximum flow through the breach being in the range from about 50 to several hundred cubic feet per second, which was less than the hydraulic capacity of the creek.

MISSISSIPPI FAILURES

The three failed dams are located in a general area of highly dispersive clays in north-central Mississippi (9).

The details and circumstances surrounding the three breaching failures were quite similar to those described above for the Oklahoma dams with two main exceptions: (1) there is no hard bedrock in the area and a much lesser tendency for differential foundation settlement; and (2) all the failures occurred along the outlet conduits, Figure 9. Also there was more damage to the Mississippi dams by rainfall erosion tunnels than to the Oklahoma dams (9).

HYDRAULIC FRACTURING IS PROBABLE CAUSE OF INITIAL LEAKS

In the analysis of the cause of failures, one of the most pertinent observations is the fact that in all cases the leak appeared at the downstream side of the dam within a very short time (within a few days, or even less than a day) after the reservoir raised rapidly behind the dam. Because of this it was apparent that the initial leakage water was traveling through a concentrated channel, such as a crack. This is true because the dams were all constructed of compacted clays, with coefficient of permeability of the order of 1.0 to 0.1 feet per year. It would have been impossible for the leakage water to pass through the embankment in such a short time except through an open leakage channel.

In the last few years it has been concluded by several investigators in various parts of the world, who have had the problem of studying difficult-to-explain leaks through well constructed dams, that such concentrated leaks have developed through cracks which were not open before the reservoir rose, and which were never seen. Such cracking can develop only if the minor principal stress acting within the embankment along the path of the potential leak is lower than the reservoir water pressure. There have been only a few published opinions on this subject (5,12); however, the writers believe there is no doubt that such a phenomenon can occur, and that it is the only reasonable explanation for some of the leakage problems which have occurred (8).

Although the initial crack may be very narrow, perhaps not even visible, water from the reservoir can penetrate it. As a result the stress acting on the plane of the crack is changed abruptly from a low compressive stress, or even tension, to a compressive stress which would approach the reservoir head if the crack did not extend through the core. This causes an increase in the crack width from readjustment of the embankment to the new stress condition. In this action the water may enter an existing crack which had been previously closed or a new crack may be formed. The term "hydraulic fracturing" has been employed to cover both situations. The internal stress conditions theoretically necessary for hydraulic fracturing can be present on the first reservoir filling or they can develop later.

As a result of the study of the Oklahoma dam failures, the writers have concluded that it is overwhelmingly probable that the initial leaks developed as a result of hydraulic fracturing. The internal embankment stresses were caused to be low either as a result of differential foundation settlement or drying and shrinkage of the embankment during or after construction, or a combination.

Embankment Stress Computations

In order to study the influence of foundation settlement on the internal stresses, finite element analyses were made by Drs. J. M. Duncan and G. Lefebvre (see Acknowledgements) using recent techniques developed at the University of California (6,7). To the writers' knowledge these calculations are the first effort to study the stress distribution in the special case of a low earth dam consisting of stiff clay, in which the internal stresses due to the weight are generally lower than the unconfined compressive strength of the embankment clay. Under these conditions the minor principal stresses acting in the lower parts of the dam can easily be zero, or tension if the material could withstand tensile stress.

Figure 3 shows the mathematical model and the basic assumptions used for the computations. Case A, Figure 4, shows the computed distribution of minor principal stresses within the embankment assuming that the foundation settles a maximum of 0.34 feet, and that all the settlement occurs during construction of the dam. Cases B and C show distribution of the minor principal stresses assuming that the maximum foundation settlement is 0.34 feet during construction and that the foundation continues to compress after construction to give total maximum foundation settlements of 0.88 and 1.42 feet, respectively. As seen in Cases B and C the computations indicate that large zones of tension develop near the abutment, with tensile stresses of more than 1.0 ton per square foot. Case D shows the results of another set of calculations, in every way similar to those of Case C except for the assumption that the embankment cannot take tension. In this case the shaded area, approximately the same size and location as in Cases B and

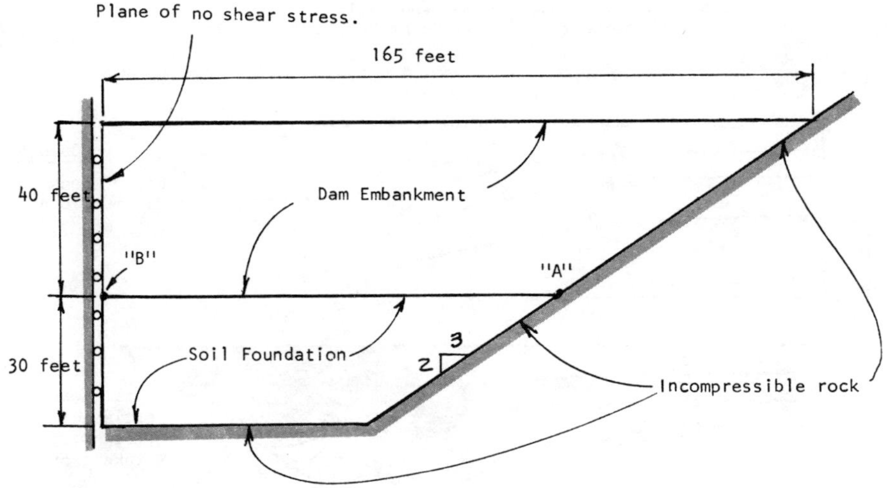

Assumptions:

1. Dam is considered a two-dimension body with above shape and boundary conditions. (Dam foundation settlement varies from zero at Point "A" to a maximum value at Point "B".)

2. Embankment is a stiff clay with non-linear stress-strain curve which varies as a function of the stress level. (Unconfined compressive strength about 4.0 tons/ft^2)

3. Embankment density = 112 pcf.

4. Dam is built in eight horizontal layers of equal (5 foot) thickness.

5. Computations are made for varying assumptions of compressibility of the soil foundation and two assumptions (1) that all foundation settlement occurs during construction; and (2) that the large portion of the foundation settlement occurs after the dam is constructed.

6. Computations are made for the two assumptions: (1) the embankment can withstand tension; and (2) no tension can develop.

7. Computations are for stresses caused by the weight of dam only: no influence of the reservoir is included.

MATHEMATICAL MODEL AND ASSUMPTIONS Figure 3

Note: In all cases, maximum foundation settlement occurs at Point "B".

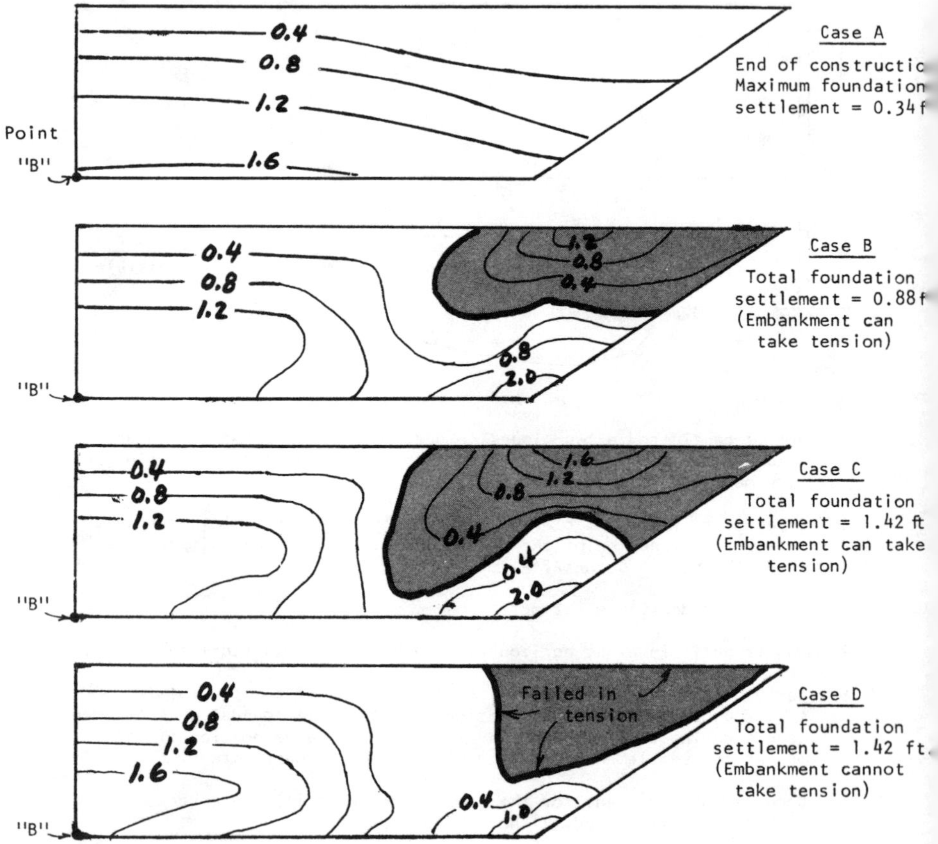

Notes:

1. Stress contours in tons/ft^2.
2. Shaded areas in cases B & C are in tension.
3. See Assumptions, Figure 1
4. In all cases, maximum foundation settlement occurs at Point "B".

MINOR PRINCIPAL STRESSES COMPUTED ASSUMING THE FOUNDATION SETTLES 0.34 FEET DURING CONSTRUCTION AND THEN CONTINUES TO SETTLE AFTER CONSTRUCTION

Figure 4

C, is the portion of the embankment which has "failed in tension".

The main result of these calculations is to show that using a mathematical model with dimensions similar to the typical failed Oklahoma dam, and reasonable assumptions of the stiffness of the embankment clay, substantial zones of tension would be expected to develop. These tension zones occur adjacent to rock abutments when the total foundation settlement at a point approximately 100 feet from the rock abutment is of the order of 12 to 18 inches. It is, of course, difficult to be confident that the results of such calculations reliably reflect stresses which develop in an actual embankment. It is apparent, however, that the results of calculations using the newly developed techniques support the hydraulic fracturing explanation for the development of the initial leaks.

Embankment Drying

It is frequently necessary to stop construction of a dam for several days during time of hot, dry weather, and it is not unusual for the construction surface to be stopped for weeks. Invariably, shrinkage cracks develop at such times in clay embankments, unless special precatuions are taken to prevent drying, Figure 5. Drying and shrinkage occurs also on the slopes of embankment closure sections and in natural clay surfaces exposed by excavation in cutoff trenches, reshaped channel banks and abutments.

When the construction recommences, the surface is often rewetted and scarified. Commonly, however, no investigation is made of the depth of the cracks and they may not be considered important. Probably in most cases the cracks may not be open to any appreciable width; however, the section of embankment influenced by severe drying, which may extend several feet below the surface, probably continues to have relatively low stresses on vertical planes. After construction is completed, the localized embankment section which has been subjected to drying and shrinkage probably continues to constitute a zone of low minor principal stress, where hydraulic fracturing could occur.

Figure 6a is a cross-section through a low homogeneous clay dam on an incompressible foundation. The cross-section is assumed to be at a location in the dam where the height is constant so that there is no influence of differential settlement. The embankment clay has an unconfined strength which is greater than the maximum value of compressive stress in the low dam. During construction the work was stopped when the dam reached the elevation of Point b as shown. During the work stoppage the clay surface ebf, was subjected to drying and shrinkage. The drying had maximum influence to a depth equal to bc, and lesser influence below Point c.

Figure 6b shows the distribution of compressive stresses acting on a vertical plane transverse to the longitudinal axis of the dam on a vertical line through the center of the dam crest. This is generally the intermediate principal stress (σ_2), but is also approximately equal to the minor principal stress (σ_3) which acts on the longitudinal vertical plane through the centerline of the crest. If there had been no influence of drying during construction, the stress (σ_2) would increase approximately linearly below the crest ($\sigma_2 = kh\gamma$), and at the level of Point b would be $\sigma_2 = rm = kD\gamma$. Because of the drying, the stress at the level of Point b, is reduced to a considerably smaller value ($\sigma_2 = rp$). Since the vertical stress acting on the horizontal plane through Point b, which is the major principal stress, is considerably less than the un-

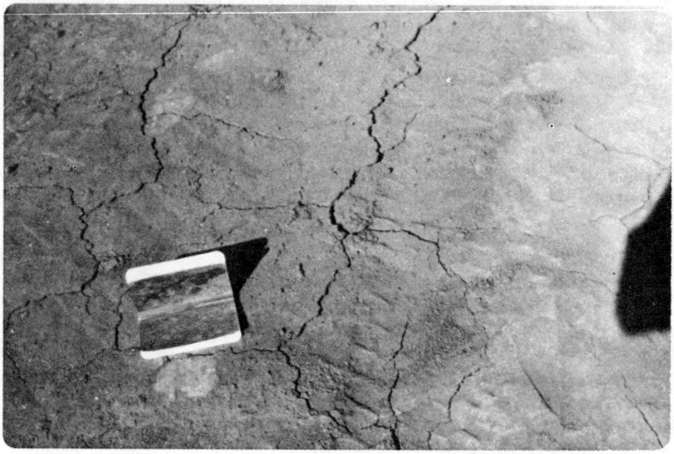

DRYING CRACKS IN CONSTRUCTION SURFACE OF COMPACTED DAM EMBANKMENT OF CLAY OF MEDIUM PLASTICITY IN DRY, HOT WEATHER SEVERAL HOURS AFTER COMPACTION.

Figure 5

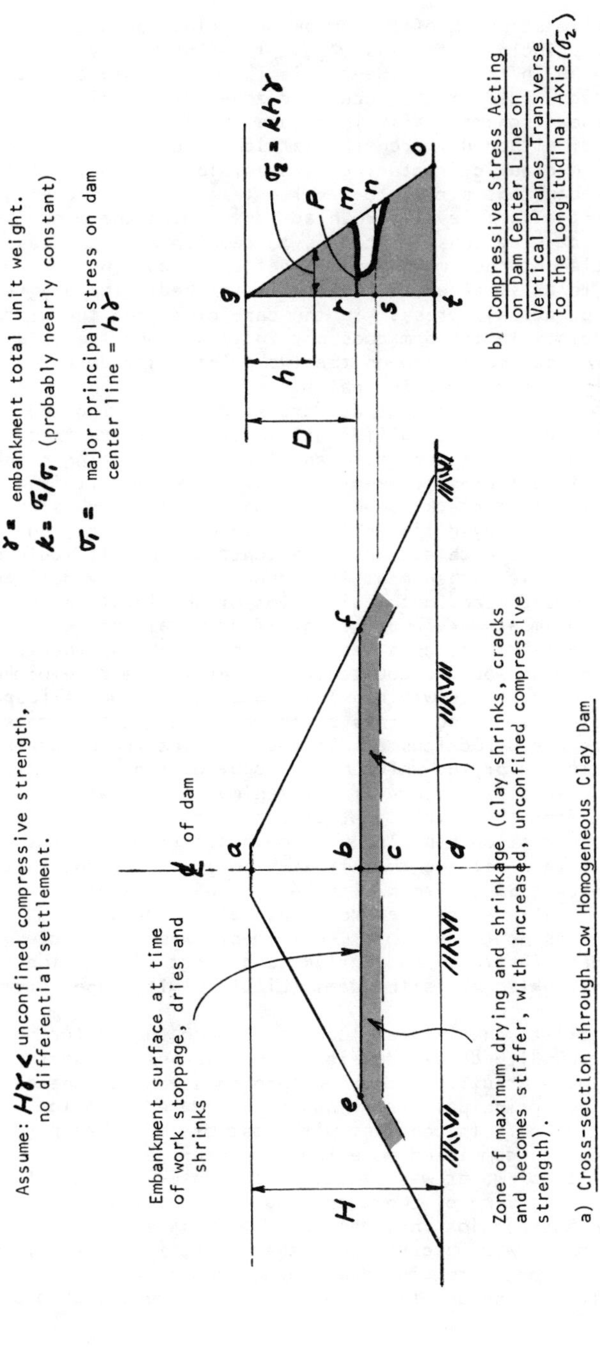

INFLUENCE OF DRYING DURING CONSTRUCTION ON STRESSES ACTING ON VERTICAL TRANSVERSE PLANES

Figure 6

confined compressive strength of the embankment clay, both the intermediate and minor principal stresses (σ_2 & σ_3) can be nearly zero, or conceivably could be in tension. Hence, it is very probable that the drying and shrinkage would create localized zones of low minor principal stress as shown schematically in Figure 6b.

The actions of either differential settlement or drying can create a situation where hydraulic fracturing is possible by reducing the minor principal embankment stress to a value less than the reservoir water pressure at the same level. Both actions reduce the minor principal stresses by causing tensile strains to develop. In the case of differential settlement the embankment is deformed, causing the critical section to be strained in tension (stretched), thereby reducing the minor principal stress. In the case of drying the distance between any two points in the dam does not increase, but the volume of the clay-water embankment between the two points is reduced by shrinkage, reducing the minor principal stress.

The writers have observed that dam cracks due to drying and those due to tensile strain caused by differential settlement are frequently related, in the sense that the pattern and location of drying cracks are similar to differential settlement cracks. Sometimes it is difficult to determine whether cracks developing on the crest and slopes of low clay dams are caused by differential settlement or by drying. Undoubtedly, in some cases both actions have contributed. It would be expected that when a clay embankment is caused to shrink by desiccation the drying cracks would open in the direction of the least resistance; that is, if an embankment were undergoing differential settlement and tension cracking was imminent as a result, then embankment volume shrinkage due to drying would undoubtedly accelerate the development of the crack and increase its width. This combined action is responsible for the fact that drying cracks often occur in low clay dams in the same locations and directions as differential settlement cracks.

In addition to any drying which might occur during construction of the dam causing localized zones of low minimum principal stress, there is evidence from this study that drying of the surface of a low clay dam after construction can also be a main contributing factor. In this regard the writers believe that it is highly significant that the embankment of Cherokee Sandy Creek Site 8A, the only one of the eleven Oklahoma failed dams in which the breaching failure occurred at a point where there was no tendency for settlement, was constructed from clay of much higher plasticity and shrinkage properties than the clay in any of the other Oklahoma failed dams (Liquid Limit ranged from 55 to 90 percent).

In all probability the initial leak leading to the failure of the Cherokee Sandy Creek Site 8A Dam developed in a series of interconnected cracks by hydraulic fracturing through a zone of the embankment where the minor principal stress had been reduced to a low value by drying and shrinkage. In contrast with most of the other projects the dam embankment was completed more than one year before any appreciable quantity of water was stored in the reservoir and it is believed likely that drying of the completed embankment during this period could have had a major influence on the failure. Following failure of the dam, it was observed that there were frequent drying cracks in the embankment extending down to unknown depths, with some of them as wide as 1 to 2 inches at the surface. Subsequent laboratory investigations show that typical specimens of the embankment clay compacted near optimum water content (about 25%) would shrink 10 to 15% of their volume when allowed to air dry from optimum water content to the shrinkage limit.

Investigators of leaks through other low clay dams have concluded that drying cracks were the probable cause. In a well-constructed low homogeneous clay dam in Australia, piping tunnels developed on the first reservoir filling in 1968 at an elevation where the construction surface had been stopped for three months during dry weather, and the investigators concluded that the most probable cause of the leak was drying cracks. At this dam also the clay was determined to be moderately susceptible to dispersive piping (4).

Another strong piece of evidence supporting the conclusion that drying cracks can lead to piping failure is the widespread damage to the surface of these dispersive clay dams by tunnel erosion from rainfall as described in Reference 9. There can be no doubt that these surface tunnels are caused by rain runoff seeping into shrinkage cracks and gradually eroding tunnels of roughly circular cross-section.

UPPER CLEAR BOGGY SITE 50

The details of the most recent of the eleven Oklahoma dam failures, Upper Clear Boggy Creek Site 50, are typical. As seen in Appendix A, the dam was completed in May 1970. The reservoir was filled for the first time during a st m on October 9, 1970, and the failure occurred immediately thereafter. There were no eyewitnesses and failure date is not known within 5 or 6 days. At the time of failure the water level was about 5 feet below the crest of the dam (maximum depth of about 20 feet) and the volume of water released was about 700 acre-feet. After the failure there was no visual evidence of a flood downstream, so that the water must have been released fairly slowly.

As shown in Figure 7 the failure breach was about 50 feet wide and was located roughly 125 feet from the right end of the dam. At the breach the bedrock was very near the original ground surface on the right side. To the left of the breach there was about 30 feet of alluvium overlying the bedrock. Hence, the failure occurred just at the location which would have been predicted as a zone of tension by the computations summarized in Figure 4.

In studying this example, it should be kept in mind that the dam has a length of 1500 feet, and that the dam height was about the same for most of its length. Hence, the fact that the breaching failure occurred at the point of maximum probable differential foundation settlement would be statistically a very rare coincidence if the differential settlement were not an important contributary cause. Further, when it is considered that most of the Oklahoma failures occurred at a point of probable high differential foundation settlement, the statistical evidence becomes overwhelming.

Figure 8 shows photographs of the conditions exposed in the left wall of the failure breach, three months after the failure. As seen in the closeup of Figure 8b, the face of the breach was checked with cracks, with one main vertical crack centered on the small erosion tunnel. It is believed that these cracks were probably closed during construction and were caused to open primarily by drying after the failure, but that they are representative of the types of incipient cracks which exist in clay embankments and can be penetrated by

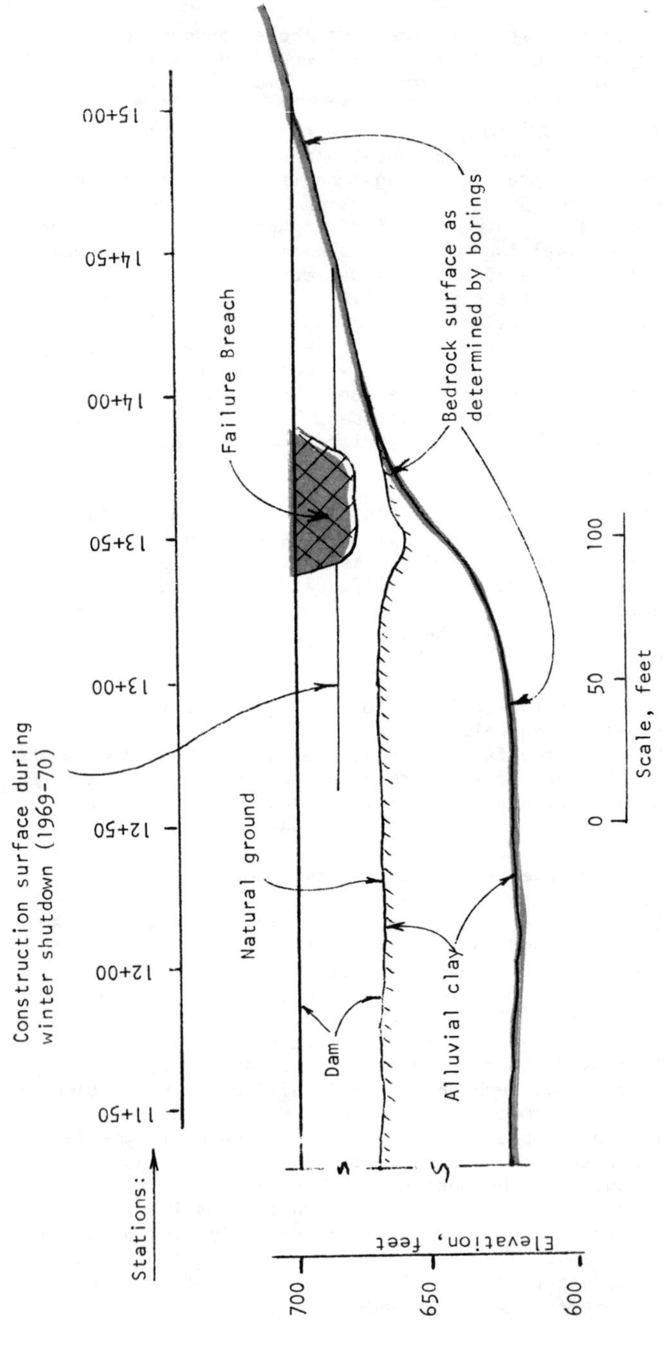

LONGITUDINAL SECTION THROUGH RIGHT END OF UPPER CLEAR BOGGY SITE #50 DAM (OKLAHOMA) SHOWING LOCATION AND DIMENSIONS OF FAILURE BREACH (Oct. 1970)

Figure 7

8a. Left wall of breach. Note 24-inch diameter erosion tunnel.

8b. Erosion tunnel in left wall of breach; note cracks.

FAILURE BREACH IN UPPER CLEAR BOGGY CREEK SITE 50
(Condition 3 months After Failure)

Figure 8

reservoir water by hydraulic fracturing.

The dam was built in two construction periods. As seen in Figure 7, the bottom of the breach, and particularly the 2-foot diameter piping tunnel observed in the wall of the breach (Figure 8), were located near the elevation of the surface of the first stage embankment, which had ample opportunity to dry and shrink. Hence, the low embankment stresses leading to hydraulic fracturing could have been caused by a combination of the actions of differential foundation settlement and shrinkage due to desiccation.

INFLUENCE OF CONDUITS

All these low flood control dams have spillways consisting of a tower at the upstream toe and a conduit through the base of the dam, usually in the range between 18 and 42 inches diameter. As discussed above, of the total of 15 piping tunnels or breaches in the 12 low Oklahoma dams, 7 were located at or near the plane of the conduit. All three of the Mississippi failures occurred directly adjacent to conduits.

Of the seven Oklahoma dams, two had corrugated sheet metal conduits, which had been backfilled on all sides only with compacted earth (Upper Red Rock Sites 42 and 48). The other conduits were of reinforced concrete pipe, provided with cast-in-place concrete cradles.

It is apparent that it is very difficult to compact soil beneath and get a good bond with the lower half of a corrugated metal conduit. Hence, conceivably the failures which occurred along these metal conduits could have been due to an initial leak which developed only because of inability to compact the soil underneath the conduit. Primarily because of difficulty getting a seal in the lower half and because the metal conduits themselves frequently were not watertight, the SCS has abandoned corrugated metal conduits, except for relatively unimportant and small structures.

Of the piping tunnels which developed near the plane of the conduits, eyewitnesses in two cases observed that the initial leaks developed through the embankment at about mid-height of the dam, many feet above the conduit (Little Wewoka Creek Site 17, Figure 2, and Leader Middle Clear Boggy Creek Site 29, Appendix A). While none of the other examples had eyewitnesses, there can be little doubt that the initial leak which led to the failure probably traveled along a path which was generally parallel with the conduit and not very far from it. In one extreme case in Mississippi a piping tunnel developed along the conduit during a storm which occurred before the dam was completed, Figure 9.

The writers believe that the high percentage of the erosion tunnels which developed along conduits is consistent with and provides other evidence for support of the hydraulic fracturing explanation as the cause of initial leaks. This is true because it would be expected that zones of the embankment immediately adjacent to conduits would have low internal stresses, both from arching and stress transfer caused by differential settlement and from drying and shrinkage.

For the Oklahoma dams it was common practice to place the conduit near one of the abutments. Also, prior to about 1965 a design detail was commonly used in which a trench was excavated into the natural foundation material along the full length of the conduit, with the

9a. Collapse of roof of piping tunnel caused sinkhole on construction surface.

9b. Downstream slope showing piping tunnel emerging on both sides of conduit.

POTACOCAWA CREEK SITE 3 DAM, MISSISSIPPI. Failed by piping along the reinforced concrete conduit before embankment construction was completed when reservoir level raised during rainstorm to an elevation which was only 4 feet above the conduit invert at the upstream end, August 1962. (H. C. Huey Photos)

Figure 9

bottom of the trench extending several feet below the concrete cradle. This trench was then backfilled with compacted earth up to about 6 inches above the top of the conduit. Then, as shown schematically in Figure 10, a narrow, vertical-walled trench was excavated into the compacted embankment, the conduit installed, and the concrete cradle placed underneath without forms. All the Oklahoma dams which failed by piping along conduits were constructed in this way. The main purpose of the compacted earth filled trench under the cradle was to provide a uniform support for the conduit.

Since the compacted earth fill under the conduit was generally less compressible than the adjacent natural foundation soil, this design feature created a condition where stress transfer from differential settlement would be expected directly adjacent to the conduit. This is particularly true because oftentimes the bottom of the trench under the conduit extended down to bedrock. Because of this and also because the reinforced concrete conduit, cutoff collars and cradle create "hard points" within the embankment, it is very probable that the minor principal stresses existing in localized strips of the embankment directly adjacent to the conduit would be lower than the stresses which would have existed if no conduit had been installed. The design detail of Figure 10 is no longer considered good practice where conduits are located on compressible soil foundations. In current practice no earth filled trench is placed under the cradle, and the cradle sides are formed with a positive batter.

In addition, drying and shrinkage cracks are especially likely to develop adjacent to conduits because: (1) it was common practice to stop the embankment construction as the conduit was being placed, and (2) backfill around the conduits must be hand placed. Unless special precautions are taken in the embankment construction around the conduit, the embankment in the immediate vicinity of the conduit is generally exposed to potential drying and shrinkage for a longer period of time than the average embankment construction surface. (Based on these experiences, good practice should include construction methods and soil types which eliminate or minimize concentrated localized desiccation of the backfill immediately surrounding the conduit.)

For these reasons, it would be expected that the minor principal stresses acting within strips of the embankment located directly adjacent to the conduits would be lower than in comparable locations at a distance from the conduits.

With respect to the locations of the failure, the experiences with the low flood control dams in Mississippi are considerably different than the Oklahoma experiences. In Mississippi, all the breaching failures occurred along the conduit, whereas in Oklahoma in approximately half the failures the conduits were not involved. Also, while the evidence is not conclusive, it is believed that on the average the embankment clays comprising Mississippi failed dams are more dispersive and erodible than the clays of the Oklahoma failed dams. The main evidence for this is that more of the Mississippi dams were damaged by rainfall erosion tunnels and the average dam was more severely damaged by rain than in Oklahoma. Hence, there are two apparent differences between the Oklahoma and Mississippi experiences; (1) in Mississippi all the failures occurred along the conduit whereas in Oklahoma the conduit was not involved in half the failures and (2) there was a greater number and higher percentage of failures in Oklahoma than in

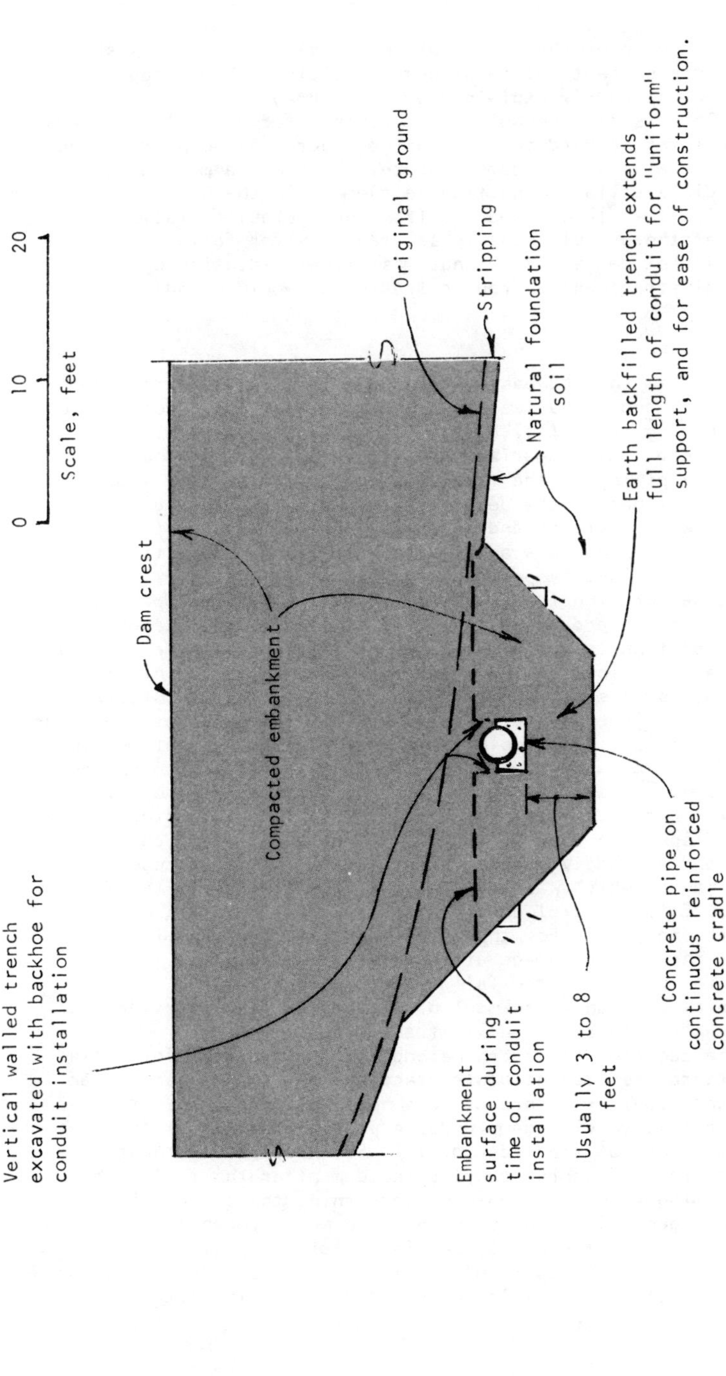

CONDUIT SUPPORT DETAIL WHICH PROBABLY CONTRIBUTED TO CONDITIONS PERMITTING HYDRAULIC FRACTURING (METHOD NOW ABANDONED)

Figure 10

Mississippi, in spite of the fact that the Mississippi clay is suspected to be more susceptible to dispersive piping. These apparent differences can be readily explained by the hydraulic fracturing hypothesis. This is true because the abutments for the Oklahoma dams generally consisted of hard bedrock whereas there was no hard bedrock under any of the Mississippi dams. Hence, there was a much greater tendency for differential foundation settlement in the Oklahoma dams. Because of this it would be expected from the hydraulic fracturing hypothesis that there would be a lesser tendency for failure on the average Mississippi dam and also that most likely location of low minor principal stress and hydraulic fracturing would be adjacent to the conduit.

WISTER DAM

The Wister Dam in southeastern Oklahoma is a fairly large, homogeneous clay dam, about 100 feet high and 5700 feet long, constructed by the Corps of Engineers (1,2,8,10). It retains a reservoir of 430,000 acre-feet. As summarized briefly in Appendix A, the dam nearly failed by piping on the first reservoir filling 1949. During the current investigation the details surrounding the leakage and piping problem were reviewed and an inspection was made of the dam.

The downstream slope has an excellent growth of protective grass which is well maintained by permanent personnel at the site. In spite of this, vertical erosion tunnels from rainfall have developed each year, with appearance and dimensions very similar to the frequent rainfall erosion tunnels which developed on the low Oklahoma and Mississippi dams as described in the companion paper (9). In this investigation four soil samples were taken at widely spaced intervals on the downstream slope of the Wister Dam, from the walls of the rainfall erosion tunnels. All samples were very similar clay of medium plasticity (CL) which was shown to be highly dispersive by both the chemical tests and the SCS Laboratory Dispersion Test, Figure 11.

The Wister Dam is located in Oklahoma. It is constructed of the same kind of dispersive clay as the low SCS Oklahoma flood control dams which failed and was badly damaged by piping on the first reservoir filling in a manner which was very unusual, but similar to the failures of the low SCS flood control dams. Also, frequent rainfall erosion tunnels occur each year. For these reasons it is overwhelmingly probable that the piping damage at the Wister Dam resulted from the same fundamental causes as the failures of the SCS dams; that is, by piping in an embankment of dispersive clay starting with an initial leak caused by hydraulic fracturing.

Since the adoption of modern methods of construction for rolled earth dams, there has been no close precedent for such a large leak and piping condition: -- a more or less horizontal leakage path of 700 feet through a massive clay dam, under a pressure of only a few feet of water and a gradient of the order of 1 in 50. Because of these surprising facts, investigators studying the dam after the failure had difficulty forming a definite opinion concerning the cause. Finally, however, it was generally concluded that the most probable and reasonable explanation was differential settlement cracking (1,2). This conclusion was reached in spite of the facts that no cracks were observed before or after the piping occurred and that the total settlements were not very large.

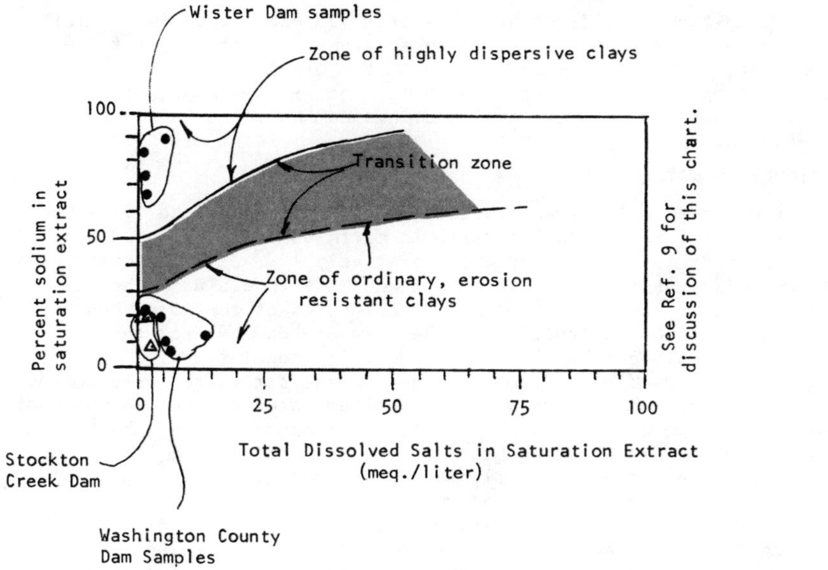

SCS Percent Dispersion	Liquid Limit	Plasticity Index	
%	%		
59	32	10	Wister Dam
44	32	11	
63	32	12	
67	33	12	
16	35	19	Washington County Dam
9	37	20	
9	37	22	
19	29	15	
2	47	27	
18	31	8	Stockton Creek Dam
15	32	11	
22	29	7	

LABORATORY RESULTS FOR WISTER, WASHINGTON COUNTY AND STOCKTON CREEK DAMS Figure 11

The writers consider that the Wister Dam experience is a relatively important part of the total evidence available for analysis of the current problem. This is true because it shows that the same trouble could develop in a dam which is a much larger and more important structure than the SCS flood control dams, and in a dam which was built by a different group of engineers.

CONSTRUCTION WATER CONTENT

It has been well established in past studies that the likelihood of differential settlement cracking in earth dams is increased when the embankment is constructed with relatively low water content. As a part of this investigation, an effort was made to evaluate the influence of this factor by reviewing the results of the water content tests made during construction of the Oklahoma dams. From this effort it was not possible to arrive at any definite conclusions. The construction records did not show that any of the structures were compacted excessively dry relative to Standard optimum water content (ASTM D698); however, in some of the post-failure studies apparent dry embankment material was observed. Undoubtedly, some sections of some of these dams were placed at a relatively low water content and very possibly this fact contributed to some of the failures.

It is not believed, however, that excessively low construction water content was a major or consistent factor contributing to the average of the failures. A part of the evidence supporting this conclusion is the experience of the Upper Clear Boggy Creek Site 50, the most recent of the Oklahoma failures. For this dam, it was discovered by tests during construction that the embankment clay was more highly susceptible to dispersion than had been realized from the preconstruction investigations. This discovery caused considerable concern during construction. The problem was carefully studied and everyone was aware that it was necessary to be especially careful with this dam. From a study of the records of the moisture-density control tests during construction it appears that the average water content was fairly uniform and near optimum.

The Wister Dam experience also supports this main conclusion. More than 900 tests of the construction water content were made showing that the average was very close to Standard optimum, and that there was not a great spread in the water content (10).

These experiences indicated that the failures can and did occur in spite of the fact that the embankments were constructed near optimum water content. It can be speculated that the likelihood of failure would have been less if the average construction water content had been several percent above Standard optimum (ASTM D698). The higher the water content the less rigid the embankment material and the lower the unconfined compressive strength, and the less likely the development of embankment zones of low minor principal stress from differential settlement. On the other hand, the higher the average construction moisture content the more precautions which would have to be taken to minimize the undesirable effects of desiccation.

OTHER DAMS IN OKLAHOMA

Several hundred other flood control dams were constructed in the same part of Oklahoma as the eleven failed dams during the same period of time, with similar designs and construction. Many of these are constructed of similar dispersive clay.

The reason that the eleven dams failed and the others did not must be attributed to the fact that the conditions were not as severe as in the others. The embankment stresses may not have been reduced to a sufficiently low value to allow an initial leak to develop by hydraulic fracturing. Alternatively, if small initial leaks did develop, the embankment clay forming the walls of the leakage channel may have swelled more rapidly than it eroded, therefy closing off the leak.

In western Oklahoma, where the soils do not have the unusual dispersive properties, several hundred similar SCS dams were constructed during the same period of time, often from silty sands and sandy silts with little or no cohesion. One of the problems with the dams in this area was that differential settlement cracks sometimes appeared on the crest and the slopes. Most of the more serious cracks were filled by mud grouting. None of these dams failed. The difference in performance is undoubtedly due to the difference in soil types. The fine fraction of the soils in western Oklahoma is not dispersive, is not highly susceptible to drying and shrinkage and often does not have the cohesion needed to support the roof of a leakage channel without collapsing.

SUMMARY OF EVIDENCE FOR HYDRAULIC FRACTURING HYPOTHESIS

There is no available means to prove with complete certainty that hydraulic fracturing was the cause of most of the initial leaks. The conclusion is not amenable to being proved with complete confidence. It could not be proved even if a great deal more information were available concerning individual failures. If all of the failures had been observed by competent eyewitnesses, and if the failed dams had been provided with extensive instrumentation, it would still be impossible to determine the cause of the initial leak with certainty. Therefore, it is necessary to draw conclusions on the basis of inference, statistical likelihood and circumstantial evidence.

The main factors leading to the conclusion that it is overwhelmingly probable that the initial leaks were the result of hydraulic fracturing are summarized below.

1. All the failures developed rapidly from an initial leak, which must have been flowing in a concentrated channel, such as a crack.

2. Most of the Oklahoma breaching failures occurred at a location where the foundation bedrock was near the original ground surface, and where the depth to bedrock was changing rapidly.

3. Computations with a mathematical model simulating the typical physical conditions of the average Oklahoma dam, showed that a large zone of tension can be expected at the point where the bedrock is near the surface on one side and the depth to bedrock varies considerably over short distances.

4. The only one of the 11 Oklahoma failed dams for which the failure tunnel did not occur at a point of high potential differential settlement was the one constructed of clay of highest plasticity, with greatest tendency for shrinkage by drying.

5. Because of the relationships between the height of the relatively low dams and the unconfined compressive strength of the clay embankment material, the minor principal stress in the lower portions of the dam can approach zero, or even be in tension. Consequently, there is no doubt that the pressure of the reservoir water can be

greater than the embankment stresses acting on vertical planes.

6. The hypothesis readily explains the large percentage of failures which occurred near the conduits.

7. The fact that similar piping developed at the Wister Dam, even under more extreme and difficult-to-explain circumstances, provides evidence that the initial leaks were not caused by some aspect of the design and construction practices employed for the great number of low SCS flood control dams.

8. Finally, there is no other hypothesis available which offers an explanation of the known facts.

Other Causes of Initial Leaks

While the writers believe that it is overwhelmingly probable that the principal cause of the initial leaks has been hydraulic fracturing, for some of the dams it is probable that there were other important contributing factors. In some cases the initial leak from the reservoir to the downstream exit points may have traveled through embankment cracks for part of the distance and found other leakage channels for other parts of the path. For example, at the Caney Coon Creek Site 2 (Appendix A) the right abutment was comprised of a rock formation which had open cracks and fissures. The post-failure study showed clearly that the main erosion tunnel at the right abutment was located just above the rock surface and that some of the eroded clay had been carried by the seeping water into the rock cracks. Hence, the initial leak may have travelled through the rock cracks for part of the length.

The failure of Leader Middle Clear Boggy Creek Site 33 Dam was different from the average experience in several main aspects: (1) the failure occurred approximately five years after the construction even though the reservoir had been nearly full several times during the five years; (2) two independent tunnels developed, apparently at the same time about 90 feet apart; and (3) the embankment slopes had been very severely damaged by frequent vertical rainfall erosion tunnels as described in Reference 9, with the worst damage located near the point where the dam finally breached. From these facts it seems probable that the rainfall erosion tunnels contributed to the failure. The initial leakage water may have entered one of the erosion tunnels, then traveled through a fissure opened by hydralic fracturing, and finally exited through an erosion tunnel on the downstream face. Conceivably also the gradual erosion of the rainfall tunnels could have reduced the average minor principal stress acting on potential leakage surfaces through the dam, increasing the likelihood of hydraulic fracturing.

Owl Creek Site 7 Dam is another example which makes it clear that the initial leak did not have one single cause. As described briefly in Appendix A, it was found that the leaking water entered the natural ground forming the right abutment about 80 feet upstream from the upstream toe and traveled within the natural ground until it hit the compacted clay cutoff under the dam. Then the erosion tunnel turned about 90 degrees and graveled along the upstream face of the foundation cutoff for about 60 feet where it encountered the outlet conduit, after which the tunnel traveled parallel with the conduit to the downstream toe. In this case it was apparent that the upstream leg of the leak developed along a crack in the fractured rock of the abutment, which created a favored seepage path.

STOCKTON CREEK AND WASHINGTON COUNTY DAMS

Washington County Dam in Illinois (3) and Stockton Creek Dam in California (8) are well constructed homogeneous dams of clay, which failed by rapid piping directly following the first reservoir filling. In neither case was the cause of the failure reliably determined. In both dams the failure occurred near relatively incompressible abutments and investigators concluded that differential settlement cracking was the most probable cause, although no open cracks were seen before or after the failure. The Washington County Dam is located in a part of Illinois which is known to have highly dispersive surface clays (13).

In this investigation, samples of the typical embankment materials from both dams were tested for dispersibility. For both dams, the tests on all the samples indicated the clays to have the properties of common non-dispersive clay, Figure 11.

While the cause of these failures can never be determined with complete confidence, the writers believe that it is probable that they were caused in the same way as the failures of the Oklahoma and Mississippi dams described herein. Neither of the embankment materials is a tough or highly plastic clay. Probably, the initial leak caused by hydraulic fracturing was sufficiently large and erosive to progress to failure by piping in spite of the fact that the embankment material was not susceptible to dispersive piping. Very possibly also the Washington County Dam embankment contained a zone of dispersive soil which was not sampled in this investigation.

In Figure 11 it is seen that all the samples from these two dams had relatively low total dissolved salts, generally less than 10 meq/liter in the saturation extract. The general study which led to the classification of dispersive soils (9) was based on limited data and experience. Possibly future studies will show that certain clays with low total salts are relatively erodible even though the sodium content in the saturation extract is less than 30% and the reaction to the SCS Laboratory Dispersion Test is low.

SOME GENERAL OBSERVATIONS ON PIPING AND FILTERS

Piping Mechanics

Literally thousands of small dams, with heights of 20 to 50 feet have been built all over the world. There can be little question that many of these have been subjected to foundation settlement and drying of generally similar magnitude and severity to the conditions which existed at the Oklahoma and Mississippi failed dams. Only few of these thousands of dams have failed by piping and this general experience supports the conclusion that it is reasonable practice to build low dams as homogeneous embankments of clay.

The writers speculate that it is probable that the main reason that there have not been a greater number of failures of such dams, by hydraulic fracturing and piping, is that when small concentrated leaks develop by hydraulic fracturing in the dam of ordinary clay, the erosive tendency of the leak is too small to cause piping. Subsequently, the embankment swells slightly, increasing the minor principal stresses and closing the cracks. Probably in most cases this action occurs before any appreciable amount of water seeps through the dam. Therefore, it appears that for small homogeneous dams built of most clays, under the common range of foundation settlement and the common tendency for drying cracks during construction, the embankments are sufficiently self-healing to prevent progressive erosion of leaks through cracks.

The Oklahoma and Mississippi dispersive clays have properties which are sufficiently different than those of ordinary clays (9), so that this general experience in the profession with the performance of low homogeneous dams is not completely applicable. For these dispersive clays, rapid turbulent leaks are not needed to cause erosion. The individual clay particles tend to go into suspension even in still water and to stay in suspension. Therefore, even the smallest leak, flowing completely without turbulence, can pick up the clay particles and carry them along. Also, the highly dispersive sodium clays are very impervious and, while they have a fairly high swelling potential, they absorb water and swell at a slower rate than ordinary clays. Hence, it is speculated that in dispersive clay embankments cracks would swell shut more slowly and that there would be more time for erosion to occur.

It should be pointed out that the nature of piping which has caused failure in the Oklahoma and Mississippi dams is fundamentally different from that which occurs in cohesionless soils, as commonly described in the soil mechanics literature. In the classic explanation of piping in cohesionless soils, the concentrated leak emerging at the downstream side is caused by water flowing through the pores of the soil. The erosion starts first at the discharge end of the leak, causing a local enlarged tunnel-like leakage channel, which creates a local concentration of seepage and erosion forces. As the piping process develops the leakage channel is gradually extended upstream by progressive erosion until it reaches the water source supplying the leak, at which time a rapid catastophic failure may result.

In the failures of the Oklahoma and Mississippi dams, the initial leak is already traveling through concentrated leakage channel from the moment of its inception, probably in the form of an individual narrow crack, or a zone of the embankment which has a series of small cracks. There can be no appreciable contribution to the volume of the leak from seepage through the pores of the compacted clay embankment. The erosion of the walls of the leakage channel undoubtedly occurs along the whole length at the same time.

Piping Failures Statistically More Frequent in Clay Dams

In the past it has generally been considered that earth dams constructed of fine cohesionless soils, such as silts and sandy silts, are statistically more likely to be damaged by piping than clay dams. The experiences summarized herein suggest the reverse may be actually true, at least for low homogeneous dams without filters. Of the 1500 dams in Oklahoma, the only piping failures were in clay dams, in spite of the fact that many of the other dams were constructed of soils with fines of low cohesion, and some of the latter developed large open cracks on the crest and slopes from differential settlement.

Small leaks through and under earth dams are fairly common and statistically only a small percentage of such leaks threaten or cause failure by piping. In a broad sense, two main conditions must be present if piping is to occur:

1. The leakage water must be capable of eroding the soil particles from the walls of the leakage channel and of carrying them out of the dam, and

2. The soil comprising the walls and the roof of the leakage channel must be sufficiently cohesive to support an open pipe and to

prevent the roof from collapsing and closing off the leak.

Neither cohesive nor cohesionless soils commonly provide both these conditions. Embankments of fine-grained cohesionless soils, which are generally easily eroded, will not allow open leakage channels to exist for any length of time. On the other hand, embankments of cohesive soils, which have the strength to support the roofs of piping tunnels, or to resist closure or settlement cracks, normally are sufficiently erosion resistant so that the erosive action of a small initial leak cannot cause piping.

One main source of the commonly held idea that piping is most likely to damage dam structures in cohesionless soil is the experience which led to the establishment of the Bligh "creep ratio", and, later the Lane "weighted creep ratio" theory. This theory was developed from studies of failures of concrete dams on soil foundations. Under the theory, cohesionless silts and very fine sands are the materials which require the longest seepage path to avoid piping: a weighted creep ratio of 8.5 or more is needed for dams underlain by very fine sand or silt, whereas a weighted creep ratio of 2.0 is adequate for foundations of medium clay (11). It would appear, therefore, that the experience underlying the creep ratio theory indicates that piping of a given leak along a given seepage path is many times more likely to occur in fine sand than in clay. It is also apparent, however, that the conditions which are needed to cause piping of a leak passing through a soil foundation directly under a concrete dam are wholly different than those which are most likely to cause piping inside an earth embankment, because the concrete dam provides a roof for the leakage channel which cannot collapse. Hence, the conclusion that piping failures in homogeneous earth dams may be statistically more likely in embankments of clay than in embankments of cohesionless soils does not conflict with the weighted creep ratio theory. Completely different mechanisms are involved in the two cases.

Filters for Dispersive Clay

All the Oklahoma and Mississippi breaching failures occurred as a result of an initial leak which emerged freely at the downstream side of the embankment with no influence of a filter. Hence, from the records of the dams included in this investigation, no direct information is available from which to judge whether or not sand filters would have had an important influence in controlling the piping.

It has been speculated by some of the Australian investigators that the individual particles of clay carried in suspension, in the water of a piping leak in dispersible clay, are so small (0.1 to 0.01 microns) that they will not be caught in the voids of even a fine sand filter. "One method of construction control often proposed is to provide an inverted filter at points of seepage outflow. However, filters built to conventional rules cannot prevent the passage of a suspension of deflocculated clay." ... "In the authors' opinion, therefore, no reliance should be placed upon filters in these circumstances."(14).

The writers are inclined to believe that if the leaks developing through the Mississippi and Oklahoma dams had exited through sand filter zones, the performance experience probably would have been quite different. It is speculated that none of the dams may have failed. The reasons for this conclusion are generally as follows:

1. Obviously the very finest colloidal clay particles suspended in the leakage water would be carried through the voids of the sand filter.

2. All the clays in the failed dams had a relatively large percentage of silt size particles, in the range between 5 and 100 microns. In order for an initially small leak to increase in volume rapidly by piping, the silt size particles and fine sands must also be carried out of the dam; however, these particles would not pass through the voids of a conventional sand filter.

3. Therefore, if the initial leak exited through a sand filter, and if the dispersed clay particles passed through the filter, the silt size particles would back up behind the filter and remain in the dam.

4. The silt size particles backing up against the filter would tend to keep the leakage channel filled with relatively impervious material so that the volume of the leak could not become very large, and would probably decrease gradually with time.

5. Even if the leak did not gradually disappear because of the accumulation of silt size particles behind the filter, there can be little doubt that the volume of the leak would increase much more slowly because of the existence of the sand filter than it would have without it. This being the case, the compacted clay embankment material forming the walls of the leakage channel has more time to swell and to squeeze off a leakage channel.

The situation existing when a piping leak through a dispersive clay dam discharges in a sand filter is very similar to the condition which exists when a drill hole (or slurry trench), filled with bentonitic drilling mud, is put down in a sand deposit. In the first moment after a step of excavation is made in the bottom of the hole, the drilling mud has a tendency to flow out into the voids of the surrounding sand. In the next moment particles of silt and fine sand, inevitably carried in suspension in the drilling mud, become wedged in the pores of the sand formation, thereby progressively reducing the sizes of the openings through which the drilling mud can flow out of the hole. When the hole is in medium sand, this action progresses very rapidly to the point where the openings are so small that the clay particles cannot get out. After this time, some water still leaves the hole and the suspended clay particles are left behind, plastered on the hole walls, forming the thin "bentonite cake" which always develops on the walls of mud-filled borings or slurry trenches.

It is well known from experiences with slurry trench construction that the bentonitic mud inside the trench will flow out of the trench easily into pervious deposits of gravel and coarse sand. There are many experiences to show, however, that the mud will not penetrate medium sand more than an inch or so before reaching equilibrium by the action described above.

While the conclusion must be considered somewhat speculative, for all the above reasons the writers believe filters of fine to medium sand, designed with adequate thickness and confinement so that the filter itself is stable, would probably be effective in the control of piping in dams of dispersive clay.

This conclusion is limited to embankments of dispersive clay of the general type of the Oklahoma and Mississippi failed dams described herein. These are fine-grained soils which do not contain any appreciable content of coarse sand, and have no gravel particles. The conclusion is considered valid for these fine soils for the two main reasons: (1) the mass of silt size particles which would accumulate behind the sand filter, and fill the leakage channel, would be relatively impervious itself, and (2) the clay embankment forming the walls of the channel would swell and tend to shut off the leak.

If the impervious embankment was comprised of a mixture of sand and gravel, with a relatively small percentage of dispersive clay fines, this sealing action might not occur. This would be true first because the material caught behind the filter, after the dispersed clay particles pass through, would consist largely of pervious sand and gravel, so that the leakage channel could remain open in the form of a sand and gravel drain through the dam. Secondly, the tendency for the embankment to swell and close off the leak would be less effective in a dam of sand, gravel and clay, because the gravel particles would "prop open" the leakage channel and resist the tendency for it to be closed by swelling. Also, other things being equal, the compacted embankment of sand, gravel and clay would have a considerably smaller swelling tendency than a dam of fine-grained clayey soil.

This tentative conclusion that a compacted impervious embankment consisting of a mixture of sand, gravel and clay may be more susceptible to erosion damage, other things being equal, than an embankment section of the clay alone, is supported by the reasons described above. Other evidence for this conclusion is given by the fact that difficult-to-explain, important erosion damage has occurred in four recent dams where the cores consisted of mixtures of sand, gravel and fines, in spite of the fact that the leak exited in sand filters designed according to conventional practice (5,8,12).

The newly improved understanding of dispersive clay piping makes it highly desirable that laboratory investigations be carried out for the purpose of establishing reliable filter design criteria. Laboratory investigations are needed to measure piping quantitatively in compacted specimens of clay with varying chemical properties as a function of the pressure and gradient of flowing water, the pressure acting within the clay specimen, the exit filter conditions, the grain size distribution of the compacted clay specimen and the chemical properties of the piping water.

MAIN CONCLUSIONS

1. It is overwhelmingly probable that the initial leaks leading to breaching failures in a number of low homogeneous dams of dispersive clay in Oklahoma and Mississippi occurred as a result of hydraulic fracturing.

2. Hydraulic fracturing is believed to have occurred where the minor principal stress acting within the dam on the plane of the potential leak was reduced to a low value, or to tension, as a result of either differential foundation settlement, drying and shrinkage during or after construction, or a combination of these two actions.

3. Stress computations show that tension can be expected to develop in low dams of stiff clay as the result of moderate differential foundation settlement. Post construction settlement of less than one foot, occurring at a point 100 feet away from a rock abutment, causes a large zone of tension in the dam embankment.

4. While damage or failure by piping from initial leaks starting

by hydraulic fracturing is far more likely to occur in homogeneous dams of dispersive clay, it is believed that under extreme conditions similar failures have occurred in homogeneous dams of ordinary, non-dispersive clay.

5. Laboratory investigations are needed to study the influence of filters in protecting dispersive clay from piping from concentrated leaks coming through cracks.

6. There is strong evidence for the conclusion that piping failures in low homogeneous dams are statistically more likely to occur in dams built of clay than in dams built of silty sands and sandy silts which have little or no cohesion in the presence of water.

7. In the same geographic areas as the failed Oklahoma and Mississippi dams there are many homogneous earth dams built of highly dispersive clay which have not failed.

ACKNOWLEDGEMENTS

The writers are grateful to the Soil Conservation Service for permission to publish the results of this study. Mr. M. M. Culp, Chief, Design Section, Washington, D. C., gave considerable assistance in reviewing the manuscript in various stages and provided valuable comments on some of the main technical points. The writers are particularly indebted to Professor J. M. Duncan and Dr. G. Lefebvre at the University of California at Berkeley, who made the finite element calculations summarized in Figures 3 and 4, as a part of their more extensive studies of this new method of calculation and supplied the results freely in a true spirit of scientific inquiry and cooperation. The writers are grateful also to the Corps of Engineers, Vicksburg Waterways Experiment Station, who sponsored the studies of the finite element method at the University of California and graciously shared the results.

APPENDIX A: BASIC DATA - OKLAHOMA FAILED DAMS

DAM NAME 1) Max. Height 2) Date of completion 3) Date of first flood on stream 4) Date of failure	FAILURE DESCRIPTION (All dams are homogeneous clay embankments)
UPPER CLEAR BOGGY CREEK SITE 50 1) 28 feet 2) May 1970 3) Oct. 9, 1970 4) Oct. 15, 1970	No eye witness. Failure date not known within 5 or 6 days. Since the bottom of the breach is well above the foundation, it is sure that the failure was wholly confined to the embankment. Conduit was not involved. Vertical walled breach, 50 feet wide at crest.
LEADER MIDDLE CLEAR BOGGY CREEK SITE 33 1) 23 feet 2) Sept. 1963 3) 1964 4) Oct. 14, 1969	No eye witness. It is almost sure that the initial leaks broke through the dam just below the high water level and gradually eroded downward. All the investigators concluded that foundation leakage did not contribute to the failure. Failure occurred following rapid reservoir filling on Oct. 13, 1969. Two independent tunnels developed, apparently at the same time, about 90 feet apart. The larger tunnel was 8 feet high and 15 feet wide. The conduit was not involved. Very severe rainfall erosion tunnels.
LEADER MIDDLE CLEAR BOGGY CREEK SITE 15 1) 25 feet 2) June 1965 3) May 26, 1965 4) May 26, 1965 (first) June 1, 1968 (second)	The dam failed twice in locations 100 feet apart. No eye witnesses to either failure. In the 1965 failure an arch-like tunnel about 3 feet high and 15 feet wide extended through the base of the dam, apparently travelling at the original ground level, passing through or under the 8 ft. deep cutoff trench. In the 1968 failure, a 3 ft. diam. circular tunnel passed through the bottom of the dam. At the entrance and exit, the inverts were 5 ft. and 2 ft. above natural ground, respectively, and the tunnel appeared to be straight through the dam, but could have dipped into the foundation in the central portion of the dam. The conduit was not involved. Very severe rainfall erosion tunnels.

Appendix A - Cont'd

DAM NAME 1) Max. Height 2) Date of completion 3) Date of first flood on stream 4) Date of failure	FAILURE DESCRIPTION (All dams are homogeneous clay embankments)
UPPER RED ROCK CREEK SITE 42 1) 23 feet 2) Dec. 1966 3) June 20, 1967 4) June 20, 1967	At 5 p.m., June 20, 1967, a small leak (about 1 in. diam.) was observed emerging just below the bottom of the discharge end of the conduit. No other eye witness observed events until 24 hrs. later when the reservoir was found to be empty from the failure. An erosion tunnel about 3 ft. high and 6 ft. wide average, ran along the right side of the conduit upstream of dam, then crossed over and ran along the left side. The conduit was exposed and washed clean over most of its length, but in certain sections it was still surrounded with intact compacted embankment. The erosion tunnel was confined to the compacted embankment; that is, it did not enter natural soil of foundation.
CANEY COON CREEK SITE 2 1) 50 feet 2) Nov. 1964 3) Nov. 17, 1964 4) Nov. 19, 1964	Eye witnesses present during entire development of the piping failure. Two independent piping tunnels, about 125 feet apart, appeared at the downstream toe. Tunnel No. 1 was first observed at 10 a.m. on Nov. 19, 1964, with a small leak emerging from a 6 in. dia. area. Apparently travelling at the contact between the dam and rock abutment along the line of the municipal water supply conduit, it rapidly increased in size, reaching 18 in. diam. by 5 p.m., at which time it was carrying an estimated volume of 25 cfs. The second leak (Tunnel No. 2) appeared at 1:30 p.m. Nov. 20, 1964 and was 10 feet in diameter during next 2 days.
UPPER RED ROCK CREEK SITE 48 1) 22 feet 2) Nov. 2, 1964 3) Nov. 16, 1964 4) Nov. 17, 1964	No eye witness. Reservoir emptied through erosion tunnel before failure was discovered on Nov. 18, 1964, probably about 24 hrs. after initial leak. Tunnel was about 10 ft. in diameter and located a few feet to right of the conduit. Conduit was exposed and washed clean at both the upstream and downstream ends. During reconstruction it was found that the conduit was completely surrounded by intact compacted embankment over the central 50 feet of its length.

Appendix A - Cont'd

DAM NAME 1) Max. Height 2) Date of completion 3) Date of first flood on stream 4) Date of failure	FAILURE DESCRIPTION (All dams are homogeneous clay embankments)
UPPER CLEAR BOGGY CREEK SITE 53 1) 26 feet 2) Jan. 1963 3) May 6, 1964 4) May 15, 1964	No eye witness. Reservoir emptied through erosion tunnel before first inspection about 24 hrs after initial leak. Tunnel entered the upstream slope at the left side of the conduit, just above its top, then crossed over the conduit and emerged downstream on the right side. Near the dam axis, the tunnel was about 10 ft. in diameter and located with its axis 8 ft. to 12 ft. to the right of conduit. Conduit exposed nearly full length but was not undermined, moved or damaged. Erosion tunnel passed through compacted embankment for full length.
CHEROKEE SANDY CREEK SITE 8A 1) 27 feet 2) Jan. 1963 3) May 6, 1964 4) May 15, 1964	Eye witnesses present from early development of trouble. Initial leak entered upstream slope near high water line, then travelled downward to near original ground surface, then travelled along base of dam horizontally, and then flowed upward again emerging on the downstream slope about 5 ft. above the toe. A tunnel was gradually eroded following the path of this initial leakage, reaching a final diameter of about 8 ft. The reservoir level was at a high point about 48 hrs before the leak was observed, and another 48 hrs were required for the release of the main volume of the reservoir through the gradually eroded failure tunnel. The conduit was not involved.
LITTLE WEWOKA CREEK SITE 17 1) 40 feet 2) May 1960 3) May 19, 1960 4) May 21, 1960	Eye witnesses present from beginning of initial leak. Small initial leak, about 6 in. diam. with "trickle" of water, was discovered at 8 a.m. on the downstream slope on the line of the conduit, about 15 ft. vertically above the conduit. The head on the initial leak must have been 5 ft. or less. By midday the leak had eroded a tunnel 10 ft in diam. and by evening the reservoir was empty, and the erosion had progressed downwardly until the conduit was exposed. The top of the tunnel collapsed leaving a breach with near vertical walls and width of 47 ft. at the crest. The conduit was exposed full length in bottom of breach.

Appendix A - Cont'd

DAM NAME 1) Max. Height 2) Date of completion 3) Date of first flood on stream 4) Date of failure	FAILURE DESCRIPTION (All dams are homogeneous clay embankments)
OWL CREEK SITE 13 1) 24 feet 2) Jan. 1957 3) May 17, 1957 4) May 19, 1957	No eye witness. Failure tunnel travelled along the right side of the conduit. At the dam the tunnel was about 18 ft. in diameter, with its bottom near the invert of the conduit. Conduit exposed full length.
OWL CREEK SITE 7 1) 24 feet 2) March 1957 3) May 17, 1957 4) May 19, 1957	No eye witness. Reservoir was emptied by the failure before first inspection on May 20, 1957. The erosion tunnel entered the natural ground of the right abutment about 80 ft. upstream from the dam, travelled in downstream direction until reaching the cutoff under the dam, then turned 90° and travelled parallel and immediately adjacent to the cutoff trench for about 60 ft. where it encountered the outlet conduit, after which the tunnel travelled to the downstream toe, crossing over the conduit from the right to the left side. In the natural ground upstream, the tunnel was about 4 ft. in diam. At the longitudinal axis of the dam, the erosion tunnel was about 15 ft. in diam. and extended 3 ft. below the conduit invert. Conduit was exposed over downstream half.
LEADER MIDDLE CLEAR BOGGY CREEK SITE 29 1) 23 feet 2) Sept. 1962 3) 1963 4) 1970	Dam has not failed. Since construction the reservoir has been at high level only in 1969 and 1970. In Oct. 1970, at time of high reservoir, a small leak of dirty water was found on the downstream slope coming out of a small hole (about 3" diam.) on the plane of the conduit at an elevation of 10 ft. to 12 ft. above the conduit. After flowing a short time the leak stopped. It is believed that the leak stopped because the reservoir level dropped. On visit of Jan. 1971, it was difficult to distinguish the leakage tunnel from the many rain erosion tunnels on the slope.

Appendix A - Cont'd

DAM NAME 1) Max. Height 2) Date of completion 3) Date of first flood on stream 4) Date of failure	FAILURE DESCRIPTION (All dams are homogeneous clay embankments)
WISTER DAM (U. S. ARMY CORPS OF ENGINEERS) 1) 100 feet 2) Nov. 1948 3) Jan. 1949 4) Jan. 30, 1949	Soon after topping out of dam embankment, heavy rains raised water level for the first time in a few days to Elev. 494 ft. (max. depth about 65 feet.). Two days later (Jan. 30, 1949), a concentrated leak broke through the dam not far below the reservoir level. (The main leaks entered the upstream slope at about Elev. 485 ft. and emerged downstream at about Elev. 475 ft.). The leakage water was obviously very muddy. Estimated volume increased from about 5 cfs to 20 cfs by Feb. 3, 1950. By Feb. 5, 1949, the reservoir was lowered, by discharge through outlet conduits to Elev. 477, after which the leaks nearly stopped. Subsequently the reservoir was lowered and extensive repairs were made including a sheet pile wall, grouting and berms on both slopes.

APPENDIX B - REFERENCES

1. Bertram, G. E., "Experience with Seepage Control Measures in Earth and Rockfill Dams," Transactions 9th Congress on Large Dams, Istanbul, 1967, Vol. 3, p. 91 (Description of serious cracking and piping in Wister Dam, Oklahoma.)

2. Casagrande, A., "Notes on the Design of Earth Dams," Journal of the Boston Society of Civil Engineers, October 1950 (Describes piping at Wister Dam, Oklahoma, "Case A", and gives opinion that differential settlement cracking was probable cause, although no cracks were seen.)

3. Hanson, W. E., "Deformation Problems in Small Earth Dams", The Military Engineer, March-April 1969, p. 119. (Describes piping failure of Washington County Dam in Illinois, believed to be caused by differential settlement cracking although no open cracks were seen before or after failure. Failure details were very similar to those of typical Oklahoma Dam.)

4. James, J. P. and Wickham, R. T., "Failure and Rectification of Morwell No. 2 Fire Service Reservoir", Bulletin of the Australian National Committee on Large Dams, July 1970, p. 20.

5. Kjaernsli, B., and Torblaa, I., "Leakage Through Horizontal Cracks in the Core of Hyttejuvet Dam", Publication No. 80 of the Norwegian Geotechnical Institute, Oslo, 1968, p. 39.

6. Kulhawy, F. H., Duncan, J. M., and Seed, H. B., "Finite Element Analyses of Stresses and Movements in Embankments During Construction", University of California, Office of Research Services Report No. TE69-4, November 1969.

7. Nobari, E. S., and Duncan, J. M., "Movements in Dams Due to Reservoir Filling"; Paper presented to the Conference on Performance of Earth and Earth-supported Structures, ASCE, Purdue University, June 1972.

8. Sherard, J. L., "Embankment Dam Cracking". A paper to be included as a part of a book, Contributions to Embankment Dam Engineering (A. Casagrande Memorial Volume), to be published by John Wiley and Sons, New York, 1972.

9. Sherard, J. L., Decker, R. S. and Ryker, N. L., "Piping in Dams of Dispersive Clay," Paper submitted for the Conference on Performance of Earth and Earth-supported Structures, ASCE, Purdue University, June 1972. (Provides information on soil properties of failed Oklahoma dams described in this paper.)

Appendix B - cont'd

10. U. S. Army Corps of Engineers (1959), "Review of Soils Design, Construction, and Performance Observations, Wister Dam, Oklahoma", Waterways Experiment Station Technical Report No. 3-305, 1959, Vicksburg, Mississippi.

11. U. S. Bureau of Reclamation, Design of Small Dams, U. S. Government Printing Office, Washington, D. C. (Contains summary of weight creep ratio theory, p. 242.)

12. Vaughan, P. R., Kluth, D. J., Leonard, M. W., and Pradoura, H.M.M. "Cracking and Erosion of Balderhead Dam and the Remedial Works Adopted for Its Repair", Transactions, 10th Congress on Large Dams, Montreal, 1970, Vol. 1, p. 73.

13. Wilding, L. P. et al (1963), "Source and Distribution of Sodium in Solonetzic Soils in Illinois", Soil Science Society of America, Vol. 27, No. 4, p. 432.

14. Wood, C. C., Aitchison, G. D. and Ingles, O. G. (1964), "Physico-Chemical and Engineering Aspects of Piping Failures in Small Earth Dams"; Colloquium on Failure of Small Dams, Commonwealth Scientific and Industrial Research Organization, Syndal, Victoria (Australia), November 1964.

Introduction to Paper 6

Embankment Dam Cracking

by James L. Sherard
Embankment-Dam Engineering, New York: Wiley, 1973, pp. 271-353

Prepared by Ronald C. Hirschfeld and Steve J. Poulos

James Sherard's doctoral thesis, **Influence of Soil Properties and Construction Methods on the Performance of Homogeneous Earth Dams,** was one of the first systematic treatments of the problems of cracking and piping of embankment dams. It was, incidentally, the only doctoral thesis completed at Harvard University under the supervision of Professor Karl Terzaghi, who taught there for two decades. At the time, the idea of cracking of dams was so new that most embankment dam designers did not consider it in their designs. Throughout his long and distinguished career, Dr. Sherard continued his research on piping. "Embankment Dam Cracking," one of his most important publications, was an invited contribution for **Embankment-Dam Engineering,** a volume prepared to honor Professor Arthur Casagande, another of Sherard's teachers, on the occasion of his retirement from active teaching at Harvard University.

EMBANKMENT DAM CRACKING

JAMES L. SHERARD*

CONTENTS

*Consulting Engineer, Berkeley, California.

1. INTRODUCTION

In 1963 the writer and coauthors summarized their ideas about the nature and importance of cracking in earth dams (Sherard, Woodward, Gizienski, and Clevenger, 1963). The present chapter contains results of a more intensive examination of the problem with emphasis on a review and evaluation of experience with cracked dams during the last few years. Theory and design trends are not covered extensively. An annotated bibliography is included.

2. HISTORICAL SURVEY

Although cracks in earth dams have caused trouble since the earliest days of dam building, little published information on the subject was available before 1950. Twenty years ago reference works for dam designers discussed cracking of concrete core walls and facings but generally did not mention cracking of impervious earth embankments. This situation has changed rapidly during the last few years so that at present it is well known that cracks have developed in the impervious sections of many dams, and that unusually large settlements are not always needed to create cracks. Recent trends in embankment dam design have been influenced to an important degree by designers' efforts to minimize the hazard of potential cracking.

Arthur Casagrande had a major influence in calling attention to the potential hazard of cracking due to differential settlement. Twenty years ago Casagrande (1950) described near failures of two dams of moderate height. One of these was a homogeneous dam of clay about 80 ft high. A large concentrated leak developed through this dam on the first reservoir filling. Although no cracks were observed, Casagrande hypothesized that the leakage was occurring through differential settlement cracks:

> I would have considered this dam perfectly safe if I had been asked about it before the appearance of piping. I am seriously disturbed by this case since it demonstrates that clay which is compacted at optimum moisture content does not necessarily possess the ability to follow even small differential settlement without cracking.

Since 1950, more attention has been paid to cracking, both by researchers and practitioners and articles have been written about cracking in other dams. Sherard et al. (1963) listed only six articles, which contained most of the available information o[n] cracking. The number of articles has increased [at] least tenfold subsequently (see Bibliography).

Such failures fall into three general categorie[s:] (1) open cracks appearing on the crest or slopes; ([2)] concentrated leaks developing through suspecte[d,] but unseen, cracks; (3) loss of drilling fluid in bo[re] holes in suspected cracks.

The ASCE Committee on Earth and Rockfill Da[ms] recently prepared a list of ten problems concer[n]ing the design and construction of earth and rockf[ill] dams for which more research is urgently need[ed] (ASCE, 1967). The problem of cracking was rat[ed] second highest in priority. Similar concern over t[he] need for improved understanding of the subject w[as] manifested by the members of ICOLD at its 197[0] conference. A large part of one of the general repor[ts] (Lowe, 1970) and a considerable number of th[e] papers and discussions were devoted to cracking.

3. HYDRAULIC FRACTURING

There have been a number of well-studied cases i[n] which dams have failed or been damaged by co[n]centrated leaks for no apparent cause. In some o[f] these experiences investigators concluded that di[f]ferential settlement cracks were the probable caus[e] even though no cracks were seen on the surface. I[n] these examples it was not determined whether th[e] crack was open before the reservoir filled or whethe[r] it might have opened afterward.

In a number of the case histories described in thi[s] paper a concentrated leak appeared abruptly at th[e] downstream side of the dam after the reservoir wa[s] filled—perhaps several hours or several days later—indicating that there was no large open crack exis[t]ing before the reservoir was raised. This is one o[f] the major pieces of evidence for the conclusion tha[t] under certain conditions, the reservoir pressure ac[t]ing on the upstream face of the dam can cause exis[t]ing closed cracks to open or can create new cracks.

The concept that cracks can be formed in so[il] deposits by fluid pressure has been understood an[d] applied for more than 15 years by grouting special[]ists, particularly in Europe. In the most commo[n] method employed for grouting alluvium under dam[s] (the so-called Soletanche "sleeve" grouting pro[]cedure) very high pressures are employed. Explora[]tory test pits put down in grouted soils have show[n] clearly that in many cases the injection pressure[s] were sufficiently great to open cracks in the groun[d] urrounding the grout holes forming thin seams of ure hardened grout.

Probably the first recognition and substantial roof that cracks could be caused to open in dam ores by the pressure of circulating drilling fluids n boreholes was in 1965, following analysis of the vents at Djatiluhur Dam, described later. Subsequently, this action has been demonstrated to have ccurred at many dams.

The concept that water pressure acting on the upream side of an earth dam can cause cracks to pen through the core has not been suggested often. o the writer's knowledge there are only three cases n the literature in which investigators concluded at concentrated leaks probably developed in racks that were caused to open by the pressure of e reservoir water. These are the Balderhead Dam nd the Hyttejuvet Dam, described later, and a eries of small dams in Oklahoma and Mississippi Sherard *et al.*, 1972a).

As discussed in more detail in the following case istories, the writer believes it has been proven eyond reasonable doubt that concentrated leaks ave broken through cracks in earth dams, which ere not open before the reservoir raised. This conlusion is supported both by the considerable number of experiences that are difficult to explain otherwise and by theoretical considerations.

In past studies of cracking, only the two extremes ave been considered: (1) open cracks develop hrough the core; and (2) the core remains uncracked. An intermediate condition must also exist in hich cracking is imminent. As the embankment is eformed by differential settlement, the minor principal stress in the zone of potential cracking dereases and may approach zero provided that the major principal stress is less than the unconfined ompressive strength. When the minor principal tress becomes zero (or even negative if the soil can vithstand tensile stress) a crack is imminent and vill open if further deformation occurs. Although he initial crack may be very narrow, perhaps not ven visible, water from the reservoir can penetrate t. As a result, the stress acting on the plane of the rack is changed abruptly from zero to a compressive stress that would approach the reservoir head f the crack did not extend completely through the ore. The result is an increase in the crack width. The water may enter an existing crack which had een previously closed or a new crack may be formed. In this paper the term "hydraulic fracturing" is employed to cover both situations.

From a practical standpoint, a crack may be caused to open on a given plane if the *effective* stress acting on the plane goes to zero; that is, if the *total* stress on the plane is equal to or less than the water pressure (provided the soil cannot withstand tensile stress). Therefore hydraulic fracturing can occur even though the total minor principal stress is compressive.

Hydraulic fracturing can occur even in a theoretically homogeneous embankment, but the probability of its occurrence is much higher if the material is not homogeneous with respect to deformability and permeability. Nonuniform deformability increases the probability that low total minor principal stresses will occur locally.

Concentrated leaks through a dam by hydraulic fracturing can occur whenever total minor principal stresses at a certain elevation in the core are sufficiently low with respect to the reservoir water pressure at the same elevation. The internal stress conditions necessary for hydraulic fracturing can be present on the first reservoir filling or they can develop later as the result of continuing differential settlement.

One of the conclusions from studies of open cracks appearing on dam crests is that such cracks often appear suddenly (in a few hours or less) with a width of a centimeter or more. This sudden development of the crack probably results from an abrupt redistribution of the embankment stresses adjacent to the crack. It is probable that unseen cracks which appear after the reservoir is filled may sometimes also open abruptly to an appreciable width. This action may explain the relatively large concentrated leaks which have sometimes occurred under very low gradients through suspected, but unseen, cracks.

4. INSTRUCTIVE EXPERIENCES – EARTH DAMS

Stockton Creek Dam

Stockton Creek Dam in central California is an 80-ft-high, essentially homogeneous embankment with a thin upstream zone of screened, pervious gravel and cobbles, a downstream rock toe, and a drain in the valley floor under the downstream slope (Fig. 1). The foundation and abutments are sound, hard, relatively impervious schist. The average abutment slopes were not excessively steep but there was a near vertical "step" with a height of about 20 ft in the right abutment at about midheight (Fig. 1).

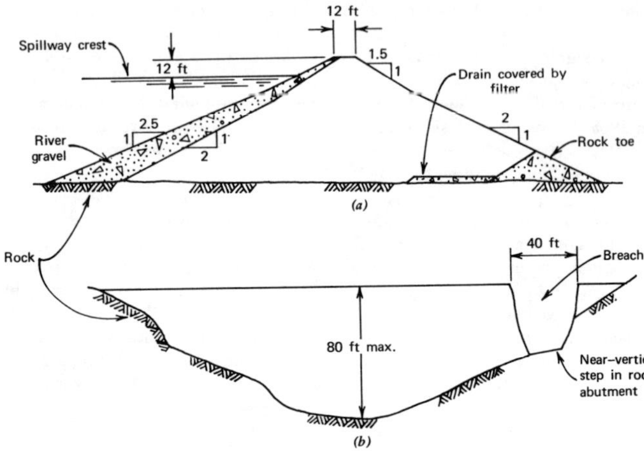

Fig. 1. Stockton Creek Dam. (*a*) Cross section. (*b*) Longitudinal section looking downstream.

The embankment material was a clayey sand of low plasticity (SC-SM) derived from *in situ* weathering of the schist bedrock (Fig. 2).

The dam was constructed with what would be considered good modern practice. The material was wetted both in the borrow pits and on the dam by sprinkling with hoses from a pond stored just upstream of the dam. The embankment was placed in 6-in. compacted layers with a sheepsfoot roller, and routine control of the water content and dry density was carried out. The rock surface forming the foundation and abutments was thoroughly cleaned and the embankment material was compacted against it in a moist state with careful hand compaction. The construction was carried out during the spring and summer of 1950. In the fall of 1950 water was retained in the reservoir to a depth of about 40 ft for several months and no seepage appeared below the dam.

On November 18, 1950, a rainstorm caused the reservoir to rise rapidly to the spillway level for the first time. The dam was inspected several times as the reservoir rose during the afternoon of November 18 and nothing unusual was noticed. By 2 PM the water level was about 8 ft below the spillway crest and at nightfall the spillway started to flow. A last inspection of the site was made about 8 PM by a representative of the owner who walked across the

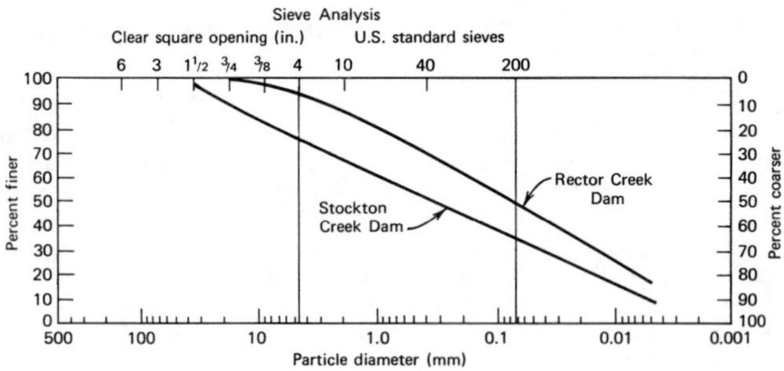

Fig. 2. Representative gradation curves of residual soils at Stockton Creek and Rector Creek dams.

crest and saw nothing unusual. At the next inspection, 8 AM the following morning, it was found that a section of the dam on the right end had washed out (Fig. 3), and a small flow of water was passing through the breach. The reservoir volume was small (about 300 acre-ft) and the resulting flood damage downstream was unimportant.

As shown in Fig. 1, the breach occurred at a point where the dam was only about 40 ft high. The width of the breach at the bottom was about 20 ft and the embankment was left standing on near vertical slopes on both sides. The reservoir had risen to El. 2297, or approximately 8 ft below the top of the dam, indicating that the maximum depth of flow in the spillway had been about 4 ft and that there was no possibility that the dam was overtopped.

Since there was no eyewitness and because subsequent investigations found nothing unusual, the cause of the failure will never be known with certainty, even though a reasonably extensive post-mortem examination was made. The writer, who had the opportunity to study the failure in some detail, concluded that the most probable cause of the failure was rapid progressive piping of an initial concentrated leak through a differential settlement crack adjacent to the near vertical "step" in the abutment rock surface. The main factors that lead to this conclusion are as follows:

1. The failure occurred directly at the vertical step in the rock abutment.
2. The embankment material exposed in the walls of the breach appeared to be very hard and rigid, and hence susceptible to differential settlement cracking.
3. No more satisfactory explanation for the failure was found. On the contrary, there was every indication that reasonable care had been taken to build the dam properly. The embankment exposed in the walls of the breach appeared well compacted and impervious (no sign of penetration of seepage water into the embankment was observed). The appearance of the section of embankment material remaining on the right abutment gave every indication than an excellent, well-compacted bond had been achieved between the dam and the rock abutment.

Fig. 3. Stockton Creek Dam. Photos taken soon after failure in November 1950. (*a*) General view of upstream face showing breach on right abutment. (*b*) Looking downstream into breach from near upstream toe. (Photos courtesy C. J. Cortright.)

The writer also suspected that the nature of the embankment material may have caused it to be especially susceptible to cracking. The soil deposit from which it was excavated had been formed by *in situ* weathering of the bedrock. When compacted in the embankment, the material had about 30 to 40% finer than the No. 200 sieve and a maximum size of 1 to 2 in. (Fig. 2). All the coarser particles were composed of the parent rock in varying states of weathering. Below about $\frac{1}{8}$ in. in diameter the particles were frequently soft enough to be crushed with the fingernail. These weathered sand and gravel-sized particles apparently adhered to each other and imparted rigidity to the compacted embankment.

The average water content of the compacted embankment material was relatively low, despite the fact that there was no shortage of construction water. Subsequent studies of all the test results show that the average water content was probably about 2 to 4 percentage points below Standard AASHO optimum.

After the failure the State Division of Dam Safety investigated the unconfined compressive strength of samples of the embankment material compacted at various water contents. The material became appreciably more rigid and brittle when compacted dry: samples compacted about 3% below Standard

AASHO optimum failed abruptly in unconfined compression at axial compressive strains of about 1% and at compressive stresses of 3 to 7 kg/cm^2. The low failure strain was considered unusual and an indication that the soil may have been especially susceptible to cracking.

The washed-out section of the dam was reconstructed in 1952 and the reservoir has been in satisfactory operation ever since. Except for a few short-lived hairline cracks on the crest, there has been no cracking or other trouble with the dam.

It seems apparent that the embankment was not highly erodible, compared with other earth embankments, because about 30 ft of water was retained in the reservoir after the failure. If the material had been highly susceptible to erosion, the water flowing through the breach would probably have continued eroding the embankment laterally down to the lowest point in the stream bed. Moreover, during the reconstruction in 1952 the partially completed embankment was overtopped during a storm. Water ran over the construction surface to a depth of several inches for a few hours and flowed down the downstream slope, causing only moderate and easily repaired erosion gullies. In addition the visual appearance of the embankment forming the 40-ft-high left wall of the breach—dense, hard, coarse, and slightly cemented (Fig. 3*b*)—gave confidence that the material would not erode rapidly.

It seems clear from the observations that no continuous open crack or other path existed through the dam before the reservoir was raised. If it had, concentrated leakage would have occurred through the crack sometime during November 18 when the reservoir was being raised and would probably have been noticed during the inspections of the afternoon and evening on that date. Therefore it is very likely that the initial leak which led to the failure developed several hours after the reservoir filled. The writer believes that the most probable cause of failure was the opening of a crack by hydraulic fracturing as described in the previous section. Since the embankment was not highly erodible and the dam was completely breached in a few hours, it seems likely that the crack opened abruptly to an appreciable width.

Wister Dam

The Wister Dam in southeastern Oklahoma is a fairly large earth dam, about 90 ft high and 5700 ft long (U.S. Army Engineers, 1959; Bertram, 1967) retaining an important reservoir (about 450,000 acre-ft). On the first reservoir filling a large concentrated leak broke through the dam and complete failure was barely avoided. Although extensive postfailure studies were made, no unusual conditions were found and the cause of the leak was not definitely determined. It was generally concluded, however, that differential settlement cracking was the most probable cause. The main details are as follows:

1. The dam is an essentially homogeneous embankment with berms of random fill both upstream and downstream (Fig. 4). The main homogeneous section is of silty clay of low to medium plasticity compacted in 6-in. layers with 8 passes of a sheepsfoot roller to an average density of 97% of Standard AASHO maximum dry density. The average measured compaction water content from about 900 tests was very close to the average Standard AASHO optimum. The random fill berms were comprised primarily of shale and sandstone from the required excavations, placed in 3-ft, uncompacted layers.
2. The dam embankment was constructed between 1946 and 1948. During the first two years the river was allowed to flow in its original channel which cuts diagonally across

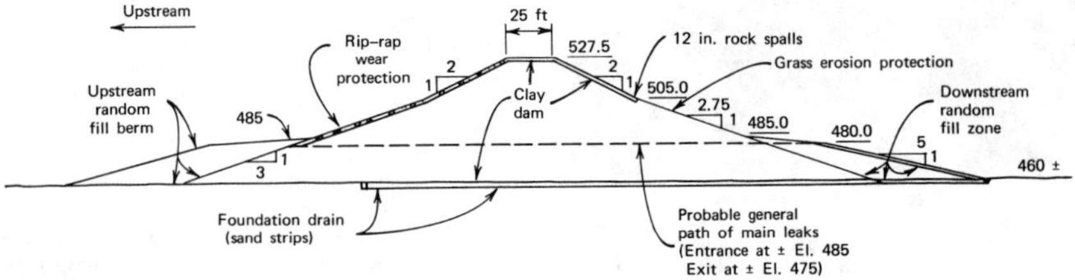

Fig. 4. Wister Dam. Section showing probable location of major leaks (After Bertram, 1967; U.S. Army Engineers, 1959.)

the dam axis (Fig. 5). The portion of the embankment across the river section was constructed last (closure section) and was completed in October 1948.

3. Over most of the length of the dam the foundation consists of 30 to 35 ft of medium stiff clay alluvium overlying shale bedrock (Fig. 6). The shale is exposed on the surface in the river bed and on the right bank.
4. The outlet works consisted of a double-barreled conduit (14 × 16 ft).
5. An unusually heavy rainstorm between January 23 and 27, 1949, caused the reservoir to rise rapidly to El. 494 on January 28 (the flood volume was estimated at nearly 400,000 acre-ft and peak flow at about 90,000 cfs). In the following two weeks the reservoir level gradually dropped to approximate El. 460 (Fig. 7) by uncontrolled discharge through the outlet conduits (maximum of about 15,000 cfs).
6. On the morning of January 30, 1949, after the reservoir level had already dropped a few feet, a small concentrated leak was observed a few feet above the toe of the downstream random berm in the vicinity of the old river channel. By the afternoon of January 31 a series of individual leaks with an estimated total volume of 5 cfs was observed emerging from the toe over a length of about 600 ft. During the next few days the leakage volume increased and the point of discharge gradually rose in elevation. On February 3 the measured total leakage was about 20 cfs and the main volume

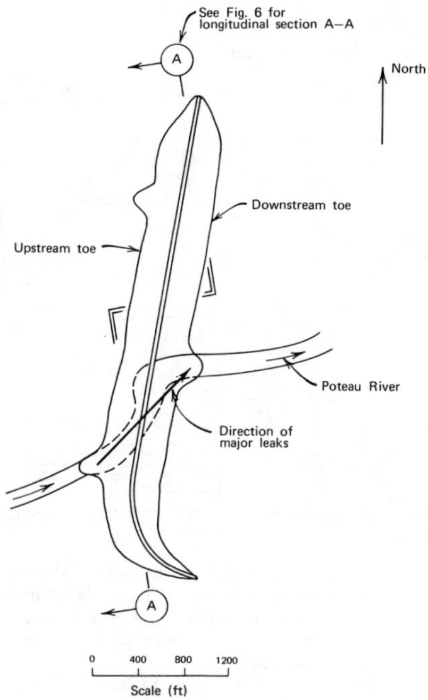

Fig. 5. Wister Dam. Plan showing location and direction of major leaks. (After Bertram, 1967.)

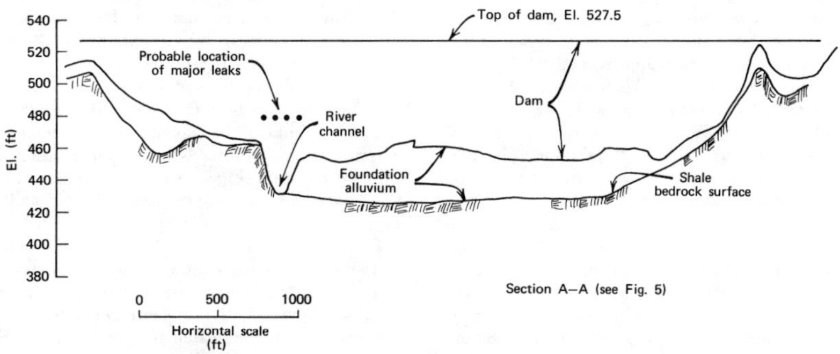

Fig. 6. Wister Dam. Longitudinal section showing thickness of foundation alluvium over bedrock and probable location of major leaks. (After U.S. Army Engineers, 1959.)

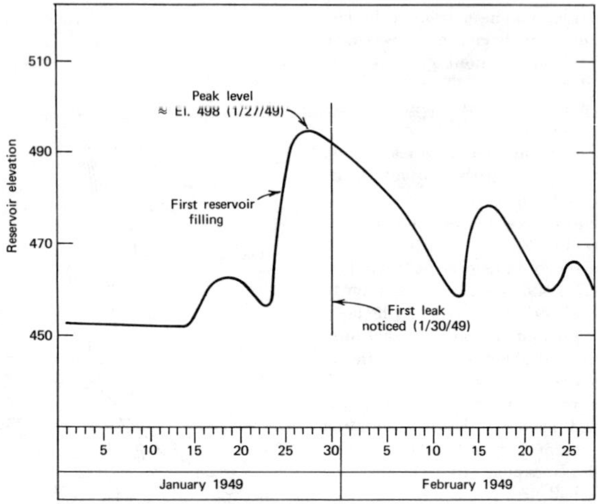

Fig. 7. Wister Dam. Reservoir elevations during first filling and period of leakage and piping. (After U.S. Army Engineers, 1959.)

of flow was exiting from the slope of the random berm at approximate El. 475. At all times during the development of the leak the water was muddy in appearance and obviously carrying suspended solids.

7. On February 3 a vortex appeared in the reservoir surface near the embankment slope, showing that the water was entering a main erosion tunnel in the embankment surface at approximate El. 485. Dye was introduced into this tunnel and found to emerge about 13 minutes later in one of the main leaks on the downstream slope at approximate El. 475. The dye followed a diagonal path through the dam, more or less coinciding with the old river channel (Fig. 5). In this test the dye traveled approximately 715 ft straight-line distance between the entrance and exit, under a head difference of about 10 to 15 ft.
8. Between February 4 and February 7 as the reservoir level continued to drop, several additional erosion tunnel entrances were located on the upstream slope at approximate El. 485 over a length of several hundred feet, more or less centered on the old river channel. By excavating into the upstream slope, erosion tunnels about 2 ft in diameter, were seen extending horizontally into the embankment.
9. As the reservoir level continued to drop in February, it fell below these tunnels and the seepage essentially stopped. The entrances to the tunnels were plugged and remedial measures immediately embarked upon consisting of a steel sheet pile wall, extensive grouting, and additional upstream and downstream berms and drains.
10. After completion of the remedial work in the fall of 1949, the dam has been in operation continuously with no trouble.

From the observations of leakage and subsequent investigation with borings it seemed apparent that the main volume of the leakage was passing through a series of erosion tunnels which passed diagonally through the dam (Fig. 5) with the tunnels entering the upstream slope at approximate El. 485 and exiting on the downstream slope at about El. 475.

From the studies in the design stage, the estimated settlement of the foundation under the central part of the dam, from compression of the 35 ft of alluvium, was about 1.3 ft. Measurements with a few settlement plates indicated that the average settlement at the base of the dam during construction had been of the order of 0.8 ft. No measurements were made of the settlement of the shale comprising the foundation in the river section, which is a relatively incompressible material.

In spite of the relatively low total settlements and the fact that the geometrical configuration of the dam and abutments was not unusual, the investigators concluded that differential settlement cracking was the probable cause. No facts could be found to support any other hypothesis.

No cracks were seen on the crest or slopes at the time of the leakage. During the remedial grouting of the embankment, open cracks with maximum width of about $\frac{1}{4}$ in. appeared on the crest and upper slopes. These were usually parallel with the longitudinal axis of the dam and apparently vertical. Although very low injection pressures were used, the cracks did appear during grouting and sometimes grout flowed out of them. It was definitely concluded that they were caused by the grouting activity.

Subsequent evidence has shown that the embankment clay is particularly susceptible to piping. The main part of the downstream slope of the dam above the random berm is protected with an excellent growth of Bermuda grass. The grass cover is maintained by mowing several times a year and by fertilizing at frequent intervals. Nevertheless, a peculiar type of erosion from rainfall has developed on the slope. As shown in Figs. 8 and 9, vertical erosion tunnels with diameter of several feet have developed. These tunnels extend downward 6 to 8 ft and then run to the slope at a lower elevation. The holes are repaired by filling with sand and covering the surface again with sod and grass; however, several dozen tunnels develop in new locations each year. It should be emphasized that the protective grass cover could not be better. There are no erosion gulleys and there is no space on the slope of more than a fraction of an inch which is not covered with grass roots (Fig. 9). The erosion tunnels undoubtedly are started by rainfall entering drying cracks.

This peculiar type of erosion on embankment slopes is rare; however, in certain locales it is common. There is a large geographic region in southern Oklahoma, not far from the Wister Dam site, where the surface clayey soils are known to be especially "dispersive" and susceptible to such erosion (see further discussion, Section 8). In this area tunnel erosion is common in both natural soils and embankments. The U.S. Soil Conservation Service has built several hundred relatively low flood control dams in this area of Oklahoma and many are plagued with tunnel erosion like that of the Wister Dam. In this same geographic area there have been about a dozen failures of small, but well constructed, earth dams by spontaneous piping, usually occurring soon after the first rapid reservoir filling. In most of these, the failure was very similar to that at Wister Dam.

It is most probable that the leak in Wister Dam broke through a differential settlement crack. It is not known whether a very narrow crack existed before the reservoir filled, or whether the pressure of the reservoir had an important part in the crack development. The latter is probable. The following seem to be the main factors contributing to the crack development:

1. The diagonal path of the river channel across the site.
2. The construction of the embankment in the river channel as a closure section after the dam had been already completed on both sides.
3. The 30-ft-high bank of relatively incompres-

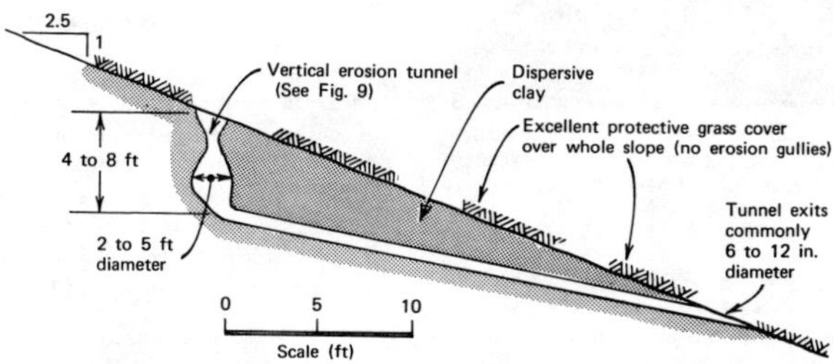

Fig. 8. Wister Dam. Common dimensions of peculiar rainfall erosion tunnels on downstream slope. *Note:* Similar erosion tunnels have developed on many other dam embankments in southeastern Oklahoma in spite of good grass protection.

Fig. 9. Wister Dam. Typical appearance of peculiar vertical rainfall erosion tunnels on downstream slope. Note excellent growth of protective grass (J. L. Sherard photo, January 1971.)

sible shale forming the right wall of the diagonal river channel.

4. The 30-ft-thick layer of clayey alluvium forming the foundation to the left of the closure section.
5. Rapid initial reservoir filling.

None of the individual conditions is rare, or even unusual; hence it is difficult to explain why this dam failed and many similar dams have not. The uncommonly high susceptibility of the embankment material to erosion, as manifested by the peculiar erosion tunnels on the downstream slope, very probably was a major part of the conditions responsible for the failure (Sherard *et al.*, 1972a).

Complete failure was averted only because the reservoir level dropped below the entrance to the leakage tunnels. If the reservoir level had remained above El. 490 for a longer time, a complete failure in which the entire reservoir volume was released could easily have occurred. This important experience, and many others (such as Stockton Creek Dam), support the conclusion that homogeneous dams without internal drains or filters to control concentrated leaks are not completely safe. Although they may be designed and constructed carefully by competent engineers, there is a small statistical risk of failure.

In current practice, homogeneous dams without internal drains or filters extending vertically up into the embankment are still commonly employed for dams of low to moderate height. They are generally satisfactory but a small percentage fails. The risk of failure can be reduced, but not wholly eliminated, by careful construction. The risk can be reduced substantially by the inclusion of a sand chimney, which acts as a "crack stopper" as well as a filter and drain, because it is cohesionless and cannot support an open crack. The writer believes that the time has come when it should be considered prudent to require at least a thin chimney of filter material in low homogeneous dams, except in circumstances where failure would not cause loss of life or important property damage.

East Branch Dam

In the great majority of experiences where differential settlement cracking is suspected to be the cause of concentrated leaks through dams, the trouble developed immediately following the first reservoir filling (within a few hours or days). The East Branch Dam in Pennsylvania is one of the few in which the failure occurred after several years. The details are briefly as follows (Bertram, 1967):

1. It is a well-constructed, modern dam with a maximum height of 56 meters.
2. In May 1957, $4\frac{1}{2}$ years after the dam had been placed in operation, muddy leakage of about

300 liters/sec emerged suddenly from the downstream rockfill toe drain.

3. It was considered that an emergency situation existed and the outlet gates were opened to lower the reservoir as rapidly as possible.
4. Borings indicated that there was an irregular cavity (3 by 6 meters) in the clayey silt core just above the bedrock near the right abutment (Fig. 10).
5. As in most such experiences, the investigators were not completely confident about the cause of the leakage; however, the cavity was directly adjacent to a horizontal bench on the rock abutment which had been used as a construction road, leading to the conclusion that, "in all probability discontinuity in the abutment slope produced sufficient differential settlement of the core to create a crack."
6. The cavity was filled and the leak brought under control by sand-cement grouting.

Although most experience shows that differential settlement cracks develop soon after the first reservoir filling, it is theoretically possible that continuing settlement can result in the opening of cracks after many years. For this reason the East Branch Dam experience, in which competent investigators concluded that differential settlement cracks probably appeared after a delay of more than four years, is particularly interesting. Also it is one of many examples where leaks through dams directly adjacent to horizontal steps and sharp irregularities in rock abutments have been attributed to differential settlement cracks.

Rector Creek Dam

Rector Creek Dam, completed in California in 1946, is an instructive early example of a dam built by methods consistent with modern practice in which important transverse cracks developed but in which simple repairs were adequate to prevent leakage. The dam is an essentially homogeneous embankment with a height of 200 ft and with thin zones of rockfill on both slopes.*

As in the case of the Stockton Creek Dam, the embankment material was a residual soil derived from a partially metamorphosed, fine-grained igneous rock. It was a silty or clayey sand, with visual appearance, gradation, and Atterberg limits very similar to those of the Stockton Creek Dam (Fig. 2). Many of the sand-sized particles were soft enough to be crushed with the fingernail.

The embankment was compacted in 6-in. layers

*Additional details of construction and measurements of the crest movements are given in Sherard et al. (1963, pp. 221–223) and Leonards and Narain (1963).

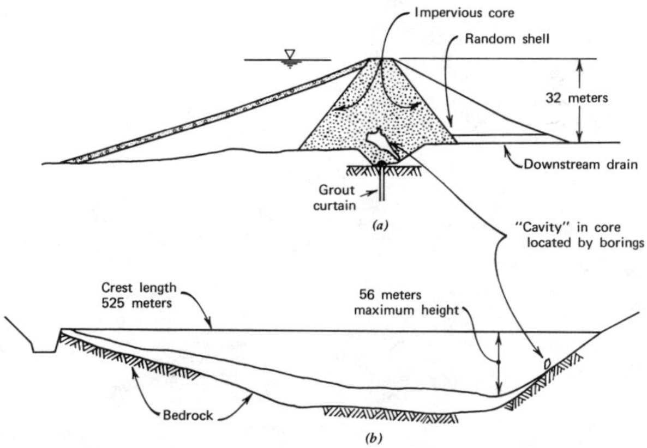

Fig. 10. East Branch Dam. (*a*) Cross section. (*b*) Longitudinal section looking downstream showing approximate location and dimensions of cavity eroded in core by leakage. (After Bertram, 1967.)

with 10 passes of a heavy sheepsfoot roller. Every effort was made to compact the embankment at Standard AASHO optimum water content and the soil was wetted by irrigating the borrow pits.

The crest of the dam settled about 3 ft in the first 10 years of operation, or about 1.5% of the height, which is more than would have been expected; however, many dams of similar height have experienced more than three feet of crest settlement without cracking. Since the dam was founded on hard rock, essentially all of the settlement was caused by compression of the embankment. Measurements of the longitudinal distance between points on the crest 100 ft apart showed gradually increasing compression in the center of the valley and longitudinal extension over the abutments, with a maximum extension of about 9 in. in 100 ft in 10 years (3/4 of 1% in 10 years).

A $\frac{3}{4}$-in. wide, essentially vertical crack developed diagonally across the crest in February 1947 during the first reservoir filling. The crack was about 100 ft from the end of the dam over the left abutment, which was steeper than the right abutment. A 24-ft-deep test pit was excavated to trace the crack down into the dam. At the bottom of the test pit the crack was about $\frac{1}{2}$ in. wide and extended below to an unknown depth. After several months the crack had not increased in width and it was filled with cement grout. In the next several years a few additional similar, but smaller cracks were noted. The reservoir has been used without restriction and no leakage or other indication of trouble from the cracks has been observed at any time.

El Isiro Dam

The most spectacular longitudinal crack that the writer is aware of in an earth dam developed on the downstream slope of the El Isiro Dam, Venezuela. The details of this case history are roughly as follows:

1. There are two dams separated by a rock knoll (Fig. 11). The main dam is about 30 meters high and its longitudinal axis has a radius of curvature of 212 meters (convex upstream). The lower dam is 25 meters high and has a straight longitudinal axis. Both dams have the same zoned cross section (Fig. 12): a central clay core upstream and downstream shells of well-graded, gravelly sand with small percentage of clayey fines (see gradation, Fig. 13), and vertical chimney and horizontal drains of a specially selected, pervious sand and gravel.
2. The dams were constructed during the summer and fall of 1963. The clay core was compacted with sheepsfoot rollers at a water content near Standard AASHO optimum. Due to a shortage of water none was added to the clayey gravel shells, which were compacted with a minimum of six passes of a 50-ton rubber-tired roller. The

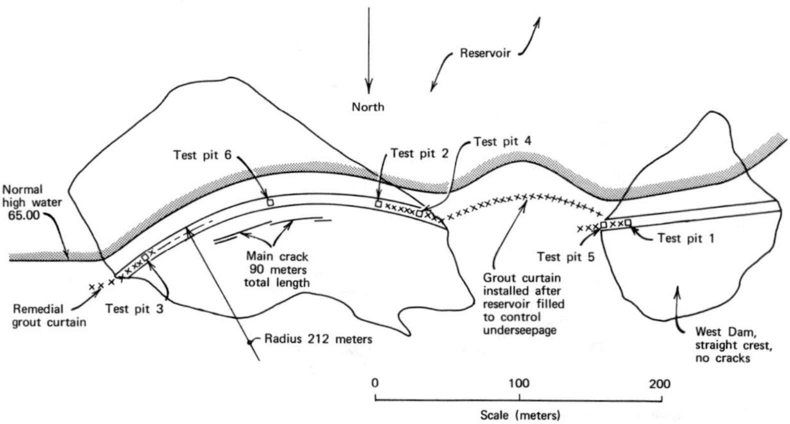

Fig. 11. El Isiro Dams. Plan showing cracking and locations of tests pits and remedial grout curtain.

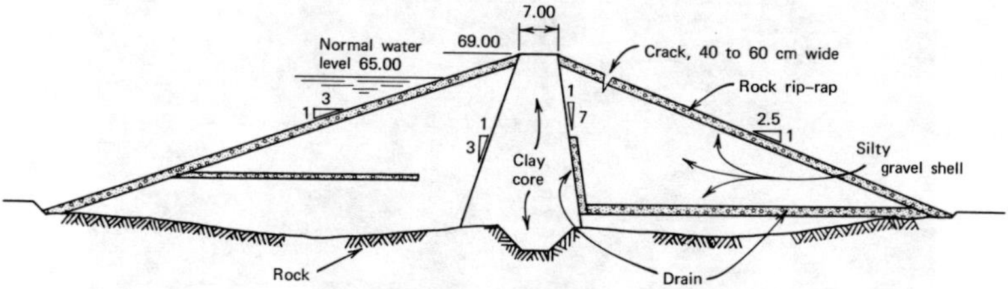

Fig. 12. El Isiro Dam. Cross section Main Dam.

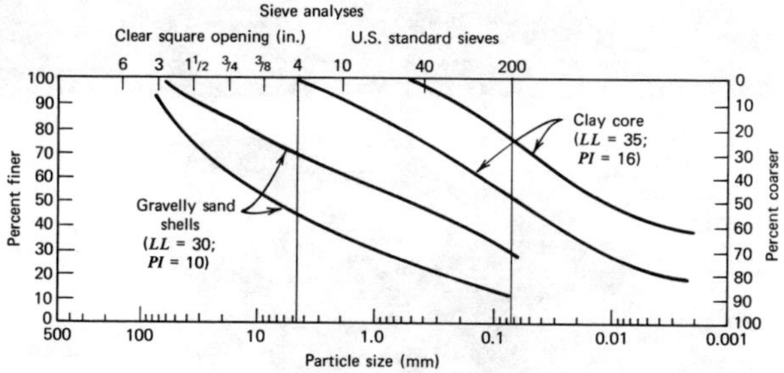

Fig. 13. El Isiro Dam. Soil properties of clay core and of gravel shells.

average measured water content of the shells was about 4%, as compared with a Standard AASHO optimum between 9 and 11% (for the portion of the soil finer than ¾ in.).

3. The reservoir rose slowly during 1964 and reached maximum El. 59* in January 1965. The water level stood between El. 58 and 59 during the spring and early summer of 1965.
4. In the last part of July 1965 a large crack was noticed on the downstream slope of the higher dam. As shown in Figs. 11 and 12, the crack had a length of about 90 meters and width between 40 and 60 cm.
5. Although the crack was very wide on the surface, the width decreased rapidly below the surface and the crack disappeared at a depth of only 2 meters, as seen in the walls of test pits. On the asphalt surfacing at the crest and on the small parapet walls of stone masonry at the upstream and downstream edges of the crest there were a number of hairline cracks which appeared to have been caused by horizontal movement of the crest toward the reservoir.
6. The foundations for both dams consisted of thinly bedded sandstone and shale, striking generally parallel to the longitudinal axis and dipping about 13° upstream. Although the rock was not very hard, it is unlikely that movements of the foundation contributed to the cracking.
7. Afer the large crack was first discovered in July 1965, no further opening of the crack or movement of the dam was apparent.
8. No cracking developed in the lower dam.

After considerable study, the investigators concluded that the large crack was caused by settlement of the clayey gravel upstream shell when it became wetted for the first time. The settlement of the upstream shell caused the crest to deflect upstream.

*Dimensions in meters.

Fig. 14. El Isiro Dam, 1965. Views of main crack on downstream slope. (Photos courtesy G. Perez Guerra.)

The downstream shell stayed dry and was too rigid to follow the movement.*

The crack may have been in existence several months before it was discovered in July 1965, since it was obscured by the thick blanket of quarried rock on the downstream slope (Fig. 14). The fact that no similar crack developed on the shorter dam was probably due partially to the fact that it was not quite as high and that the movements were not quite as great. It also seems likely to the writer that the curvature of the higher dam contributed in some way to the difference in behavior.

In June 1966 a vertical inclinometer casing was installed through the crest near the midpoint of the main dam. It was about 31 meters long and pene-

*As discussed later, DeFries (1968) showed that the location of the crack was consistent with the stresses computed by the finite element method, assuming reasonable movements in the upstream shell.

trated 2 meters into bedrock. Readings during the first $3\frac{1}{2}$ years show that the crest moved downstream gradually about 1 in, and at a depth of about 40 ft below the crest there was a movement of about 1 in. upstream.

Some small leaks developed through the rock abutments of both dams. Although small, this leakage was of concern because (1) it contained considerable dissolved gypsum leached from the foundation rock, and (2) it caused some small landslides in the clay overburden on the downstream slopes of the abutments. To reduce the leakage a new grout curtain was installed in the rock of both abutments of the main dam and across the rock knoll separating the two dams (Fig. 11). The grout holes were cased through the embankment; the casings were sealed well into the bedrock and every effort was made to keep the grout in the foundation. Nevertheless, the grout flowed upward and created grout-filled cracks in the core.

After grouting, four test pits were excavated through the crest of the higher dam and two were excavated through the crest of the lower dam at the locations shown in Fig. 11. With one exception, grout-filled cracks were encountered in the walls of all the test pits and the following general observations were made:

1. Most of the grout-filled cracks were vertical and parallel to the longitudinal axis of the dam (Fig. 15).
2. Almost all cracks were filled with grout, but occasionally the grout filling stopped and an open vertical crack continued beyond the grout seam for a short distance. In some places there were individual grout-filled cracks and in other places there were zones several inches wide of apparently disturbed material in which there was a series of parallel veinlike, grout-filled cracks.
3. Usually the cracks were a few millimeters wide or less, though a number of cracks ranged between 20 and 30 mm wide. In one case, in test pit 3 at a depth of 13.7 meters, a more or less horizontal, grout-filled crack, 100 mm wide, was found.
4. In all cases the grout-filled cracks gave the general appearance of having been opened and filled by the vertically rising grout. In some of the test pits there were no cracks or veins of grout in the upper part of the core of the dam. In others, the cracking and grout penetration extended nearly to the crest of the dam.

Fig. 15. El Isiro Dam, 1966. Typical, near-vertical, longitudinal, grout-filled crack seen in wall of test pit 3 at depth 11.60 meters. (Photo courtesy C. Flores E. Calcaño.)

5. In test pit 6, which was excavated to a depth of 21.4 meters in the center part of the main dam (Fig. 11) at a considerable distance from the abutments and from the grouting, no cracking or grout was found.
6. It was not possible from observations in a few test pits to determine with certainty whether the core was cracked before grouting or whether the grouting caused the cracks to open; however, the latter seemed by far the more probable. The investigators who saw the walls of the test pits concluded that the grout had flowed vertically upward from the foundation into the core of the dam.
7. There was a definite indication that in some places the grout had flowed horizontally from the abutments toward the center of the dam in vertical, longitudinal cracks. In test pit 2, for example, on the left abutment of the main dam there were vertical, longitudinal, grout-filled cracks in the upper 4 or 5 meters of the walls of the test pit, but no cracks below this depth. Therefore the grout must have flowed horizontally. The investigators guessed that the grout

encountered in the cracks of test pit 2 may have flowed horizontally as much as 40 meters.

Shek Pik Dam

As discussed in more detail in Section 7, when borings are put down through an impervious embankment, drilling water or mud is sometimes lost. This loss has frequently been attributed to open, differential settlement cracks, although the evidence was not conclusive. One of the more interesting examples on a major project is that of the Shek Pik Dam, constructed during 1961-1963 for the Hong Kong city water supply.* It is a zoned earth dam (maximum height about 180 ft and crest length 2300 ft) with a thin, inclined chimney drain and a horizontal drainage blanket under the downstream slope. The core material is a residual soil derived from rhyolite and consists of a sandy, clayey silt with an average of a little more than 50% finer than the No. 200 sieve and about 20% finer than 0.002 mm (liquid limit = 38; plasticity index = 12). Although the core material was quite impervious and was compacted at an average water content more than 2% higher than standard AASHO optimum, the pore pressures in the core measured during construction were negligible. Over half of the length of the dam the bedrock at the site is overlain by a heterogeneous mixture of alluvial boulders, gravel, sand, and silt, varying in thickness up to a maximum of 40 ft. This material was left in place and a cutoff was constructed under the core consisting of two parallel walls of overlapping concrete piles; the soil between the walls was grouted. Over the rest of the length of the dam the core was founded on bedrock which was generally weathered at the surface to a considerable and variable depth. The upper weathered zone of bedrock under the core was grouted with three lines of grout holes using the Soletanche method for grouting soil. The grouting was carried out as the dam was being built by drilling down through the partially completed embankment. It was specified that at least 20 ft of embankment material should be in place before grouting, and much of the early grouting was done through 50 ft of embankment. "Injection pressure was kept at about 140 psi and the quantity of grout limited to 3.5% of the volume of the ground between the grout pipes" (Carlyle, 1965).

*A description of the design and construction of this dam and complete details of the water loss, cracking, and grouting problem are given by Carlyle (1965, 1966). See also Little and Beaven (1963) and Little (1967).

When the embankment had reached a height of about 50 ft in 1962, the engineers were surprised and concerned to find that drilling water was lost in some of the grout holes being put down through the dam core. At one location "a system of interconnecting transverse fissures was explored by boreholes and subsequently grouted with clay-cement grout" (Carlyle, 1965). Two test trenches put down to a depth of 50 ft allowed examination of the effectiveness of the grouting. In the walls of these trenches grout-filled cracks were found at random spacing and orientation, and the core material was uniform and hard, with no evidence of soft or brittle layers or open cracks.

At first it was thought that the region of cracking was isolated, but, as the construction continued, water losses occurred in an increasing number of the grout holes in the core of the dam. A full-scale investigation was begun when the dam had reached a height of about 130 ft. Thirty-one holes were drilled along the full length of the dam and in ten of these, at random locations, water was lost; the largest water loss was 12,000 gallons per day. The point of exit could not be determined by observation at the upstream toe or by an increase in the underdrain flow downstream.

In the summer of 1963, when the dam was essentially completed, the reservoir raised rapidly for the first time to a depth of 80 ft. "No unusual features were observed and the leakage was negligible" (Carlyle, 1965). Measured settlement of the foundation at three points was about 1 ft, and most of this occurred during construction.

Subsequently 32 more holes were drilled through the core of the dam at 50-ft centers along the full length to close the fissures by introducing water. The borings were drilled on the centerline of the core to a depth of approximately 150 ft, so that the bottoms of the holes were still within the core material, and a 2½-in. perforated pipe was placed in each hole. Water was poured into alternate holes to keep them full. "Immediate connection between two holes 50 feet apart occurred only in one case. There was, however, a gradual response in half of the intermediate holes (Carlyle, 1965). As seen in Fig. 16, the rate of inflow in the holes typically decreased considerably in the first 10 to 15 days and then reached an equilibrium value. In the hole which took the greatest quantity of water (chainage 1500 ft, Fig. 16) the rate of flow into the hole was initially about 22,000 gallons per day (about 15 gallons a minute) and it decreased continuously until after 15 days it

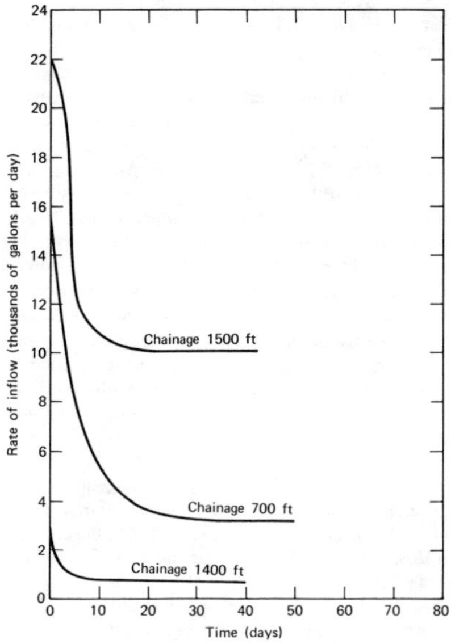

Fig. 16. Shek Pik Dam. Borehole water losses at constant head in long time tests decrease greatly after 10 to 20 days. (After Carlyle, 1965.)

reached an approximately constant value of 11,000 gallons per day. At chainage 1500, where the greatest loss was measured, five nearby piezometers responded rapidly to changes in the water level in the test hole.

As a result of these tests it was decided to grout the core over its full length before refilling the reservoir during the 1964 rainy season. Grouting was done with a series of holes on 25-ft centers, using clay-cement grout in uncased borings with only sufficient pressure to provide a free return of the grout to the surface. Where grout acceptance was high, intermediate holes were drilled to reduce the spacing to 12 ft and finally to 6 ft. Grout takes were generally less than 0.2 ft^3/ft of hole with a maximum of 2.75 ft^3/ft.

"An additional cause of differential settlement was the heave of the core due to grouting the underlying weathered rock. Half the length of the dam was affected by this operation and although heave was limited as far as possible, local variations were considerable, maximum values being 0.3 feet (Carlyle, 1965). At one location on the downstream embankment slope on the left abutment, this heave caused an open crack. This was the only crack seen on the embankment surface.

The reservoir was filled completely by August 1964. The maximum flow from the foundation drains was 0.16 cfs and nothing unusual occurred. The maximum measured crest settlement was only about 0.3 ft by April 1966. Piezometer readings in the core showed a normal seepage pattern, with no indication of the influence of cracks or of the grout.

In spite of the rather extensive investigations and observations made, the cause of the water losses in the boreholes and the grout-filled cracks remained puzzling to the investigators. They concluded that although the total settlements of the foundation and the embankment were not great, open cracks due to differential settlement probably occurred. It was believed that the residual soil of the core was unusual, as manifested by the fact that no pore pressures developed during construction even though the material was compacted wet. The responsible engineer also concluded that if he had it to do over again he would probably not grout the upper zone of weathered bedrock by drilling through the partially completed embankment but would either excavate the material or grout in advance of embankment placement (Carlyle, 1966).

Postconstruction studies of the mineralogical properties of the core material showed that iron oxide hydrate is present in the weathered rhyolite and "may act as a cementing agent in the recompacted soil, making it brittle and so liable to crack if subjected to differential settlement" (Little 1963).

As with most of the case histories described in this paper, it is difficult to be completely confident about the validity of any explanation for the trouble that developed. When the events are examined in light of the experiences summarized in this paper, the writer believes that the following conclusions are reasonable:

1. It is possible that there were some open cracks in localized zones within the embankment and that some of the observed boring water losses were due to flow into these cracks.
2. It is overwhelmingly probable, however, that most of the losses of water and grout from the boreholes occurred in cracks that were not in existence before the holes were drilled and that

were opened and held open by the pressure of the fluid in the borehole.

3. If the majority of the water losses in the borings were due to preexisting cracks, open cracks would also have been observed on the embankment surface, because cracks are more easily formed and kept open near the surface.
4. The rigidity of the two concrete foundation walls, combined with localized heaving of the base of the dam by grouting, probably caused localized zones of very high compressive stress and consequent neighboring zones of relatively low stress. The minimum principal stress in the zones of low stress was undoubtedly less than the fluid pressure in the boreholes with the result that hydraulic fracturing occurred.

Brazilian Experience

In the south-central region of Brazil there are few deposits of sand and gravel available from which to build pervious zones of earth dams. Therefore major structures are built with essentially homogeneous cross sections and thin, vertical, chimney drains, usually only about 1 meter wide and composed of a relatively uniform, medium sand. This basic design has been employed for many dams retaining large and important reservoirs. The embankment materials for many of these dams are residual clayey sands and sandy clays.*

These dams were constructed in thin layers with heavy rollers, following common modern practice. To prevent high pore water pressures during construction the dams were generally compacted dry of optimum. In many of the dams cracks were seen on the surface or were suspected because water was lost in boreholes, but in none of the dams have serious leakage, piping, or other important problems developed. Vargas and Hsu (1970) describe briefly the experiences with six of these dams, all constructed of sandy clay derived from the weathering of gneiss. All developed cracks.

The first of these homogeneous dams with the narrow vertical chimney were the two dams built for the Vigario reservoir, completed in 1951, and designed with the assistance of Terzaghi. As shown in Fig. 17, the Vigario Dike was about 45 meters high and had a rockfill downstream toe, a horizontal drainage blanket under the downstream slope, and a 1-meter-wide vertical sand chimney. The dam fills a narrow canyon and has a crest length of roughly 200 meters.

The reservoir filled for the first time between December 1951 and March 1952. In 1953 a vertical, transverse crack, with a maximum width of about 1 cm was found on the crest about 60 meters from the right end of the dam (Fig. 17*a*). An exploratory trench revealed that the crack narrowed rapidly to a few millimeters in width and disappeared at a depth of 6 meters. Up to the date the crack was noticed, the crest had settled about 10 cm, approximately 0.25% of the dam height. The crack was repaired by trenching and backfilling with clay and there has been no subsequent cracking or trouble with the dam. Vargas and Hsu (1970) describe five other dams of similar design in which similar vertical cracks developed, the majority of which "were small enough so that they did not cause concern of those responsible for the job."

At the Graminha Dam, also a homogeneous embankment with a thin vertical sand chimney, 55 meters high, (Vargas and Hsu 1970), a boring was put down through the crest at the end of construction to install a piezometer. At a depth of about 33 meters all the drilling water was lost abruptly (about 400 liters in a few minutes), presumably through a crack. Subsequently a second boring was drilled at a distance of approximately 4 meters from the first and again all the boring water was lost abruptly at a depth of 37 meters. From these observations it was concluded that there was probably a crack in the embankment, inclined at approximately 45° (Fig. 18). No further studies were made nor was remedial action taken, and the dam has been in operation since 1965 with no leakage or other indication of trouble. There have been several other experiences since 1960 in Brazil in which water loss in borings caused some temporary concern, but where no repairs were made and no leakage through the dam occurred.

At the Euclides da Cunha Dam, a similar homogeneous dam, 60 meters high, constructed of residual sandy clay and having a thin vertical chimney (Vargas and Hsu, 1970), one small spring appeared, approximately 2 months after the first filling of the reservoir, on the downstream slope approximately 35 meters vertically below the crest of the dam and about 10 meters from the abutment contact. The leak was directly opposite a local irregularity in the left rock abutment, which had a steep face. An exploratory pit about 2 meters deep was excavated at

*The problems of design and construction of some of these dams are described by Hsu (1963) and Vargas, Silva, and Tubio (1965).

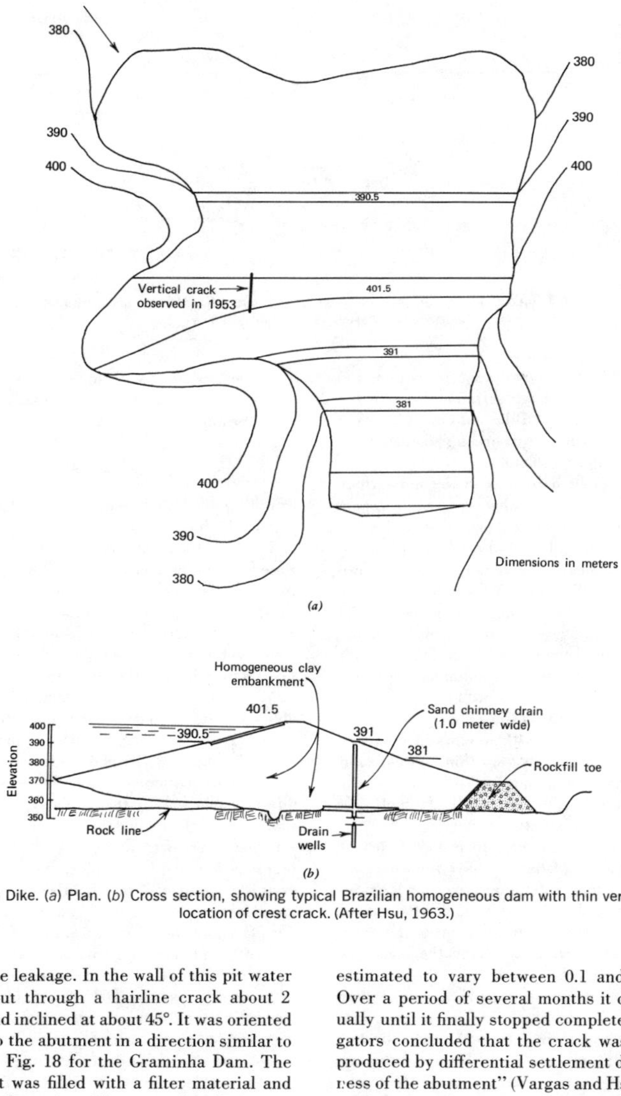

Fig. 17. Vigario Dike. (*a*) Plan. (*b*) Cross section, showing typical Brazilian homogeneous dam with thin vertical chimney and location of crest crack. (After Hsu, 1963.)

the point of the leakage. In the wall of this pit water was coming out through a hairline crack about 2 meters long and inclined at about 45°. It was oriented with respect to the abutment in a direction similar to that shown in Fig. 18 for the Graminha Dam. The exploratory pit was filled with a filter material and drained through a pipe. The rate of leakage was estimated to vary between 0.1 and 0.3 liters/sec. Over a period of several months it decreased gradually until it finally stopped completely. The investigators concluded that the crack was "undoubtedly produced by differential settlement due to the steepness of the abutment" (Vargas and Hsu, 1970).

These Brazilian experiences indicate that:

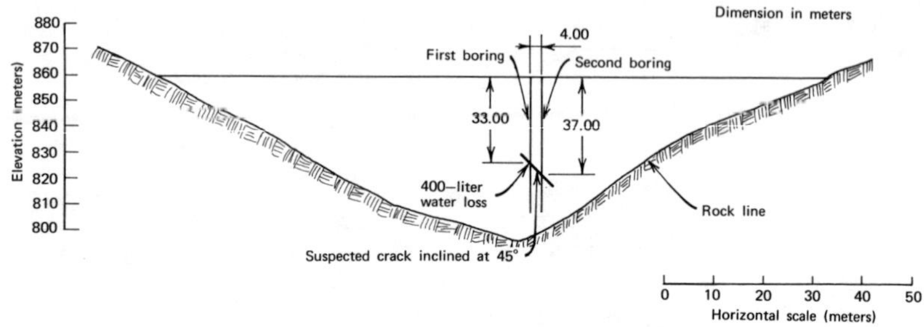

Fig. 18. Graminha Dam. Longitudinal section showing water loss in two adjacent borings attributed to inclined crack. (Courtesy S. J. C. Hsu, Themag Engenharia Ltda.)

1. Narrow transverse crest cracks are common in their dams even though settlements were not large. At the Vigario Dike a 1-cm-wide crack developed and the corresponding postconstruction settlement was 10 cm.
2. Loss of water in boreholes in the impervious embankments has been observed on several occasions and has been interpreted as indicating the existence of differential settlement cracks.
3. Cracking has been observed or suspected more frequently in the major Brazilian dams than would be expected from experience in the United States. Since the dams are commonly constructed of residual soil there is reason for speculation that the embankments are especially rigid and susceptible to cracking because of the nature of the soils.
4. Homogeneous dams with heights of 60 to 70 meters and having only very thin vertical sand chimney drains have performed satisfactorily in spite of the apparent relatively high frequency of embankment cracking.

The Euclides da Cunha Dam is one of the few cases in which investigators have reported seeing direct leakage water emerging from a crack. Almost surely the observed crack was caused by differential settlement adjacent to the nearby abutment irregularity. Since the leak was not observed until 2 months after the reservoir filling, it seems reasonable to conclude that water, traveling through cracks in the rock abutment (by-passing the chimney drain), gradually built up pressure in the embankment zone near the crack, finally causing the crack to open. The cessation of the leak could have been caused by subsequent closing of the crack due to swelling (or softening and collapse) of the adjacent embankment material.

Virginia Ranch Dam

The Virginia Ranch Dam, constructed in northern California during 1961-1963, has a maximum height of 160 ft and a crest length of about 3000 ft. It is a zoned earth dam constructed following good modern practice. The core is generally a clayey sand or sandy clay, with a liquid limit between 28 and 37 and a plasticity index between 8 and 15. The average water content of the compacted core material was near Standard AASHO optimum.

At the right end of the dam, the embankment abuts a reinforced-concrete counterforted spillway wall (Fig. 19). The height of the embankment at the vertical spillway wall is about 50 feet.

The short section of dam at the extreme right end, directly adjacent to the spillway, was built fairly rapidly after the main dam had been completed. The working space was limited, and, since it was not practical to use a large roller, most of the core material was compacted in thin layers by laborers using hand-operated compactors and by truck traffic. The density and water content were carefully controlled.

Crest monuments were installed at 100-ft intervals along the 3000-ft crest. Settlement measurements between December 1963 and February 1967 indicated nothing unusual. The maximum settlement occurred at the point of maximum embankment height and was about 0.4 ft, or roughly 0.25% of the height. The maximum horizontal crest deflection was about 1 in. upstream.

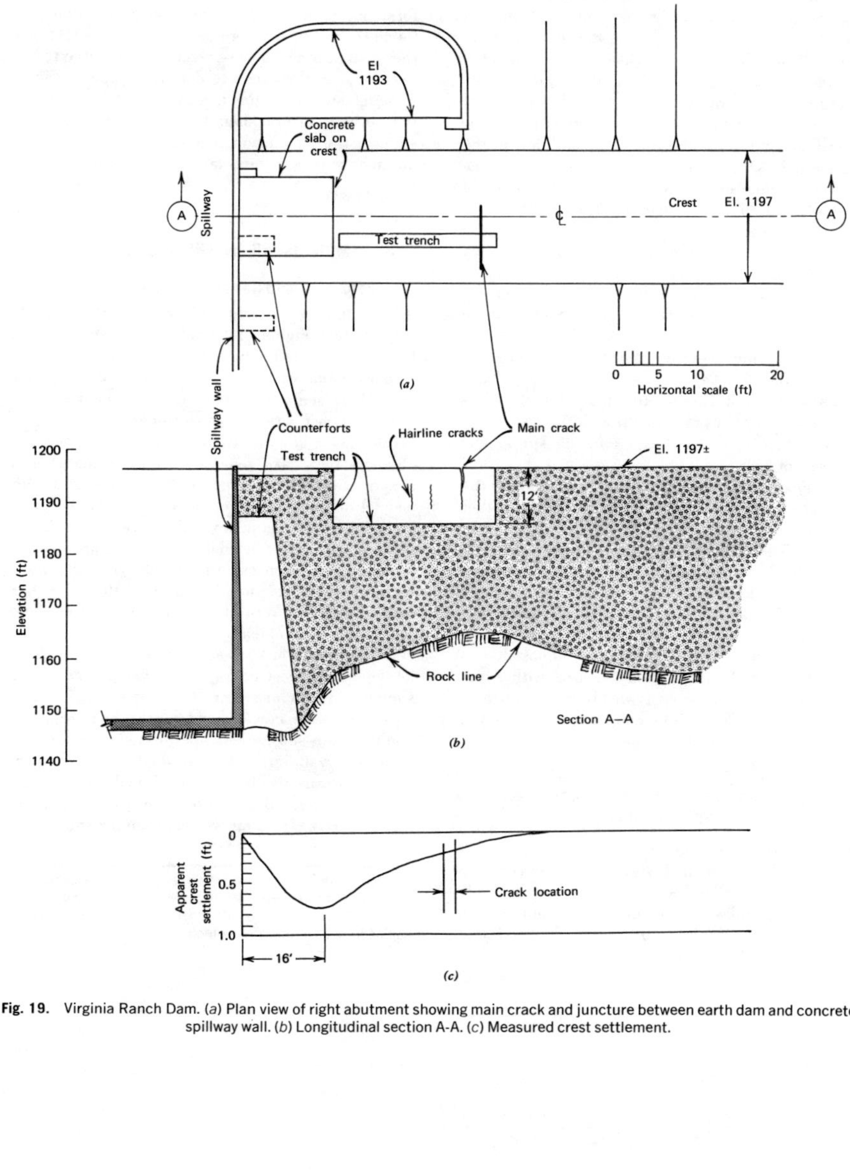

Fig. 19. Virginia Ranch Dam. (*a*) Plan view of right abutment showing main crack and juncture between earth dam and concrete spillway wall. (*b*) Longitudinal section A-A. (*c*) Measured crest settlement.

In the first week of November 1966, after the dam had performed with no trouble for 3 years, a single transverse crack, about 1 in wide, was found running across the crest about 40 ft from the spillway wall (Fig. 19).

After discovery of the crack, measurements of the crest elevation were made over the first 60 ft from the spillway wall (Fig. 19*c*), showing that the crest apparently settled a maximum of about 9 in. at a distance of 16 ft from the edge of the concrete spillway wall, which indicates that the hand-compacted portion of the embankment settled considerably more than the rest of the dam.

A 12-ft-deep test trench was excavated through the crest following down the crack. In addition to the main crack which appeared on the surface, three vertical hairline cracks were found in the walls of the pit (Fig. 19*b*). The main crack had a width of 3 in. at the top; from 2 to 6 ft below the crest it had a width of about $\frac{1}{4}$ in. and was filled with soft clay which had evidently been washed in by rain; below 6 ft it was barely recognizable as a hairline crack, and below 9 ft it could not be traced. No horizontal cracks were found.

Several borings were put down through the crest between the spillway wall and the crack. Water losses were small. Piezometers placed in the holes indicated a normal gradient through the core when the reservoir was raised again in the spring and summer of 1967.

No repairs were made, except for backfilling the exploratory trench with compacted clay. No further cracking developed.

In this case the differential settlement cracking was probably caused primarily by the large post-construction settlement of the hand-compacted section adjacent to the rigid concrete structure. Regardless of the care exercised in construction, the writer believes it is practically impossible to build a 50-ft-high embankment section with hand-held compactors in such a way that the postconstruction settlement will be comparable to that of a section compacted with heavy rollers.

Junctures Between Earth and Concrete Gravity Dams

On large rivers concrete gravity dams are often built in the main river channel with earth dams connecting on each end. The connections are usually made simply by extending the concrete dam into the earth dam, and wrapping the embankment slopes around the end of the concrete dam on both sides. The embankment dam is generally compacted directly against the concrete dam with heavy rollers; therefore high embankment settlements would not be anticipated. In some cases the dams are fairly high at the junctures, creating a situation with potential for differential settlement cracking.

The writer made inquiries concerning the performance of this kind of structure. No example was found in which cracking was observed. The fact that the upstream and downstream faces and the transverse end of the concrete dam are plane surfaces, although steeply inclined, is probably the reason for the absence of cracking. This experience tends to support the conclusion that the steepness of rock abutments is less important than abrupt irregularities.

5. SMALL DAMS IN THE UNITED STATES*

Frequency and General Nature of Cracking

A large number (probably 150–300) of earth dams, 20 to 75 ft high, retaining relatively small reservoirs, are built in the United States each year. Many are homogeneous dams with little or no internal drainage for controlling seepage. Often they are constructed at minimum cost without the careful attention given to higher, more important dams. However, in the last 15 years they have generally been reasonably well constructed in thin layers with a roller, and with control of the water content and density. Reliable statistics on the frequency and nature of cracking in these dams are not available, but from his own experience and discussions with colleagues who have built a considerable number of them, the writer has formed a general view of the importance, frequency, and details of cracking.

In the last 15 or 20 years a large number of these low, minimum-cost dams have developed cracks of sufficient width and extent to cause considerable concern. These cracks have usually been repaired and the dams subsequently performed satisfactorily. In narrow valleys these small dams exhibit typical crack patterns as shown schematically in Fig. 20. Transverse vertical cracks frequently develop in the distance between 15 and 50 feet from the ends of the

*The small dams discussed in this section were designed and built under the supervision of professional engineers since about 1950. Experiences with small farm dams built without engineering attention are not included.

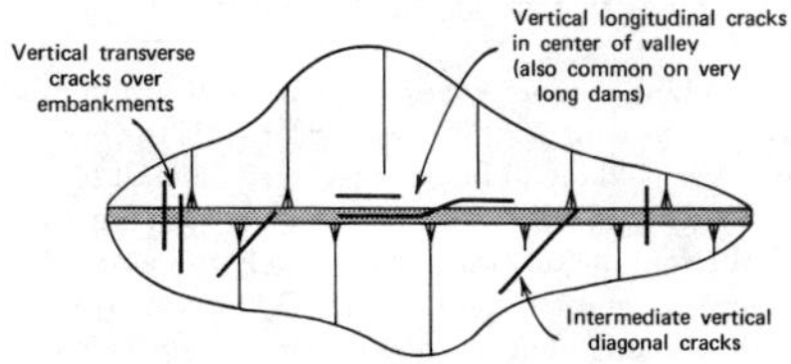

Fig. 20. Common crack patterns in small dams.

crest. In the central portion of the valley, where the dam is highest, vertical longitudinal cracks often develop on or near the crest as shown. Occasionally intermediate diagonal cracks develop. A summary of this general experience follows:

1. The transverse cracks generally vary in width from a hairline to an inch or so but are commonly from $\frac{1}{8}$ to $\frac{3}{4}$ in. They are usually of greater width than the diagonal or longitudinal cracks. There may be a single transverse crack, or several parallel cracks, a few feet apart.
2. The transverse cracks may develop only on the crest, or they may run down the slopes, in some cases extending all the way to the toe. (In a few extreme cases where large foundation settlements have occurred after the first wetting, cracks as wide as 6 in. at the crest have developed and they have occasionally extended completely through the dam and into the foundation to a depth of 10 ft or more.)
3. In the test pits cracks often disappear at a depth of several feet but some are open to depths of 20 ft or more.
4. If both abutments have about the same steepness, and if the compressibility of the foundation is uniform, all three types of cracks shown in Fig. 20 sometimes develop. In other cases, only a single crack may develop.
5. The cracks commonly develop by about a year after completion of the dam. There are cases, however, where cracks of significant width have developed abruptly for the first time as much as five years after completion of construction.
6. Large postconstruction crest settlements are not required to cause cracking. Dams 50 ft high, with crest settlements of the order of 6 in., have developed transverse crest cracks $\frac{1}{2}$ to 1 in. wide.
7. Commonly, most of the postconstruction settlement that causes cracking results from fairly rapid compression of unsaturated soil foundations when the reservoir fills for the first time.
8. In many cases it has been concluded that cracks developed because the embankments were constructed at an excessively low water content. Yet cracking has also developed in some dams where every effort was made to place the fill at Standard AASHO optimum.
9. Cracks of appreciable width are easily obscured. A crack that is observed on one inspection may not be seen on subsequent inspections and may then reappear. They may be obscured by traffic on the crest, embankment swelling, and other actions.
10. Cracks are more common in dams in which the reservoirs remain at low levels for long periods of time, such as dams acting wholly as flood control structures with ungated outlet conduits.

Typical cracking (Fig. 20) reflects the stress distribution commonly existing in the various portions of the dam. As the embankment is subjected to progressively larger differential settlements, the minimum principal stresses in the critical zones gradually decrease, and they reach zero at the time of cracking. It can be assumed that the cracks show the directions of minimum principal stresses acting in the deformed embankment. As would be expected, the dam is stretched over both abutments in the longitudinal direction, and the minimum principal stresses in these portions of the dam act on vertical transverse planes. In the center of the valley the minimum principal stresses act on vertical longitudinal planes. These longitudinal cracks are also influenced by the fact that the foundation settles different amounts under the crest than under the shells, especially if there is a cutoff trench through the foundation soil. In the portions of the dam between the center of the valley and the abutments, the direction of minimum principal stress gradually changes and cracks may open on the diagonal planes.

In some cases the longitudinal, diagonal, and transverse cracks develop simultaneously and the result is a more or less continuous curved crack resembling the scarp of an incipient shear slide. Particularly in the central portions of long dams, away from the influence of the abutments, differential settlement cracks patterns are governed by the variation in thickness and compressibility of the foundation soil, and cracks often follow continuous curved lines of apparently random shapes. At some dams considerable time and investigation have been required to demonstrate that the cracks were caused by differential settlement and were not a manifestation of a deep-seated shear slide.

Embankment shrinkage due to drying is undoubtedly an important factor in the development of cracks, especially in homogeneous dams of clayey soil with no thick surface layers of gravel or rock slope protection, and in dams which sit with empty reservoirs for considerable lengths of time in dry climates. In a number of cases, investigators have not been able to conclude definitely whether cracks which developed were caused by differential settlement or drying. It would be expected that drying cracks would develop in the direction of least resistance; that is, if a portion of the embankment is undergoing tensile strain and open differential settlement cracks are imminent, embankment shrinkage due to drying undoubtedly would accelerate the development of the crack and cause it to be larger.

Selective Experience

The percentage of dams that crack is much higher in some parts of the country than in others. Engineers building low dams in some geographic areas express the opinion that small differential settlement cracks almost always develop. In other areas cracking is rare. A higher percentage of cracking in certain areas undoubtedly results from some common element in the dams, such as embankment soils that are especially brittle and susceptible to cracking or foundations that are unusually subject to settlement due to wetting.

Cracking is more common in areas of low rainfall (15 in. a year or less) for two reasons: (1) the borrow material is likely to be dry and more difficult to process to a uniform water content near optimum; and (2) the foundation soil is more likely to be partially saturated and susceptible to compression when wetted. Dams built in areas where the surface soils are comprised of wind deposits of silt and clay (loess) have sometimes developed cracks of relatively great width.

A high incidence of cracking occurred in a very local area in southern California where four 40- to 50-ft-high and 500- to 1000-ft-long homogeneous flood control dams were built between 1953 and 1956 about 30 miles east of Los Angeles. The dams were constructed according to good modern practice with a heavy sheepsfoot roller and moisture-density control. All four dams cracked. Two of them required extensive repairs.

1. All four were constructed of a very similar soil —a silty sand derived from weathering of the granitic bedrock in the region.
2. Since the purpose of the dams was wholly flood control, they stood with empty reservoirs except for a few days following heavy rainstorms once every few years.
3. The measured settlement of the crest in the first few years after construction was 6 to 8 in. and was caused largely by gradual foundation settlement.

The gradation of the typical disintegrated granite in the embankments is shown in Fig. 21. In some borrow pits the excavated material was weathered-

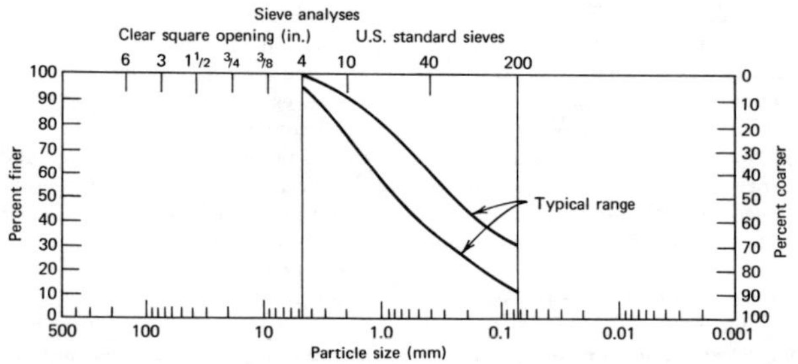

Fig. 21. Gradation of disintegrated granite in southern California (cohesionless silty sand) in low homogeneous earth dams with special susceptibility to cracking.

in-place granite; in other borrow pits the material had been transported a short distance as slope wash but had essentially the same properties. The fine portion of these soils has a plasticity index equal to zero and reacts rapidly to the shaking test. In the borrow excavations some of the material has the appearance of a cohesionless silty sand but there are some layers and zones that are cemented. The harder of these cemented zones in the borrow area were not easily excavated with scrapers and were left as remnants in the excavations. Chunks of the typical harder cemented material can be easily broken into small pieces with a geologist's pick and have an unconfined compressive strength of about 10 kg/cm^2. The cemented material is mottled with white substance of unknown composition, which may be the cementing agent.

The writer concluded from study of these dams that the average construction material probably contained at least a small amount of this unknown cementing agent, which caused the embankments to be more rigid, brittle, and susceptible to cracking than the ordinary earth dam. The fact that the reservoirs were empty undoubtedly played an important role. It seems possible that if the reservoirs had been filled, the embankments might have been softened by the seepage and may have been able to follow the moderate settlements without cracking.

Cracking Adjacent to Conduits

Concentrated leaks, apparently along outlet conduits through the dam base, have been one of the most frequent causes of troubles with small, low-cost dams. Usually these leaks have been attributed to the difficulty of making a good seal between the conduit and the impervious embankment. In a few cases, investigators suspected that the leakage took place through cracks caused by differential settlement between the conduit and the embankment, though usually no crack was seen. The writer believes it is probable that some, perhaps a large percentage of these leaks are caused by differential settlement. The leak may travel in an open crack or the water may create a crack by hydraulic fracturing in the zone of low embankment stress directly adjacent to the conduit (Sherard *et al.*, 1972b).

At sites where the dam foundation consists of compressible soil and it is not practical to found outlet and spillway conduits on rock, the choice of conduit details is one of the major design problems for these small dams. In recent years the development of superior precast concrete pipes, and particularly better connections allowing axial extension and rotation of the joints without leakage, has been an important improvement. To reduce the settlement of these pipes, a transverse foundation trench is sometimes excavated and backfilled with compacted embankment as shown schematically in Fig. 22. This detail has been used primarily where the foundation soil is expected to be especially susceptible to settlement due to wetting. In some cases where this detail has been employed, leakage has emerged on the downstream slope, apparently traveling along a path through the embankment immediately adjacent and parallel to the conduit, even though seepage collars were provided and considerable care had been taken in construction. This experience has led to the conclusion that the potential hazard of leakage through transverse differential settlement cracks caused by

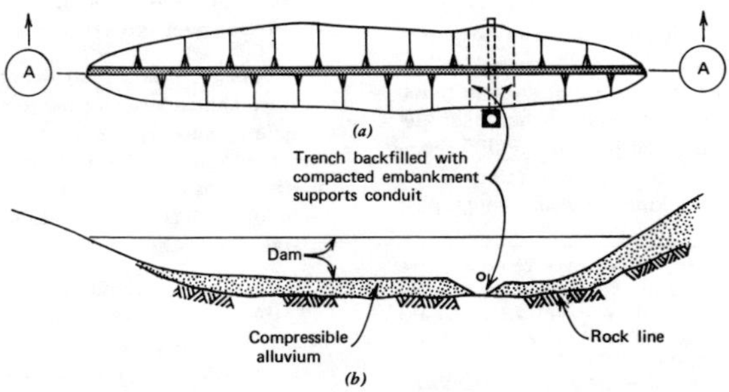

Fig. 22. Method of supporting conduit through the low dams. Considered generally undesirable because of hazard of leaks through differential settlement cracks. (*a*) Plan. (*b*) Longitudinal section A-A.

the existence of the backfilled trench may outweigh the benefit of reducing the settlement and distortion of the conduit (Sherard, Decker and Ryker 1972a, pp. 668–672).

In summary, experience indicates clearly that leaks along conduits, resulting from cracking as well as other causes, are common sources of piping in small dams. The writer believes the following conclusions are valid for conduits that must be placed on earth foundations:

1. It is particularly important that the embankment adjacent to the conduit be placed at a relatively high water content and not be a soil susceptible to piping.
2. Even in small, homogeneous dams where no chimney drain is installed it is advisable to provide a drain and filter around the conduit at its downstream end for the purpose of intercepting concentrated leaks which follow the conduit.
3. In cases where the soil foundation is thick and compressible, it is not desirable to excavate a trench under the conduit and fill it with compacted earth.

Importance of Construction Water Content

Excessively low construction water content has been a frequent cause of cracking. Dams built in arid parts of the western United States 30 to 40 years ago were frequently constructed of very dry soils. Commonly they settled several feet in the first few years of operation and cracked badly, with frequent complete failures from leaks breaking through cracks (Sherard, 1953).* This early experience was largely responsible for the development and current universal acceptance of modern methods of moisture-density control.

It is clear from experience with small homogeneous dams constructed in the last 20 years that it is dangerous to construct them at water content much below Standard AASHO optimum. It has been demonstrated that only a small percentage difference in the water content can have a large influence on the susceptibility to cracking. In some geographic areas, engineers building many of these low dams had much trouble with cracking 10 to 15 years ago but learned from experience that the frequency and magnitude of cracking could be greatly reduced if the average water content of the embankment material during construction was uniform and at or above Standard AASHO optimum.

One of the recurrent conclusions of investigators of cracks in small low-cost dams is that while an effort was made to control the water content, the embankments were compacted dry simply because of difficulty in controlling the contractor's operation in a practical manner. The smaller the dam, the more difficult it has been to achieve a uniform and satisfactory water content for the following reasons:

1. A certain amount of experimentation and time is required at any site to establish satisfactory procedures. For the small dam, the construction period is so short that the dam may be completed before the problems of raising the water content in a satisfactory manner are solved.
2. The construction surface is very small and if the material does not have a uniform and satisfactory water content when delivered to the dam, there is insufficient working space for blending it by turning it over several times with disks, which is the only practical means of creating a uniform condition after the material is dumped on the fill.
3. Commonly for small dams, the specifications for moisture control have not been stringent, so that when it became apparent during construction that better moisture processing was desirable, it was difficult to force the contractor to improve the operation without a change in the contract and extra payment.

For these reasons, stringent specifications for moisture control are as desirable (or more desirable) for small dams as for more important structures. The requirements do not have to be varied greatly as a function of the soil type. The writer believes the following basic specifications are reasonable and satisfactory for the majority of sites:

1. The water content at compaction should be in the range between one percentage point below and three percentage points above Standard AASHO optimum water content.
2. The material should be processed to a uniform water content by adding water in advance in

*The literature before 1930 contains many such experiences. For important examples, see the descriptions of Apishapa Dam, which failed completely and rapidly in 1923 when water broke through a large horizontal crack near one abutment (*Eng. News-Rec.*, 1923, pp. 357, 418, 422, 900, and 1029), and of the Seros Dam in Spain where 2-in.-wide transverse cracks appeared on the first reservoir filling (*Eng. News*, 1914, p. 536).

the borrow area. Only light sprinkling should be allowed on the construction surface to compensate for drying.

3. The material should be compacted with 10 to 12 passes of a heavy sheepsfoot roller (3000 to 4000 lb/lineal ft) in layers that are not more than 6 in. in thickness after compaction. (Certainly other methods of compaction can be used satisfactorily; however, for low dams with relatively small embankment volumes the sheepsfoot roller has many important advantages: it can be used satisfactorily for essentially all soil types and no decision has to be made as to whether it is suitable; it is the most common type of roller; it minimizes potential problems with bonding between layers; in addition to compaction, it provides a desirable supplementary blending action.

All experience indicates that with few exceptions soils can be compacted with large standard rollers efficiently at Standard AASHO optimum water content, or a little above. It is also known that if the water content is decreased only a few percentage points below Standard AASHO optimum, the stiffness of embankments constructed of most soils is increased significantly. For example, the Stockton Creek Dam was constructed at an average water content estimated to be approximately 3% below Standard AASHO optimum. As discussed later, in one of the few laboratory investigations on this subject, it was shown that the tensile strain at failure for specimens compacted at Standard AASHO optimum was of the order of three times that of specimens compacted at 3% below Standard AASHO optimum.

For these reasons, and because Modified AASHO optimum water content is commonly 2 to 4 percentage points below Standard AASHO optimum, the writer strongly believes that the Modified AASHO test should not be used for construction control of small earth dams.

Experience also clearly indicates that if the natural water content of the borrow material is below Standard AASHO optimum it should be raised by irrigation in the borrow areas, as far ahead of time as possible. For the average, clayey embankment material used in small dams it is desirable to sprinkle or pond the borrow areas several weeks in advance of excavation.

Even when borrow pit irrigation is specified, trouble may develop during construction because it is difficult on a small project to force the contractor to add the water a sufficient time in advance to achieve a uniform water content. One method that has worked well for reducing the problem has been to specify in writing in advance that the borrow pit be divided into three sections of more or less equal area and that one of these sections always be in the process of being wetted, while the second is curing and the third is being excavated.

Methods of Repair

Various methods have been employed to seal surface cracks. The most common and preferred method has been to excavate a narrow trench (3 to 6 ft wide) along the crack to the depth at which the crack disappears and then to backfill the trench with impervious, hand-compacted soil. Provided no further large settlements occur, this method has generally been satisfactory. It has the advantage that the depth and extent of the crack can be determined at the same time that the repairs are being made. Sometimes cracks have been filled with sand or mixtures of dry sand and powdered bentonite.

Occasionally the cracks reappear and are trenched and backfilled again. If the crack extends deeper than it is practical to excavate an open trench, efforts have been made to fill the lower parts of the cracks with cement grout, either by pouring the grout into the crack or drilling holes and pumping the grout in.

In a few cases continuing foundation settlement has caused recurrent serious cracking over many years. One such example is the Woodcrest Dam in Southern California,* a 40-ft-high, 900-ft-long flood control dam. Transverse differential settlement cracks appeared at the left end soon after completion of construction in 1953. In the first 15 years of operation, the reservoir was never filled completely; only a few feet of water accumulated behind the dam during storms. In the first 5 years the cracks were repaired by trenching and backfilling (and grouting the deeper parts of the cracks) several times. Within several months after each set of repairs was made, the cracks appeared again in the same general area over the left abutment. In April 1960, the largest crack to date was found, again in

*This is one of the four cracked dams constructed of disintegrated granite described earlier. The construction of the dam and cracking in the first few years of its operation are described by Narain (1962) and Leonards and Narain (1963).

the same general location on the left end. It was a vertical, transverse crack about 1.5 in. wide. (At this time the total measured crest settlement since construction was about 0.6 ft.) Repair was accomplished by excavating a wide notch through the top of the dam with a bulldozer. The crack was 12 ft deep and the trench was excavated to a depth of 14 ft. The backfill in the trench was compacted with a sheepsfoot roller. Again in 1961, $\frac{1}{4}$-in.-wide vertical transverse cracks developed; cracking of lesser magnitude occurred and repairs were made at recurrent intervals until the spring of 1969.

Heavy rains during January and February 1969 caused the Woodcrest reservoir to fill completely for the first time and a concentrated leak emerged from an erosion pipe downstream of the dam in the left abutment. The reservoir was lowered before the erosion got out of control. Investigations showed that the concentrated leak and piping had developed in the natural soil of the left abutment, about 4 ft below the bottom of the embankment. A cemented layer in the abutment formed a "roof" for the pipe, which was about 8 in. in diameter at the dam centerline. All investigators concluded that it seemed probable that the foundation piping was not related to the embankment cracking. In 1969, about half of the dam on the left end and some of the natural soil of the foundation and abutment were excavated, and the embankment was rebuilt. In the fall of 1970, small new cracks developed in the original embankment near the juncture with the reconstructed section.

From this experience and a few others, it appears that the likelihood and severity of cracking, and the hazard of a complete failure, may be greater for dams that stand with empty reservoirs during most of the year and then fill rapidly during storms. In the past, engineers have sometimes reasoned that dams designed to act wholly as flood control structures do not need conservatively designed internal filters and drains, since the reservoir is filled only for a few days and the water does not have time to seep into the dam. The writer believes that the contrary appraoch is more valid; that is, other things being equal, the need for conservatively designed internal filters and drains to control concentrated leaks is greater for flood control dams which sit with empty reservoirs for long periods than for dams in which the reservoir is kept continuously filled. This is true for two main reasons:

1. Cracking is more likely to develop because the embankment is not softened by seepage; any small crack that develops is likely to be widened and extended by drying.
2. Because of the rapid reservoir filling, concentrated leaks in cracks may cause rapidly progressive piping, whereas if the reservoir were raised slowly, the water entering the cracks might cause the cracks to close by swelling or collapse of the soil in the crack walls.

In a few cases repairs have been made with membranes of various kinds.* The 45-ft-high Harrison Street Dam is one of the southern California flood control dams described earlier; it was constructed of residual silty sand derived from disintegrated granite, and had a history of transverse cracking. The cracks were not wide, generally $\frac{1}{4}$ to $\frac{1}{2}$ in. or less. In 1965, 12 years after completion of construction, a longitudinal trench was dug with a backhoe to a depth of about 12 ft below the crest for the full length of the dam. It was hoped that this trench would extend below all the cracks and that the dam could be satisfactorily repaired by simply backfilling the trench with suitable impervious compacted soil. It was found that the cracks continued below the bottom of the trench. In addition to backfilling the trench with a well-compacted mixture of the disintegrated granite and about 2% of powdered bentonite, a membrane was placed on the entire area of the upstream face of the dam using following procedure:

1. Excavate about 2 to 3 ft of material from the upstream face and compact to a smooth, hard surface.
2. Spray the surface with a sterilizing chemical to prevent subsequent growth of vegetation.
3. Place on this surface a layer of coarse woven-jute matting of the type used to control erosion on soil slopes.
4. Spray the surface with asphalt emulsion.
5. Spread about $\frac{1}{4}$ in. of sand on the surface by hand.
6. Spray with a second coat of asphalt to bind the sand, jute matting, and asphalt into a membrane.
7. Cover the membrane with a 3-ft-thick layer of protective soil.

*See *Eng. News-Rec.*, Feb. 14, 1963, p. 45, for a description of the use of a plastic membrane to repair the upper cracked portion of the 70-ft-high Mast Dam in northern California in 1963. After the repairs described, this dam has retained nearly full reservoir each year without problems (1970).

As of this writing (November 1970) the reservoir has not retained very much water since the repairs, and no further cracking has been observed.

Starting in 1953, the U.S. Soil Conservation Service has repaired about a dozen badly cracked, small dams in Oklahoma, Texas, Utah, and Wyoming by filling the cracks with a slurry consisting primarily of natural clayey soil and water.* Commonly the slurry is mixed in a pit dug on the crest of the dam, about 6 ft square and 6 ft deep, using a rotary drillrig for the mixer (Fig. 23*c*). The slurry is then pumped with a low head pump through a hose (about 2 in. diameter) and discharged into the cracks under gravity head. Sandbags and small dikes are used to prevent the slurry from flowing out of the cracks in the lower elevations of the embankment (Figs. 23*a* and *b*).

This method has been used particularly for cases of extreme cracking in which large foundation settlements have caused open cracks with widths of 1 to 6 in. on the dam crest and slopes, and often in circumstances where the cracks extend through the whole height of the dam and downward into the foundation material.

A local clayey soil is used for the slurry. Materials with a fairly high content of medium to fine sand have been considered preferable and an effort has been made to select materials with a plasticity index less than 10, for the purpose of reducing shrinkage on drying. In a few cases, sand has been excavated from a separate source and mixed with the clay, in a ratio of about 1 part sand to 3 parts clay. The final slurry is mixed with as much sand as possible, while still keeping it in a sufficiently fluid state so that it can be pumped and will flow readily into and fill the cracks. Each batch (generally 4 to 8 yd^3) is mixed for about 30 minutes before it is pumped into the cracks and the slurry in the sump is continuously mixed during the pumping. In some cases additional sand has been poured into the mud in the cracks after they have been filled.

Before pumping any mud slurry, the cracks are

*Information courtesy of Mr. Jasper L. Holland, U.S. Soil Conservation Service, Engineering and Watershed Planning Unit, Portland, Oregon, who has closely followed the development and application of this method of repair.

Fig. 23. Repairing severe cracking in small dam by pumping in slurry of native soil and water. Cedar City, Utah, 1969. (*a*) and (*b*) Recently filled cracks in upper part of dam. (*c*) Mud sump and rotary mixer. (Photos courtesy J. L. Holland.)

first washed with water. Usually the pump on a drill-rig has been used and the water is jetted into the cracks. Using a ½- or ¾-in. pipe, about 10 ft long, on the end of a hose, and starting in the lower elevations of the dam, water under pressure is injected into the cracks as far as the pipe can be inserted. The water washing has two main purposes: cleaning loose debris from the crack and wetting the embankment material in the walls.

Mud-filled cracks are allowed to dry for several days or a week. If appreciable shrinkage of the slurry occurs, the process is repeated. Finally the crack is sealed at the surface by trenching and recompacting to a depth of 1 to 2 ft.

The most important advantage of the method is economy. A very badly cracked dam can be repaired for a fraction of the cost of excavating and rebuilding the cracked portion of the embankment. Another advantage is that if large foundation settlements continue, the mud-filling process can be repeated economically, whereas if the embankment is rebuilt it may crack again. Another advantage is that cracks extending to a considerable depth in the foundation can be filled. The main disadvantages are that the crack filling may shrink and crack and that some cracks may be missed. Shrinkage may occur as a result of evaporation from the surface of the crack, or as a result of water being sucked into the adjacent dry embankment.

The engineers of the Soil Conservation Service have had mixed feelings about this method of repairing cracks. It is used generally where the cracks are very wide and deep and where it would not be practical to seal them by trenching to their full depth and backfilling. Some engineers have been concerned that the slurry crack filling would dry out and leave the crack only partially filled. In most of the dams repaired there has been no reappearance of the cracks on the surface or reports of other trouble. Moreover, after nearly 20 years and a dozen experiences, the Soil Conservation Service is currently (1970) using the method.

6. INSTRUCTIVE EXPERIENCES—ROCKFILL DAMS WITH EARTH CORES

Differential settlement cracking has caused problems at a considerable number of the major rockfill dams constructed in the last 15 years. Rockfill dams with earth cores are often well adapted at sites where high dams are needed. These sites generally have relatively steep rock abutments which are essentially incompressible. Also the various embankment zones usually have appreciably different compressibility. For these reasons the upper portions of such dams are subjected to high differential settlements and the risk of cracking cannot be wholly eliminated.*

The trouble has ranged from minor and unimportant crest cracks, which were repaired and did not reappear, to serious leaks and piping through the core. The case histories described below illustrate the range and details of the problems and include four cases in which serious damage from piping occurred.

Cougar Dam

Cougar Dam, Oregon, is a 520-ft-high rockfill dam with a central earth core sloping slightly upstream. The crest length is 1600 ft and the rock abutments are relatively steep. The performance and the cracking are well described by Pope (1967). The details are summarized briefly as follows:

1. The embankment was completed in the fall of 1963 and the reservoir was filled for the first time in the summer of 1964.
2. Vertical and horizontal components of the movement of the upstream and downstream edges of the crest from 1964 to 1966 are shown in Fig. 24*a*. Note that the width of the crest had increased about 0.7 ft by the end of 1964.
3. In June 1964, when the reservoir was filled for the first time, a transverse crack opened across the crest at a point about 40 ft from the left abutment. The crack had a maximum width of about 1.5 in. At the center of the crest it was about ¾-in. wide, and in a test pit it was found to disappear about 5 ft below the top of the dam. The crack was repaired and did not open again.
4. A few days after the transverse crack opened, a series of longitudinal cracks appeared abruptly on the crest, over almost the full length of the higher section of the dam. The main longitudinal crack developed at the center of the crest more or less along the boundary between

*As an example of current opinion on this subject, Casagrande (1969) states, "I believe that for very high dams with steep abutment slopes, it is not possible to prevent entirely the formation of substantial tension zones and transverse cracks in the top of dams in the vicinity of abutments, no matter what materials we use in the dam. Therefore, we must defend ourselves against the effects of the cracks."

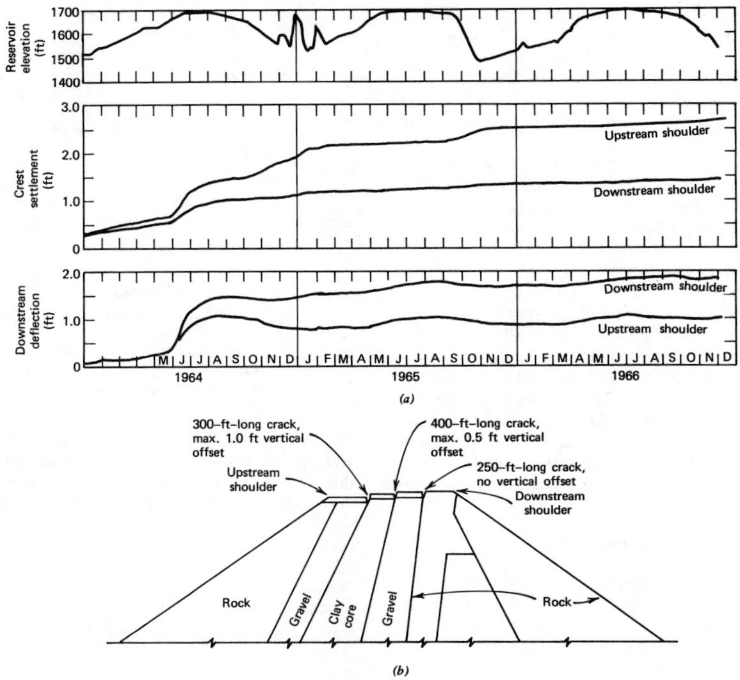

Fig. 24. Cougar Dam. (*a*) Settlement and downstream crest deflection during 1964-1966. (*b*) Longitudinal crest cracking. (After Pope, 1967.)

the downstream edge of the core and the upstream edge of the gravel filter. The width of this crack was less than ½ in. and remained fairly constant for several months with full reservoir during the summer and early fall of 1964. When the reservoir was lowered during the late fall and winter of 1964, movements started again and additional longitudinal cracks appeared. By January 1965, the width of the main crack at some points had become as great as 6 in. and the upstream shoulder of the crest had settled as much as 1 ft with respect to the downstream shoulder, giving the crest a "stepped" appearance as shown in Fig. 24*b*.

5. Widening of the crest and differential settlement between the upstream and downstream edges of the crest increased only slightly during the next year (1966) and then reached equilibrium (Fig. 24*a*).
6. The cracks were filled with sand and the crest was regraded. No further crest cracking has been observed up to 1970, and there has been no indication from leakage or other observations that unseen cracks exist.

The increase of the crest width and longitudinal cracking at Cougar Dam are similar to experiences at several other recently constructed high dams of the same type.* Generally in the central part of the

*For example, the 150-meter-high Gepatsch Dam completed in Austria in 1964 had crest movements and longitudinal cracking which were very similar to those at Cougar Dam except that no transverse crack appeared (Schober, 1967, 1967a). In the first two years, the crest width increased about 22 cm (a stretch of about 2.5%).

dam the crest widens considerably during the first months after construction and longitudinal cracks develop. The upstream rockfill shell settles considerably during the first raising and the first lowering of the reservoir, and transverse cracks develop a short distance from one or both abutments, usually where the embankment is about 50 ft high, and often directly above recognizable irregularities on the rock abutments. Moreover, longitudinal crest cracks develop very rapidly—they may not be observed one day and the next day they may be several hundred feet long. These longitudinal crest cracks usually develop along the boundaries between the core and the upstream and downstream transition sections.

In retrospect, it is generally agreed that the cracking of Cougar Dam did not create a serious threat to its integrity. During the months when the major movements of the crest were taking place, however, there was concern as to when the movements might stop and as to whether the final magnitude of the movements might be very large. Also, as discussed later, the writer believes it is quite possible that internal cracks which were never seen may have developed.

Round Butte Dam

Round Butte Dam, Oregon, completed in the fall of 1964, is a rockfill dam with a central earth core sloping slightly upstream, and has a geometry very similar to that of Cougar Dam. The core material was a well-graded, cohesionless silty sand with 30 to 40% passing the No. 200 sieve. The dam has a maximum height of 440 ft and a length of about 1400 ft at the crest. The movements and cracking (both longitudinal and transverse) were very similar to those of Cougar Dam, and are described by Patrick (1967).

The transverse crack appeared approximately 70 ft from the right end of the dam and was about 1 in. wide at the top and nearly vertical. Probing with a stiff steel tape indicated that it was at least 16 ft deep and that the width decreased with depth. Although this 16-ft-deep crack extended 5 ft below the level of the water in the reservoir at that time, no free water was found in the crack. It was later filled with a mixture of fine dry sand and powdered bentonite. Probably the crack was closed at the upstream face of the core because of softening and collapse of the cohesionless silty sand in the presence of the reservoir water. The transverse crack was located just over a prominent irregularity in the rock abutment (Fig. 25).

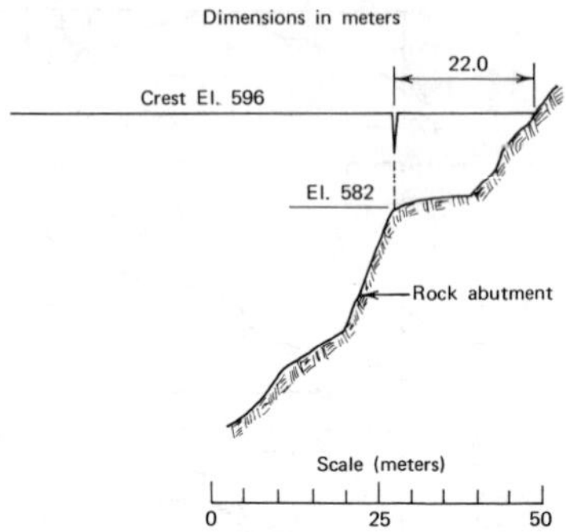

Fig. 25. Round Butte Dam. Transverse crest crack appeared directly over sharp irregularity in rock surface in upper part of right abutment. (After Casagrande, 1965.)

Djatiluhur Dam

The Djatiluhur Dam in Indonesia, completed in 1965, is a 112-meter-high rockfill dam with clay core sloping slightly upstream (Fig. 26), generally similar in geometrical design to the Cougar and Round Butte dams. The foundation consisted of a shale-sandstone formation which was relatively incompressible and

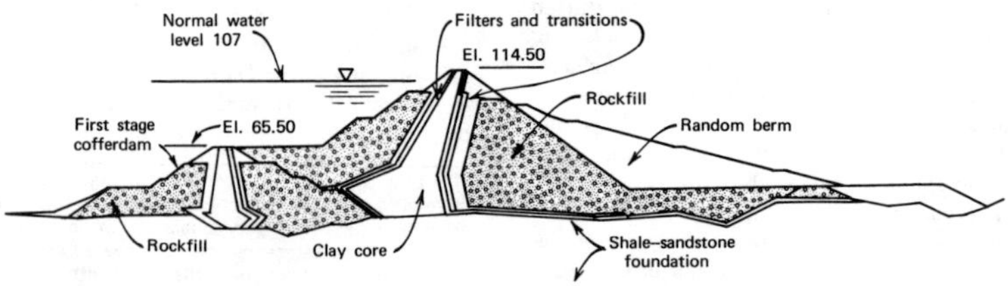

Fig. 26. Djatiluhur Dam. Cross section. (After Hamon and Buellet, 1966.)

did not contribute to differential settlement. The dam is 1200 meters long and the abutment slopes are uniform and not steep. The rockfill shells were constructed by dumping in high sluiced lifts. The core is a clay of high plasticity. Longitudinal crest cracks developed similar to those at Cougar and Round Butte dams. In test pits, however, open horizontal cracks were discovered in the upper part of the core. A detailed description of the measurements of post-construction settlements and cracking was given by Farhi and Hamon (1967) and is summarized here.*

The first cracks developed during construction of the embankment. In January 1965, when the embankment construction surface reached El. 103,† about 13 meters below the crest, a longitudinal crack appeared along the boundary between the core and the downstream fine filter. It first appeared near the midpoint of the dam and then extended rapidly in both directions, reaching a total length of more than 500 meters, or half the length of the dam. Construction on the main section of the dam was stopped for several months to study the problem and an extensive system of surface reference monuments was installed. Figure 27 shows the measured settlements and horizontal movements in the upstream-downstream direction of typical points on the construction surface at El. 103 from January 10 to April 13, 1965. Measurements showed that the core was settling with respect to both shells. A few water tests in boreholes gave no evidence of cracks in the core. During this time, water was stored in the reservoir at approximate El. 70.

In the spring of 1965 the crest movements and longitudinal cracking at Cougar and Round Butte dams were already known. It seemed clear that the movements and cracking at Djatiluhur Dam,

*Further descriptions of the design and construction are given in Hamon and Buellet (1966) and Bohn and Hamon (1967).

†Elevations in meters.

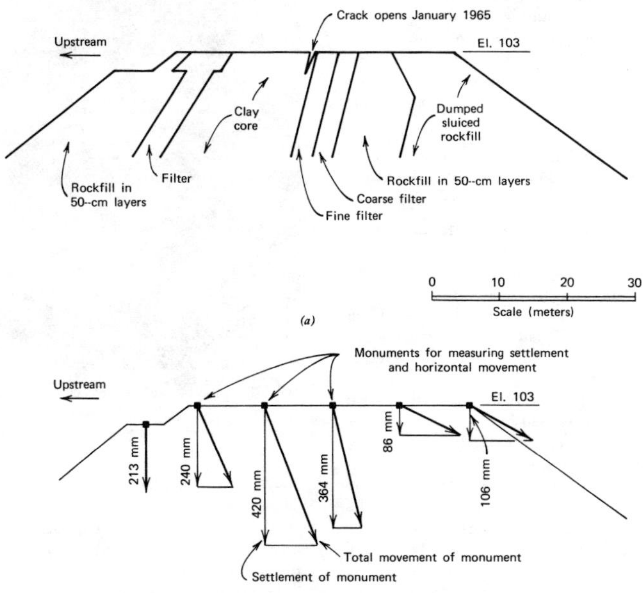

Fig. 27. Djatiluhur Dam. (*a*) Location of crack observed during construction, January 1965. (*b*) Movement of monuments on construction surface at El. 103, January–April 1965.

although they developed during construction, had the same general cause. Construction was resumed and careful measurements of the deformations were continued. The main portion of the dam was completed to its final elevation (114.50) at the end of August 1965, and the downstream berm was completed shortly thereafter.

Soon after construction was completed a longitudinal crack approximately 300 meters long appeared on the crest in the central portion of the dam. The maximum width of the crack ranged generally from 1.0 to 1.5 in. Photographs of the crest cracking in October 1965 are shown in Fig. 28. By February 1966 the main crack had widened slightly but the general nature and extent had not changed appreciably.

Between October 1965 and June 1966 the upstream edge of the crest settled about 50 cm with respect to the downstream edge. During this period the rate of settlement decreased rapidly and was very low by June 1966. The horizontal distance between the monuments on the upstream and downstream edges of the crest increased only a few centimeters during this period and also stabilized.

The core consisted of a clay of high plasticity (CH), derived from *in situ* weathering of the claystone bedrock and having the following average properties:

Liquid limit	65%
Plasticity index	38
% Finer than 200 sieve	60
% Finer than 0.005 mm	35

Water was added to the clay both by irrigation of the borrow area before excavation and by sprinkling during the excavation process. Excavation was by bulldozers pushing the material into piles from which it was placed into trucks with loaders. When compacted, the material had a very homogeneous appearance. Its water content was uniform, averaging about 2% above Standard AASHO optimum.

To investigate the possibility that unseen internal cracks might have developed in the core, a number of borings and two test pits were put down during the last part of 1965 and the first part of 1966. The borings were inclined, to follow the core. Some of the borings extended as much as 75 meters below the crest. In almost all of the borings, drilling water was lost at one or more elevations, presumably through cracks in the core. The rate of water loss ranged up to 150 liters/minute, which was the capacity of the pump being used.

A consistent pattern was found for these water

Fig. 28. Djatiluhur Dam. Crest cracks, October 1965. (Photos courtesy Coyne and Bellier.)

losses in the boreholes: water loss occurred only when the elevation of water in the borehole was above a certain value. During the drilling of a given hole, for example, there would be complete return of the circulation water until at some depth there would be an abrupt partial or total loss of circulation. By stopping the drilling operation and pouring water into the hole from the top, so that the pressure head in the hole was equal approximately to the elevation of the crest, it was found that the hole would accept flow at a constant rate, e.g., 100 liters/minute. When no more water was added to the hole, the water level did not drop to the bottom of the hole, but consistently dropped to a level between 7 and 12 meters below the crest and then stabilized, even when the water losses (and apparent cracks) had been encountered at depths varying from 30 to 70 meters below the crest (Fig. 29).

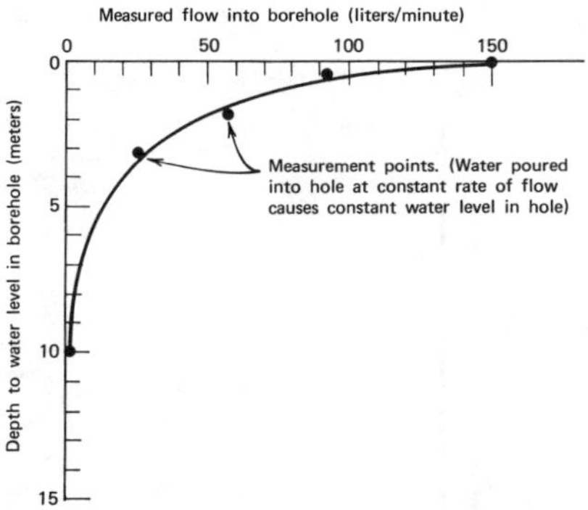

Fig. 29. Djatiluhur Dam. Typical results of experiments to determine the quantity of lost water in a borehole through embankment crack as function of pressure in the hole. (After Farhi and Hamon, 1967.)

Finally, piezometers were installed in several of the boreholes where water losses were encountered. These consisted simply of small diameter pipe, perforated for 20-ft lengths, surrounded by sand and sealed with grout above the sand. The water pressures recorded in all piezometers stabilized rapidly and all showed a pressure head of roughly El. 100, which was about 20 meters higher than the level of the reservoir at the time (approximate El. 80).

After much study of the results of the exploratory borings, the water losses, and piezometer readings, the investigating engineers concluded that cracks were probably being opened in the core by the "jacking" pressure of the water in the boreholes.* It seemed probable that the water pressure in the lower part of the borings was sufficient to open cracks. These cracks could be new fractures or old cracks which had been closed before being penetrated by the boring. Subsequently, when the water level was allowed to drop in the holes, it seems likely that the crack widths gradually became smaller as the water pressure in the lower part of the hole was reduced, in the same way that the thickness of a "flat jack" becomes smaller as the internal pressure is reduced. Finally, as the water level in the borings dropped to approximate El. 100, the pressure was no longer great enough to keep the crack open and no more water left the boring.

A small amount of cement grouting with no pump pressure was tried in some of the holes. In a few cases, when the same elevation in the same hole was grouted several times in succession, the rate of grout take increased, yet the grouting had no influence on the water loss observed in nearby borings. It was apparent from these experiments that a considerable volume of grout could be put into the core, even using gravity pressure alone, and that much of the grout would flow out into the core through fissures which either were closed or did not exist before grouting.

The test pits were excavated very carefully to avoid obscuring the cracks. In the first test pit (about 1.5 meters^2), put down in November 1965 to a depth of 12.5 meters at a location where maximum crest movements had been measured, numerous horizontal cracks were observed. There were more cracks on the downstream wall of the test pit than on the other sides but in some cases the crack extended around all four sides (Fig. 30). The small number shown opposite each crack in Fig. 30 is the average width of the crack in millimeters. At a depth of about 7 meters there was a 10-mm-wide horizontal crack which extended around the four sides of the test pit. At a depth of 9 meters a horizontal crack with a width of 60 mm was discovered on the downstream face of the pit only. Figure 31 shows typical cracks.

A second pit was excavated to a depth of 21 meters in January and February of 1966. It was also located in the high portion of the dam where maximum surface cracking and crest settlement had occurred, but at a distance of about 70 meters from the first pit. This second pit was put down through the rockfill of the upstream slope 9 meters upstream of the crest centerline, so that a deeper portion of the core could be examined. No sign of an open crack was observed in this second, deeper test pit.

A boring was put down on the crest of the dam at a

*As discussed in more detail later, this is probably the first case where this phenomenon was recognized and well explained.

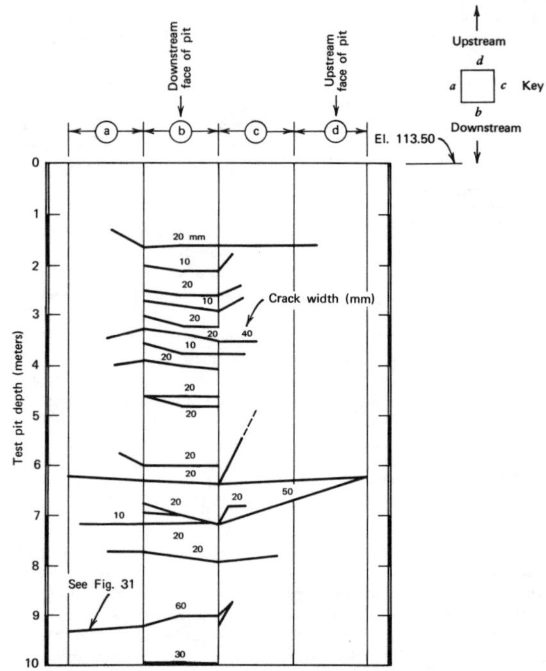

Fig. 30. Djatiluhur Dam. Developed elevations of walls of first test pit showing locations and widths of cracks seen. (After Farhi and Hamon, 1967.)

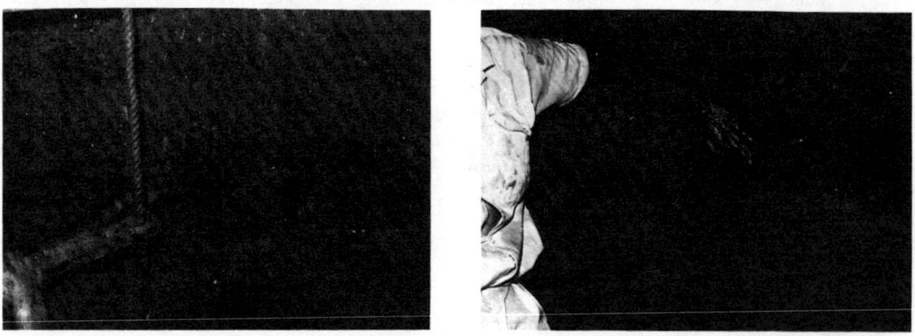

Fig. 31. Djatiluhur Dam. Looking at the 60-mm-wide crack at approximate depth 9 meters in first test pit. (Photos courtesy Coyne and Bellier.)

location and inclination such that at El. 101 (13.5 meters below the top of the dam) the boring was about 19 meters horizontal distance from the center of this second test pit. When the boring had penetrated to El. 101, drill water was lost abruptly. It was possible to pour 150 liters of water per minute into the boring when the hole was kept full. After water had been poured into the boring for some time, a small stream of muddy water appeared on the wall of the test pit, also at El. 101. As observed in the test pit wall the crack was essentially closed (the thickness could not be measured); however, it was obviously a horizontal crack. The leak emerged from three sides of the pit in several very small, concentrated flows along a horizontal line. When no more water was poured into the boring, the water level in the boring gradually dropped, at a decreasing rate, and stabilized at El. 105 at the end of 19 hours, after which no more water entered the test pit.

When these investigations were being carried out, in the spring of 1966, the water in the reservoir was at approximate El. 80, or about 27 meters below the maximum normal water level (El. 107). During this time there was very little leakage observed at the downstream toe. Because of this and the results of the water tests in the borings, it seemed likely that there were no open cracks of important width below El. 80 at least.

Finally, in February 1966 it was necessary to decide whether it was safe to continue filling the reservoir. Serious consideration was given to grouting the core in the upper elevations before allowing the reservoir level to rise. It was decided finally not to grout unless a concentrated leak developed when the reservoir was raised. The main reasons for this decision were:

1. The good quality and thickness of the transition zones upstream and downstream of the core.
2. The rapid decrease in the rate of crest settlement since the end of construction.
3. The probability that the horizontal core cracks seen in the November 1965 test pit were closed in the upstream portion as the result of the settlement of the upstream shell causing a downdrag on the core.

In the spring and summer of 1966 the reservoir was raised gradually in steps to El. 107, while careful observations of the performance were made. No problems developed. The settlements continued, as expected, at a decreasing rate. By making some special provisions it was possible to measure essentially all the seepage water emerging downstream from the dam, including seepage through the core and seepage through the shale foundation. The total seepage for the nearly 1-mile-long dam when the reservoir was full was less than 300 liters/minute; that is, there was essentially no leakage through either the core or the foundation. Subsequently the reservoir has been used without restriction and has been full or nearly full each year. The leakage has not increased and no other problems have developed (1970).

This experience is one of a number providing strong evidence in support of the general conclusion that unseen open cracks on horizontal planes in the upper elevations of high dams probably develop much more often than has been suspected from the frequency of appearance of cracks on the crest. It is likely that open cracks similar to those found in the first test pit at the Djatiluhur Dam (Figs. 30 and 31) may frequently develop temporarily during construction of high dams at the interfaces between zones and near abutment irregularities.

The Djatiluhur experience also illustrates another point which has not been widely considered: the probability of visible cracks during construction depends on the rate of increase in elevation of the construction surface. At a given dam if embankment placement were to be stopped for a certain period of time, cracking might appear. In contrast, under the same circumstances, if the construction were to continue at a uniform rate it is possible that no crack would be seen, since the construction surface would be progressively covered by new layers of embankment material and the cracks would be masked.

A somewhat similar experience occurred at the Mattmark Dam in Switzerland (Gilg, 1970).* Following a disastrous construction accident (avalanche falling on construction camp), at which time the 100-meter-high dam had been completed to approximately midheight, work on the embankment was stopped in August 1965. In October 1965 differential settlement cracks appeared near and parallel to the right abutment. These ran completely across the core and down the upstream slope, and caused serious concern. After observation for nearly two years, during which time the construction continued on the main section of the dam, the cracks were repaired by trenching and backfilling in June 1967. Subsequently,

*Another recent example of important cracking appearing during construction is the Duncan Dam (Gordon and Duguid, 1970).

the remainder of the dam in the area of the cracks was constructed fairly rapidly at a uniform rate in the summer and fall of 1967. No further cracking was observed during the remainder of the construction period. The reservoir has been full or nearly full four seasons; no cracking has developed on the crest and there has been no indication from leakage or displacement measurements of internal cracks (1970). It seems quite probable in this case that if it had been possible to construct the dam without a work stoppage no cracks would have been observed.

Matahina Dam*

The Matahina Dam, constructed in New Zealand in 1965-1966, is a 200-ft-high rockfill dam with a central earth core (Fig. 32). The crest length is roughly 1300 ft. When the reservoir was nearly full for the first time, a large, concentrated leak of dirty water broke through the core adjacent to a steep-faced bench on the right rock abutment in the upper part of the dam. After a few hours of muddy flow the leak decreased greatly in volume and subsequently ran clear. Details of the design and construction, the leakage, the investigations and the repairs are described briefly.

In the center of the valley the foundation consists of a relatively incompressible and impervious Tertiary sedimentary deposit (silty clays, clayey gravels, and silty sands). The abutments consist of a generally hard igneous flow (called ignimbrite) resting on top of the Tertiary sediments. The abutments are generally of hard rock.

The core is a gravelly clay derived by deep, *in situ* weathering of a greywacke. As is common with some types of *in situ* residual soils excavated with scrapers, the gradation of the material varied widely (Fig. 33). The Atterberg limits of the undried material typically were liquid limit, 30 and plasticity index, 10. Lower plasticity values were obtained on specimens which were air dried or oven dried before testing.

Both transitions were constructed from unprocessed materials excavated from weathered sections of the ignimbrite. The inner transition was material that could be excavated with scrapers without ripping. This material was cohesionless and had gradation shown in Fig. 34. The coarser particles ranged from soft, easily crushed material to hard rock.

The outer transition zone consisted of material excavated from the weathered ignimbrite in areas where it was necessary to use a ripper but no blasting. As seen in Fig. 34, this material was a well-graded gravel; from a maximum size of about 12 in. down to a few percent passing the No. 200 sieve, and with an average size (D_{50}) typically in the range between 0.5 and 3 in.

The shells consisted of hard, quarried ignimbrite with very few fines, placed and compacted in layers. The ignimbrite flow comprising the rock abutments has a slight dip from the right side to the left side with the result that the columnar joints in the rock on the right side are more open than on the left side. The abutment surface over the whole core area on the right side was covered with a concrete slab and

*The design and construction and a brief description of the crest settlement, cracking, and piping are given by Galloway (1967). Further details were supplied subsequently through the courtesy of Messrs. W. M. Duncan and J. H. H. Galloway of the New Zealand Ministry of Works.

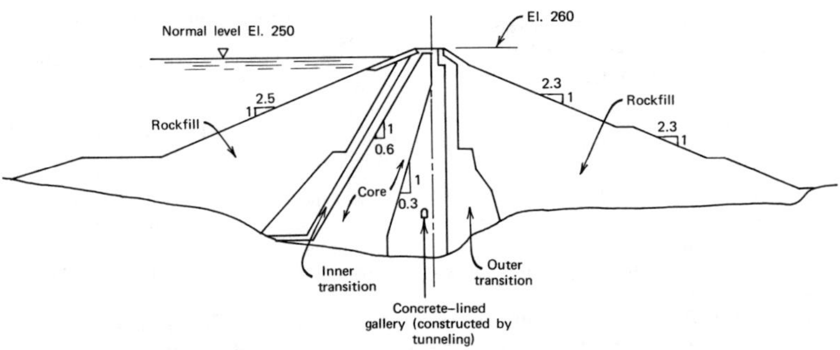

Fig. 32. Matahina Dam. Cross section.

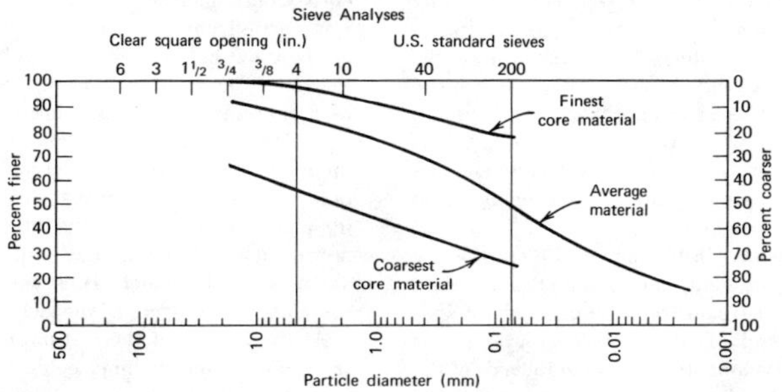

Fig. 33. Matahina Dam. Weathered greywacke core gradation. (Courtesy of New Zealand ministry of works.)

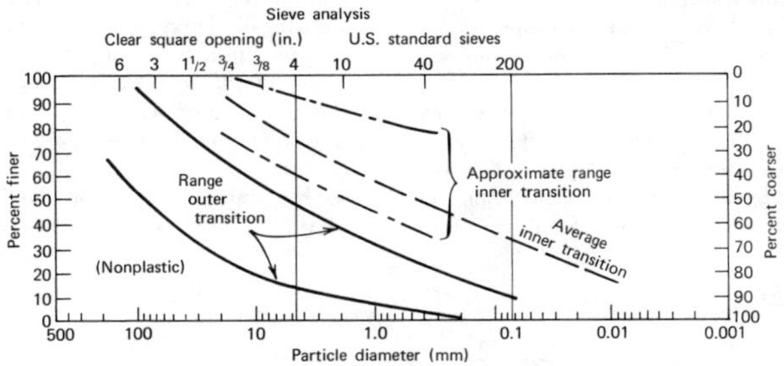

Fig. 34. Matahina Dam. Gradation of transition zones. (Courtesy New Zealand Ministry of works.)

massive grouting was carried out (150,000 sacks of cement were used for 50,000 ft² of grout curtain.) Most of the grouting on the right abutment was done in advance of the fill placement. A small amount of grouting on the right abutment was done by drilling through the dam crest after completion of the embankment.

Since the open joints extended deep into the right abutment, it was anticipated that the leakage around the end of the grout curtain might be in the range between 3 and 30 cfs.

The spillway has 40-ft-high gates with a fixed concrete crest at El. 210 and a normal full-reservoir lake level at the top of the gates at El. 250.*

Starting in the first part of December 1966 the reservoir level rose rapidly for the first time to El. 210 and was held at this level for about two weeks. During this time the total flow in the drainage outlet was constant at about 2.5 cfs, including leakage emerging from the right abutment.

There were no large settlements at any time. During construction of the embankment, measuring plates placed in the foundation showed a uniform settlement of about 0.6 ft across the floor of the valley. During reservoir filling the crest of the dam in the central portion moved downstream about 0.4 ft and settled 0.15 ft. No unusual settlements were observed at the right abutment. A monument was established on the downstream slope during construction at El. 203, approximately 70 ft horizontally from the right rock abutment. This monument settled about 0.7 ft as the dam was being constructed from El. 203 to El. 260 and then settled less than 0.1 ft between the end of construction (Nov. 1966) and May 1967.

Starting early in January 1967 the reservoir was raised above El. 210 fairly rapidly (approximately 5 ft/day) and on January 13, with the reservoir level

*Elevations in feet.

at approximate El. 235, 15 ft below normal operating level, the flow from the drainage outlet increased abruptly to approximately 20 cfs and the water became "slightly cloudy." Within a few hours the total seepage decreased to approximately 9 cfs and within 24 hours the water was clear. No further signs of trouble were observed and the reservoir was allowed to rise to approximate El. 240 (the minimum level for power generation).

In the last few days in January 1967 a subsidence crater appeared suddenly on the crest of the dam immediately downstream from the core, about 75 ft from the right end of the dam. This crater was initially about 20 ft long, transverse to the axis of the dam, 10 ft wide, and 10 ft deep, with an estimated volume of 50 yd^3. Subsequent collapse of the cavity increased this volume several fold.

The lake level was lowered rapidly to approximately El. 213 and held there while the investigations and repairs were made.

A vertical-walled, timber-braced test pit was excavated as a series of contiguous shafts on the center of the subsidence crater with overall plan dimensions at the crest of 28 by 20 ft and with the upstream face of the test pit located on the downstream face of the core. Continuous sheeting was not required. The pit was carried down to El. 207 at the deepest point, or about 6 ft below the water level of the reservoir at the time. The lower part of the test pit followed the rock abutment from El. 221 down.

As seen in Fig. 35, a horizontal bench in the abutment surface was found at approximate El. 221. This bench had been used during construction as a grouting platform. It had been the intention to smooth the abutment by either filling the filet with concrete or shooting off the bench. However, only a small concrete filet was placed, as shown in Fig. 35.

At the bottom of the test pit, on the downstream face, an erosion channel could be seen where the material from the downstream transition zones "had been washed into the rockfill shell between a large boulder and the abutment rock" (New Zealand Ministry of Works, 1970). The eroded channel was followed until it disappeared behind a large rock (about 1.0 yd^3) which was resting directly against the abutment. There was a clear opening between this large rock and the rock abutment of about 9 in. at El. 205.

The total amount of material in the downstream transition zones which was disturbed by the leak

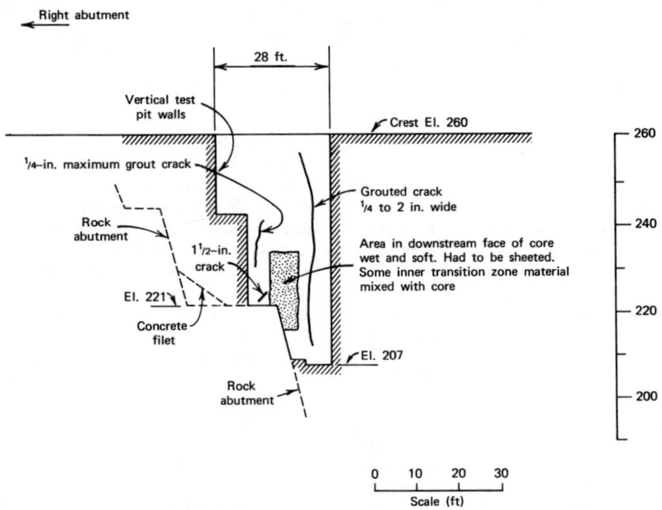

Fig. 35. Matahina Dam. Major features observed in test pit, looking upstream at downstream face of core. (Courtesy of New Zealand Ministry of Works.)

and subsequent collapse leading to the crater on the crest was estimated to have been about 800 yd^3, and the amount washed out was estimated to have been no more than 50 yd^3 (New Zealand Ministry of Works, 1970).

Directly opposite the bench on the rock abutment at El. 221, the downstream face of the core was wet and soft and had some transition material mixed with it. To prevent this face of the test pit from caving, nearly continuous sheeting was required. Except for this soft zone, and a few small cracks which may have been caused by stress relief during the excavation of the test pit, the downstream face of the core appeared uniformly stiff and intact.

Grout seams, generally vertical and transverse to the longitudinal dam axis, with a maximum width of about 2 in., were found running continuously through the downstream inner transition zone.

The soft spot in the downstream face of the core zone was patched with "a sand-gravel mixture heavily plasticized with bentonite" and the test pit was backfilled with "zones of clean washed sand and gravel to form a filter and a strong hydraulic sink. Observation wells left in the backfill and water level readings in these and flows from the dam drainage system show that the repair is completely satisfactory" (New Zealand Ministry of Works, 1970). No grouting or other remedial work was carried out in the core or upstream transitions. The reservoir was then raised to El. 250 and has been kept full for the last 2 years without further incident.

There was no doubt in the minds of the engineers associated with the project that the trouble had been caused by a concentrated leak through a differential settlement crack in the core directly opposite the bench on the abutment. It was speculated that there might have been an open crack before the reservoir was raised. Maximum leakage of about 15 cfs probably passed through the crack in the core under a head of about 20 ft.

It seems probable that some of the upstream inner transition material was carried into the crack, thus reducing and stabilizing the leakage and preventing any large amount of rapid, progressive erosion of the core. The quantity of upstream transition material that entered the core is unknown, but it could not have been very large since there was no subsidence of the upstream edge of the crest. About 50 yd^3 of the downstream transitions were washed into the downstream rockfill shell through an opening which had been created by several large rocks directly against the rock abutment. Progressive collapse of the roof of the cavity, during about 2 weeks, resulted in the appearance of the subsidence crater on the crest.

Several other circumstances may have been important contributing causes:

1. The reservoir rose rapidly (5 ft/day) at the critical level (from El. 210 to 235).
2. The presence of the grout seams may have increased the rigidity of the downstream inner transition and consequently reduced its ability to deform and to heal the leak (these seams were undoubtedly caused by leakage from grout holes cased through the body of the dam or from leakage around the concrete slab, which was 12 meters wide at this elevation).
3. Similar grout seams might have existed in the upstream inner transition, conceivably having a reinforcing action which would decrease the ability of the material to be washed into core crack. This may also explain the fact that no subsidence crater developed at the upstream edge of the crest; that is, grout seams may have created a structural skeleton within the upstream inner transition. In the worst case a concentrated leak could develop through the transition zone underneath a grout seam, which would act as a roof to prevent the material from collapsing and sealing the leak. Such action must have occurred to permit this large volume of leakage to pass through the upstream transition under the only 20 ft of head.
4. The transitions consisted of partially weathered rock (i.e., they were not constructed of processed alluvium or crushed hard rock as normally employed for such zones). Although it was a completely cohesionless soil, the compacted embankment formed had some cementation and unconfined compressive strength. This was observed particularly when constructing the drainage gallery in this zone, which was installed by tunneling through the completed embankment (Fig. 32). In this work it was noted that the soft crushed-rock inner transition material appeared to revert to its original state (i.e., a homogeneous soft rock) under pressure. This gave the material sufficient cohesion so that there was no trouble with roof support or lost ground during the tunneling.
5. The core material was a residual gravelly clay derived from *in situ* rock weathering, which may have been especially rigid and susceptible to cracking.

It was not believed probable that the rather unusual geology of the dam site (the igneous rock abutments overlying unconsolidated sediments) had any appreciable influence on the problem. It was also quite clear that the total and differential settlements responsible for the trouble were not unusual or greater than those experienced in many similar dams.

It is pertinent that no cracking appeared on the crest before the development of the dirty leak. Hence in this case history we have another experience which shows clearly that, for this kind of dam in any event, serious internal core cracking can exist without being manifested on the crest. This is the only experience on record to the writer's knowledge where there has been definite evidence of cement grout penetrating zones of transition or filter material in seams.

This is also the only experience to the writer's knowledge where it is known that trouble was caused in a modern rockfill dam by the existence of large rock and large voids in downstream rockfill zones. At Matahina Dam there is no doubt that the placement of the large rocks with large voids directly downstream from the downstream transitions had a major influence on the trouble which developed. It is quite conceivable that the core and transition would otherwise have been self-healing.

Finally, there can be little doubt that the sharp irregularity on the rock abutment, with the horizontal step (Fig. 35), had a major part in the problem. It is interesting to compare the location of this abutment step and the adjacent damaged zone in the core with the location of the washed-out section of the Stockton Creek Dam in Fig. 1. In both cases the abutment irregularities were located at a depth of about 40 ft below the crest. At this depth the weight of a column of overburden (40 ft of embankment) would be expected to be of the same order of magnitude as the unconfined compressive strength, so that open cracks could be sustained without being forced closed by the internal pressure.

Another common feature of the Stockton Creek and the Matahina dams was the core materials, both being coarse residual soils with very similar properties. Both were excavated with scrapers from deposits of soil formed by *in situ* rock weathering. Commonly the total thickness of usable soil in such deposits is not great and the excavation is carried down in each area of the borrow pit to a level at which the soil is judged to be too coarse and pervious for the core, at which level this area of the borrow pit is exhausted. When excavated with scrapers, the material obtained near the surface may be a fine-grained soil with only a small content of coarse sand or gravel-sized particles, whereas the material obtained at the lower part of the pit may be a gravel with barely enough fines to make it usable for an impervious core. Inevitably in these circumstances the resulting core is comprised of alternate layers of considerably different properties, and particularly of different percentages of gravel-sized particles. This accounts for the wide range of gradation for the Matahina Dam core material (Fig. 33). From a recent examination of the borrow pits at Stockton Creek Dam, the writer has no doubt that a comparable range of properties existed there.

It could of course be simple coincidence that these two dams, one of which failed and one of which was badly damaged from concentrated leaks breaking through the core on the first reservoir filling, had very similar core materials. The writer believes it is reasonable to conclude, however, that the nature of the core material may have contributed an important part to the set of conditions in the two dams which allowed the water to break through the cores abruptly. The laminar characters of the cores could very well make the embankment more susceptible to the development of localized zones of low minor principal stress and high susceptibility to hydraulic fracturing as discussed earlier.

It should be noted that many *in situ* residual soils are quite different from those of the Stockton Creek and Matahina dams in the sense that they have very little or no coarse sand or gravel-sized particles and do not grade gradually into the weathered bedrock below. For example, soils derived as the result of *in situ* weathering of some granites have no appreciable percentage of gravel-sized particles (Fig. 21).

Balderhead Dam*

The Balderhead Dam, constructed in England in 1961-1965, is a rockfill dam with a narrow, vertical earth core (Fig. 36) and a maximum height of 48 meters. When completed it was the highest dam in England. In April 1967, after the reservoir had been full continuously for more than 2 years, a subsidence crater appeared near the upstream edge of the crest.

*Details of the design and construction and of the subsequent problem and repairs are given by Kennard (1964), Kennard, Knill, and Vaughan (1967), Kennard, Penman, and Vaughan (1967), and Vaughan, Kluth, Leonard, and Pradoura (1970).

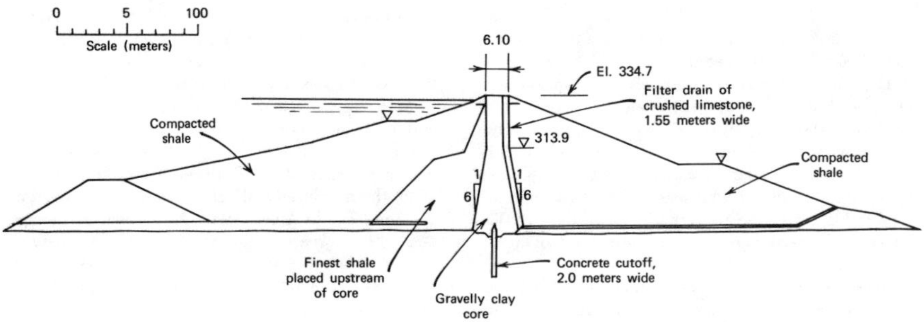

Fig. 36. Balderhead Dam. Cross section. (After Vaughan *et al.*, 1970.)

Extensive investigations led to the conclusion that concentrated leaks had developed through horizontal differential settlement cracks in the core and that a considerable portion of the core material had been damaged by progressive erosion. Following are the details of this experience.

The core was constructed of a gravelly, sandy clay (boulder clay) with the gradation shown in Fig. 37. The clay fraction was tough at the plastic limit. The liquid limit was 30 to 40; the plastic limit was 12 to 22. The material was compacted in 6-in. layers with a tamping roller at an average water content near standard AASHO optimum.

The shale comprising the shells of the dam was excavated with scrapers after ripping. The specifications required that the material be placed in 9-in. layers and compacted to 95% Standard AASHO maximum density. During the first construction season the shale was compacted in 9-in. layers with four passes of a grid roller. This was subsequently changed and most of the volume of the shale shells was compacted in 30-in. wetted layers with a heavy, smooth-wheel, vibrating roller, which was shown by tests to give higher densities. Tests in boreholes during construction showed that the shells had a coefficient of permeability of roughly 1×10^{-3} cm/sec. The most weathered shale was placed in a zone directly upstream of the core (Fig. 36).

The vertical drain at the downstream edge of the core was constructed of minus 3-in., crushed, hard limestone, compacted in 9-in. layers with a vibrating roller between steel forms placed exactly 5 ft apart to control the width accurately. This material was not processed beyond crushing and had a fairly wide range of gradation (Fig. 37). The filter material was designed to have the 15% size of the filter not more than three times the 85% size of the core, based on the gradation of the portion of the core finer than

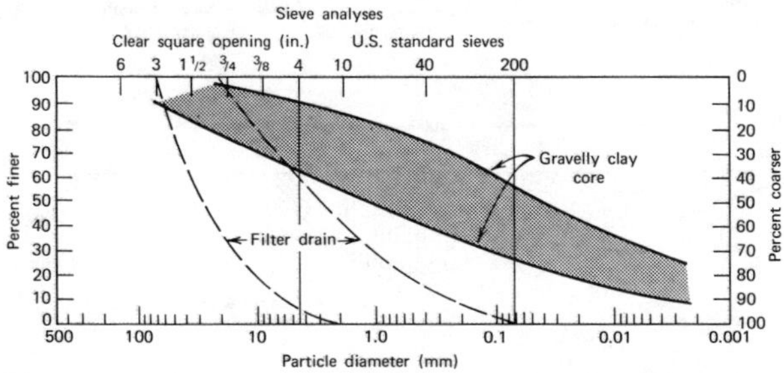

Fig. 37. Balderhead Dam. Grain size distribution of core and vertical filter drain.

about 1 in. "In the dam, due to segregation and variations in the material, this ratio may have been locally as high as 6" (Vaughan et al., 1970).

The foundation was relatively incompressible. Over most of the length it is a shale; over portions of the length the shale is overlain with a very stiff clay, with maximum thickness of about 18 meters, which was left in place under the dam (Fig. 38). A 1.8-meter-wide concrete cutoff wall was sunk into the shale foundation along the full length of the dam in the center of the earth core to a maximum depth of about 25 meters.

Reservoir filling commenced in October 1964 and the water level rose relatively slowly to its maximum elevation by February 1966 and remained there until March 1967 (Fig. 39).

Settlements during construction were not unusual. Settlement plates placed in the center of the core near midheight settled about 6 in. more than plates at the same level in the adjacent shells. Pressure cells indicated that the total pressure acting on horizontal planes within the core was less than the weight of a column of overburden, evidently because of arching of the core between the shells. By February 1967 the total postconstruction settlement of the crest at the point of maximum dam height was roughly 25 cm.

As the reservoir was rising in the fall of 1965, the total measured seepage from the main underdrain for the full length of dam was roughly 10 liters/sec (Fig. 39). In January 1966, as the reservoir reached its maximum level, the measured seepage dropped suddenly to 5 liters/sec for a few days and then abruptly increased again to about 35 liters/sec by the end of February 1966. Between March and August 1966, the measured leakage was roughly constant at 25 liters/sec. In August 1966 the measured leakage started to increase again, and to fluctuate considerably, reaching a maximum of about 55 liters/sec by the end of December 1966, when it abruptly fell to about 30 liters/sec.

At the end of January 1967 a small depression was discovered on the downstream edge of the crest at Station 266.* At the beginning of April 1967, a larger subsidence crater, about 3 meters in diameter and 2½ meters deep, developed on the upstream edge of the crest at Station 317 (Fig. 38*a*).

Following discovery of this crater the reservoir was lowered about 9 meters during the first half of April 1967. As the reservoir was lowered the measured leakage fell immediately from 45 to about 10 liters/sec and another subsidence crater, similar

*Stationing and elevations in meters.

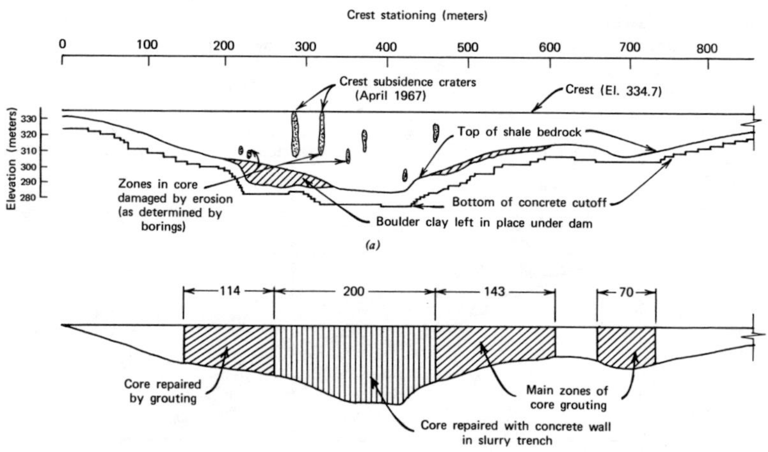

Fig. 38. Balderhead Dam. Longitudinal sections. (*a*) Location of zones of erosion damage in core. (*b*) Methods employed for repair of core damage. (After Vaughan *et al*., 1970.)

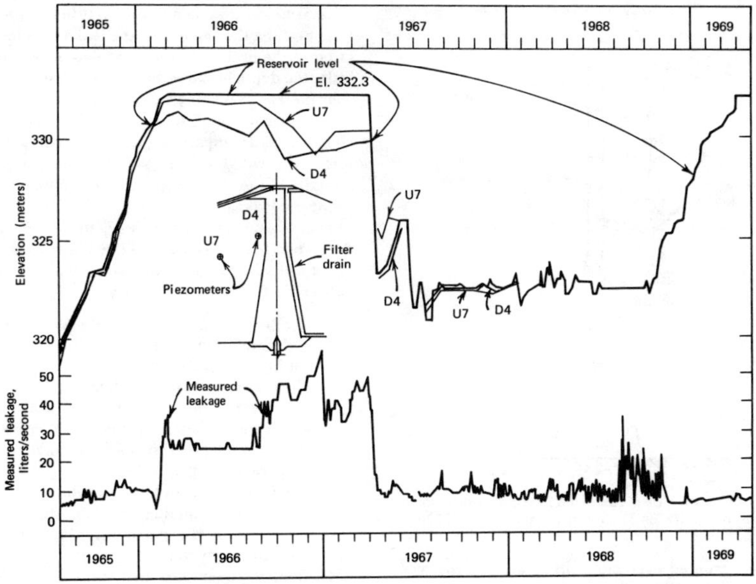

Fig. 39. Balderhead Dam. Reservoir level, measured leakage, upstream piezometer readings. (After Vaughan *et al.*, 1970.)

to that at Station 317, developed on the upstream edge of the crest at Station 287. During the reservoir drawdown the crest in the central part of the dam settled an additional 10 to 15 cm.

A year earlier, starting in the spring or early summer of 1966, when the reservoir was full, some of the piezometers located in the upstream shell indicated a gradually reducing water pressure (piezometers D4 and U7 in Fig. 39). It was apparent in retrospect that these piezometers were measuring a head loss in the semipervious upstream shell and were giving a first indication of concentrated leaks through the core.

Seepage emerging from the main underdrain had been alternately slightly cloudy and clear for about 4 months preceding the discovery of the subsidence crater. After drawdown in April 1967, the water was continuously clear.

A vertical-walled, braced test pit was put down to a depth of 12 meters through the downstream drain at Station 317, exposing the downstream face of the core (Fig. 40). "No evidence of significant damage was found, although some softened clay was noticed in the center of the exposed face" (Vaughan et al., 1970). The drain material encountered in the test pit was coarser than the average material used in the dam.

A series of very careful borings was made with a double-tubed diamond core barrel using drilling mud. Nearly complete recovery of the core material was achieved, even in the damaged areas.

Numerous zones of erosion damage were encountered in the core (Fig. 38). In these damaged zones the samples obtained consisted primarily of the coarser particles of the core material, showing that leakage through the core had washed out the fines. Figure 40 shows the investigators' best interpretation from the borings of the extent of the damaged zone at Station 317.

The reservoir was held at approximately El. 322 until the fall of 1968 while investigations and repairs were being made. In the summer and fall of 1968 the

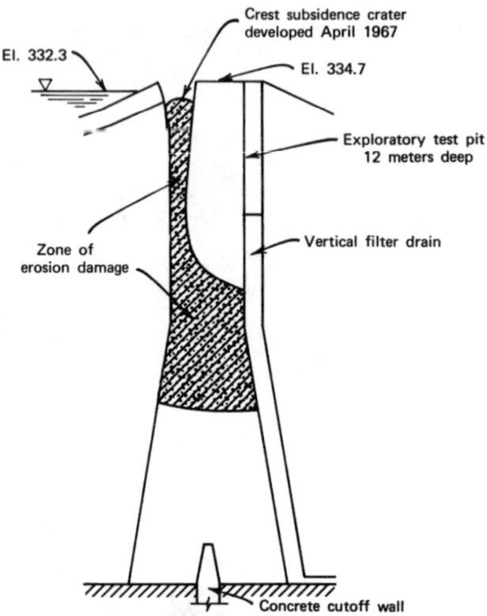

Fig. 40. Balderhead Dam. Estimated zone of erosion damage in dam core at Station 317. (After Vaughan *et al.*, 1970.)

core was grouted over more than half the length of the dam and a 200-meter-long cutoff wall was constructed in the core (Fig. 38*b*). The grouting and construction of the diaphragm wall are described by Vaughan et al. (1970).

Starting in October 1968 the reservoir level was allowed to rise again and reached the maximum level in May 1969, when the total measured leakage was of the order of 7 liters/sec and had a turbidity of less than 1 ppm. After the reservoir was refilled all piezometers upstream of the core showed water levels within 10 cm of the reservoir level.

It was assumed that there was a possiblity that additional cracks and leaks might occur, as the pressures generated in the core by the grouting dissipated. In addition to monitoring the quantity and turbidity of the leakage carefully, 69 piezometers were installed in inclined boreholes in the shale fill just upstream of the core, for the purpose of detecting any reduction in pore pressure below the reservoir pressure, which would indicate a leak through the core.

In March 1970 new leaks developed between Stations 600 and 650, where no grouting had been done (Fig. 38). These leaks were inferred primarily from the fact that the adjacent upstream piezometers measured a head about 2 ft below the reservoir level. At the same time the leakage being monitored in the drain outlet downstream increased very little (the depth of water passing through the V-notch weir increased roughly from $\frac{3}{8}$ to $\frac{1}{2}$ in.) and there was no measurable increase in the turbidity of the water. Hence it was apparent that the upstream piezometers provided a more sensitive measure of the integrity of the core than could be obtained by monitoring the leakage.

The engineers who investigated the problem developed the following hypothesis for the mechanics of the failure, which the writer believes is probably correct:

1. The first leaks probably developed in near horizontal cracks through the core which were not in existence before the reservoir was filled.
2. The cracks were opened by the pressure of the water acting on the upstream face of the core, which was possible because arching had caused the total pressure on horizontal planes in the core to be low.
3. The first cracks and leaks probably developed in the core at midheight of the dam or lower, and probably did not develop until the reservoir was nearly full.
4. Since the upstream shell of the dam is not very pervious, the total volume and velocity of the leakage passing through the cracks was not high.
5. The leakage gradually eroded the soil, carrying the finer particles with it into the downstream drain, depositing some of the fine material in the drain itself, and carrying some of the clay-sized particles out of the drain. The coarser particles of the core material were moved some distance by the leakage but remained in the core. Progressive collapse of the roof of the leakage channels gradually worked upward and finally caused the subsidence craters on the crest and the extensive zones of erosion damage.

The fact that the erosion damage was distributed over several hundred meters of the length of the dam makes it probable that the difficulties were not the result of a "construction accident," which would more likely be localized.

Although there is little doubt about the cause of the trouble, it is not wholly clear why the problem developed. The settlements were not very large and the dam is not high. Although such a narrow vertical core is not a common design, there is no fundamental reason from past experience why the design should not have been satisfactory. Since there is no evidence that any one aspect of the design or construction was primarily responsible for the trouble, it is reasonable to assume that several aspects combined unfavorably to create the unprecedented condition. The writer believes that several of the following circumstances must have had an important influence on the trouble:

1. The very narrow core, with vertical faces.
2. The nature of the core material with its wide gradation from cobble-sized particles down to clay and the character of the clay fraction which did not have sufficient swelling potential to seal the cracks but which was tough and hard and would erode slowly.
3. The existence of the concrete cutoff wall in the foundation protruding up into the bottom of the narrow core.
4. The coarseness and wide range of gradation of the crushed rock used for the vertical drain.

Yard's Creek Upper Reservoir Dam*

The Yard's Creek pumped storage project in New Jersey was constructed in 1963–1965. The upper reservoir is a circular lake of about 5000 acre-ft, retained by a series of peripheral rockfill dams having a total length of about 9000 ft and a maximum height of about 80 ft. On the first reservoir filling dirty leakage emerged at the downstream toe of the dam.

The dam has a symmetrical section with a narrow vertical earth core, rockfill shells, and thin transitions between the core and shells (Fig. 41). The core extends down to a fairly impervious quartzite bedrock and, in general, the shells rest on a layer of dense glacial moraine that covers the bedrock.

The core was constructed of a gravelly, clayey sand (glacial moraine) and the fine and coarse filters were processed sand and gravel, with the gradations shown in Fig. 42. Atterberg limits of two samples of the core material were liquid limit, 25 and 33 and plasticity index, 6 and 10. The rockfill shells consisted of hard quartzite which was placed without water in 2-ft layers and compacted by the hauling equipment and a minimum of two coverages of the tread of a heavy crawler tractor.

The reservoir was filled for the first time at the end of July and early in August 1965. As seen in Fig. 43, starting on July 31, the reservoir was raised about 21 ft in 2 days (approximately between El. 518 and El. 539). During July there had been no appreciable leakage. On August 1, 1965, leakage of several hundred gallons per minute appeared abruptly at the downstream toe of the dam over a length of about 300 ft centered at approximately Station 55, at a point where the dam was generally 50 to 60 ft high.

The surface of the rock foundation under the core over this length of the dam was irregular, 3 to 6 ft high near-vertical steps, running transversely across the core (Fig. 44). Very little concrete had been used during construction to smooth up this surface. Over most of the rest of the length of the dam, the bedrock surface under the core was more uniform and smooth.

Since the pumped storage could be emptied in a few hours, it was considered that there was no great hazard of complete failure and the reservoir was raised to the maximum water level on August 11, in spite of the dirty leakage. But in the early part of August a number of other concentrated leaks emerged downstream of the dam at various other locations on essentially all sides of the reservoir. Except near Station 55, all these leaks were apparently flowing clear water. On August 13, 1965, at full reservoir the total volume of leakage emerging at seven individual areas at the downstream toes of the dams was estimated to be about 1400 gpm, including 400 to 500 gpm in the section of the dam centered at Station 55. The leakage emerging in the vicinity of Station 55 alternately ran very dirty (like "chocolate milk") and clear, in cycles of about 1 or 2 days.

In early August a crack parallel to the longitudinal axis of the dam appeared at the upstream edge of the crest over the upstream filters at approximately Station 54. It had a width of about ½ in. and a length of about 25 ft. This crack was soon obscured by traffic on the crest. No further cracks developed.

In the middle of August 1965 the reservoir was lowered about 30 ft and was not filled again until the end of September (Fig. 43). At the lower reservoir level, during late August and early September, the estimated total volume of seepage emerging down-

*The owner maintained that the cause of the leakage was faulty construction. There was litigation in 1968 and the reader is cautioned that the opinions given are only those of the writer, who investigated the cause of the failure on behalf of the contractor. A description of the general layout of the project and some construction photos are given in *Eng. News Rec.*, June 25, 1964, p. 32.

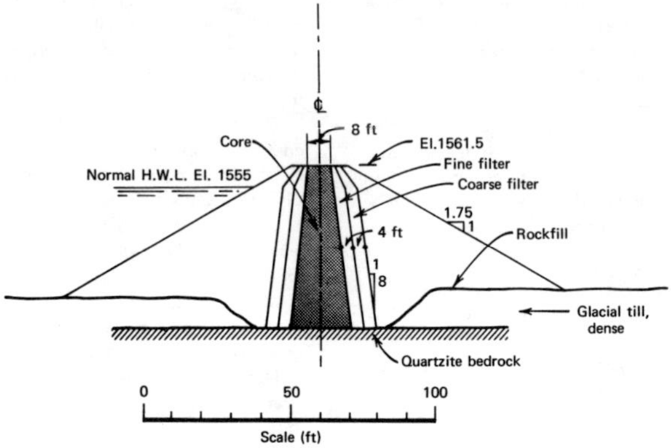

Fig. 41. Yard's Creek Upper Reservoir Dam. Cross section.

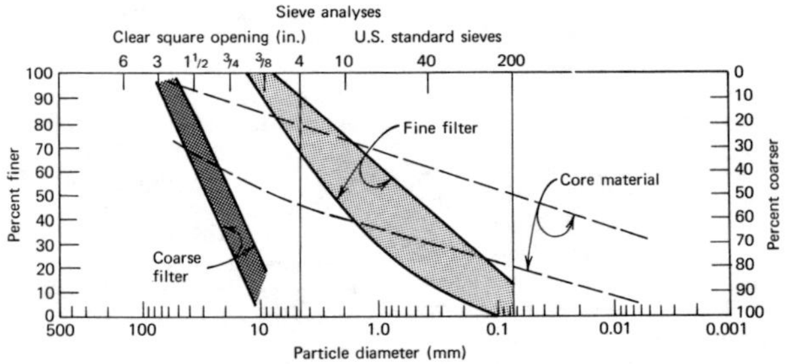

Fig. 42. Yard's Creek Upper Reservoir Dam. Gradation of core and filters.

stream of the dams at various places around the reservoir had decreased from 1400 gpm to roughly 200 gpm, about half of which was coming out in the vicinity of Station 55.

When the reservoir was filled again in the latter part of September and the first part of October, the total estimated leakage from the reservoir was about 1000 to 1200 gpm. This included about 400 gpm emerging downstream from the damaged area in the vicinity of Station 55, which continued running alternately dirty and clear. It was apparent that a considerable volume of fine material was being washed out of the core but no reliable estimate of the quantity was obtained.

In the fall of 1965 a large number of exploratory borings were put down through the core in the area of the dirty leakage. In nearly all of these water was lost in the lower half of the core, especially just above the bedrock. Dye inserted in the holes usually emerged at the downstream toe within a few hours. Efforts to obtain relatively undisturbed samples of the lower part of the core with 5-in. diameter double-tube, rotary core barrels and with thin-walled drive tubes were not successful, since the core contains an appreciable percentage of coarse gravels. It was apparent, however, from this exploratory program that there were extensive zones, especially in the lower portion of the core (generally between El. 1500 and 1525) in which the material was completely disturbed by erosion and in which the fines had been

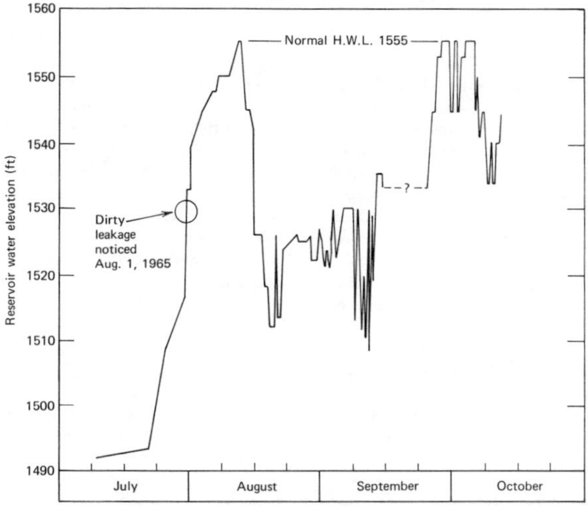

Fig. 43. Yard's Creek Upper Reservoir Dam. Reservoir hydrograph July-October, 1965, during period of leak discovery and studies.

Fig. 44. Yard's Creek Upper Reservoir Dam. Views of rock surface under dam core in general area of dirty leakage, October 1963. (Photos courtesy of D. Ferrer.)

washed away leaving the coarser soil particles behind.

Exploratory borings were also made in other sections of the dam to see if water would be lost. Water was lost in many of these holes. In at least one case it was reliably observed that dye placed in one of these exploratory borings appeared in the reservoir upstream and did not appear at the downstream toe. In a few locations other than the section of the dam at Station 55, leaks emerging at the downstream toe, which were formerly clear, turned temporarily cloudy as exploratory borings in the core were made nearby.

In the spring of 1966 cement-fly ash grouts, with a small percentage of bentonite, were pumped into the core in the zone of worst leakage. The fine filters apparently prevented any large quantity of grout from leaving the core.

On May 31, 1966, in the general vicinity of Station 51, where there had been no leakage for some months following the completion of the grouting, a small leak started. It increased rapidly, reaching a maximum of about 20 gpm within a few hours, and it was very muddy. Within a day or so, a small subsidence crater was found at Station 51 at the upstream edge of the crest directly over the fine filter. This crater was about 2 ft deep, 6 ft long, and 3 ft wide. In the next day or so the leak decreased to only a few gallons per minute of completely clear water. No further subsidence of the crater developed.

In the two years following the summer of 1966, the leakage was clear and the reservoir has been used without restriction. Maximum total leakage at full reservoir was less than 1000 gpm.

Observations in 1965 Test Pits. In the fall of 1965 seven vertical-walled, braced test pits were put down through the core. Only one of these was in the section where muddy leaks developed near Station 55 and it was only a few feet deep. Hence essentially all the information obtained from the test pits was from areas where no muddy leaks had occurred. Generally the test pits were in locations where it was possible to reach bedrock at not excessive depth or where it was desired to investigate some special condition, such as water loss in a borehole. The deepest pits were about 25 to 30 ft deep.

Where the dam was relatively low, the core material appeared homogeneous and dense. In the walls of the test pits it was not possible, for example, to push a sharp pointed steel rod by hand into the typical compacted core material more than a small fraction of an inch. In some of the deeper test pits, however, there were horizontal layers of the core that were much softer. It was possible to push a sharp steel rod into these softer layers several inches without great difficulty. They were approximately 6 to 12 in. thick and 3 to 6 ft apart. These softer layers were quite impervious and appeared in every way identical to the rest of the embankment material except that the water content was higher. Although these softer layers had obviously been wetted selectively by seepage from the reservoir, they were relatively impervious and there was no evidence that concentrated leakage had passed through them.

Generally when the reservoir level was raised above the bottom of a test pit, water seeped in only along the contact between the bedrock surface and the core. Water entered in concentrated streams at the rate of several gallons a minute. No water broke through the walls of the test pits even though in some cases the reservoir level was raised a considerable distance above the bottom of the test pit and the thickness of the core upstream of the pit was only a few feet.

Probable Piping Caused by Boring Water Loss. One of the test pits, located a considerable distance from the section of the dam where the muddy leakage developed, was excavated at the location of an exploratory boring. As this exploratory boring was being put down in the center of the core it was observed that there was a temporary loss of drilling water. This occurred when the casing was approximately 10 ft deep and the bottom of the hole was 12 ft deep. Subsequently, when the drill rods were removed it was found that it was possible to pour about 14 gpm of water into the hole when the casing was kept full (i.e., when the water pressure head was at the elevation of the dam crest). At least several hundred gallons were poured in.

In the walls of the test pit, centered on the location of this boring at a depth of about 12 ft (Fig. 45), there was a horizontal layer of material about 2 to 3 in. thick which could be seen on all four sides, through which muddy leakage had apparently traveled. The layer contained numerous small "pipes," with maximum size of the order of 2 or 3 mm in diameter, which had a lining of soft, wet mud of the same color as the clayey fines of the core material. Directly above and below this layer the core material was dense and impervious.

It was apparent that not only had a concentrated leak passed through this thin horizontal layer, but the flowing water had carried the clay fines with it, creating the small open flow channels observed. As in most of the other individual experiences, it is not

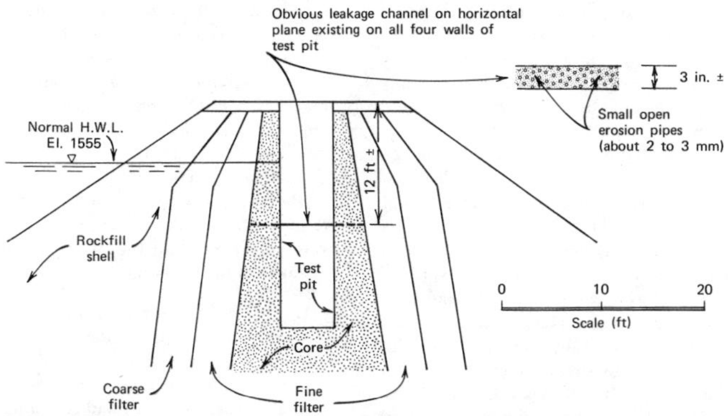

Fig. 45. Yard's Creek Upper Reservoir Dam. Concentrated leak on horizontal plane through core visible in test pit November 1965. Suspected piping in tension crack. *Note:* Test pit was centered at location of boring which had lost water at depth 12 ft.

possible to be completely confident regarding the cause of this leak and piping. The writer believes, however, that the leak can be explained as follows:

1. The vertical earth pressure in the core at the elevation of the leak was considerably less than the weight of a 12-ft column of overburden because of the tendency of the narrow core to settle more than the adjacent filters and to arch and transfer a portion of its weight to the filters and shells. The vertical stress may have been as low as 200 to 500 psf.
2. The pressure of the drilling water in the borehole was considerably in excess of the earth pressure acting on the horizontal plane. Because of this the borehole water created a fissure by hydraulic fracturing and flowed out radially from the hole on a horizontal plane.*
3. When circulation of the drill water was stopped, the width of the horizontal fissure (which may have been jacked open several millimeters under the pressure from the drillrig pump) decreased but remained open wide enough to allow the 14 gpm concentrated leak observed.
4. During the subsequent observation period when several hundred gallons of water were placed in the hole, the water left the boring through the crack and traveled both to the upstream and downstream filters carrying with it some of the fines, which must have gone into suspension in the water and been carried out through the fine filter.
5. Subsequently, when the reservoir level was above the elevation of this fissure, water seeped through the core in the seepage path formed as a result of the boring operation.

It was also conceivable that the leakage and piping observed (Fig. 45) could have been caused by the fact that this horizontal layer had not been compacted, and hence that it was pervious because of its low density. The writer concluded that this was extremely improbable. The layer did not have the appearance of being poorly compacted. Moreover, the core material was sufficiently impervious so that if it had been simply dumped in a pile with no compaction, it is unlikely that a concentrated leak would have developed as a result of the water pressure acting on the upstream face, which was a maximum of about 6 ft at this level when the reservoir was completely full.

*Fluid losses in boreholes caused by hydraulic fracturing have been a common phenomenon, as discussed in considerable detail later. In this case the borings were being made with a drillrig equipped with a positive displacement pump so that when the drill bit was advanced faster than the returning water could wash out the cuttings, the water pressure in the bottom of the hole quickly rose to a value which could approach the maximum pump pressure (several hundred psi). Therefore, it is probable that the pressure head of drill water at the bottom of the hole reached 30 to 50 ft of water, or even considerably more. One of the drillers employed in making the borings through the core at the Yard's Creek Dam made the observation that he could cause a loss of circulating water at any time in any hole by feeding the drill bit downward at a fast rate.

Influence of Soil Type. From the observation of piping seen in the test pit described previously (Fig. 45), it was clear that once a concentrated leak had started through the core the fines of the widely graded core material could be selectively eroded and carried along with the seepage water through the filters, causing segregation and progressive damage to the core similar to that which had been observed at Balderhead Dam. It is remarkable that the core materials at the Balderhead and Yard's Creek dams are very similar, both being very widely graded soils (from coarse gravels to clayey fines). Moreover, the fines in both cases have similar plasticity and both materials are of glacial origin. Heretofore, materials of this kind have been generally considered reasonably self-healing and stable from the standpoint of piping resistance. These two cases (and the Hyttejuvet Dam case described later) indicate that soils of this type might be unstable in the sense that a concentrated leak even of small quantity can wash the fines out of the granular matrix. The soils of these dams are not "well graded" in the sense that this term refers to a soil with gradation curve that is concave upward on the typical semilogarithmic plot, where grain size distribution has the proper proportion of the various particle sizes to be "self-filtering" and to result in maximum density. The soils of the Yard's Creek and Balderhead dams are more properly described as "widely graded." They have a gradation curve that is essentially linear on the conventional semilogarithmic plot (Figs. 37 and 42), and having such a considerable quantity of fines that the compacted density is dominated by the fines and is not very high.

Probable Failure Details. The evidence available did not allow a definite conclusion regarding the cause of the initial concentrated leaks which led to the severe erosion damage. The writer believes that they probably were caused by one or both of the following:

1. The difficulty of obtaining a satisfactory seal between the narrow core (less than 25 ft) and the irregular rock surface.
2. Arching of the core between the upstream and downstream filters or from differential settlement adjacent to steps in the rock surface which caused horizontal cracks in the core. These cracks were probably opened at least partially by hydraulic fracturing.

The records of the fluctuating volume of leakage, which alternately ran clear and dirty, show that erosion damage probably worked its way gradually upward within the core, as the roof of the individual leakage channels progressively collapsed. The mechanics of the erosion damage to the core probably were very similar to that at the Balderhead Dam. At Yard's Creek Dam there was little doubt that the fines were carried through the filter in suspension, as they were at Balderhead Dam, in spite of the fact that the fine filter at Yard's Creek Dam was considerably finer than the drain at Balderhead Dam, because of the following:

1. The leakage water carried a large quantity of fines in suspension for such a long time that it was inconceivable they came from any source other than the dam core.
2. From water losses and sampling in exploratory borings, it was shown that the lower part of the core was damaged by erosion, apparently by removal of the fines.
3. Erosion and segregation of the fines were seen to be occurring in the one leak observed in the test pit described earlier (Fig. 45).
4. Since the average of the fine filter material has a D_{15} size of the order of 0.2 to 0.3 mm, it would not be expected to retain individual clay-sized particles in suspension in the water.

It is also possible that a large portion of the total volume of muddy leakage escaped from the core by passing underneath the downstream filter, as shown schematically in Fig. 46. Since the downstream fine filter was only 4 ft wide and the rock surface was irregular, it would have been difficult to obtain a perfect contact between the fine filter and the rock surface at every point.

The litigation concerning responsibility for the damage deterred free discussion and served to crystallize the differences of opinions among the various interested engineers. In spite of this situation the writer believes that the following observations and conclusions would not be strongly disputed:

1. There had been almost no leakage on the first reservoir filling until the water level reached approximate El. 1535; furthermore, the erosion damage was generally limited to the lower part of the core (between elevations 1500 and 1525), indicating that the initial leakage channels were probably closed or nearly closed before the reservoir filled.
2. The rapid raising of the reservoir during the

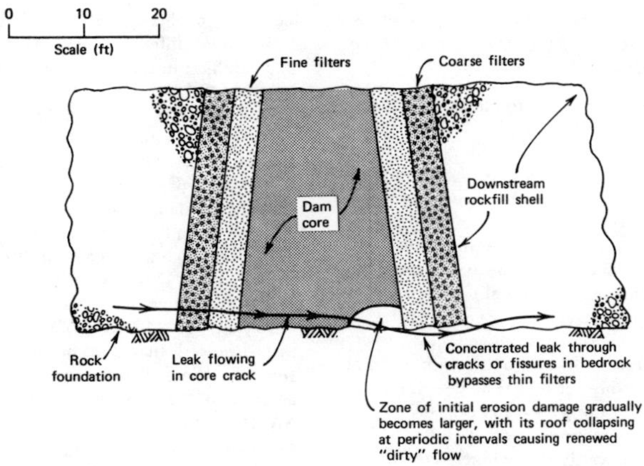

Fig. 46. Yard's Creek Upper Reservoir Dam. Possible piping by leak traveling under filter.

first filling may have played an important role in causing the trouble. It seems likely to the writer that if the reservoir had been filled more slowly, the trouble might not have developed because the core might have been able to adjust to the strains imposed on it without cracking.

3. Although rockfill dams with central earth cores having widths in the range between 30 and 50% of the water head are widely used for high dams in relatively narrow, steep-walled canyons, this core-width ratio is not satisfactory for low dams. Although the hydraulic gradient is not high, the absolute width of the core is so narrow that it is difficult to insure a satisfactory contact over the entire length of the dam.
4. It is not clear how effective the upstream filter may have been as a "leak stopper"; however, the upstream fine filter could have played a major role in preventing more serious damage than occurred. The small crest crater which appeared over the fine filter in April 1966 is evidence that at least some of the material must have been carried into the leakage channels in the core.

Microcracks. The periodic horizontal softened layers observed in the test pits may provide a clue to an aspect of the behavior of thin central core dams which is not presently recognized. The core material in these layers was softer and wetter than would have been expected if the wetting action had been due simply to gradual seepage of water through the soil pores; that is, the material was softer than would have been expected for a compacted laboratory specimen subjected to a long time permeability test.

These layers could have been caused to soften simply because they were not well constructed (too thick, too dry, or poorly compacted). For a number of reasons, however, the writer believes that it is unlikely that the softened layers resulted from construction which was greatly different from the average practice employed for dams: (1) construction was supervised by experienced engineers; (2) there was a continuous program of moisture-density control, the results of which all show adequate compaction; (3) the site is in an area of relatively high precipitation (40 in./year or more) which is distributed fairly uniformly throughout the year so that there is no season of prolonged dry weather; (4) the natural water content of the borrow material was not far from standard AASHO optimum; (5) standard heavy hauling units and rollers were used to place and compact the core material

The writer believes it is likely that the periodic softened layers were caused by the same action as that which caused the dirty leakage and damage in the vicinity of Station 55, differing only in the widths of the cracks that opened. If this is the correct explanation the mechanics of the development of the softened layers were as follows:

1. Before the reservoir filled, arching of the narrow core caused a large portion of its total weight to be transferred to the adjacent filters and shells, so that the average vertical pressure acting on horizontal planes through the core was a small fraction of the weight of a column of overburden.
2. After the reservoir filled the internal stresses in the core changed; however, the average vertical total stress on some horizontal planes was less than the water pressure acting on the upstream face of the core at the same elevation.
3. Water penetrated the core on these horizontal planes in very small cracks. The penetration of water increased the stresses in the core until a condition of equilibrium was reached, after which the crack widths did not increase further.
5. At this equilibrium condition, two types of action may happen. On the one extreme, if the crack is wide enough and the volume of leakage great enough, internal erosion and piping commences. At the other extreme, the crack width is very narrow and the volume of flow through it is so small that piping does not get started; the embankment material adjacent to the crack swells slightly or softens and collapses into the crack.

There is no theoretical or practical reason to doubt that very narrow cracks (microcracks) develop in earth dam embankments. These may have widths only several times greater than the average pore diameter of the compacted soil, and they may not be visible to the human eye. Nevertheless, the width may be sufficient to provide a privileged path of seepage and allow the entry of water under pressure. Because of the narrow width, the velocity of the penetrating water is low and there is no tendency for piping.

Hyttejuvet Dam

Hyttejuvet Dam, about 90 meters high and with a crest length about 400 meters, was constructed in 1964–1965 in Norway. It has a thin, vertical-sided central earth core, rockfill shells, and thick gravel transition zones (Fig. 47). During the first reservoir filling in the summer of 1966 a concentrated leak of dirty water emerged abruptly at the downstream toe when the reservoir was nearly full. Investigations led to the conclusion that the leak developed through a horizontal differential settlement crack, resulting from arching in the core. The details are described briefly here.*

The core was constructed of a gravelly, clayey sand (glacial moraine) with a gradation very similar to that of the core materials at Yard's Creek and Balderhead dams (Fig. 48). The Atterberg limits were liquid limit, 21 and plasticity index, 6. The core was placed in 25-cm-thick layers and compacted at a relatively high water content with five passes of the treads of a 15-ton crawler tractor, following a common practice in Norway and Sweden (see Sherard et al., 1963, pp. 622–629). Construction control tests showed that the placement water content generally ranged between 1.0 percentage point below and 2.0 percentage points above Standard AASHO optimum. The dry density ranged generally between 97 and 103% of Standard AASHO maximum density, with an average of 99%.

At the end of the 1964 construction season, when the dam was about at midheight, the design was changed and the core made much thinner, as shown in Fig. 47. The change was made because the construction pore pressures being measured in the core were higher than had been anticipated.

The reservoir was filled rapidly the first time starting in May 1966. As seen in Fig. 49, very little leakage (less than 2 liters/sec) appeared at the downstream toe until the middle of August 1966 when the reservoir level had reached to within 7 meters of the high water level (El. 737).† As the reservoir was raised above this level in the last 15 days in August the leakage increased abruptly, reaching a maximum of about 63 liters/sec by the end of August, when the reservoir had reached approximate El. 740. The leakage water, metered with a weir at the downstream toe (Fig. 47), was gray colored and contained approximately 0.1 g of fines per liter. At the beginning it was not clear that the dirty flow was caused by piping of the core, since the water could have been picking up fines as it passed through the rockfill before emerging at the downstream toe. Later, primarily because of the persistence of the dirty leakage, the investigators concluded that most of the turbidity was caused by piping of the core.

*Details of the design and construction and of the subsequent problem and repairs are given by Kjaernsli and Torblaa (1968). See also Bjerrum (1967). Supplementary details were supplied through the courtesy of Mr. B. Kjaernsli of the Norwegian Geotechnical Institute.

†Dimensions in meters.

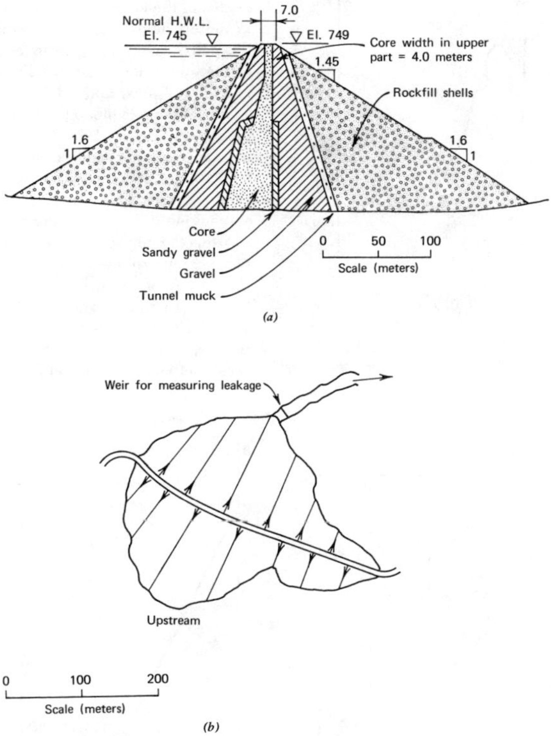

Fig. 47. Hyttejuvet Dam. (*a*) Cross section. (*b*) Plan. (After Kjaernsli and Torblaa, 1968.)

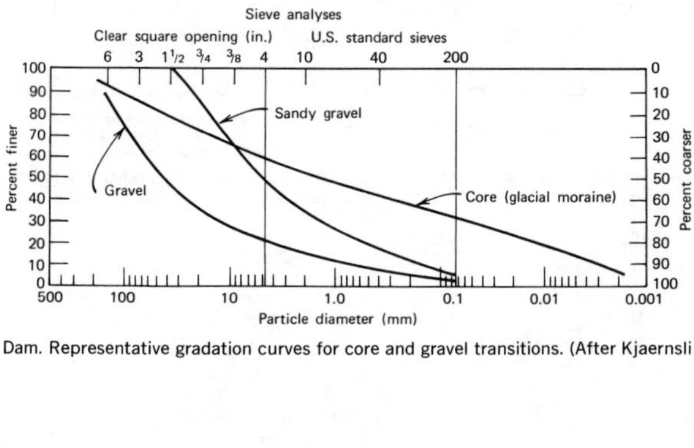

Fig. 48. Hyttejuvet Dam. Representative gradation curves for core and gravel transitions. (After Kjaernsli and Torblaa, 1968.)

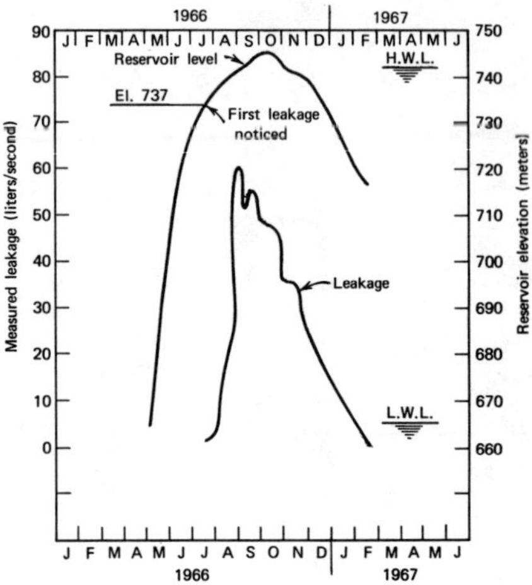

Fig. 49. Hyttejuvet Dam. Reservoir level and measured leakage 1966-1967. (After Kjaernsli and Torblaa, 1968.)

During September and early October 1966, the reservoir rose gradually another 5 meters to the high-water level. During this time measured leakage decreased from about 62 to 45 liters/sec (Fig. 49).

A single earth pressure cell was installed during construction to measure total vertical stress in the center of the core 21 meters below the dam crest, or 17 meters below high-water level. As seen in Fig. 50, this cell measured a pressure of about 17 tons/meter^2 at the end of construction. The pressure decreased to about 14 tons/meter^2 by June 1966 and increased again to about 23 tons/meter^2 when the reservoir was full in mid-October 1966. These measurements made it seem probable that an appreciable portion of the total weight of the core has been transferred to the shells by arching. (Since the weight of 17 meters of embankment is of the order of 40 tons/meter^2.)

After the abrupt development of the leakage a number of exploratory borings were made through the center of the crest at various points along the 350-meter length of the dam. In most of these, water was lost from the holes into the core, especially in the range of depth from 10 to 20 meters below the crest (between El. 730 and 740).

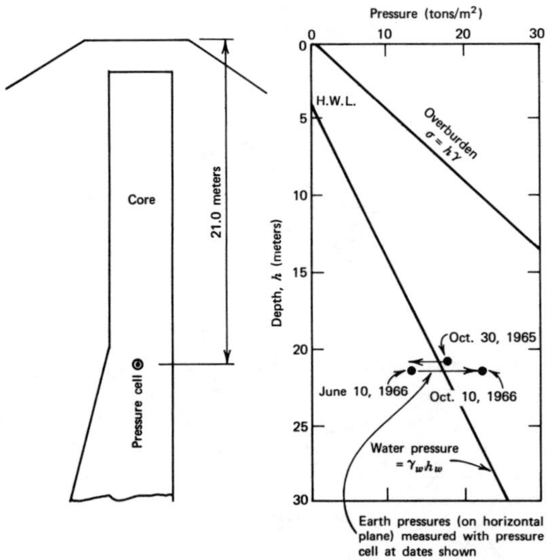

Fig. 50. Hyttejuvet Dam. Measurement of earth pressure in core shows probable stress relief due to arching. (After Kjaernsli and Torblaa, 1968.)

From the leakage observations, the low values of vertical total stress measured by the pressure cell and the fluid losses in the borings, the investigators concluded that there was a reasonable likelihood that the leakage was passing through open horizontal cracks in the core, and that grouting might be a satisfactory remedial measure. Some tests were made to see if it would be possible to inject grout into the core without grouting the gravel transition zones upstream and downstream. This was done by drilling holes in the gravel zones and testing them first with water pressure and then by attempting to pump in grout. These tests confirmed that the holes would take a large volume of water but no grout. From these tests, and from the gradation of the gravelly layers (Fig. 48), it was concluded that if the core were grouted the adjacent transition layers would act as a filter to prevent any large quantity of cement grout from leaving the core.

The core was grouted during the summer of 1967 from holes drilled to bedrock or to a maximum depth of 30 meters. Primary holes were spaced 50 meters apart, usually with one or two secondary holes between. The holes were grouted starting at the bottom working upward in stages of 1 to 3 meters. Pressures of 1 to 5 kg/cm^2 (measured at the top of the dam) were used, with the higher pressures used for the greater depths. Some grout was taken in all holes; however, the largest takes occurred over the right half, where the dam was 20 to 30 meters high, and especially just above the bedrock surface.

The total grout consumption in the core consisted of 67 tons of Portland cement, 33 tons of bentonite, 308 tons of sand, and 535 tons of water. If this volume is reduced by 40% to account for consolidation, possible waste, and loss of water, the calculated volume of grout in the core becomes 420 $meters^3$. Assuming a width of core of 5 meters and a length of 350 meters, this volume would fill horizontal cracks corresponding to a combined thickness of 24 cm throughout the core.

Monuments on the crest settled approximately 18 cm between the fall of 1965 and the fall of 1967. During the summer of 1967 the grouting caused some of the monuments to heave slightly (maximum of about 1 cm).

The reservoir was full or nearly full during the period of grouting in the summer of 1967 and the leakage gradually decreased. When the grouting was completed in the fall it was no longer possible to measure the leakage since the weir was covered with snow. At full reservoir in the summer of 1968, the rate of leakage was 15 liters/sec and the water was clear.

In August 1970, as the reservoir was being lowered from the maximum level, the measured leakage abruptly increased to about 20 liters/sec and became visibly turbid again. Studies of the condition are currently (1970) under way.

Although no test pits had been put down in the dam and no cracks were seen on the surface, the engineers investigating the problem formed the strong opinion that the most likely explanation for the trouble was that leakage had broken through a horizontal crack made possible by arching of the core between the upstream and downstream gravel zones. In common with most of the other individual experiences, it is not possible to know the cause of the piping with complete certainty. The writer believes, however, it is probable that the leakage developed through a horizontal crack which was caused to open by the pressure of the water acting on the upstream face of the core. A number of aspects of the leakage would be very difficult to explain from any other cause:

1. The total measured leakage was less than 2 liters/sec until the reservoir reached El. 738 and then increased abruptly to more than 60 liters/sec by the time the reservoir reached El. 740. This could only be accounted for by either the existence of a large open leakage channel through the dam core (or perhaps the rock abutment) at approximate El. 738 or by the abrupt opening of a crack in the core below El. 738.
2. The decrease in measured leakage during September and the first part of October, as the reservoir was raised to El. 745, would be difficult to explain if the leakage were passing through the rock abutment but would be explained readily by partial collapse of a crack in the core.
3. The leakage continued to flow appreciably until the reservoir was lowered below El. 716. This can only be explained by the existence of a leak through the core at or below El. 716. Hence the crack must not have been in existence when the reservoir was raised above El. 716 for the first time.

It is not easy to explain why the crack developed. The total postconstruction settlement of the crest was relatively low. It seems likely that the most important factor contributing to the cracking was the narrow width and vertical walls of the core. The relatively rapid rate of reservoir filling (about 65 meters in 60 days in May and June 1966) was probably an important contributing factor.

Furthermore, it may be significant that the core was constructed of a glacial moraine with a wide grain size distribution (extending from coarse gravels to clay-sized particles). This soil is a very similar soil to those used for the cores of Yard's Creek and Balderhead dams. This provides additional support for the conclusion that these soils are not self-filtering and may be susceptible to a type of piping in which the fines are carried away by the leakage water, leaving the coarser particles behind.

The fact that a large amount of the total grout take was just above the bedrock contact in holes at the right end where the dam height was less than about 25 meters makes it seem likely that most or all of the initial leakage and piping may have been at this location. If this is true, the problem is very similar to that of the Yard's Creek Dam described earlier. The experience at the Hyttejuvet Dam supports the conclusion that thin (less than 50% of the water head), near-vertical cores are not desirable for dams less than 75 ft high.

7. FLUID LOSSES FROM BOREHOLES

Experiences

The practice of drilling holes into completed sections of impervious embankment zones to install piezometers and for other purposes has become increasingly frequent in recent years. Total or partial loss of drilling fluid has often occurred in one or more holes, indicating that the fluid has escaped into the surrounding embankment through a crack or other channel. As in the examples of the Shek Pik and Graminha dams, explorations to determine the cause of the fluid loss are frequently inconclusive and the dam subsequently displays no sign of distress when put into service.

Sufficient experience is now available to conclude that such water losses are common in well-constructed dams, in which there were no open cracks before the boring was made. In these cases the fluid flows out from the boring into a crack which was caused to open by the fluid pressure. This may be a new crack or an existing, but closed, crack. The term "hydraulic fracturing" is used for both cases.

In the following section a number of such experiences are described briefly to show that the phenomenon is widespread and to illustrate the range of conditions under which it has been observed. Also a brief theoretical treatment is given which shows that such losses may be expected to occur when the fluid pressure in the borehole exceeds the minimum principal total stress in the adjacent embankment, which is not a rare condition even in well-constructed dams.

The amount of fluid lost from the borings varies considerably. A few gallons per minute is common and frequently 30 to 50 gallons or more per minute have been poured into a hole for several hours without building up pressure. In some cases as much as several thousand gallons of fluid have been put into a single hole in less than a day and subsequent investigations with test pits failed to reveal any clues as to where the water went.

The Garrison Dam on the Missouri River is a typical early example where holes were drilled through the partially completed embankment in 1951 and 1952 for the purpose of grouting the underlying formation: "A high percentage of the holes exhibited a major loss of drill water within the embankment.... test pits to 25 feet deep in these areas failed to show any cracks" (Lane, 1963).

Another example is the Lovewell Dam in Nebraska, completed in 1957. After the dam was finished, an exploratory hole was drilled through the embankment near the downstream edge of the crest for the purpose of testing the strength of the soft foundation clay:

> The engineers were surprised to find that a considerable amount of drilling water was lost through the hole into the impervious embankment at a level of about 5 feet above the clay foundation. Water was poured into the hole at a rate of 100 gpm and completely disappeared. Later, 35 additional holes were drilled through the embankment on the center line of the crest at various points in the same general area, and water was lost in 25 of them in the lower 3 to 5 feet of the embankment. The loss varied from a minor quantity to a maximum of 160 gpm; one hole took 7700 gallons in 24 hours without filling" (Sherard, Woodward, Gizienski, and Clevenger, 1963, p. 199).

Later a test pit was put down at the location of one of the exploratory holes where a considerable volume of water had been lost and the material exposed in the walls of the pit was carefully examined. At all points the embankment appeared uniformly well compacted, without cracks, and with good bond between the layers.

Dayte (1968) describes a "heavy loss of drill water" in borings in the impervious zones of the Girna and Vir dams in the state of Maharashtra, India. At the Vir Dam the loss occurred in the vicinity of a 20-ft-high "step" in the foundation surface, and Dayte states, "it is not conclusively established whether the loss of drill water was in cracks in the embankment due to differential settlement."

To the writer's knowledge, the first recognition and substantial proof that borehole fluid caused cracks to open was in 1965 by the engineers of Coyne and Bellier, following analysis of the events at the Djatiluhur Dam (Fig. 29 and accompanying description). At the time of this experience, the investigating engineers, Post and Londe, were also involved in basic studies of the similar action by which fissures are opened in rock formations during water pressure tests in boreholes and high pressure grouting. The experiences at Djatiluhur Dam left no room for reasonable doubt that water in the boreholes was jacking cracks open in the clay core. Also it was shown conclusively that the action was "elastic" in the sense the crack remained open only when the hole was kept full of water and the crack width varied with the fluid pressure.

Wilson and Squier (1969) referred to water losses in boreholes at the Aksombo Dam in Ghana and at the La Villita Dam in Mexico. In both cases the borings were put down to install instrumentation soon after completion of the dam. At the La Villita Dam, the experience was very similar to that at Djatiluhur Dam:

> This phenomenon was a matter of concern until subsequent investigations at La Villita revealed the probable mechanism. Here a boring was being accomplished with a rotary drilling technique using a heavy bentonitic drilling fluid. At a depth of about 40 feet, there was an abrupt loss of fluid. At approximately the same time, malfunctioning of instruments at about the same elevation but some 100 feet distant in a direction parallel to the axis indicated that fluid, presumably drilling mud, entered the instrument system. The borehole was flushed out with clean water and the rate of drop of the water surface observed with time. It was found that the water surface in the borehole stabilized within minutes at an elevation such that the water pressure at the bottom of the hole was nearly identical with that recorded at a nearby pressure cell in the core material, oriented such that its face was in a vertical plane parallel to the axis. It was evident that the coefficient of horizontal earth pressure in the vertical core against this plane was of the order of 0.40, which is less than the fluid pressure of the drilling mud at the same depth. The low value of earth pressure resulted from lateral spreading of the upstream and downstream shells, which were granular. In any event, it is apparent that the pressure of the drilling mud, which locally may have been increased by pumping pressures and the up and down motion of the drill bit, exceeded the earth pressure against a vertical plane parallel to the axis. Once a crack started, it quickly spread longitudinally, but once the fluid pressure reduced to that of the soil itself, the crack immediately closed. It is believed that the temporary opening of such a crack, although undesirable, has no harmful effect on the imperviousness of the core, and as the reservoir is filled the earth pressure probably will increase. Fortunately, the earth pressure against transverse vertical planes is greater (except possibly near the abutments); therefore, any such cracks that may be opened up by the drilling operations will spread in a longitudinal direction only.

The experience at El Isiro Dam described earlier, where grout penetrated into the core, is particularly interesting in this context because of the observation that the grout-filled cracks were generally vertical and parallel to the longitudinal axis of the dam. The investigators were quite confident that their observations in the test pits could most reasonably be interpreted as showing that the grout flowed considerable distances horizontally through near-vertical, longitudinal cracks which were caused to open by the fluid grout. Hence the action of "hydraulic fracturing" in causing vertical longitudinal cracks was similar to that believed to have been caused by drilling mud at La Villita Dam by Wilson and Squier (1969).

As part of the repairs at the Wister Dam, described previously, a considerable number of holes were grouted, generally by gravity pressures only, or with pressure of less than 5 psi at the top of the hole. As the grouting progressed, many cracks appeared on the crest and upper part of the slopes. The cracks were almost all parallel to the longitudinal axis of the dam, appeared to be nearly vertical, and had widths generally varying from a hairline to about $\frac{1}{4}$ in. Adjacent to holes being grouted under pressure, grout sometimes emerged at the surface through the cracks. Hence in this experience we have another example showing hydraulic fracturing occurring primarily in vertical and longitudinal planes, which should generally be perpendicular to the direction of minimum principal stress, in the upper elevations of the valley section of a dam.

As discussed earlier, vertical grout seams, transverse to the longitudinal dam axis, were exposed in the test pit in the downstream inner transition at the Matahina Dam. Since the test pit was located over the steep part of the abutment, the embankment in this zone would have been expected to stretch in the longitudinal direction, and the minimum principal stresses would have been expected to act on vertical planes transverse to the longitudinal axis of the dam. Hence, if grout were going to force its way upward into this portion of the embankment, the creation of vertical, transverse fissures and grout seams is easily explained.

The investigators of the leakage at the Hyttejuvet Dam were particularly cognizant of the fact that some of the water losses in their boreholes in the core may have been through cracks opened by the pressure of the fluid in the borehole:

> It should be mentioned that it is quite possible that hydraulic splitting of the core took place as a result of the tests in the boreholes. The pressure in the bottom of the hole when the casing was entirely filled with water, and even more so with grout, would be higher than the pressure created by the water in the reservoir at the same elevation. The results of the water tests should not, therefore, be used to conclude that the tests proved the existence of open cracks. It is more likely that the tests produced cracks (Kjaernsli and Torblaa 1968).

The fluid pressure in a borehole during drilling can be very much higher than the weight of the static column of fluid. High pressure caused by partial obstruction of the return flow of the drilling fluid has been the suspected cause of hydraulic fracturing in a number of cases.* At the Livingston Dam near Houston, Texas, completed in 1969, a large number of borings were put down through the clay core during construction to install piezometers in the foundation. Water losses during drilling were experienced in more than half of the borings. In some of them all the drill water was lost, especially in the lower part of the impervious clay core just above the clay foundation. In one boring a total of 5500 gallons was lost. At the Livingston Dam it was observed that the speed of advancing the drill hole had an important influence on the probability that water would be lost. When a rotary fishtail bit was advanced downward at a more rapid rate than the soil cuttings could be washed out of the hole, a "wall pack" of cuttings formed around the drill stem near the midheight of the hole, restricting the return water flow and causing the development of high pressures:

> In some cases, packing develops to the extent that the entire annular space between the borehole and the drill stem is filled with tightly packed clay cuttings. Rotation of the drill stem in this tightly packed hole can be continued and the drill stem can be moved up and down without diffculty.... In many cases it is necessary to pull the drill string and manually clean the sticky clay from the pipe, and then slowly redrill the lower 10 to 15 feet of the borehole.*

In several cases where a wall pack developed and where water was lost from the boring after the hole was completed, the water loss was much less than the loss during drilling.

At Livingston Dam it was assumed that there were no open cracks and that the water losses were being caused by hydraulic fracturing. The reservoir is almost full and the seepage pattern is normal (1970).

In some cases, fluid losses are encountered in many adjacent borings. Occasionally large losses have been experienced in one boring but not in nearby borings. Following completion of the Hartwell Dam on the Savannah River in Georgia in 1961 a boring was put down in the crest of the essentially homogeneous earth dam on the right abutment at a point where the dam was about 100 ft high. During the drilling of this hole all the drill water was lost at a depth of approximately 20 ft. The hole was carried down to a depth of about 55 ft, and loss of drill water was experienced continuously between 20 and 55 ft. Subsequently steel casing was installed to the bottom of the hole (55 ft), and as soon as it was washed out, water was lost again:

> We then attempted to seal the crack by pumping heavy drilling mud into the hole. Four pits of mud holding approximately 50 cubic feet each were pumped in the hole. After the first two pits of mud were pumped in without regaining circulation, we thickened the mud considerably in the last two pits by adding a small amount of cement. These were also pumped into the hole without regaining circulation. We then extracted the casing from the hole and attempted to grout the hole with cement and gravel. Ten 100-lb. sacks of cement were mixed with an equal portion of gravel. This was poured down the open hole. Use of the cement and gravel mixture finally plugged the crack and the hole was backfilled. We then moved 10 feet away and drilled another piezometer. No crack was encountered in this hole.†

About six other piezometers were subsequently installed in the embankment not far from the first and no fluid loss was observed. No further investigations

*An experience has even been reported in which a hole being drilled with compressed air to blow out the cuttings lost circulation in the impervious section of a dam. It is not difficult to imagine that the air pressure in the bottom of the hole approached the maximum pressure of the compressor if the return flow were blocked by squeezing of the hole or by an accumulation of cuttings (Johnson, 1970).

*Personal communication with Mr. R. F. Reuss, National Soil Services, Inc., Houston, Texas.

†Personal communication with Mr. William V. Conn, South Atlantic Division, U.S. Army Corps of Engineers.

or repairs were made and the dam has subsequently performed satisfactorily. Piezometers show a normal seepage pattern through the dam.

A number of engineers have suspected that water was lost from borings at various sites for reasons other than differential settlement cracking and hydraulic fracturing. Among these reasons are the following:

1. Embankment zones which dried and cracked in the sun during construction and were subsequently covered up and left in place.
2. Embankment zones consisting of alternate layers of appreciably different rigidity, resulting either from differences in material properties or construction water content or both. When a more brittle layer is sandwiched between relatively plastic layers, any deformation of the entire mass is liable to produce localized fissures in the brittle layer.
3. Inadequate compaction of individual layers or poor bond between layers.
4. Horizontal shear planes caused by a heavy roller traveling on an embankment that has a high water content. At the Livingston Dam, where water losses were experienced in more than 50% of the borings in the clay core, the material was compacted well above Standard AASHO optimum and extensive systems of these shear planes were created by the roller (see photo of typical shear plane of this type in Sherard et al., 1963, p. 101.)

The writer has little doubt that all four may increase the probability of hydraulic fracturing in some cases.

Particularly in recent years as the wheel loads on earth-hauling equipment have increased, cracking of the construction surface from travel of the equipment has become more frequent and more severe. These cracks are accentuated by drying, e.g. see Fig. 51. The cracks may be eliminated by scarification of the surface before placing the next layer. Usually no investigation is made of the depth of the cracks and they are generally considered unimportant. It seems likely that in some cases these cracks extend downward at least several inches below the depth to which the material is scarified or reworked before placing the next layer. Undoubtedly, as the embankment is constructed to greater heights any such open cracks are squeezed shut; however, they still may constitute a "weak point" (zone of low minimum principal stress) where hydraulic fracturing can most easily occur.

Drying cracks have sometimes been very severe on transverse slopes between two sections of a dam that are constructed at different times, such as on the slopes of closure sections. At the Ilha Solteira Dam, a 60-meter-high homogeneous clay dam currently (1970) under construction for a major hydroelectric project in Brazil, drying cracks on the transverse slope between two sections of the dam were found to extend to a depth of at least 2 meters and resulted in deep erosion gullies from rain runoff. In order to minimize this problem on one of the higher transverse slopes, a sprinkling system was established to keep the surface continually wet.

Also at the Ilha Solteira Dam, a large water loss was encountered in a single boring put down in the clay embankment for the purpose of installing a piezometer. The water loss, at a depth of 11 meters, was very near the depth at which the boring passed through a 4:1 transverse juncture between two embankment sections constructed at different times. This boring was located only about 3 meters away from a vertical telescoping pipe settlement measuring instrument, around which the embankment had been hand compacted. It was estimated that about 20 meters^3 of water from a water truck was poured into the hole in the first hour or so and more water was pumped in on the subsequent days.

Extensive investigations of the water losses were made, including 60 additional borings on an 8-meter by 10-meter grid in the adjacent embankment, three test pits, and the use of radioactive isotopes to trace the flow of water. No water loss was experienced in any of the additional 60 borings. The test pits encountered some erosion pipes with diameter of the order of 10 cm at the depth where the water loss occurred in the boring. Exploratory tunnels were opened horizontally along these erosion pipes for a few meters until they disappeared. In spite of the careful investigations it was not possible to conclude definitely that there was one single cause of the trouble. It was believed that the water loss probably resulted from a combination of drying cracks and poor compaction in the localized zone around the settlement measuring pipe. It seems probable to the writer that this is another case of hydraulic fracturing. The water probably broke out of the boring into the clay embankment at a point of localized low minimum principal stress, which may have been related to drying cracks or to the hand compaction around the settlement measuring instruments, or both. Since water was lost only in the one boring and not in the other 60, it must be concluded that the

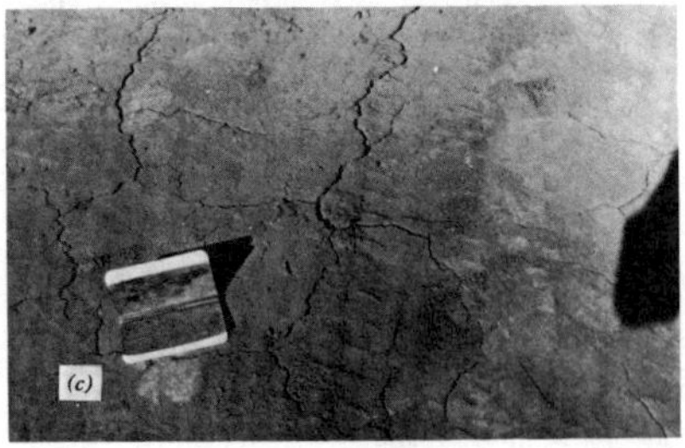

Fig. 51. Cracks commonly seen on construction surfaces of impervious zones of well constructed modern dams. (*a*) Dam core consists of well-graded silt, sand, gravel mixture, cohesionless. Terrace alluvial deposit. Cracks from rapid travel of heavy trucks, probably accentuated slightly by drying. (*b*) Dam core from silty sand, cohesionless. Disintegrated granite. Cracks from truck travel not influenced by drying. (*c*) Dam core from medium plastic clay. Drying cracks in hot weather after about an hour. (Photos, J. L. Sherard.)

stresses in the embankment generally were high enough to prevent hydraulic fracturing and the single water loss must have been a result of the localized zone of the low stress.

A common characteristic of these experiences has been that several thousand gallons of fluid have been poured into the holes with no visible return, raising the question, "Where does this fluid go?" Such large quantities of fluid cannot be absorbed by swelling of the core material but must finally find an exit in one of the pervious zones of the dam. After exiting in a pervious zone, such as a chimney drain or the upstream or downstream pervious shells, the water flows down to the foundation. Since the foundation surface is usually irregular, even a large volume of fluid, such as 10 or 20 thousand gallons, can be trapped in closed depressions and held within the base of the dam without appearing in the form of a leak at the dam toes.

It is impossible to say how common the conditions may be in modern dams which would allow water loss in boreholes, since there are no reliable statistics available on the number in which fluid was lost and since boreholes are not drilled in most dams. The writer believes it is probable that if we were able, hypothetically, to drill several holes into the impervious sections of a large number of our modern dams chosen at random, losses of circulating water during the drilling would be observed in a large percentage, say 25% or even more. This guess is based primarily on the frequency with which the problem has been observed in the last few years and also on theoretical considerations of the fluid pressure needed in a borehole to cause hydraulic fracturing as discussed below.

Analysis

Hydraulic fracturing of rock formations to increase the yield of oil wells has been common practice for more than 25 years.* It is done simply by subjecting boreholes to fluid pressures which are high enough to open cracks and then pumping sand in to prop the cracks open. The pressures needed to open cracks in a formation at a given level are often much less than the weight of a vertical column of overburden. The following conclusions have been drawn from an analysis of hydraulic fracturing, assuming the rock to be an elastic medium (Hubert and Willis, 1967):

1. The fractures will always open on the plane perpendicular to the minimum principal total stress.
2. If the ratio of maximum to minimum principal stresses in the formation before drilling the hole is in excess of approximately 2, the borehole pressure required to create the crack must be equal or slightly greater than the minimum principal stress. If the ratio of principal stresses is less than 2, the fluid pressure in the borehole must be somewhat greater than the minimum principal stress to get the crack started.
3. The borehole fluid pressure needed to hold open and extend the fracture is equal to or slightly in excess of the minimum principal total stress, regardless of the principal stress ratio.

*It has been common European practice to grout rock foundations at pressures which are sufficiently high to cause deliberate fracturing of the rock.

These conclusions apply equally to a porous rock and to a hypothetically solid rock without pores.

A rough idea of the quantity of water that will leave a borehole through a crack of a given width can be obtained from theory developed to evaluate the permeability of rock formations. Assuming that a single crack with a constant width e intersects a borehole with a radius r, the quantity of water loss (laminar flow) from the borehole q into the crack will be (Sabarly 1968)

$$q = \frac{\pi}{6\eta \log_e (R/r)} Pe^3$$

where η = viscosity of the fluid, P = the borehole pressure, and R = a distance from the borehole at which the fluid pressure in the crack is equal to 0.

As seen in Fig. 52, in which this equation is plotted for a borehole pressure of 1.0 kg/cm^2, the ratio R/r has relatively small influence on the water loss compared with the crack width, and a flow from a borehole of more than 100 liters/minute can be expected in a crack with a width of less than 1 mm.

Figure 53 shows the distribution of horizontal stresses in the upstream-downstream direction acting on the centerline of a 100-ft-high homogeneous embankment with 2.5:1 slopes, computed by the finite-element method using reasonable assumed elastic properties and density (Clough and Woodward, 1967). As seen in this plot, the horizontal stress, which is the minimum principal stress, is less than the hydrostatic pressure of water in a borehole drilled through the crest over most of the height of the dam.

Although the foregoing theories have serious limitations when applied to water losses in an embankment dam, it is clear that the results do not conflict with the experience described in this paper.

Hydraulic fracturing cannot occur where the pore water pressure in the surrounding embankment is greater than the fluid pressure in the borehole, since the minimum total principal stress in the embankment is always greater than the pore water pressure. Furthermore, a dam core is likely to be more plastic and deformable if the pore pressures are high, and less likely to sustain localized zones of low total stress. From this reasoning we can conclude that the lower the construction pore water pressures, the more likely it is that water losses will occur in borings.

In embankments where the pore pressure is very

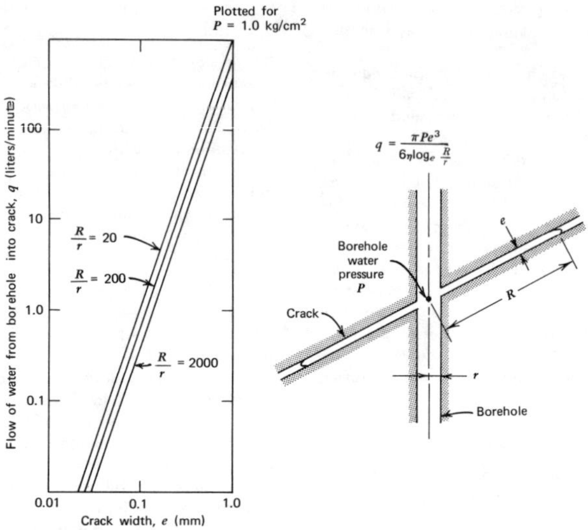

Fig. 52. Calculated rate of borehole water loss into a crack. (After Sabarly, 1968.)

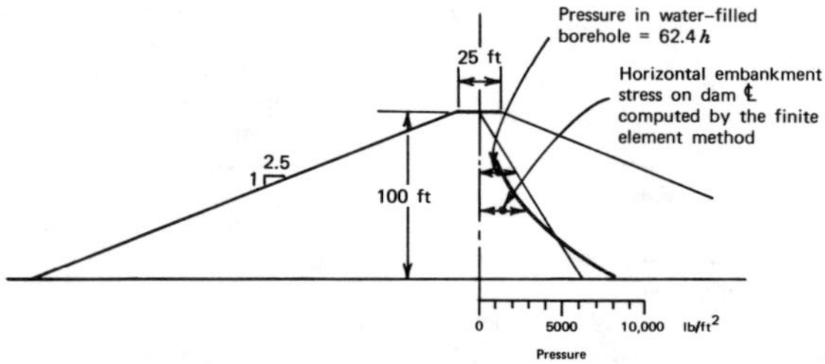

Fig. 53. Comparison of computed minor principal stress on vertical axis of symmetrical elastic dam with water pressure in boring drilled through crest. Assumptions (from Clough and Woodward, 1967): (1) embankment density = 135 pcf; (2) Poisson's ratio = 0.40; (3) no strain in dam foundation.

low or negative, the high fluid pressure inside a deep borehole will create a high hydraulic gradient in the first instant as the water commences to leave the hole through the newly opened crack. As the fluid-filled crack is extended to progressive distances from the borehole by hydraulic fracturing, the gradient progressively decreases, approaching an equilibrium condition. If the crack extends to a pervious zone, an equilibrium condition will be reached in which the quantity of water that can be poured into the hole is constant and less than the rate of flow of the initial water loss (assuming no piping occurs). The progressive decrease in the rate of boring water loss at constant head experienced at the Shek Pik Dam (Fig. 16) is probably caused by this action.

Recently it was proposed in England (Vaughan,

1970a, 1972) that it might be practical to measure the minimum principal stress acting within the embankment by purposely inducing hydraulic fracturing.* Figure 54 shows the results of a research test made in a dam core in which the pressure in a simple piezometer was gradually raised (with a tank of gas under pressure) until a fracture developed. Very little flow occurred in this test until the pressure exceeded about 52 ft of water; however, when the pressure was increased only a small amount above this threshold, the flow increased abruptly, undoubtedly through a small crack opened by hydraulic fracturing. As the pressure was subsequently decreased, the flow decreased, reaching zero at a pressure of 20 ft of water. At this point the crack was evidently closed again. The results of this test indicate that the minimum principal embankment stress in the vicinity of the piezometer approximates the weight of a 20-ft water column. In some similar tests, the pressure causing the initial crack was much closer to the pressure at which the crack closed again.

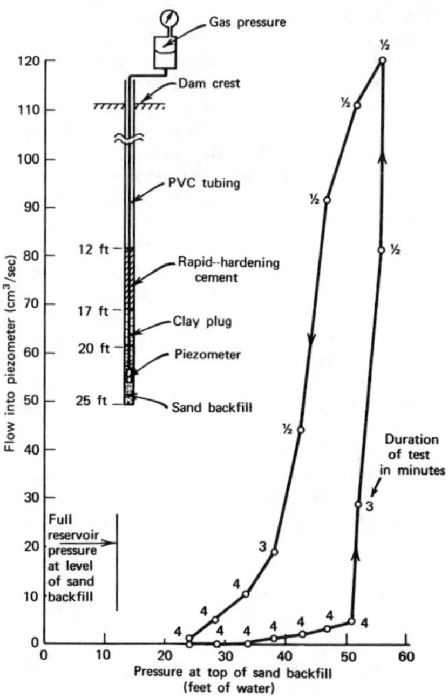

Fig. 54. Research test in which hydraulic fracturing is artificially induced in dam core by applying back pressure to a sealed borehole section. (After Vaughan, 1972.)

*See also Bjerrum and Andersen (1972) and Bjerrum *et al.* (1972) for descriptions of hydraulic fracturing purposely caused by raising borehole fluid pressures.

Evaluation

Losses of drilling fluid are commonly experienced in borings drilled during or just after construction in the impervious sections of earth dams. The fluid loss characteristically occurs abruptly during the drilling, when the hole reaches a certain depth.

Both experience and theory indicate that in the majority of cases there probably was no open crack in existence before the hole was drilled and a crack was formed by the pressure in the borehole fluid. If cement grout is placed in open holes drilled in a dam embankment, the greater density of the grout (and any added pump pressure) increases the likelihood of hydraulic fracturing.

It is not difficult to imagine that the cracks through which water has left a boring would not be found in test pits, because the crack may only be 1–2 mm in width. By the time that the test pit is excavated, which would be at least several days after any water had been poured into the borehole, the thin crack would have closed and would probably not be detectable. Furthermore, it would be difficult to distinguish between a very narrow crack through which the water had flowed and the fine cracks which always exist in the walls of test pits due to drying and stress relief. For example, at the Djatiluhur Dam, water poured into a nearby borehole entered the walls of a test pit through a horizontal crack, which was almost invisible and certainly would not otherwise have been detected.

If the rate of water loss out of a boring is large, and if water is poured into the hole for many days, erosion pipes might develop, depending on the piping resistance of the soil and the gradient to the point of exit of the leak (which will be high inevitably if the core is narrow). Therefore it is not desirable to pour large quantities of water into such a borehole. At the Yard's Creek Dam, for example, there was strong evidence that drilling exploratory borings caused nearby leaks appearing at the downstream toe to change from clear to dirty water.

Since water losses in borings are common in dams

that do not leak, it is apparent that fluid in borings can cause hydraulic fracturing much more easily than reservoir water acting against the dam, for a number of reasons:

1. The reservoir water acts against the whole upstream face of the dam, whereas the borehole fluid acts only on the relatively small peripheral surface of the hole. Thus the reservoir water exerts a large total force, increasing the compressive stresses which close cracks. In contrast, the pressure in the borehole fluid, before hydraulic fracturing occurs, has negligible influence on the stresses in the embankment surrounding the borehole, and acts only to open cracks.
2. Seepage from the reservoir water acting on the whole face of the dam tends to increase the water content of the core as the water penetrates, reducing the likelihood of hydraulic fracturing by two actions. First, most well-compacted core materials have at least a moderate tendency to swell when the water content is increased, with a resultant increase in the internal embankment pressure. Second, the increase in water content of the upstream portion of the core causes it to be less rigid and brittle and better able to follow the differential settlements without cracking. In contrast, the fluid in a borehole can only raise the water content of a very small volume of the core material directly surrounding the borehole.
3. Even in reservoirs that are filled rapidly, the pressure of the reservoir water is applied to the core much more slowly than the pressure of the water in a borehole. With some exceptions, reservoir levels are not raised faster than 5 to 10 ft/day, whereas a 100- to 200-ft-deep boring may be drilled in a few hours. Hence the core commonly has some time to adjust to the pressure of the reservoir water and to seal cracks by swelling, whereas the fluid pressure in a boring is imposed almost instantaneously.
4. Many of the cracks through which water is lost from boreholes are probably vertical and parallel to the longitudinal axis.
5. When water is lost from a borehole by hydraulic fracturing in a crack which is generally transverse to the core, the water does not travel through the whole width of the core but only from the borehole to the easiest exit, either at the upstream or downstream face.
6. The pressure head in the borehole fluid is higher than the maximum reservoir level since the holes are usually drilled from the dam crest. In addition, the pressure in the borehole may be much higher than the static pressure of the fluid column because of the added pump pressure needed to cause circulation in the hole and remove the cuttings.

On the basis of the experience that has been described here, the following conclusions are reasonable:

1. There is no way to determine from losses of borehole fluid in a given dam whether or not it is possible that reservoir water could open a crack through the core.
2. Borehole water losses would be expected in dams which have cracks and in dams where cracking is imminent; however, they are also common in well-designed and constructed dams, in which there is no possibility of leaks developing through the dam.
3. If the engineer suspects open cracks may exist in a dam, he can only hope to confirm his suspicion by making test pits. The knowledge that borehole fluid losses occur does not give any practical assistance in the problem of evaluating the safety of the dam.
4. If holes are to be drilled in the core of a dam, it is reasonable for the engineer to avoid the problem by employing methods which do not use drilling fluid in uncased holes, such as with hollow stem augers or by driving and cleaning out casing.
5. It is conceivable that hydraulic fracturing by borehole fluid can cause internal erosion and create a dangerous condition in the dam where none previously existed, especially in thin-core dams and in cases where large quantities of water are poured into the hole for investigation purposes.

8. RESEARCH CONTRIBUTIONS

Laboratory Studies of Cracking in Compacted Soil

The only concentrated laboratory research on cracking was done by Narain and Leonards at Purdue University (Narain, 1962; Leonards and Narain, 1963. A wholly novel approach was taken, in which soil samples were compacted into prisms with rectangular cross section and tested in bending, as a beam. The tensile strain at which the first crack developed was measured carefully. This ranged from a little more than 0.1% for specimens compacted near Modified AASHO optimum water content to a little more than 0.3% for specimens compacted at or above Standard AASHO optimum. One result that might not have been anticipated was that the tensile strain causing cracking increased considerably as the water content was increased from 3% below Standard AASHO optimum to optimum, but a further increase of the water content on the wet side of optimum had little influence.

Tests of this kind were performed on soil samples taken from several dams which had cracked. The laboratory values of tensile strain at which cracking occurred were reasonably consistent with the best estimates of longitudinal crest elongation that had caused cracking in the dams.

To the writer's knowledge nothing further has been done in the United States using this beam testing procedure developed at Purdue, either in the form of continued research or application to the design of a specific project. The technique has been used, however, to study the potential for cracking in the 75-meter-high Sklope Dam, a rockfill dam with very thin central core in a narrow, steep-walled valley constructed in Yugoslavia in 1966–1968 (Nonveiller, 1968). The test results were generally similar to those obtained at Purdue: the tensile strain causing cracking increased from about 0.1 to 0.3% as the water content was increased from 23 to 28%. They were different from the Purdue results, however, in that the maximum value of tensile strain at failure occurred at a water content considerably above Standard AASHO optimum, which was about 20%.

The designers at Sklope Dam made a definite effort to compact the core at an especially high water content in limited zones at the top of the dam, where transverse open crest cracks most frequently have been observed. Figure 55 is a longitudinal section showing two areas at the top of the dam, located over the steepest part of the abutments, in which the core was constructed using clay of higher plasticity than the average at water content in the range between 3 and 5% above Standard AASHO optimum. These zones were compacted with a special lightweight sheepsfoot roller. According to the laboratory tests, still higher placement water contents would have resulted in higher ductility, but it was considered impractical to place the core material at higher water contents since the construction surface would

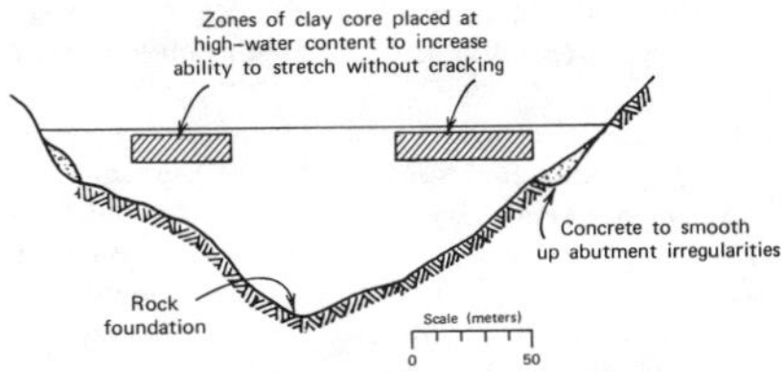

Fig. 55. Sklope Dam. Longitudinal section. (After Nonveiller, 1963.)

have been too soft to support equipment. Measurements of the horizontal distance between crest monuments showed longitudinal compression in the center part of the valley and extension over the abutments in the normal pattern. By November 1969, after 2 years of reservoir operation, the crest had settled a maximum of about 30 cm, longitudinal crest compression in the center part of the valley was approximately 0.15%, and the crest extension over the right and left abutments, respectively, reached a maximum of 0.11 and 0.06%. No cracking developed.*

Finite-Element Analysis

Besides the research at Purdue the only theoretical approach to the cracking problem has been the application of the finite-element method to computation of the stresses and strains within embankments. Before this development the mathematics were so cumbersome that it was practically impossible to compute the stresses in an elastic embankment model, even with the simplest boundary conditions. Clough and Woodward (1967) demonstrated the use of the finite element method by computing stresses and deformations in a simple, homogeneous earth dam model, including the influence of building the dam in a series of horizontal layers. Subsequently the method has been applied by a number of investigators for studies of stress distribution in dams. Lee and Shen (1969) show the results of finite-element calculations made for the El Infiernillo Dam in Mexico in which the computed postconstruction crest movements approximated measured movements remarkably well. DeFries (1968) applied the method of the analysis of the El Isiro Dam, showing that with reasonable assumptions concerning the settlement of the upstream slope the location and

*Personal communication with Professor E. Nonveiller, Geotechnical Institute, Zagreb, Yugoslavia.

direction of the large downstream slope crack might have been predicted from the computed internal stresses.

Subsequently Covarrubias (1969) computed stresses and strains for many models of homogeneous and zoned dams, varying the geometry and compressibility of the dam and foundation. Making the assumption that the dam embankment could withstand tension, Covarrubias studied the influence of the various factors on the zone of tension that developed. He also placed cracks in the mathematical models in zones of maximum tension and recomputed the stresses. He compared his calculations with measurements made at some recent dams including the Gepatsch and El Infiernillo dams and showed that the cracks occurred where the computed tensile stresses were maximum.

Kulhawy and Duncan (1970) showed a comparison between the calculated and measured stresses and movements in the Oroville Dam (California). Finite element calculations were made assuming the dam to be constructed in stages using different moduli of compressibility for each stage in an effort to approximate a nonlinear relationship between stress and strain. The calculated movements show a remarkable correlation with the measured movements, but the comparison between the calculated stresses and the stresses measured with buried pressure cells was not good.

One of the most important applications of the finite-element method is to compare the internal stress and strain patterns within dams of various possible designs. Such studies allow a choice of a final design which has maximum freedom from localized zones of low stress in the core, caused by arching between zones and stress transfer near foundation irregularities. However, the quantitative results of such computations vary considerably, depending on the assumed embankment properties such as the bulk modulus and Poisson's ratio. For example, Covarrubias (1969) computed the stress distribution with a 100-meter-high dam with vertical central core using two assumptions for Poisson's ratio, 0.35 and 0.45. The average computed vertical stress on horizontal planes in the core near the base of the dam changed with these two assumptions from about 6 to 9 kg/cm^2.

Cracking Potential as Related to Embankment Soil Type

Only a few studies have been made of the influence of the embankment soil type on cracking potential. The information available does not provide much practical assistance to the dam designer.

As a result of an earlier study of 60 relatively low dams in the western United States, the writer concluded that most of the dams in which severe cracking had occurred had been poorly compacted, usually at a low water content, and that there was also a tenuous correlation between the incidence of cracking and the embankment soil type (Sherard, 1953). In a similar study of the performance of a number of dams in Mexico, some of which cracked, Marsal (1960) concluded that the main cause of cracking was relatively high postconstruction settlements resulting from inadequate compaction and low construction water contents, and he found little or no correlation with soil type.

Experience indicates that dams constructed of widely different types of soils have cracked; these soils include: clays of high plasticity (Djatiluhur Dam), widely graded glacial moraine deposits consisting of gravel-sand-clay mixtures (Balderhead, Yard's Creek, and Hyttejuvet dams), and cohesionless silty sand (Round Butte Dam). From the evidence currently available it therefore appears that earth dam cores consisting of almost all soil types have suffered differential settlement cracks. Nevertheless, it seems that there must be a considerable difference in the ability of various compacted soils to withstand tensile strain without cracking. Future laboratory research, along lines parallel to the investigations initiated by Narain and Leonards, but probably using techniques other than bending an unconfined beam, could provide information of potential value to the dam engineer in guiding the choice of the core material.

Recently Biarez, Bordes, and Londe (1970) defined two complex parameters, one relating to the settlement potential (including measures of dam height, embankment slopes, thickness of compacted layers, and grain size distribution of the embankment material), and the other relating to the "brittleness" of the core (including Atterberg limits, compaction water content, and density). These parameters were computed for 23 dams, 7 of which had developed longitudinal crest cracks. The results showed a possible correlation.

In a supplement to their paper Biarez, Bordes, and Londe (1970) reported recent laboratory research indicating that the ratio of the compaction water content to the plasticity index may be a significant parameter which characterized the "crackability" of compacted soils. Figure 56 is a plot of this parameter as a function of the ratio of crest settle-

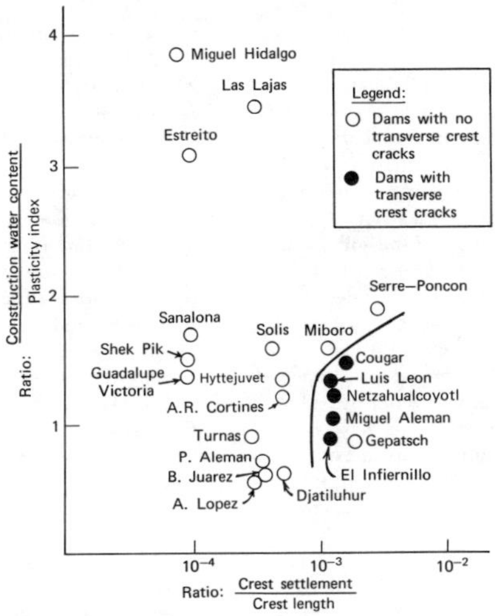

Fig. 56. Statistical study of transverse cracking. (After Londe, 1970.)

ment to crest length for 22 dams, 5 of which had developed transverse crest cracks. With one exception (Gepatsch Dam), a boundary can be drawn in this plot between the dams that cracked and those that did not. The authors are cognizant of the many serious limitations of this rough empirical correlation and they present the material only because the first results indicate that "there may be a research path to follow." Figure 56 is included here for the same general reason and also because it shows that there is a threshold in the ratio of crest settlement to crest length of about 0.001, above which there has been a much higher frequency of transverse cracking.

It is interesting to note in this regard that Vargas and Hsu (1970) report that the residual soils in south central Brazil, used for dams which have a high frequency of cracking, consistently have plastic limit in the range of 8 to 10% higher than Standard AASHO optimum. Possibly the relationship between plastic limit and optimum water content might be shown by research to be a better parameter to define brittleness.

There seems to be enough evidence available from the experiences at the Stockton Creek, Rector Creek, Woodcrest and Shek Pik dams, and others described here, to justify the conclusion that embankments constructed of some kinds of residual soils are especially brittle and susceptible to the development of cracks under moderate settlements. This general conclusion has little practical value since the broad designation "residual soils" comprises materials of a wide range of properties and at present there is no way to identify those that may be most susceptible to cracking. The suspect soils appear to have an internal cementation that causes hardening with time. Possibly this cementation phenomenon is not limited to soils of residual origin. Research on the existence and nature of this cementing action would be of value.

"Dispersion" Piping Influenced by Clay Chemistry

Some original research has been carried out in Australia in the last few years relating the incidence of piping in small clay dams (for farm ponds) with the chemical characteristics of the clay. The results provide an important increase in our understanding of piping of clayey soils and may also provide the key to understanding some of the experiences with dam leakage and piping described in this paper which are currently difficult to explain.

Aitchison and Ingles (1964), Aitchison and Wood (1965), and Rallings (1966) report studies of many low dams (usually less than 20 ft in height) in various parts of Australia. These are generally homogeneous embankments without filters or drains constructed without moisture control and compacted only by the travel of hauling equipment. Because of the prevailing climatic conditions the embankments were usually built dry of optimum. Of more than 3000 such structures counted, roughly 8% failed by rapid progressive erosion of spontaneous leaks. Nearly 100 of the failures were studied. Tests of the soil chemistry as well as of the engineering properties of the soil were made. The investigators concluded that there is a strong correlation between the likelihood of failure and the presence of soils with high percentages of exchangeable sodium cations, especially in the case where the reservoir water has a low content of soluble salts. Similar results have been reported from a study of 10 small dams in Israel, 5 of which failed by piping (Kassiff and Henkin, 1967). Some of the main conclusions and observations from these studies are:

1. The piping of the clay is caused by a process called "dispersion" (or "deflocculation"). This occurs when the repulsive forces (electrical surface forces) between the individual clay particles exceed the attractive (van der Waals) forces so that when the clay mass is in contact with water, individual clay particles are progressively detached and go into suspension. If the water is flowing, the dispersed individual clay particles are carried away. Among various soils with clayey fines there are great differences in the tendency toward dispersion piping.
2. The tendency for dispersion piping is governed by a number of properties of the clay, including the exchangeable sodium percentage (ESP),* the pH, the soil type, and the content of dissolved salts in the water. The sodium acts to increase the thickness of the diffuse double layer and to increase the repulsive forces between particles, making it easier for individual clay particles to be detached from the mass.
3. When a concentrated leak starts through a clay embankment with high ESP, one of two actions can occur: (1) if the velocity of flow is sufficiently low, the clay surrounding the flow channels swells and progressively seals off the leak; or (2) if the initial velocity is sufficiently rapid, the dispersed clay particles are carried away, enlarging the flow channel at a faster rate than it is closed by swelling and the dam fails by piping.
4. The initial leaks may be through channels built into the clay embankment by inadequate compaction (macropores between soil chunks) or through drying cracks.
5. Soils with an ESP of 7 to 10 are moderately dispersive and were associated with piping when the reservoir had water with a low content of dissolved salts. Soils with ESP exceeding 15 have serious piping potential.
6. There is no good relationship between the piping potential and the results of ordinary index properties tests for clays (Atterberg limits, particle size, shrinkage). Soils with high contents of exchangeable sodium range from cohesionless silts (ML) to clays of high plasticity (CH).
7. High ESP values and piping potential are generally found in soils where the clay fraction is composed largely of montmorillonite. Some illites have high ESP values. High ESP kaolins are rare. Moreover, kaolins tend to cause less trouble because the dispersed particle size is much larger than montmorillonite and a larger initial leakage channel is required for piping to start.
8. A simple dispersion test ("crumb test") was found to give good correlation with the incidence of piping failure. Briefly the test consists of placing air-dried crumbs of soil in a 1-meq/liter solution of sodium hydroxide. If, after a period of 1 hour, a colloidal cloud of dispersed clay exists around the soil crumbs, then the soil is considered to be dispersive. "The test is considered to be technically superior to the chemical analysis as it automatically accounts for the ESP and clay type. Soils which disperse in this test cannot be used safely in farm dam construction without compaction and moisture content control" (Rallings, 1966). In 25 of 26 dams which failed, Rallings found that the soil showed dispersive characteristics in this test.

Aitchison and Wood (1965) give a summary of studies of 20 dams (of which 14 failed) in 4 states of Australia chosen to represent a variety of climatic

*Rallings (1966) plotted his results in terms of ESP. Aitchison and Woods (1965) and Kassiff and Henkin (1967) used another measure of the quantity of exchangeable sodium cations called the Sodium Adsorption Ratio (SAR). For many clays the numerical values of the ESP and SAR are not greatly different.

and soil conditions. Kassiff and Henkin (1967) show a similar study of 10 dams in Israel (of which 5 failed). Both investigators plotted the SAR on one axis to an arithmetic scale and total cation concentration in the reservoir water (a measure of the dissolved salts) on the other axis to a logarithmic scale. As seen in Fig. 57, in which the results of the two studies are combined, a boundary can be drawn, as shown by the dotted line, separating the dams which failed and those which did not. The example marked 11A and 11B represents an Australian case history in which the reservoir was initially filled with a groundwater relatively high in dissolved salts. The dam held water with only slight seepage for several years (11A), and then failed within 3 days after the reservoir was filled with water low in dissolved salts.

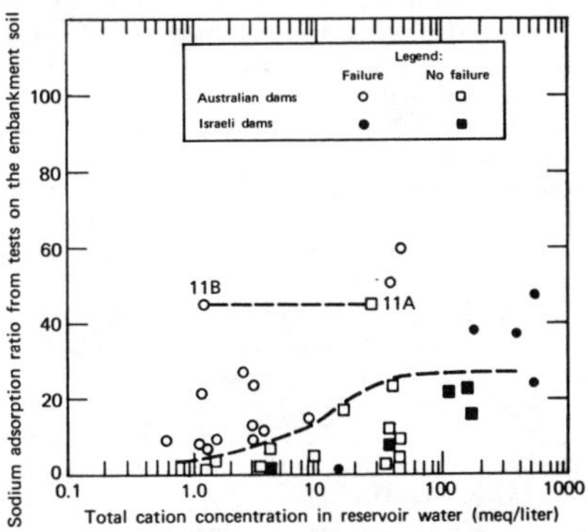

Fig. 57. Influence of soil and water chemistry on piping in small earth dams. (After Aitchison and Wood, 1965; Kassiff and Henkin, 1967.)

The studies described here were based on the performance of small farm-type dams, which were generally constructed with tractors and scrapers with no roller and without engineering control. In their early papers the Australian investigators speculated that piping in clay cores had not occurred in well-constructed dams, even when the soil had a high ESP and the reservoir water was low in dissolved salts.

Recently, however, some Australian engineers have concluded that piping of highly dispersive clay cores may have developed even in dams constructed with rollers and moisture-density control. Ingles et al. (1968, 1969) describe the failure of Flagstaff Gully Dam, retaining a small municipal water supply reservoir (50 million gallons) for the city of Hobart, Tasmania. The dam, 55 ft high and 690 ft long, is a rockfill dam founded on rock with a central clay core, comprised primarily of montmorillonite, with upstream and downstream filters (Fig. 58). The clay comprising the lower half of the core had the following average properties:

Liquid limit 40
Plasticity index 25
Particle size distribution:

Size	Average percent finer
2 mm	90
200 sieve	65
0.002 mm	30

The upper elevations of the clay core, taken from another borrow pit, had a liquid limit of 75 and a plasticity index of 60. ESP in the lower portion of the core was 17 and the upper portion 10. The core was apparently well constructed and at a water content near Standard AASHO optimum.

The reservoir was filled with water having a low

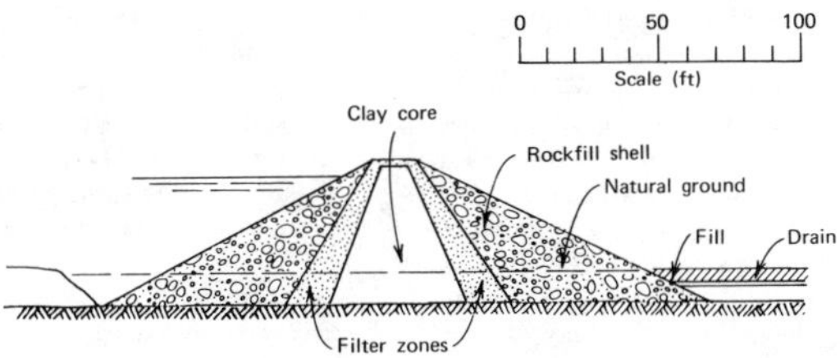

Fig. 58. Flagstaff Gully Dam. Cross section. (After Ingles *et al.*, 1969.)

dissolved salt content from a source 20 miles away. (No local runoff was permitted to enter.) The total of all dissolved salts in the water was about 1.0 meq/liter. In July 1963, three weeks after the first filling, a large dirty leak appeared at the downstream toe. Nearly all of the water in the small reservoir (about 40 million gallons) was lost. Several samples of the dirty leakage water were tested; the content of suspended clay particles was about 1000 ppm and the average size of suspended particle was about 0.6 μ. In subsequent explorations it was found that there were two erosion tunnels through the core just above the bedrock surface with diameter of about 2 to 3 ft. As in most such cases, the cause of the failure was not determined reliably; however, because of the coincidence of the high exchangeable sodium and the very pure reservoir water, the investigators concluded that dispersion of the clay could have been a contributing cause.

The Flagstaff Gully Dam experience is similar to that of the Yard's Creek Dam. Both are relatively uncommon designs in the sense that it has been rare to construct such low dams with rockfill shells and a narrow central core founded on bedrock. At Flagstaff Gully Dam the width of the contact between the core and rock surface at the point of failure was about 36 ft; at Yard's Creek Upper Reservoir Dam it was less than 25 ft.

James and Wickham (1970) describe the failure of another small dam (Morwell Dam) in 1968 in Victoria, Australia. It was a homogeneous dam of clay of medium plasticity (CL-CH) and a height of about 30 ft, constructed according to modern practice with moisture-density control. The average construction water content was about 2% below Standard AASHO optimum and the average density was about 98% of Standard AASHO maximum dry density. When the reservoir filled rapidly for the first time in December 1968, a leak of about 100 gpm broke out at the downstream toe through an erosion pipe about 1 ft in diameter in the clay embankment. The small reservoir was rapidly lowered to prevent more extensive damage.

When the embankment was only a few feet high, work was stopped for 3 months. During this period the weather was very dry. The leaks and erosion pipes subsequently developed at about this level. The rapid development of piping in a well-constructed, impervious clay dam cannot be easily explained unless the material was highly susceptible to piping. Investigations showed that the clay did not have very high ESP (5 to 7), but it did nevertheless have some reaction to the Rallings dispersion test. The reservoir water was found to have a total cation concentration of about 5 meq/liter.

As part of the remedial measures, gypsum (calcium sulfate) was spread on the upstream slope of the dam, in a layer about 1 in. thick (about 30 tons). "As the reservoir water is used and replenished, and the salt concentration falls, the gypsum blanket will still be relied upon to increase the concentration of divalent salts in the seepage water entering the embankment (and prevent dispersion) until the clay swelling closes off any pervious channels" (James and Wickham, 1970).

Although clays generally are considered to have fairly high resistance to erosion, it has long been noticed that there is a great difference between the erodibility of different clay soils in nature. Some natural clays are so resistant to erosion that they form stable beds for rapidly flowing streams; others erode so rapidly that the action is not unlike sugar dissolving in water. These differences in erodibility are not measured well by ordinary soil mechanics tests. Until publication of the Australian research,* no explanation for these differences had been proposed. In many cases where dirty leakage developed through dams there is no really satisfactory explanation of why these dams failed, whereas similar dams did not.

At the present (1972) the civil engineer dealing with earth dams in the United States and most other parts of the world is not generally familiar with the concept of dispersive clay piping. Recently similar studies have been made in the United States and Venezuela which confirm completely the main conclusions of the Australian investigators and show that dispersive clays are not rare in nature. A number of failures of low clay dams in Oklahoma and Mississippi were shown conclusively to have been caused by dispersive clay piping, probably resulting from initial leaks in cracks caused by hydraulic fracturing in zones of low embankment stress (differential foundation settlement, drying, and shrinkage). These experiences and a more extensive discussion of this subject are given by Sherard, Decker and Ryker (1972 and 1972a). Basic research is needed on this subject; however, it is now clear that the Australian investigators have made an

*Ingles and Aitchison (1969) contains a review and extension of the Australian research.

important contribution to the understanding of earth dam performance. We already know enough about the subject to conclude that tests of clay dispersibility should be made as a routine part of studies for earth dam design.

9. RELATIONSHIP BETWEEN CRACKING AND EARTHQUAKE DESIGN

One of the most probable and hazardous effects of earthquake shaking on embankment dams is settlement and cracking, very similar in pattern and location to the cracking commonly caused by differential settlement (Fig. 59). Design measures to increase the safety of dams against concentrated leaks in differential settlement cracks also act to make the dam safer against cracking due to earthquake shaking, and vice versa.

A dam that settles slowly is less likely to develop large cracks than one that settles rapidly. For a dam that is undergoing relatively large but slow differential settlements, even moderate earthquake shaking could be sufficient to cause abrupt opening of cracks. A combination of moderate differential settlement and a moderate earthquake could result in large cracks.

The construction of large reservoirs not only can increase the general level of seismic activity in the area of a dam but can trigger major earthquakes. For this reason, the possible effects of earthquakes on cracking must be considered even in areas that are not seismically active.

The first time that seismic activity was definitely attributed to the construction of a reservoir was the filling of Lake Mead behind Hoover Dam in the mid-1930s. Little or no earthquake activity had been recorded during the 15 years immediately preceding its construction. About 100 tremors were felt during the first year of reservoir filling and the greatest frequency of small tremors was reached when the full reservoir level was reached for the first time, two years after construction (Carder, 1968).

Shasta Dam in northern California, constructed in the mid-1940s, at which a swarm of small earthquakes occurred during the first reservoir filling, was the only other such experience reported until about 1965. In the last few years there have been a considerable number of experiences reported in various parts of the world which leave little doubt that the

(a)

(b)

Fig. 59. Crest cracks caused by earthquakes. (*a*) Transverse crack in long, low dam near Anchorage, Alaska, April 1964. (Prof. H. Bolton Seed.) (*b*) Transverse crack near abutment, Upper Van Norman Dam, Los Angeles, California, March 1971. (Photos, J. L. Sherard.)

construction of reservoirs can cause earthquakes.* In December 1970, an international group of specialists convened at the UNESCO Headquarters in Paris to review this problem, concluded: "During the past few years, the impounding of certain reservoirs has been found to be responsible for triggering seismic phenomena, irrespective of the seismicity of the reservoir region. Characteristic examples are associated not only with recent tectonics and high seismicity but also with older and more stable masses of very early tectonics" (UNESCO, 1970).

The latest and strongest piece of evidence is the experience of the Koyna Dam in India (Mane, 1967; Ambraseys and Sarma, 1968), which was constructed in an area where it is reported there was no seismic activity for hundreds of years and where geologists believed the region was stable (Guha et al., 1968). The dam retains a reservoir of about 3000 million meters^3, which is relatively large but not greater than many others built in recent years. Immediately after the dam was completed and the reservoir commenced to fill in 1962 many earth tremors were reported and the frequency of these tremors began increasing after the middle of 1963. The strongest tremors in the first few years rattled windows and disturbed utensils. On September 13, 1967, two relatively strong shocks (Richter magnitude 5.0–5.5) occurred with epicenters very near the dam. Subsequently on December 11, 1967, the whole of western India was shaken by a strong earthquake (magnitude about 6.5 and a "felt radius" of about 600 km), with the epicenter about 3 km from the dam. The shaking measured by a strong motion instrument inside the dam showed a peak acceleration in excess of 50% gravity and duration of very strong motion of about 5 seconds (Housner, 1969). About 180 people were killed and more than 2000 injured. The dam, a concrete gravity structure, was cracked sufficiently to cause considerable concern, and extensive repairs by grouting and prestressing were made (Berg et al., 1969; Murthy and Thomas, 1970).

No fault movements at the ground surface have been reported in any of these earthquakes. However, it is not unreasonable to foresee that fault movement could occur in some future reservoir-induced earthquake.* Circumstantial evidence indicates the possibility that the strong West Yellowstone (Montana) earthquake of 1959 may have been related to the presence of the Hebgen Reservoir (Sherard et al., 1963). During this earthquake the earth's crust over an area of several dozen square miles suddenly subsided 10 to 15 ft with extensive vertical fault displacements. The subsidence area was almost exactly centered on the Hebgen Reservoir (300,000 acre-ft). The dam itself dropped about 10ft, and the maximum measured subsidence of the bedrock in the area was 22 ft.

*Endersbee (1969), Jaeger (1969), Carder (1968), and Rothé (1969) are recent papers summarizing literature on this subject and describing many reservoirs in various parts of the world at which increased seismic activity has been reported following first reservoir filling.

*The failure of the Baldwin Hills Dam in 1963 was caused by movement of a fault. The writer and other investigators (Hamilton and Meehan, 1971) believe that fault movement was probaby caused by pumping water into deep boreholes.

10. SUMMARY: MAIN CONCLUSIONS AND OBSERVATIONS

Frequency and Importance of Cracking

Differential settlement cracks have been common in all types of embankment dams with earth cores. In the great majority of cases no leaks have developed, the cracks have been repaired by simple means, and the dam has subsequently performed satisfactorily.

In a small number of dams, concentrated leaks in cracks have caused serious piping damage or complete failure.

In many cases, severe cracking in modern, well-constructed dams has been caused by compression of soil foundations. Cracking has been common in dams that were compacted at an average water content several percentage points below Standard AASHO optimum. Important cracking has also developed in a considerable number of dams where the foundation settlement was not large and the embankments were well constructed, with water content at or above Standard AASHO optimum water content. In many of these cases extensive postfailure investigations did not reveal conditions which were greatly different from the conditions existing at many other sites.

Complete failures, in which the entire reservoir released through a breach or erosion tunnel, have usually occurred only in homogeneous, relatively low dams which have no internal vertical filter drain (chimney) or zoning to control concentrated leaks.

In one case (Wister Dam) complete failure of an important dam with large reservoir was probably avoided only by good luck.

Concentrated leaks and serious erosion damage, believed to have been caused by cracks through the core, have occurred in a number of important rockfill dams with vertical, or near vertical, earth cores since 1960. Probably these dams have not been in danger of complete failure.

For dams of low to moderate height, the likelihood of cracking can be reduced to a very low value by conservative measures to reduce the differential settlement. The problem is not completely understood, however, and it is not possible to be completely confident that open cracks could not develop through almost any dam.

For high dams, particularly with thin cores and steep abutments, cracking must be anticipated regardless of the method of construction.

Leaks in Unseen Cracks and Hydraulic Fracturing

In the majority of cases where cracking is suspected to have been the cause of the initial leak leading to failure or erosion damage, no cracks were seen on the crest or slopes of the dam before the leakage started.

For dams which have failed or been badly damaged by piping, where cracking was suspected to be the cause, it has been commonly observed that concentrated muddy leaks develop abruptly some time after the reservoir has risen above the level of the leak. Probably in many of these cases a differential settlement crack opened only after the reservoir was filled, and the crack was caused to open in part by the water pressure (hydraulic fracturing).

At one extreme, hydraulic fracturing is believed to have caused complete failure of homogeneous dams (e.g., Stockton Creek Dam) or important erosion damage to the core of zoned dams (e.g., Balderhead Dam). At the other extreme, hydraulic fracturing may cause the opening of very narrow cracks (microcracks), through which no appreciable concentrated leakage takes place and no damage occurs.

Leaks believed to have been caused by cracks have developed abruptly after several years of satisfactory performance (e.g., East Branch Dam and the second appearance of troubles at the Balderhead and Hyttejuvet dams).

It is believed that hydraulic fracturing probably is also responsible for some concentrated leaks at the contact between the earth core and rock foundation surfaces or concrete structures, such as conduits.

In three rockfill dams, the fines in widely graded, coarse core material were picked up by the leaking water and carried in suspension through the filters, leaving the coarse fraction behind (Balderhead, Yard's Creek, and Hyttejuvet dams).

Rockfill Dams with Central Earth Cores

Longitudinal cracks with widths of several centimeters are common on the crests of high rockfill dams with central, vertical (or near vertical) cores. Frequently the width of the crest (the horizontal distance between monuments on the upstream and downstream edge of the crest at the same station) increases appreciably, perhaps 6 to 12 in.

From the general experience described in this paper, and particularly from the experience at the Djatiluhur Dam, it is reasonable to suspect that these longitudinal crest cracks in dams with thin central cores may be accompanied in some cases by unseen horizontal cracks at depth; that is, if 30 to 50-ft-deep test pits had been put down in these dams at the right place and at the right time, open horizontal cracks would have been found.

The initial leak leading to the damage at Yard's Creek, Balderhead, Hyttejuvet, and Matahina dams probably passed through a near horizontal crack in the core opened by hydraulic fracturing. It is probable that unrecognized leaks have developed through horizontal cracks in the upper part of many other high rockfill dams.

In none of the central core dams that were damaged by piping believed to have originated in cracks was the core compacted at a low water content. At the Djatiluhur Dam, which is the only dam in which open horizontal cracks in the core were seen, the core consisted of a clay of high plasticity, well compacted at an average and uniform water content about 2 percentage points above Standard AASHO optimum.

Core widths in the range of 30 to 50% of the water head are commonly considered adequate. Experience indicates that for low, central core dams with shells of quarried rock greater widths are desirable, especially when the core is compacted directly against the rock surface.

The Matahina Dam experience illustrates the trouble that can develop when very large rocks (3 ft diameter or larger) are placed in the main rockfill zone immediately adjacent to the downstream transition.

The Balderhead Dam experience shows that piezometers located in a semipervious upstream shell can provide a more accurate indication of the development of concentrated leaks through the core than can be obtained by measuring the increase in the volume of seepage emerging at the toe.

The Matahina Dam experience illustrates a potential hazard of grouting the foundation through completed embankments: grout can penetrate filter and transition zones by hydraulic fracturing. Grout-filled seams can form a reinforcing skeleton in the filters, which can provide a structural roof for concentrated leakage channels.

Small Dams

Cracking in small dams is very common. The frequency is much higher in some geographic areas than others. It is especially common in areas of low rainfall and in areas of compressible soil foundations, such as loessial deposits.

Sometimes it is impossible to differentiate between cracks caused by drying and by differential settlement. If small differential settlement cracks are imminent, shrinkage due to drying undoubtedly accelerates the development of the cracks and increases their width.

Transverse cracks with a width of 1 cm or more have appeared on the crest of dams that have had postconstruction settlement of 15 cm or less.

A common cause of cracking has been compression in foundation soils that soften when they become wet after the first reservoir filling. In some cases transverse cracks have opened completely through the embankment and extended 10 to 15 ft, or more, into the foundation.

For small dams with low embankment volume, stringent control of the construction moisture is equally or more desirable than for more important structures.

For small dams of relatively low embankment volume, it is especially important to require that the water content of the embankment material be processed in the borrow area in advance to a uniform and satisfactory value.

For homogeneous dams there is a certain statistical risk of complete failure by progressive erosion of leaks through cracks. Some of the dams that developed piping were well built by competent engineers, at sites without particular problems (Wister, Stockton Creek, East Branch dams and many smaller structures). It is believed that even low, homogeneous dams should include at least a thin vertical chimney of filter material as a "crack stopper," except in those cases where the consequences of a failure are not great.

Soil Properties

Cracking has developed in cores constructed of almost all types of soils. There is no reliable guide from the experience currently available as to which soils may be especially susceptible to cracking.

The experiences at the Yard's Creek, Balderhead, and Hyttejuvet dams indicate that some widely graded soils, characterized by a nearly straight gradation curve on a conventional semilogarithmic plot and ranging between 5 and 15% of coarse gravel sizes (3-in. cobbles) and 10 to 22% clay sizes, are susceptible to a type of piping in which the fines are selectively eroded by leaking water and are carried out of the coarser granular matrix.

There is considerable evidence to support the conclusion that embankments constructed of certain kinds of residual soils and possibly soils of other origin contain a "cementing agent" which make them especially susceptible to cracking (Stockton Creek, Woodcrest, Rector Creek, Shek Pik, and Brazilian dams). Except at the Shek Pik Dam, where tests indicated the presence of iron oxide hydrate, no studies have been made of cementing agents. Research on this subject is needed.

Recent research shows that dam cores from soils with fines having a high percentage of exchangeable sodium cations are susceptible to "dispersive piping." Further research is needed.

Core materials with radically different resistance to "dispersive piping" may have identical properties when classified according to conventional soil engineering procedures; that is, they may have the same Atterberg limits and grain size distribution. Laboratory tests of the types used by soil chemists should be adopted by the civil engineer for evaluation of the piping potential and "dispersibility" of dam core materials.

There have been a number of experiences in which rapid piping has seriously damaged well-constructed dam cores consisting of clays of medium to high plasticity, showing clearly that some clays have uncommonly low resistance to piping. These experiences indicate that the erosion resistance of clays varies over a wide range, and probably depends on the relative tendency of the clay to "disperse."

It is easily possible that some of the experiences described here in which dams were damaged or failed by piping were caused primarily by the fact that the embankments were constructed of clays having susceptibility to dispersive piping.

Fluid Losses in Boreholes

Fluid losses from boreholes drilled in the impervious sections during or after construction have been common. In the great majority of cases, there was probably no open crack in the core before the hole was drilled, but the crack was opened by hydraulic fracturing.

Fluid losses from boreholes can be caused by hydraulic fracturing when the minimum total principal stress acting within the embankment is less than the fluid pressure in the hole.

Fluid pressure acting in a borehole can cause cracks to open by hydraulic fracturing much more easily than the pressure of the reservoir water acting on the upstream face of the core.

When drilling holes in the cores of dams, procedures that do not employ fluid in uncased holes are desirable.

Internal erosion caused by fluid losses in borings can damage the core and reduce the safety of the dam, especially in thin-core dams.

General Experiences

Cracks on the crests of dams sometimes appear abruptly with a width approximately equal to the ultimate width. For example, no crack may be observed during one inspection, and within a day or two a crack several centimeters wide may be found. Often the crack does not increase in width appreciably in the subsequent months.

Cracks which have sometimes appeared during construction shutdowns might not have been seen, or might not have opened, if construction had continued at a uniform rate.

Concentrated leaks have frequently developed through embankment cracks located directly adjacent to rock abutments at points of abrupt change in slope, and particularly adjacent to horizontal steps used for haul roads, grouting platforms, and so on. Investigators concluded that such features probably caused cracking at the following dams: East Branch, Matahina, Round Butte, Stockton Creek, Yard's Creek Upper Reservoir, Euclides da Cunha (Brazil), and Vir (India).

From the foregoing experiences, and from the knowledge that many dams have been built against very steep rock abutments and near vertical concrete structures without experiencing leakage through cracks, it seems reasonable to conclude that the steepness of the rock abutment is considerably less important from the standpoint of cracking potential than is the existence of pronounced surface irregularities.

There is every reason to believe from experience and theory that the likelihood of leaks breaking through cracks on the first reservoir filling is much greater if the reservoir is filled very rapidly.

Foundation grouting carried out by drilling through completed or partially completed embankments is undesirable from the standpoint of cracking. Even if great care is taken to seal the casing into the bedrock surface and to keep the grout in the bedrock, it can flow upward in rock cracks and create high pressure on the base of the dam. These uplift forces can result in localized zones of low embankment stress where cracks can be opened by hydraulic fracturing. At the Shek Pik Dam, foundation grouting during construction caused a measured embankment heave of 0.3 ft and an open crack on the slope about ½-in. wide.

There are a number of examples where grout curtains in rock foundations under earth dams were constructed by drilling through the completed embankment and where the grout penetrated upward into cracks within the embankment in spite of careful efforts to keep the grout only in the underlying rock (e.g., El Isiro Dam).

Commonly differential settlement cracks develop on vertical transverse planes over the abutments and on vertical longitudinal planes in the central portion of the dam. Many other patterns of cracking have developed, depending largely on the geometry and compressibility of the foundation and abutments. In some cases differential settlement cracks have developed with generally circular trace, resembling the type of crack that would occur in the incipient stages of a shear slide.

Dams designed as flood control structures which will stand with empty reservoirs most of the time have at least as great a need for internal zoning, drainage, and filters to control leaks through cracks as dams designed to retain continuously full reservoirs.

Much experience shows that the rigidity of compacted embankments, and the likelihood of cracking, increases rapidly with decreasing water content below standard AASHO optimum. For this reason, water contents close to modified AASHO should not be *generally* used (or specified).

No cases are known where cracks developed at the juncture between an earth dam and a concrete gravity dam, in spite of the fact that there is a considerable potential for differential settlement. This is probably due to the fact that the faces of the concrete dam are plane surfaces without steps or breaks in slope, even though they are steeply inclined (but not vertical).

Cracking is likely to develop adjacent to concrete structures in circumstances where the working space is confined so that a standard heavy roller cannot be used.

The El Isiro Dam experience shows that very wide longitudinal cracks in a zoned dam may develop if the upstream shell is constructed in such a way that it is especially susceptible to settlement on saturation.

The finite element method of computing stresses and strains within an embankment shows promise of providing a very useful tool for the dam designer, particularly for the purpose of comparing alternative designs.

Experience in the last few years gives strong evidence that the construction of large reservoirs has triggered major earthquakes.

Design measures to increase the safety of a dam against the hazard of concentrated leaks in differential settlement cracks also act to make the dam more secure against potential damage or failure from earthquake shaking.

ACKNOWLEDGMENTS

The writer wishes to express his gratitude to Drs. R. C. Hirschfeld and S. J. Poulos for their thorough review of the manuscript and many helpful ideas for improvement. He is also indebted to Messrs. G. Post and P. Londe for valuable suggestions and comments.

ANNOTATED BIBLIOGRAPHY

Aitchison, G. D., O. G. Ingles, and C. C. Wood (1963), "Post-Construction Deflocculation as a Contributory Factor in the Failure of Earth Dams," *4th Aust.-N. Z. Conf. Soil Mech. Found. Eng.*, p. 275. Speculation that clays in a flocculated condition at the time of construction, as determined largely by the composition of the electrolyte surrounding each particle at the time of construction, are likely to be especially susceptible to piping. After construction, a significant change occurs in the electrolyte in the pores as a result of seepage into the dam from the reservoir and from rain.

Aitchison, G. D., and C. C. Wood (1965), "Some Interactions of Compaction, Permeability, and Post-Construction Deflocculation Affecting the Probability of Piping in Small Earth Dams," *6th Int. Conf. Soil. Mech. Found. Eng., Montreal*, Vol. II, p. 442. Study of piping failure in 60 small dams leads to conclusion that incidence of failure is strongly dependent on the exchangeable sodium percentage in the embankment soil and the salt content of the reservoir water.

Ambraseys, N. N., and S. K. Sarma (1968), "Large Earthquake Forces on Gravity Dam," *Nature*, Vol. 219, Sept. 28, 1968. Brief article summarizing the measurements of December 10, 1967, ground surface shaking in the vicinity of the Koyna Dam after the filling of the large reservoir showing a peak acceleration approaching 60% gravity.

ASCE Committee on Earth and Rockfill Dams (1967), "Problems in Design and Construction of Earth and Rockfill Dams—Progress Report," *J. Soil Mech. Found. Div., ASCE*, May 1967, p. 129. A report by a group of engineers having considerable experience with embankment dams summarizing and interpreting the results of a survey among practitioners in the field concerning the current problems of design and the relative need for research in various directions; concluded that "cracking" is a problem of relatively high priority in need of better understanding.

Berg, G. V., et al., 1969, "The Koyna, India, Earthquakes," *Proc. 4th World Conf. Earthquake Eng., Santiago, Chile*, Jan. 1969 (in publication). Description of the geology of the site, the seismic activity and the damage to the dam.

Bertram, G. E. (1967), "Experience with Seepage Control Measures in Earth and Rockfill Dams," *Trans. 9th Cong. Large Dams, Istanbul*, Vol. 3, p. 91. Description of serious cracking and piping in Wister and East Branch dams of the Army Corps of Engineers.

Biarez, J., J. L. Bordes, and P. Londe (1970), "La Fissuration des Noyeaux des Barrages en Enrochments ou en Alluvion," *10th Cong. Large Dams, Montreal*, Vol. 1, p. 27. See also Londe, 1970. Two parameters suspected to be related to the cracking potential are defined and 17 dams are studied on the basis of longitudinal crest cracks. Results show possible correlation.

Bird, J. M. (1961), "Uncertainties in Earth Dam Design," *J. ASCE Soil. Mech. Found. Div.*, June 1961, p. 33. Cites an example in which a 200-ft-high dam developed a transverse crack through the narrow central core opposite an abrupt discontinuity in the rock abutment. Muddy leakage broke through the crack.

Bjerrum, L. (1967), Verbal discussion, *9th Cong. Large Dams, Istanbul*, Vol. VI, p. 456. Gives opinion that horizontal cracks can develop through narrow vertical cores, especially when compacted wet, as result of arching.

Bjerrum, L., J. K. T. L. Nash, R. M. Kennard, and R. E. Gibson (1972), "Hydraulic Fracturing in Field Permeability Testing", *Geotechnique (London)*, June 1972, p. 319. *In-situ* permeability tests of fine-grained soils in boreholes are influenced by hydraulic fracturing when pressure exceeds a low value.

Bjerrum, L. and Andersen, K. H. (1972), "*In-situ* Measurement of Lateral Pressures in Clay", *Proc. Fifth Eureopean Conf. on Soil Mech. Found. Eng., Madrid,* Vol. 1, p. 11. Artificially induced hydraulic fracturing in boreholes in clay is employed to measure earth pressures.

Bohn, M., and M. Hamon (1967), "The Djatiluhur Project," *Water Power,* Aug. 1967, p. 305. A general description of the overall aspects of the project in English. Special emphasis on the 300-ft-high, 300-ft-diameter concrete tower in the upstream slope in which are included the spillway, irrigation outlets, river diversion facilities, and hydroelectric power house.

Carder, D. S. (1968), "Reservoir Loading and Local Earthquakes," *Proc. 6th Ann. Symp. Eng. Geol. Soils Eng., Univ. Idaho,* p. 225. Detailed summary of seismic activity in vicinity of Lake Mead and a brief description of other earthquakes occurring at sites of newly constructed reservoirs.

Carlyle, W. J. (1965), "Shek Pik Dam," *Proc. Inst. Civil Eng. (London),* Vol. 30, March 1965, p. 557. Extensive water loss in borings in 180-ft-high earth dam caused considerable concern. Dam core grouted for full length using low-pressure clay-cement grout. The cause of the trouble remained unclear even after much investigation.

Carlyle, W. J. (1966), "Discussion on Shek Pik Dam," *Proc. Inst. Civil Eng.,* Sept. 1966, p. 164. Additional information on the Shek Pik Dam problem.

Carmichael, K. M. (1970), "Applying the Lessons of the 1960's to Farm Dams in the 1970's," Report No. 34, Water Research Foundation of Australia Ltd., Water Problems in Inland Southern Queensland. Review of Australian research relating piping of clay farm dams with exchangeable cations. Refers to research at Queensland University on a variation of the crumb test in which solutions of varying sodium and calcium content are employed to give a quantitative reaction.

Casagrande, A. (1950), "Notes on the Design of Earth Dams," *J. Boston Soc. Civil Eng.,* Vol. 37, Oct. 1950. Calls attention of profession to hazard of cracking in well-constructed earth dams. Examples are given of two dams in which leaks through cracks caused serious damage.

Casagrande, A. (1965), "Hohe Staudämme (High Dams)," *Mitteilungen des Institutes für Grundbau und Bodenmechanik, Technische Hochschule Wien,* Vienna, December 1965, p. 7. (In German.) Summary of the author's recent experience on high earth core-rockfill dams and of his current thinking about some of the main problems of design and construction using Cougar, Round Butte, and Furnas dams as illustrations. Emphasis on the hazard of cracking and on design measures to counteract it. States that he believes it possible for horizontal cracks to open through narrow central core rockfill dams as the result of arching of the more compressible core between the shells.

Casagrande, A. (1969), Participation in panel discussion on earth and rockfill dams, *7th Int. Conf. Soil Mech. Found. Eng., Mexico,* Vol. 3, p. 301. Transcript of an informal discussion in which general opinions are given on the hazard of cracking and preventive design.

Clough, R. W., and R. J. Woodward, III (1967), "Analysis of Embankment Stresses and Deformations," *J. Soil. Mech. ASCE,* July 1967, p. 529. Application of the finite-element method for the computation of the stresses and deformations in embankment dams and their foundations due to the load applied during construction. A pioneer effort which shows the power of this method of analysis.

Cole, D. C. H., and J. G. Lewis (1960), "Piping Failure of Earthen Dams Built of Plastic Materials in Arid Climates," *3rd Aust.-N. Z. Conf. Soil Mech. Found. Eng., 1960,* p. 93. A study of six low dams which failed by piping gave tentative indication that relatively "active" clays with a low natural water content are particularly suspect.

Covarrubias, S. W. (1969), "Cracking of Earth and Rockfill Dams," Harvard Soil Mech. Series No. 82, April 1969. A theoretical investigation by means of the finite-element method. Studied influence of abutment shape, foundation compressibility, and embankment zoning on development of zones of tension.

Dayte, K. R. (1968), "Detailing Earth Dams to Minimize Cracking Hazards," *Indian Natl Soc. Soil Mech. Found. Eng., Symp. Earth and Rockfill Dams, Punjab, India,* Oct. 1968. Paper based on practice in India. Contains detailed consideration in general terms of various aspects of embankment design and foundation preparation, as these relate to cracking problems, such as maximum desirable longitudinal slopes between embankment and sections constructed at different times, and special problems of steep-walled foundation trenches and junctures between earth dams and concrete structures. Speculates about the pattern of cracking which may develop inside a dam core adjacent to sharp irregularities in rock foundations. Generally presents the opinion that embankment cracking has presented a very real hazard to the safety of dams in India.

DeFries, K. C. (1968), "Analisis de Esfuerzos y Deformaciones en una Presa de Tierra," *Bull. Venez. Soc. Soil Mech. Found. Eng., Caracas,* April-June 1968, No. 27, p. 3. (In Spanish with an abstract in English.) An interesting description of the cracking which occurred at the El Isiro Dam with some photographs of the very wide crack on the downstream slope, the results of laboratory tests on the various embankment materials, and the study of the embankment strains made by the author using the finite element approach.

Endersbee, L. A. (1969), "Man-Initiated Earthquakes?", *Bull. Aust. Natl Com. Large Dams,* Oct. 1969, p. 25. Contains a description of 13 reservoirs at which increased seismic activity has been reported soon after filling.

ENR (1923), A series of news articles on the failure of the Apishapa Dam, pp. 357, 418, 422, 900, and 1029. When the reservoir filled for the first time, a leak with an initial estimated volume of 2000 cfs broke through a horizontal crack near the top of the dam leading rapidly to a complete failure.

Evans, D. M. (1967), "Manmade Earthquakes—A Progress Report," *Geotimes*, American Geological Inst. publication, July-Aug. 1967. Current report on the increased earthquake activity near Denver, Colorado, caused by pumping waste liquid into a deep well and a discussion of earthquakes caused in general by reservoir construction including the flurry of small shocks centered near the construction of the recently completed Cabin Creek pumped storage project in the Colorado Rockies.

Farhi, F. J., and M. Hamon (1967), "Barrage de Djatiluhur. Deformations du Remblai et Fissuration du Noyau," *9th Cong. Large Dams, Istanbul*, Vol. V, p. 457. (In French.) Detailed summary of measurements of rather large movements of the embankment and descriptions of cracking in clay core-rockfill dam in Indonesia.

Galloway, J. H. H. (1967), "Installed Instruments in Matahiña Earth Dam," *5th Aust.-N. Z. Conf. Soil Mech. Found. Eng.*, pp. 147–152 and discussions, pp. 165–168. Description of cracking, piping, and resulting crest subsidence in 280-ft-high dam with central earth core and rockfill shells on first reservoir filling in 1967.

Gilg, B. (1970), "Apparition de Fissures dans la Dique de Mattmark," *10th Cong. Large Dams, Montreal*, Vol. 1, p. 189. Description of important transverse cracking developing during construction in 1965 of the 100-meter-high Mattmark Dam in Switzerland.

Gordon, J. L. and D. R. Duguid (1970), "Experiences with Cracking at Duncan Dam," *10th Int. Cong. Large Dams, Montreal*, Vol. 1, p. 469. Unprecedented high differential settlement caused severe transverse cracking during construction. Embankment design in cracked area changed during construction. No leakage problems during first two seasons of reservoir operation.

Guha, S. K., et al. (1968), "Recent Seismic Disturbances in the Shivajisagar Lake Area of the Koyña Hydroelectric project, Maharashtra, India," *Report of the Central Water and Power Research Station, Khadakwasla, Poona, 24, Govt. of India,* March 1968. Detailed description of the earthquake activity in the vicinity of the Koyña hydroelectric project since the first impounding of the reservoir in 1962.

Hamilton, D. H., and R. L. Meehan (1971), "Ground Rupture in the Baldwin Hills," Science, April 23, 1971. Extensive analysis of the fault movement which caused the failure of the Baldwin Hills Reservoir in December 1963, leading to conclusion that the cause was the high-pressure water injection into deep boreholes for the purpose of repressurizing the oil field.

Hamon, M., and H. Buellet (1966), "Djatiluhur Dam," *Travaux, Paris,* July-August 1966. A thorough discussion of Djatiluhur Dam in 2 main parts: the design and the construction. (In French.)

Hsu, S. J. C. (1963), "Residual Clay Earth Dams Constructed by Rio Light S.A.," *2nd Pan-Am. Cong. Soil Mech. Found. Eng., Brazil.* Description of design and construction of four homogeneous earth dams in Brazil with thin vertical chimney drains of sand.

Hubert, M. K., and D. G. Willis (1957), "Mechanics of Hydraulic Fracturing," *Petrol. Trans., AIMining Eng.*, Vol. 210, p. 153. Comprehensive review of relationship between pressures of fluids in boreholes needed for penetration of fluid into cracks in surrounding rock formations and the state of stress in rock before the boring is made, from elastic theory. Results compare favorably with known experience in petroleum industry.

Housner, G. W. (1969), "The Seismic Events at Koyna Dam," Paper presented at Rock Mech. Conference, Berkeley, Calif., June 17, 1969. Review of seismic history of region and details of reservoir induced earthquake in 1967.

Ingles, O. G. (1964), "The Water-Soil Regime in Three Dams Sampled during Failure," Paper presented at the Colloquium on Failure of Small Dams, November 16, 1970, Melbourne, Australia. Contains details of post-failure studies of the Flagstaff Gully Dam described in Ingles et al., 1968.

Ingles, O. G., and G. D. Aitchison (1969), "Soil-Water Dis-Equilibrium Observations on the Reconstructed Flag-and Earth Embankments," *Symp. Land Subsidence, Tokyo,* Sept. 1969, International Association of Scientific Hydrology (IASH), Gent-Brugge, Belgium, Vol. 2, p. 342. Review and extension of research relating piping potential and soil chemistry.

Ingles, O. G., J. G. Lang, and B. G. Richards (1968), "Pre-Equilibrium Observations on the reconstructed Flagstaff Gully Dam," *Symp. Earth Rockfill Dams, Talwara (Punjab), India*, Nov. 1968, Sarita Prakashan (pub.), Meerut, India, Vol. 1, p. 162. It is proposed that failure of a 50-ft central core dam in 1963 may have been caused in part by piping due to deflocculation of a sodium rich clay by water of low salinity.

Ingles, O. G., J. G. Lang, B. G. Richards, and C. C. Wood (1969), "Soil and Water Observations in Flagstaff Gully Dam," Technical Report No. 11, Soil Mechanics Division, Commonwealth Scientific and Industrial Research Organization, Syndal, Victoria, Australia. More detailed description of the studies of dam failure described in Ingles et al., 1968.

Jaeger, J. C. (1969), "Occurrence and Effects of Earthquakes," *Bull. Aust. Nat. Comm. Large Dams,* Oct. 1969, p. 2. Extensive review of literature on the correlation of earthquakes with construction of reservoirs including some recent experience in Australia.

James, J. P., and R. T. Wickham (1970), "Failure and Rectification of the Morwell No. 2 Fire Service Reservoir," *Bull. Aust. Natl Com. Large Dams*, July 1970, p. 20. Failure of a 30-ft-high, well-constructed homogeneous clay dam by piping on first reservoir filling, suspected to have been caused by drying cracks occurring during a 3-month halt in the construction, and "dispersive" montmorillonite clay of low piping resistance. As part of repairs a blanket of gypsum was spread on the upstream face to reduce the risk of a second piping failure.

Johnson, W. E. (1970), Discussion of fluid loss in boreholes, *Trans. 10th Cong. Large Dams, Montreal*, Vol. VI, p. 390.

Kassiff, G., and E. N. Henkin (1967), "Engineering and Physico-Chemical Properties Affecting Piping Failure of Low Loess Dams in the Neger," *Proc. 3rd Asian Regional Conf. Soil Mech. Found. Eng., Haifa,* Vol. 1, p. 13. Study of 10 low clay dams, including 5 which failed by piping, shows a strong correlation between incidence of failure and the percentage of exchangeable sodium cations in the clay.

Kennard, M. F. (1964), "Balderhead Reservoir," *Civil Eng. Pub. Works Rev.* (*London*), May 1964. A description of the construction of the Balderhead Dam published before the cracking trouble developed.

Kennard, M. F., L. K. Knill, and P. R. Vaughan (1967), "The Geotechnical Properties and Behavior of Carboniferous Shale at the Balderhead Dam," *J. Eng. Geol.* (*London*). Description of investigations of the shale used for the main part of the Balderhead Dam plus a summary of construction methods and measurements of pore water pressure and settlement during the first year of operation, before the main trouble with cracking and dirty leakage developed.

Kennard, M. F., A. D. M. Penman, and P. R. Vaughan (1967), "Stress and Strain Measurements in the Clay Core at Balderhead Dam," *9th Cong. Large Dams, Istanbul,* Vol. 3, p. 129. Summary of instrumentation and measurements at the Balderhead Dam just before cracking and piping problems developed.

Kjaernsli, B., and I. Torblaa (1968), "Leakage through Horizontal Cracks in the Core of Hyttejuvet Dam," Publication No. 80, Norwegian Geotechnical Institute, Oslo, p. 39. Description of experience with large concentrated leak through 90-meter-high earth core, rockfill dam on the first reservoir filling, probably caused by a horizontal crack through the core in upper part of dam.

Kulhawy, F. H., and J. M. Duncan (1970), "Nonlinear Finite Element Analysis of Stresses and Movements in Oroville Dam," Report Te-70-2, Univ. of Calif., Berkeley, Dept. of Civil Engineering, Jan. 1970. Comparison of measured stress and movements with finite element method calculations assuming nonlinear stress-strain properties for embankment.

Lane, K. S. (1963), Discussion of paper, "Flexibility of Clay and Cracking of Earth Dams," by Leonards and Narain (1963), *J. ASCE Soil Mech. Found. Div.*, Nov. 1963, p. 109. A brief summary of the writer's opinions concerning the current status of the ability of the profession to evaluate the hazard of cracking and to design against the hazard. Includes discussion of design of Garrison Dam on the Missouri River where special precautions were taken in the central river closure section to control possible adverse effects of leaks through cracks. Major loss of drill water experienced in boreholes put down through the embankment in the limited area on right abutment. Origin never known for certain but finally considered not serious.

Lauffer, H., and W. Schober (1964), "The Gepatsch Rockfill Dam in the Kauner Valley," *Trans. 8th Cong. Large Dams, Edinburgh.* Detailed description of the design and construction problems of the high Gepatsch Dam in Austria.

Lee, K. L., and C. K. Shen (1969), "Horizontal Movements Related to Subsidence," *ASCE J. Soil Mech. Found. Eng.*, Jan. 1969. Shows theory for prediction of longitudinal strain of crest of dam based on assumed foundation settlement—checks reasonably well with few actual experiences.

Leonards, G. A., and J. Narain (1963), "Flexibility of Clay and Cracking of Earth Dams, *ASCE J. Soil Mech. Found. Eng.*, Mar. 1963, p. 47. A laboratory and theoretical investigation of the problem. Tests were made on compacted beams of soil. Results compared with calculations on basis of elastic theory and observed movements or cracking on five dams. Results of additional tests are given in the authors' closure in July 1964 *Journal*, p. 135.

Little, A. L. (1967), "The Use of Tropically Weathered Soils in the Construction of Earth Dams," *Proc. 3rd Asian Regional Conf. Soil Mech. Found. Eng., Haifa,* Vol. 1, p. 35. Further details of the problem at Shek Pik Dam as described by Little and Beavan, 1963. "The precise cause of the cracking and in particular the means of detecting soils susceptible to cracking when used as fill, remain unknown." Speculates trouble may be result of fact that soil for the core was derived from weathered rhyolite and contains minerals which act as cementing agents in the recompacted soil to make it especially brittle and susceptible to differential settlement cracking.

Little, A. L., and G. C. G. Beavan (1963), Discussion of "Flexibility of Clay and Cracking of Earth Dams," by Leonards and Narain (1963), *J. ASCE Soil Mech. Found. Div.*, Nov. 1963, p. 116. Description of water losses in boreholes in the core of the Shek Pik Dam.

Lofquist, B. (1951), "Earth Pressure in Thin Impervious Core," *4th Cong. Large Dams, New Delhi,* Vol. 1, p. 99. Summary and analysis of measurement of pressures in clay cores of two central core rockfill dams and the discussion of the potential of horizontal cracks opening as a result of arching.

Lofquist, B. (1955), Verbal discussion of Question 18, *5th Cong. Large Dams,* Vol. 3, p. 22. Brief description of an experience with a dam with a thin vertical earth core. An estimate was made that the postconstruction settlement of the core should have been about 2 ft. Actual settlements amounted only to 2 in. and stopped after a few months. The engineers were concerned that it was possible to have a horizontal crack of considerable width through the core and a steel sheet piling cutoff wall was installed before the reservoir was filled.

Londe, P. (1970), "La fissuration des noyaux", *Trans. 10th Cong. Large Dams*, Vol. VI, p. 384.

Lowe, J., III (1970), "Recent Development in Design and Construction of Earth and Rockfill Dams," *10th Cong. Large Dams, Montreal,* Vol. 5, pp. 1–28. General report on Question 36 devoted largely to the problem of differential settlement cracking.

Mane, P. M. (1967), "Earth Tremors in the Koyña Project Area," *9th Cong. Large Dams, Istanbul,* Vol. V, p. 509. Description of the early earthquakes which were generated by the reservoir construction. Published before the very large quake occurred.

Marsal, R. J. (1959), "Earth Dams in Mexico," *Proc. 1st Pan-Am. Conf. Soil Mech. Found. Eng., Mexico City,* Vol. III, p. 1294. Summary of experience with 26 dams in which 4 developed transverse cracks and 5 had longitudinal cracks. No correlation was found between soil type and incidence of cracking.

Marsal, R. J., and Ramirez, L. (1967), "Performance of El Infiernillo Dam, 1963-66," *J. ASCE Soil Mech. Found. Eng.*, July 1967, p. 265. Description of measurements of movement with detailed instrumentation in high clay core-rockfill dam, during and after construction and of development of transverse crack.

Murthy, Y. K., and K. C. Thomas (1970), "Koyña Dam," *10th Int. Cong. Large Dams, Montreal,* Vol. IV, p. 1019. Description of the damage to Koyña Dam during the 1967 earthquake.

Narain, J. (1962), "Flexibility of Compacted Clays," Thesis, Civil Engineering, PhD, Jan. 1962. A laboratory and theoretical study of the cracking problem. Most results are summarized by Leonards and Narain, 1963.

New Zealand Ministry of Works (1970), "Subsidence and Piping in Matahiña Dam, Rangitaiki River, Rotorua, New Zealand," unpublished report. Summary of crest settlement, cracking, and piping. Probably will be published in future.

Nonveiller, E. (1968), "Nasuta Brana Sklope Na Lici," published in the Yugoslav Civil Engineering Journal, *Gradjevinar,* Mar. 1968, Zagreb, Yugoslavia. (In Yugoslav with good synopsis in English.) Describes design and construction of the rockfill, central clay core Sklope Dam with height of about 75 meters, in a narrow rock gorge. In two areas over the abutments, where the greatest longitudinal stretch of the core is to be expected, special zones of clay with higher plasticity and higher water content were used.

Patrick, J. G. (1967), "Post Construction Behavior of Round Butte Dam," *ASCE J. Soil Mech. Found. Eng.* July 1967, p. 251. Description of the measurement of movements and of the cracking which developed in this central core rockfill dam during the first period of operation.

Peterson, R., and N. L. Iverson (1953), "Study of Several Low Earth Dam Failures," *3rd Int. Conf. Soil Mech. Found. Eng.,* Vol. 2, p. 273. One of the earlier discussions in the literature concerning the hazard of embankment cracking. Description of several examples in Canada.

Pope, R. J. (1967), "Evaluation of Cougar Embankment Performance," *ASCE J. Soil Mech. Found. Div.,* July 1967, p. 261. Description of the performance and rather spectacular longitudinal cracks on the crest of the dam with a single transverse crack at one end near the abutment.

Rallings, R. A. (1966), "An Investigation into the Causes of Failure of Farm Dams in the Brigalow Belt of Central Queensland," Water Research Foundation of Australia, Bull. No. 10. Investigation of causes of piping failure of small farm dams. Concludes that strong correlation exists between failure and exchangeable sodium in the soil. Good bibliography of articles relating chemistry of soils to engineering properties for dams.

Rothé, J. P. (1969), "Earthquakes and Reservoir Loadings," *Proc. 4th World Conf. Earthquake Eng., Santiago, Chile.* A summary of examples where it is believed that the construction of reservoirs has triggered earthquakes, including several in France.

Sabarly, F. (1968), "Les Injections et les Drainages de Fondation de Barrages," *Geotechnique* (*London*), June 1968, p. 229. Contains theory of flow through rock cracks.

Schober, W. (1967), "Behavior of the Gepatsch Rockfill Dam," *9th Cong. Large Dams, Istanbul,* Vol. III, p. 677. Summary of measurements from extensive instrumentation in high central core, rockfill dam with very steep side slopes.

Schober, W. (1967a), "Contribution to Question 34," *9th Cong. Large Dams Istanbul,* 1967, Supplementary Vol. 6. A supplement to his main paper describing the performance of the Gepatsch Dam, Schober, 1967, specifically describing the latest measurements of the movement of the crest of the dam and of the longitudinal cracking which developed.

Sherard, J. L. (1953), "Influence of Soil Properties in Construction Methods on the Performance of Homogeneous Earth Dams," Technical Memorandum 645, U.S. Bureau of Reclamation, Denver, Colorado, January 1953. A study of the performance of 60 old dams in the western part of the United States including records of 6 dams which developed differential settlement cracking. Includes an attempt to correlate incidence of cracking with the type of soil from which the embankments were built. The results are summarized briefly in Sherard et al., 1963.

Sherard, J. L., R. J. Woodward, S. F. Gizienski, and W. A. Clevenger (1963), *Earth and Earth-Rock Dams—Engineering Problems of Design and Construction*, John Wiley and Sons, New York. Reference book contains review of knowledge and experience concerning cracking.

Sherard, J. L. (1967), "Earthquake Considerations in Earth Dam Design," *ASCE J. Soil Mech.,* July 1967, p. 377. General summary of writer's main opinions concerning earthquake design written following a study of the subject for the Calif. Dept. of Water Resources. Contains a discussion of properties of soil used for dam cores as a function of the relative desirability of the material from the standpoint of resistance to erosion by concentrated leaks, such as through cracks.

Sherard, J. L., R. S. Decker, and N. L. Ryker (1972), "Piping in Earth Dams of Dispersive Clay", Proc. Conf. on the Performance of Earth and Earth-Supported Structures, ASCE, June 1972, Vol. 1, p. 589.

Sherard, J. L., R. S. Decker and N. L. Ryker (1972a), "Hydraulic Fracturing in Low Dams of Dispersive Clay", Proc. Conf. on the Performance of Earth and Earth-Supported Structures, ASCE, June 1972, p. 653.

State of California (1964), "Investigation of Failure of Baldwin Hills Reservoir," Report of Dept. of Water Resources, Sacramento, Calif., Apr. 1964. A comprehensive summary of the facts surrounding the failure of the reservoir caused by progressive erosion of a leak originating in a crack in the foundation.

Tamez, E., and G. Springall (1960), "The Use of Soils as Construction Materials for Earth Dams," *Proc. 1st Pan-Am. Conf. Soil Mech. Found. Eng., Mexico,* Vol. 3. A summary paper of experience with a number of dams calling attention to the potential hazard of cracking with an effort by the authors to classify soil types used for earth dam cores on the basis of susceptibility to cracking.

Terzaghi, K., and Y. Lacroix (1964), "Mission Dam, an Earth and Rockfill Dam, on a Highly Compressible Foundation," *Géotechnique,* Mar. 1964, p. 14. Description of the design and first few years of performance of a dam of a height of about 150 ft at a site where the foundation was expected to settle roughly 15 ft and the abutments were incompressible. The dam embankment used was designed to be safe regardless of the embankment distortion which might develop.

Terzaghi, K., and R. Peck (1967), *Soil Mechanics in Engineering Practice,* 2nd ed., John Wiley and Sons, New York, p. 597. Brief description of the failure of the Stockton Creek Dam in California in 1960 showing a cross section of the dam and a schematic picture of the type of failure.

UNESCO (1970), "Draft Report of Working Group on Seismic Phenomena Associated with Large Reservoirs," Paris, December 23, 1970. A group of international experts agreed that reservoir construction has caused earthquakes and suggested research and a future technical symposium on the subject.

U.S. Army Corps of Engineers (1959), "Review of Soils Design, Construction and Performance Observations, Wister Dam, Oklahoma," Waterways Experiment Station Tech. Report No. 3-505, Vicksburg, Miss. Description of design and construction, and of the piping which occurred on the first reservoir filling.

Vargas, M., and S. J. C. Hsu (1970), "The Use of Vertical Core Drains in Brazilian Dams," *10th Cong. Large Dams, Montreal,* Vol. 1, p. 599. Description of six homogeneous clay dams in Brazil, all of which had evidence of differential settlement cracks.

Vargas, M., F. P. Silva, and M. Tubio (1965), "Residual Clay Dams in the State of Saõ Paulo, Brazil," *6th Int. Conf. Soil Mech. Found. Eng., Montreal,* Vol. 2, p. 578. Experiences with homogeneous dams for important reservoirs in Brazil.

Vaughan, P. R. (1970), "Cracking of Clay Cores of Dams," *Surveyor* (*London*), Jan. 30, 1970, p. 32. An analysis of the mechanics of cracking and some new concepts for design of filters to protect against concentrated leaks through cracks. A brief report on some fundamental and original work being carried out by the author, stimulated by the problem at Balderhead Dam.

Vaughan, P. R. (1970a), "Cracking of Clay Cores of Dams," *Proc. Inst. Civil Eng.* (*London*), May 1970, p. 115. Record of informal discussion following presentation of Vaughan (1970) as a lecture. Includes discussion of the feasibility of measuring the minimum principal stress acting within an embankment by increasing the water pressure in a hydraulic piezometer to the point the water breaks out of the piezometer tip into a newly formed crack.

Vaughan, P. R., D. J. Kluth, M. W. Leonard, and H. H. M. Pradoura (1970), "Cracking and Erosion of Balderhead Dam and the Remedial Works Adopted for Its Repair," *10th Cong. Large Dams, Montreal,* Vol. 1, p. 73. Comprehensive description of dirty leakage through rockfill dam with narrow central earth core, probably resulting from horizontal cracks in the core due to arching, including repairs by grouting and construction of new concrete core wall.

Vaughan, P. R. (1972), Discussion of Bjerrum and Andersen, 1972. *5th European Conf. Soil Mech. Found. Eng., Madrid.* Describes the use of artificially induced fracturing in boreholes in the cores of earth dams for purpose of estimating the minimum principal stress.

Walker, F. C. (1970), "Prevention of Cracking in Earth Dams," *Int. Cong. Large Dams, Montreal,* Vol. 1, p. 361. Philosophical discussion of cracks of many kinds by the long-time head of earth dam design of the U.S. Bureau of Reclamation.

Wilson, S. D., and R. Squier (1969), "Earth and Rockfill Dams," *7th Int. Conf. Soil Mech. Found. Eng., Mexico City,* State of the Art Volume, pp. 137–224. Includes description of water loss in borings drilled in the impervious section of the Akosombo and LaVallita dams. At LaVallita the fluid flowed in vertical cracks parallel to the longitudinal axis as determined by instrument observations.

Wood, C. C., G. D. Aitchison, and O. G. Ingles (1964), "Physico-Chemical and Engineering Aspects of Piping Failures in Small Earth Dams," Paper presented at the Colloquium on Failure of Small Dams, Nov. 16, 1964, Melbourne, Australia. Comprehensive review of literature on piping in dams plus study of 100 small dams in Australia. Concludes that the incidence of piping in clay soils is governed by the relationship between the cation balance of the soil and the cation concentration of the percolating water.

Introduction to Paper 7

Potentially Active Faults in Dam Foundations

by J.L. Sherard, L.S. Cluff, and C.R. Allen
Geotechnique, Vol 24, No. 3, September 1974, pp 367-428.

Prepared by Clarence R. Allen[2]

My primary interaction with Jim Sherard came about because of our joint service on the Consulting Board for Earthquake Analysis of the California Department of Water Resources. In the 1960's, this Board was primarily concerned with problems of the California Aqueduct, then in planning and under construction, but the Board still serves today--with some of the same original members. During Jim's tenure, other Board members were Hugo Benioff, George Housner, Harry Seed, Bruce Bolt, Nate Whitman, and myself. Service with this group was an exciting and educational experience for all of us, as we were often treading new ground in attempting to establish criteria for vibratory ground motions near the center of large earthquakes, fault offset criteria, etc. The interplay between the geologists, seismologists and the engineers was especially vigorous, and it was a subject in which Jim Sherard had a particular interest.

During the exploratory excavation for Cedar Springs Dam, active faults were unexpectedly discovered in the foundation, and Jim was intrigued by their potential significance in the design of the dam, if, in fact, the dam was to be built at all. This, as I recall, was the primary stimulus behind the following article. Lloyd Cluff, then a junior geologist with Woodward-Clyde-Sherard, and I provided the backboard off which Jim bounced many of his ideas. The article is primarily Jim's, but preparing it was an exceptional educational experience for Lloyd and myself, and, I think for Jim as well. Although we have learned a great deal since that time, the article is, in my opinion, still somewhat unique in its broad overview of this sensitive and important subject.

[2]Professor of Geology and Geophysics, Emeritus, Seismological Laboratory, California Institute of Technology, Pasadena, CA.

SHERARD, J. L., CLUFF, L. S. & ALLEN, C. R. (1974). *Géotechnique* **24**, No. 3, 367–428.

Potentially active faults in dam foundations

J. L. SHERARD,* L. S. CLUFF† and C. R. ALLEN‡

The Paper contains information on existing dams founded on active faults, a summary of pertinent lessons learned from study of historic fault breaks and fault mechanisms, recommended practice for evaluation of active faults, and opinions concerning design of dams on active faults. While a dam site with an active fault should be avoided if possible, if a reservoir is vitally needed and a better site is not available, it is reasonable practice to construct a conservatively designed embankment dam. Concrete dams on active faults, or near some major active faults, are not advisable. For evaluation of fault activity, geological studies usually must be carried a considerable distance from the dam site, a departure from recent past practice. Experience of the last few years with many fault studies indicates that thorough geological investigations with modern techniques will usually provide sufficient evidence to allow a judgement on the activity or inactivity of a fault.

Cet Article présente des renseignements pertinents aux barrages construits sur des failles actives, un resumé historique des leçons tirées de la rupture de failles et de leurs mécanismes de rupture, des conseils pratiques pour l'évaluation de l'activité des failles, et des commentaires sur la conception de barrages situés sur des failles actives. Bien que la conception d'un barrage sur une faille active devrait être évitée, la construction d'un barrage en terre conçu d'une façon très sécuritaire est raisonable s'il est absolument nécessaire d'avoir un réservoir à cet endroit et qu'il n'y a pas d'alternative. Il est déconseillé de prévoir des barrages en béton sur des failles actives ou près d'importantes failles actives. Pour l'évaluation de l'activité d'une faille, il faut entreprendre des études géologiques généralement à une très grande distance du site, ce qui est contraire à une pratique encore récente. L'expérience de ces dernières années, avec l'étude de nombreuses failles, montre que des études géologiques détaillées, utilisant des techniques modernes, fourniront normallement suffisament d'évidences pour l'évaluation de l'activité ou de l'inactivité d'une faille.

INTRODUCTION

In earthquake regions faults are frequently found to exist in dam foundations. Evaluation of the likelihood that displacement could occur along the fault during the lifetime of the dam, and the selection of the design details to ensure safety against such possible fault displacement, are difficult problems for which there is little guidance in the literature.

In 1963, one of the Authors (with co-authors) summarized ideas about the influence of active faults in dam foundations (Sherard *et al.*, 1963). This Paper presents a more intensive examination of the subject. Emphasis is on the engineer's problem in evaluating what kind of study should be made of the fault and how the results of the study should influence the dam design. While written primarily for the engineer, the Paper contains points of interest for the geologist regarding techniques employed in evaluation of potential fault activity not widely used in other geological studies.

Any dam site located near a potentially active fault may be subjected to strong earthquake shaking. The problems of designing the dam to withstand the accelerations imposed by strong earthquake shaking and of making the dam safe against fault movement in the foundation are related; however, this Paper is directed wholly to the potentially active foundation fault which might displace abruptly during an earthquake, rupturing the dam.

The terms 'fault break', 'fault movement', 'fault displacement', 'fault slippage' and 'faulting' are used synonymously to mean the shearing movement of ground on one side of the fault against the other. Usually this faulting occurs during a strong earthquake and, in fact, it is the energy release associated with the faulting which causes the earthquake. The

* Consulting Engineer, Berkeley, California.
† Vice President and Chief Engineering Geologist, Woodward Lungren and Associates, Oakland, California.
‡ Professor of Geology and Geophysics, Seismological Laboratory, California Institute of Technology, Pasadena, California.

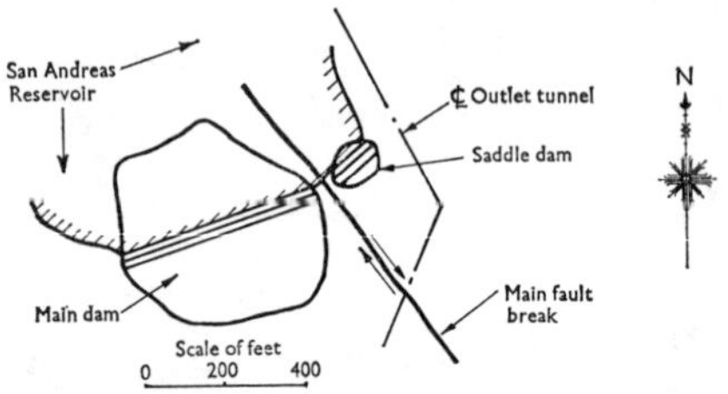

Fig. 1. San Andreas Dam—location of main fault break (after Lawson, 1908)

size of the earthquake at its source is described by the Richter magnitude M in this Paper. For background information on this and other aspects of seismology and seismic geology the reader is referred to the classic work by Richter (1958).

In some areas of frequent strong earthquakes, faults in dam foundations are the rule rather than the exception for the two main reasons that very many faults exist, and that river channels often follow faults because the rocks in the fault zone are more crushed and broken and consequently more easily eroded. Of the hundreds of dams built in regions of high earthquake potential, there are few in locations where faults were not recognized or suspected to exist.

For the purposes of this study, an active fault (or a potentially active fault) is defined as one on which there is sufficient evidence of displacement within the recent geological past to make it reasonable to anticipate that future surface displacements could occur within the lifetime of a dam (about 100 years). It must be emphasized, however, that this definition implies the exercise of judgement by geologists, and that there will be honest differences of opinion among competent geologists specializing in this field as to the likelihood of displacements on a given fault.

According to the definition adopted by the US Atomic Energy Commission (1973) for nuclear power plant site studies, an active fault (now termed by the AEC a 'capable' fault) has one or more of the following characteristics:

(*a*) displacement at or near the ground surface at least once in 35 000 years or movement of a recurring nature within the past 500 000 years;

(*b*) instrumentally measured macro-seismicity determined with sufficient precision to demonstrate a direct relationship with the fault;

(*c*) a structural relationship with an active fault having the characteristics of (*a*) or (*b*) such that displacement on the one could be reasonably expected to be accompanied by displacement on the other.

This is a reasonable and conservative definition when applied also to dams; however, unfortunately the chronological histories of recent displacements on most faults cannot be as yet established with certainty, so that it finally becomes a matter of the geologist's judgement in interpreting the evidence as to how recently and how often the fault has moved in the past.

Dam design is largely an empirical activity. While increasingly useful methods of analytical treatment are available to guide design decisions, it is clear that one builds dams with confifidence primarily because many other dams of similar design details have been built and performed satisfactorily. With regard to movements of faults in dam foundations, there is essentially no experience record to guide us, so that related performance and judgement must be relied on. Because of this and of the many uncertainties and the judgement required in fault activity evaluation, it is clear that the main decisions concerned with designing safe dams at sites with potentially active faults must be based largely on judgement, guided by the accumulated experience and knowledge in the fields of geology, seismology, and the several engineering specialities that comprise dam engineering. In this Paper the Authors, who have collaborated for a number of years on several projects, attempt to summarize the main points involved and present opinions on some of the debatable aspects of the problem. No large changes are proposed in what might be considered current conservative practice for building dams in earthquake regions at sites with foundation faults. Information and opinions are given on lessons learned from study of historical active faulting, on methods of studying fault potential activity, and on interpreting the results of such studies.

EXPERIENCE WITH FAULT BREAKS NEAR DAMS

To the Authors' knowledge there have been essentially no failures or damage to dams caused by displacements of faults during earthquakes. There are of the order of 25 000 dam structures of moderate size in existence in the world, and there are records of more than 500 which have failed or been seriously damaged in other ways. Hence, while fault displacements during earthquakes are spectactular events which could seriously damage a dam, they have been so rare that the coincidence of a large offset passing through a dam foundation has not yet taken place.

In addition to the fact that there have been no instances where faults have displaced in dam foundations during strong earthquakes, there have been only a few 'near misses'. The Authors know of only two examples.

1906 San Francisco earthquake

The San Andreas Dam is a 27 m high earth dam, built more than 100 years ago a few miles south of San Francisco, California. It comprises a main embankment and a smaller saddle dam on the east abutment separated by a narrow ridge. During the 1906 San Francisco earthquake ($M=8{\cdot}3$), the main break on the San Andreas Fault passed through the small ridge forming the common abutment between the two embankments (Fig. 1). The main embankment dam was moved bodily, approximately 2·5 m north relative to its east abutment and the saddle dam. A tunnel in the east abutment was twisted about 3 m out of line, but there was no damage to the dam. The 20 million m^3 reservoir was full and no leakage or other alarming symptoms developed (Lawson, 1908).

From a statistical point of view this experience is a special case because the dam was constructed directly across the main fault valley of the major San Andreas Fault (Fig. 2). It was done more than 100 years ago at a time when very little was known about earthquakes and faulting. At the present state of knowledge such major faults as the San Andreas can be easily recognized as active faults, and the fault valley would not generally have been considered a suitable dam site.

The Crystal Springs Reservoir of the San Francisco water supply system is also located in the main San Andreas Fault Valley, a few miles south of the San Andreas Dam. This reservoir was crossed by a road embankment (Fig. 3) which had formerly been a dam (Upper Crystal Springs Dam). At the time of the 1906 earthquake this embankment was not being used as a dam but as a roadway. During the earthquake the main fault break passed through the central portion of the embankment approximately normal to the longitudinal axis. The embankment was offset approximately 2·5 m with the west side moving north with respect to the east side. Both longitudinal and transverse cracks were observed on the crest extending to a depth of at least one meter (Sewell, 1907). Since there was water on both sides there was no tendency for a leak to develop through the fissured zone and it is not known how the embankment might have behaved had it been performing as a dam. The Lower Crystal Springs Dam, a concrete gravity dam of 43 m height, constructed in 1888, was not damaged by the earthquake or fault displacement, although located about 200 m from the fault break.

Figure 4 shows an offset (about 2 m) of yet another small dam (Old San Andreas Dam). This abandoned dam was submerged under the north arm of the Crystal Springs Reservoir at the time of the 1906 earthquake and was only exposed in 1931 when the reservoir was lowered (Louderback, 1937).

1959 West Yellowstone earthquake

The second example of a 'near miss' was at the Hebgen Dam, Montana, during the West Yellowstone earthquake of 17 August 1959 ($M=7{\cdot}1$). The 27 m high, 210 m long earth dam, constructed in 1914, retained a 370 million m^3 reservoir with a length of about 25 km. The reservoir was full at the time of the earthquake. Movements occurred on a number of major faults with primarily vertical displacement (Fig. 5). The aggregate length of all the main fault breaks was more than 75 km. The main Hebgen Fault break traversing the north wall of the reservoir, about 200 m from the right end of the dam had a vertical displacement of about 5 m at this point, with the north side of the fault going up relative to the south side (Figs 6–8).

Associated with the faulting there was a general subsidence of the bedrock underlying the reservoir, with maximum of about 22 ft (6·7 m), (Fig. 5). The bedrock under the dam (and the whole dam, spillway and outlet conduit) went down approximately 3·0 m in a remarkably uniform fashion. The 'average' subsidence of the large reservoir was about 3·2 m, so that the dam and water surface remained at the same relative elevation.

The earthquake shaking caused moderate slumping and some cracking in the dam but no threat of serious trouble developed (Sherard, 1959; Sherard *et al.*, 1963), and the fault breaks did not influence the dam. The dam was repaired at relatively little cost and has continued in operation.

All American Canal

Figure 9 shows the resulting fault breaks through a large canal in 1940, illustrating particularly that for breaks on strike-slip faults (with primarily horizontal movement) the zone of fracturing is limited to a relatively narrow zone.

BALDWIN HILLS RESERVOIR FAILURE

Failure of Baldwin Hills Reservoir, California, in 1963 was caused by progressive erosion of an initial leak along a fault (Jansen *et al.*, 1967). No earthquake was involved. The fault moved a few inches sometime between construction of the dam and failure, breaking the compacted clay reservoir lining. The reservoir bottom and natural abutments consisted of a horizontally stratified deposit of cohesionless fine sand, silt, and clay of Lower Pleistocene and Upper Pliocene age. The fault was recognized during construction, passing through the formations.

The leakage water entered the crack along the fault and emerged as a concentrated leak on the downstream sloping surface of the natural abutment. The clay layers were sufficiently cohesive to form arches over erosion tunnels and the erodible fine sand and silt was washed out. The cause of the movement at the fault is strongly debated because of nearby relatively large ground surface subsidence related to oil withdrawal (Leps, 1972; Casagrande *et al.*, 1972; Castle and Youd, 1972). The experience emphasizes the hazard of uncontrolled concen-

Photo: Sunderland Aerial Photographs

Fig. 2. Looking north along the main rift valley of the San Andreas Fault. The city of San Francisco is in upper right. Two reservoirs are shown in the fault valley; the San Andreas Reservoir in the distance and the Crystal Springs Reservoir in the foreground. The 1906 fault break went up the valley through the centre of both reservoirs. The embankment of the old Upper Crystal Springs Dam (now serving as a highway) offset in 1906 is shown running across the Crystal Springs Reservoir in the centre. The valley is typical of fault valleys of 'master', strike-slip faults seen in many parts of the world (see also Figs 10, 19, 22, and 25)

Photo: L. S. Cluff

Fig. 3. Looking west at the Lower Crystal Springs Dam. The 43 m high concrete gravity dam retains a 65 million m^3 water supply reservoir in the north–south rift valley of the San Andreas Fault. The dam closes a gorge in the east wall of the valley and is off the fault. The 1906 break on the fault passed near the centre of the reservoir, running parallel with the longitudinal axis of the dam about 200 m upstream

Fig. 4. Old San Andreas Dam showing offset of embankment caused by 1906 break on the San Andreas Fault. At the time of the earthquake the dam was abandoned and submerged in the north arm of the Crystal Springs Reservoir. The photograph was taken in 1931 at a time of low reservoir level (after Louderback, 1937)

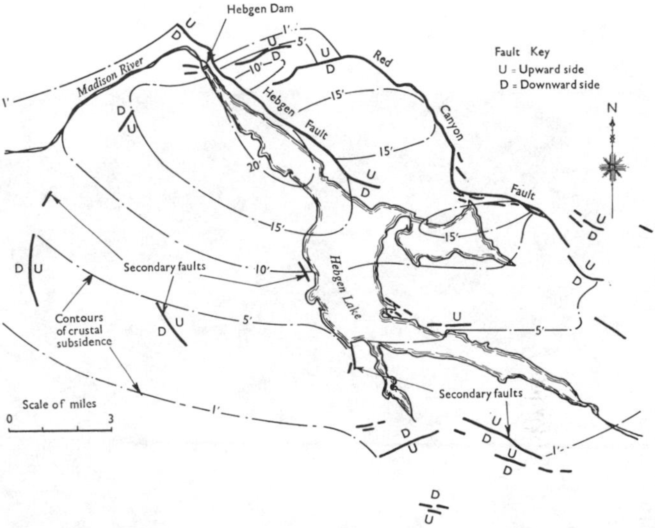

Fig. 5. Fault breaks and crustal deformation near Hebgen Lake, Montana during the August 18, 1959 earthquake (after Witkind *et al.*, 1962)

Photo: J. L. Sherard

Fig. 6. Looking upstream with Hebgen Dam in foreground, three days after the earthquake. The main Hebgen Fault break is clearly visible in the trees running along the valley wall parallel to the lakeshore on the left side of the photograph

Photo: J. L. Sherard

Fig. 7. Break on Hebgen Fault directly opposite dam; approximately 5 m vertical displacement

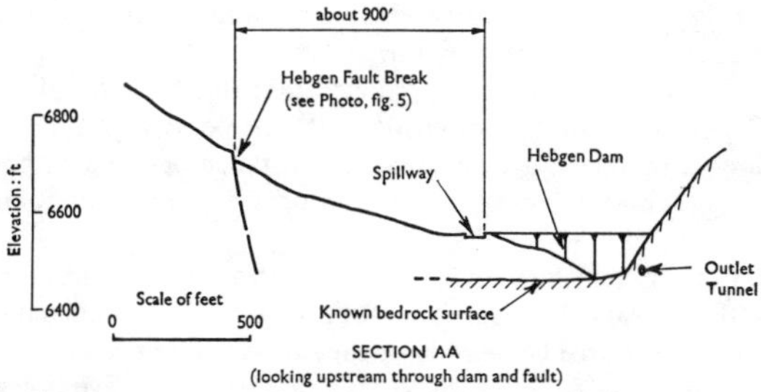

Fig. 8. Hebden Dam—longitudinal section through dam and fault looking upstream

trated leaks through foundations in abutments consisting of alternate layers of erodible, fine-grained, cohesionless soils with interbedded clay layers. The combination allows the cohesive soil to form roofs over erosion tunnels in the sand layers, so that very large leakage channels can develop. If the foundations had consisted wholly of the cohesionless fine sand, it is probable that no leakage channels could have developed, since the sand would have caved in and closed any temporary openings caused by the fault movement.

Eroded caverns ('bear caves') observed by investigators of the failure in some of the thick clay layers are difficult to explain (California Department of Water Resources, 1964). Recently it has been shown that these clays belong to the family of 'highly dispersive' (erodible) clays, whose existence has been recognized only in the last few years (Sherard *et al.*, 1972). This dispersive clay undoubtedly had some role in the failure, and conceivably the dominant one.

CURRENT AND PAST DESIGN PRACTICE

While there are notable exceptions, in common practice faults in dam foundations in earthquake regions have not been extensively studied. Usually they have been investigated by exploratory borings, pits, etc., and by detailed geological studies only in the immediate vicinity of the dam. Also, a cursory examination is commonly made of the general geology of the region (within a radius of 10–30 km), mainly by review of existing geological literature. These investigations frequently provide little or no evidence concerning the possible activity or lack of activity of faults.

For these commonly encountered faults, for which little or no evidence of activity is available, it has been considered reasonable practice to proceed with the dam construction. Underlying this practice are the tacit assumptions that it is highly improbable that the fault will break during the lifetime of the dam, and that if it should break, the amount of movement would probably not be great enough to damage the dam seriously.

In current practice where active faults are diagnosed in foundations, other sites have usually been chosen. There have been very few cases where dams have been built on sites at which active faults were believed or suspected to exist in the foundation. The Authors know of only a few cases.

Coyote Dam

The 38 m high Coyote Dam, retaining a 30 million m^3 reservoir, was constructed in 1936 near San Jose, California, at a site where the foundation is underlain by the main branch of the Calaveras Fault (Fig. 10). The nearby Hayward Fault broke last in 1868 causing an earthquake with a probable magnitude of 6·5 to 7·0.

As a part of the design studies geologists predicted that the fault would move again within the lifetime of the Coyote Dam and recommended that it be assumed the fault might move about 6 m horizontally, 1 m vertically, and that the distance between opposing sides of the fault might increase as much as 300 mm, forming an open fissure in the foundation (Louderback, 1950).

A consulting board of eminent authorities agreed that an earth dam could be safely constructed at the site, with general design details as follows: a thick, compacted earth core should be used, as opposed to a reinforced concrete, upstream-facing, concrete-core wall, or narrow clay puddle; the earth core should have a thickness equal to at least five times the water head; extra freeboard should be provided; a layer of sand and gravel should be placed against the rolled earth impervious core, so that if an opening should occur, the sand and gravel would limit the amount of water which could enter.

The dam was constructed according to these recommendations. As shown in Fig. 11, it has a crest width of 100 ft (30·5 m), freeboard of 22 ft (6·7 m) above normal water level and zones of sand-gravel and rockfill both upstream and downstream.

This design is described here primarily for historic interest; however, it is basically consistent with the writers' opinions for such a dam as described later, except that the writers would place greater emphasis on the cohesionless transitions and less emphasis on the need for such a wide core.

No fault displacement has occurred since the dam was constructed and the dam continues in service without problems.

Cedar Springs Dam

Cedar Springs Dam was completed in 1971 for an important reservoir in southern California as part of the main California Aqueduct System (USCOLD, 1970). The site is in a region of expected strong earthquakes, with the main San Andreas Fault about eight kilometres away. During the foundation explorations in 1964, several faults were discovered trending across the river valley and running parallel with the longitudinal axis of the proposed main dam, but crossing the dam from upstream to downstream on the left end (Fig. 12).

Test trenches excavated through several metres of recent alluvium down to the top of bedrock, showed that the alluvium had obviously been displaced vertically 3–5 ft (0·9–1·5 m) indicating that the fault had moved since the deposition of the alluvium (Fig. 13). It was not possible to judge if, or how much, horizontal movement was associated with the vertical movement.

The upstream fault (fault 1) was at the boundary between two geological formations. Upstream of fault 1, the bedrock was a slightly cemented sandstone (Harold Formation) that could be excavated with earthmoving equipment. Downstream of fault 1 the foundation was hard granite.

The observations in the test pits left no doubt that the faults were active. Much consideration was given to the prudence of building a dam at the site. Finally, with the concurrence of two independent boards of consultants (one for earthquake analysis and one for earth dam design) it was decided that it was reasonable to build a conservative dam at the site. The design was the responsibility of an experienced group of dam engineers of the California Department of Water Resources, working in a region and at a time when it was considered imperative that no real risk to safety could be taken. Following discovery of the faults, a

Photo: Spence Air Photo

Fig. 9. Break on the Imperial Fault displaced the All-American Canal during the 1940 El Centro earthquake (M=6·7). The fault offset was nearly completely horizontal (strike-slip), approximately 4·3 m. The canal was nearing completion of construction but not carrying water at the time of the earthquake. Note that the fractured zone between offset canal embankments is relatively narrow

Photo: L. S. Cluff

Fig. 10. 38 m high Coyote Dam built across the rift valley of the Calaveras Fault, California. Note typical narrow valley of strike-slip fault

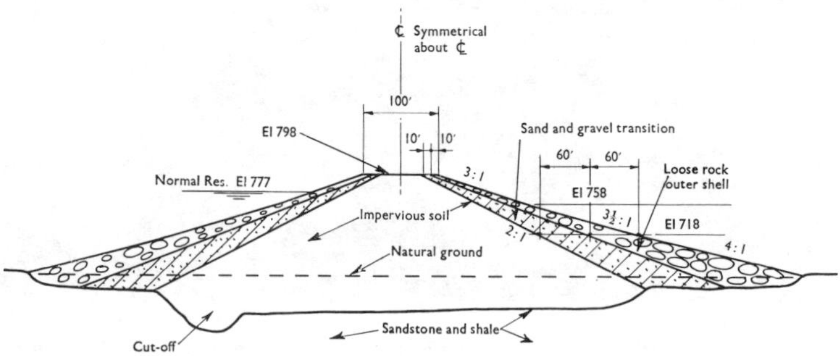

Fig. 11. Coyote Dam (after Tibbetts, 1936)

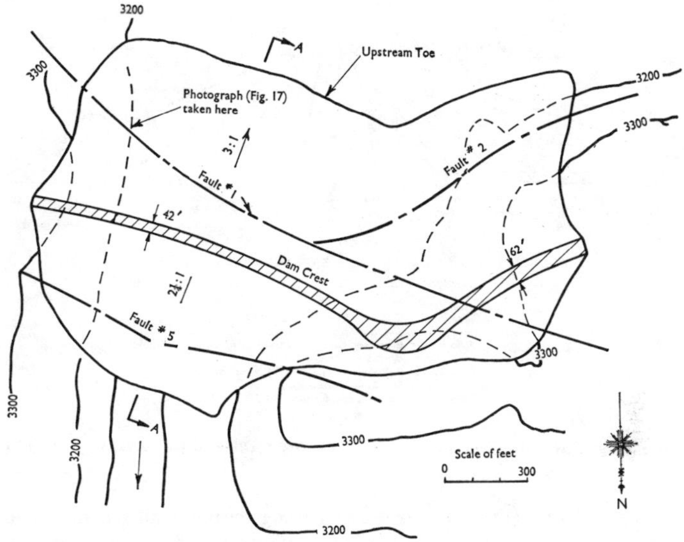

Fig. 12. Cedar Springs Dam—plan view and fault locations

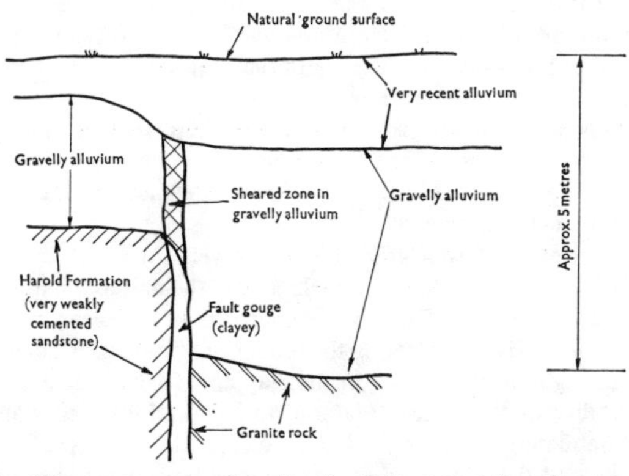

Fig. 13. Cedar Springs Dam; appearance of small active fault exposed in exploratory test trench

number of changes were made in the previously existing plans, and the main design details for the new project were as follows. The dam height was reduced to the minimum possible value consistent with the project needs, from about 300 ft (92 m) to 215 ft (66 m), reducing the reservoir storage volume from about 160 to 60 million m^3. The axis of the dam was moved so that the entire contact between the clay core and the foundation of the dam (in the main river valley) was on the granite formation. The embankment dam cross-section was changed completely from an ordinary zoned dam with narrow chimney drain, to a thoroughly zoned dam with thick exterior shells of quarried hard rock and thick transitions of well graded,

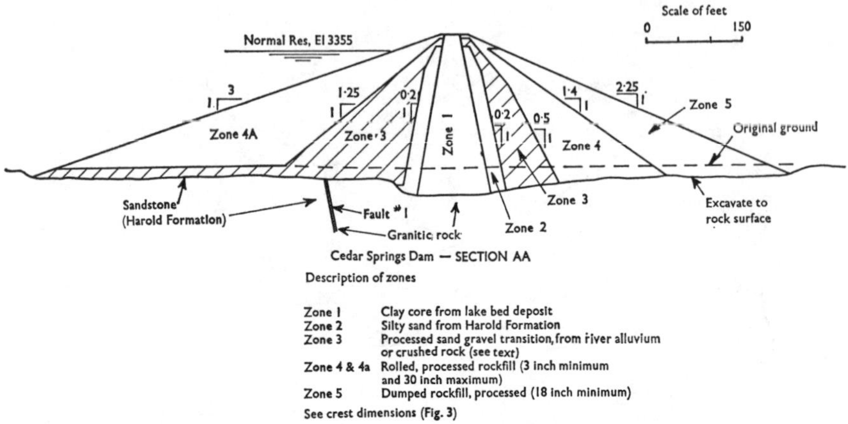

Fig. 14. Cedar Springs Dam—typical cross-section A–A (after California Department of Water Resources, 1968)

coarse, sand-gravel mixture (Fig. 14). Instead of using locally available silty sand for the core, a lake-bed clay, of especially tough and apparently erosion-resistant nature, was hauled from a distance of about 50 km. A wide crest was used for the main dam with generous widths of clay core and transition zones at the top of the dam. In the portion of the dam (west end) where the fault crosses the dam from upstream to downstream, the crest width was increased to 62 ft (19 m), or probably twice the width that might otherwise have been employed (Fig. 12).

The dam design is considered illustrative of the kind of dam which is inherently safe against damage from displacement of faults in the foundation.

The well graded, sand-gravel zone 3 is the main line of defence. Gradation specifications for this zone are shown in Table 1.

The material was composed of well graded river gravels (in which case an 18 in. maximum size was allowed) or quarried rock fines (in which case the material had to pass a vibrating screen with 3 in. square openings). This zone 3 material is too clean and cohesionless to allow an open crack to exist. Even neglecting the influence of the sandy zone 2 the amount of water that could get through an open crack in the clay core following fault rupture would be limited by the permeability of the zone 3 material to a quantity which would be controlled safely with complete confidence by the thick downstream rockfill shells.

Figure 15 shows a view of the construction surface looking downstream with the clay core (zone 1) in the foreground and zones 2, 3, 4 and 5 at progressive distances. Fig. 16 shows a

Table 1

Size	Percent finer
18 in.	100
3 in.	50–100
1·5 in.	40–90
#4 sieve	20–70
#200 sieve	0–8

Photo: J. L. Sherard

Fig. 15. Cedar Springs Dam during construction, looking downstream showing core at right and zones 2, 3, 4 and 5 progressively downstream

Photo: J. L. Sherard

Fig. 16. Cedar Springs Dam during construction, looking at the boundary between transition zone 3 and downstream rockfill zone 4. It is apparent that the permeability of rockfill zone 4 is several orders greater than that of zone 3

Photo: J. L. Sherard

Fig. 17. Cedar Springs Dam during construction showing fault No. 1 exposed in excavation in east abutment at location shown in Fig. 12

Photo: C. R. Allen

Fig. 18. Wellington Fault (arrows), New Zealand: one of the several parallel major active faults traversing the north island. City of Wellington in centre of photograph

Photo: L. S. Cluff

Fig. 19. Lower Karori Dam: 25 m high earth dam across the active Wellington Fault near Wellington, New Zealand. The reservoir is in the photograph of Fig. 18 but too small to be seen. Note the typical narrow reservoir in valley of strike-slip fault

Photo: C. R. Allen

Fig. 20. Horizontal offset of wall by creep along North Anatolian Fault, Ismetpasa, Turkey. It is reported that creep is progressing at 1–2 cm/year since the photographs were taken in 1969 (Nason, 1972). During the 1944 Bolu earthquake ($M=7{\cdot}6$) 1·5 m of horizontal displacement took place at this same location

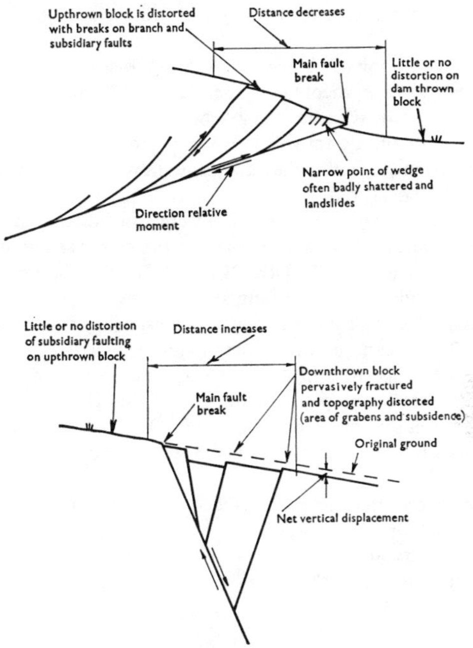

Fig. 21. Characteristics of thrust and normal faults

typical view during construction of the juncture between zones 3 and 4 of the downstream shell, which gives confidence that any amount of water that could get through the zone 3 material would percolate away into the downstream rockfill safely.

During construction the active faults as observed in the excavations were only a few inches wide and resembled the minor faults commonly seen in many dam foundations in tectonized regions (Fig. 16).

In the design stage the clay for the impervious core was chosen because its visual appearance gave confidence that it would not erode rapidly, even if a leak through a crack developed. Particularly, it was very tough and hard at water contents near the plastic limit. At the time of design there was no laboratory test available to evaluate the erosion resistance of the clay in a reliable manner. Later, after construction was under way, laboratory tests using the newly developed procedures and criteria for evaluating 'dispersive clays', indicated that the clay is in fact highly non-dispersive and resistant to erosion (Sherard *et al.*, 1972).

The dam is in operation with no problems.

Palmdale Dam

In 1891, a 10 m high earth dam, retaining a small reservoir of about 8 million m^3 for irrigation and domestic water supply, was built across the rift valley of the San Andreas Fault near Palmdale, California. No movement on the fault has occurred at the site since the dam was constructed (although displacement of many feet almost certainly occurred during the great 1857 earthquake).

Starting about 1955, the question of vulnerability to fault displacement was raised, as part of general practice of review and strengthening of older dams in California. The State Division of Safety of Dams suggested that the dam be strengthened substantially or that a new reservoir be constructed to replace it. In an interim period a storage level restriction was imposed below which sag pond depressions in the fault valley downstream would store a significant portion of the water released if a failure occurred.

Other possible dam locations and other schemes were studied to obtain storage elsewhere. No suitable alternative storage sites were available and the dam was completely reconstructed in 1969 at the same site (Cortright, 1970; USCOLD, 1967). It was the opinion of geologists and seismologists that the fault in the dam foundation could be expected to break during the lifetime of the dam, that the displacement would be primarily horizontal with magnitude probably not exceeding 20 ft (6·2 m) and that there might be a vertical component of displacement of 2–3 ft (0·6–1·0 m). The new dam has a very conservative design including a 40 ft (12·2 m) crest width, flat exterior slopes, interior zones wide enough to remain intact even after being displaced by relative movements of 30 ft (9·5 m) horizontally and 5 ft (1·6 m) vertically, transition zones of cohesionless sand and gravel and a large downstream rockfill zone.

This example is particularly interesting from the standpoint of illustrating current practice. The dam is crossed by a 'major' active fault with the likelihood of 3 m or more of movement during the lifetime of the reservoir very great. It was designed and constructed in the last few years by experienced engineers, with the approval of the California State Division of Dam Safety.

Morris Dam

Morris Dam is a concrete gravity dam about 80 m high, constructed in 1935 in southern California for water supply. During design it was concluded that a foundation fault might be active and the dam was designed with a special joint at the fault which would allow fault displacement without distress (Louderback, 1937 and 1950). During construction it was found that the alluvium covering the fault had not been displaced and it was concluded that the fault was probably not active. Nevertheless the dam was built with the joint. The dam continues in good service with no indication of fault displacement. This example is included for historic interest. The Authors have not studied this case, have not seen details of the joint in the dam and have no opinion regarding the dam, except for the general opinions concerning concrete dams and faults described later.

CURRENT PRACTICE

Current practice requires that every effort be made to avoid sites with active faults. As illustrated by the foregoing examples, however, where no alternative site is available and where a reservoir is deemed vitally important, it has been considered reasonable practice to build dams on active faults. With the exception of the Morris Dam, conservatively designed embankment dams with internal cohesionless transition zones for leakage control have been employed.

While the examples given are all in California, the conclusions drawn are considered representative of general thinking and conservative practice in the industry. In fact, dam building practice in California is generally more conservative than the typical practice in the United

States of America. Following the failure of the St Francis Dam in 1928, with about 400 deaths, the State legislature established the Division of Safety of Dams for the purpose of requiring conservative practice for dams and this organization has become increasingly more influential over the years. All proposed new dams in California are now studied independently in great detail by the Division's staff of dam engineers and engineering geologists. In general only the more conservative designs are permitted. No other states exercise comparable control over dams.

The examples of Coyote, Cedar Springs (Fig. 17), Palmdale and Morris are believed to comprise practically all the experience in the profession in which dams were constructed on demonstrably active faults and where the designer, owner and official agencies were all consciously aware of the situation. In addition, however, it can be confidently assumed that there have been many other unpublicized cases where dams have been built at sites crossed by faults for which the designing engineer suspected possible activity and where he quietly assumed the fault might move several feet and proceeded with the construction of a conservatively designed embankment dam.

Undoubtedly there are many dams in various parts of the world with foundation faults which would be shown to be active by adequate geologic study, and some cases have been discovered after many years of reservoir operation. One example is the Wellington Fault in New Zealand (Fig. 18). Two dams were built in the fault valley for water supply reservoirs about 1 km apart. The Upper Karori Dam is a concrete dam built in 1908 and the Lower Karori Dam is an earth dam built in 1872 (Fig. 19). Both dams are about 25 m high and both retain very small reservoirs (about 250 000 m^3). It is now recognized that both these old dams straddle the active fault. They are kept in service nevertheless because it was concluded that the flood from release of the small reservoirs by dam failures would be impeded by a culverted road embankment and reach Wellington harbour without hazard of loss of life or property damage.

DETAILS OF FAULT BREAKS

The main lessons of value to the dam engineer from observations of surface fault breaks during historic time are summarized in the following. Judgements as to the nature of events that are likely to happen in the future are determined almost entirely by our knowledge of the kinds of things that have happened in the recent past, including both historical observations and geological evidence of prehistoric events.

Number of historic breaks

There have probably been in the order of 100 to 200 surface fault breaks during the past 150 years that have been noticed by geologists, although most of these have not been adequately described in the published literature. Judging from the fact that a large part of these incidents has been reported during the past ten years alone, it is reasonable to assume that hundreds of additional surface breaks have occurred in the past 150 years that have gone unnoticed and unreported by geologists. It would be misleading to estimate the number of fault breaks throughout the world solely on the basis of the occurrences that have been described in the literature.

Until a few years ago, it was thought that surface fault breaks occurred only in association with very large and infrequent earthquakes, but recent intensive field studies have revealed that surface displacements are far more common than was once thought. Not only have ruptures been observed in association with relatively small shocks, but continuous slippage—known as fault creep—has recently been recognized in several parts of the world (Fig. 20). Furthermore, it is now known that earthquakes resulting from displacement and energy release on one fault can sometimes trigger small displacements on other unrelated faults many kilometres distant.

Breaks on existing faults

In almost all known instances of historic fault breaks, the fracturing has occurred along a pre-existing fault that could have been recognized by geologists prior to the earthquake, and in most cases the fault could have been recognized as active. Many historic fault breaks occurred exactly along easily recognized scarps of previous breaks which occurred in prehistoric but very recent geological time.

There have been a few recorded instances of more than one historic fault break along nearly the same fault line, such as near El Centro, California, in 1940, 1966 and 1968 (Allen *et al.*, 1972), near Parkfield, California, in 1901, 1922, 1934 and 1966 (Brown and Vedder, 1967), and probably numerous events dating back to AD 10 along portions of the North Anatolian Fault in Turkey (Ambraseys, 1970).

While it seems probable that a given fault would break again on the exact location of the last break, this cannot be concluded with certainty. In 1954, fault breaks occurred on the Gold Fault near Fairview Peak, Nevada, along the same general lines as breaks that had occurred in 1903. But in detail the 1954 faulting generally criss-crossed the earlier faulting and was frequently more than 50 m from it (Bonilla, 1970). All the 1954 breaks, however, occurred on existing faults, whether or not they had broken in 1903.

Every fault must, of course, have originated as a 'new' fault at some time in geological past, but the statistical chances of a new fault breaking previously unbroken rock within the life of an engineering structure are so remote that they can be reasonably neglected, except in unusual situations such as near the tip of the wedge of an active thrust fault (Fig. 21). Similarly, old faults are sometimes 'healed' or cemented, by metamorphic or intrusive processes at depth. Because healing probably indicates great age since the last movement, and because the healed zone is probably no longer a weak surface, it is also reasonable to ignore the possibility of movement on healed faults.

Fault types and nature of breaks

The relative movements between the walls of a fault during an earthquake may be primarily vertical (dip-slip) or horizontal (strike-slip), depending on the local stress system in the earth's crust that caused the earthquake. Historical observations as well as mechanical considerations suggest that most faults fall into one of three distinct categories.

Strike-slip faults. These involve primarily horizontal displacements. The fault planes are typically near vertical and are markedly linear in plan view, master faults of this type often appearing as linear 'scars' on aerial photographs (Figs 22 and 23).

Thrust faults. These involve vertical displacements caused by horizontal compression. The distance between points on opposite sides of the fault decreases as a result of the displacement (Fig. 24). The fault planes tend to have a shallow inclination (dip) from the horizontal —typically 20–40°. Because of the shallow dip, thrust fault traces tend to be sinuous in plan view through hilly terrain. Many major mountain systems of the world are bounded by thrust faults, such as the Andes and Himalayas, and most of the world's great earthquakes have been caused by thrusting in the oceanic trenches where the ocean floor is being thrust beneath the continental edges. The great 1964 Alaska earthquake and the 1960 Chile earthquake are typical results of breaks on such fault systems.

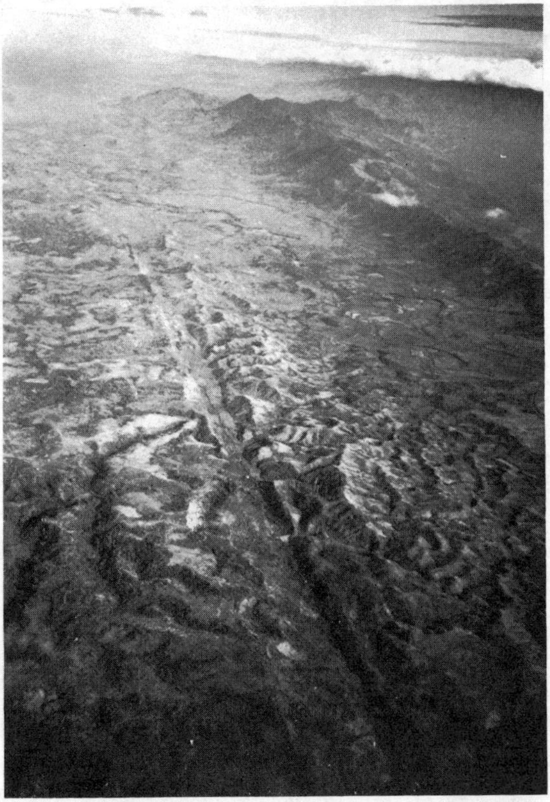

Photo: C. R. Allen

Fig. 22. Philippine Fault in northern Leyte: master strike-slip fault of more than 1000 km length. Note tendency of streams to erode along the fault valley. No historic displacements are known along this segment of of the fault, but a 3·2 m displacement occurred on the fault 400 km further north, during the 1973 Ragay Gulf earthquake (Morante and Allen, 1973)

Fig. 23. Scarps (arrows) showing recent displacements on the Boconó Fault (strike-slip) in Merida Andes, Venezuela, a master fault of more than 500 km length which probably broke during the $M=8$ earthquake in 1812. Amounts of offset of 10 000-year-old glacial moraines (large arrow) suggest $M=8$ earthquakes and fault displacements occur every 200 years. Fault in photograph is on drainage divide between two major rivers, but lies within river valleys over most of its course, passing through several potential dam sites (see Fig. 26)

Photo: L. S. Cluff

Fig. 24. 1971 San Fernando, California thrust fault break passing through concrete pavement showing buckling resulting from shortening of a distance between points on opposite sides

Photo: L. S. Cluff

Fig. 25. 1963 photograph of 1915 break on the Pleasant Valley Fault (normal fault) Nevada. Fault scarp is exceptionally well preserved in the arid climate

Photo: L. S. Cluff

Fig. 26. Main rift valley of Boconó Fault, Venezuela (master strike-slip fault) with length of at least 500 km, looking northeast from Lagrita towards Merida in the far distance

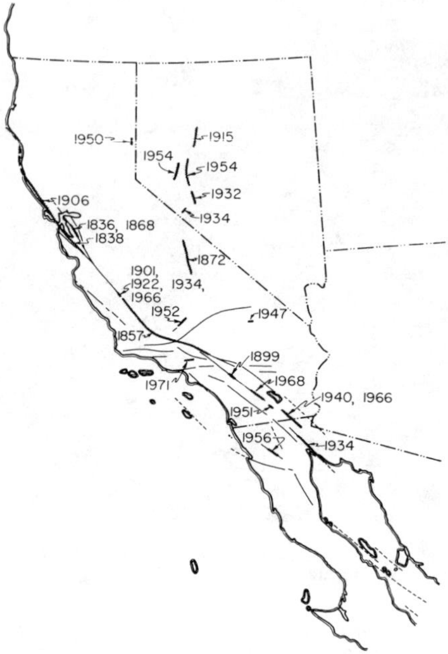

Fig. 27. San Andreas and associated fault zones in California, Nevada and northern Mexico. Zigzag lines show where surface of ground was broken during historic earthquakes and the dates of the breaks

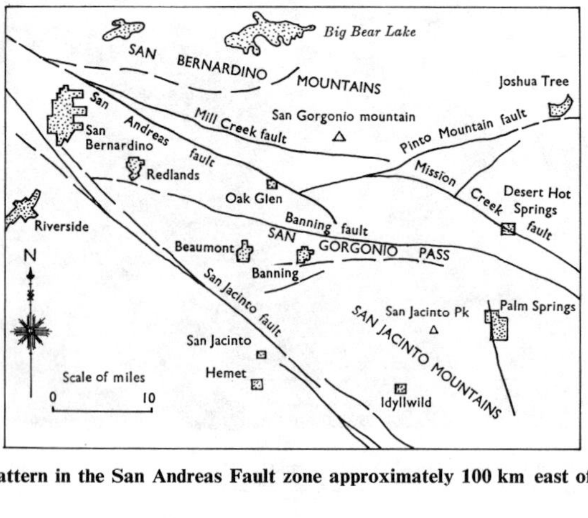

Fig. 28. The fault pattern in the San Andreas Fault zone approximately 100 km east of Los Angeles (after Allen, 1957)

Normal faults. These involve vertical displacements caused by horizontal extension (Fig. 21). The distance between points on opposite sides of the fault is increased by the fault displacement. Normal faults tend to dip steeply (typically about 60° from the horizontal) and their traces are moderately complex in plan view. They are less frequently found on the land masses than the other fault types; most of the areas of crustal extension in the world are along the mid-ocean ridges. Three areas of pervasive normal faulting in continental areas are the so-called basin ranges of the western United States (Nevada, Utah and adjacent regions), the great rift valleys of East Africa, and the Shansi Graben area of China.

The nature of ground fracturing caused by displacements on the three types of faults varies markedly and is of very significant engineering import. Since the strike-slip fault slips with primarily horizontal displacement, there is no consistent tendency for the ground on one side to be higher than on the other. Therefore, it is more common that streams may be captured by the fault and follow along it, creating a long narrow valley. For the thrust and normal faults, which generally cause a consistent change in the relative elevation of the ground on opposite sides, the fault commonly runs along a contour on the lower slope of a hill or mountain. For these faults therefore, it is not common for a stream or river to follow them for any distance. The streams coming out of the hills may cross the faults or may run parallel to the fault at a lower elevation. The main faults in the vicinity of the Hebgen Dam are normal faults running along parallel with the river on the hillside about 300–600 m from the river (Figs 5 and 6).

Fault breaks on strike-slip faults tend to be relatively simple and well defined, with fewer branching and subsidiary fractures than for the other two types. By observing physiographic evidence of where previous breaks have taken place along strike-slip faults, it is often possible to predict within a very few metres exactly where the next break will occur, inasmuch as breaks tend to repeat faithfully along the same lines.

Thrust faults represent a much more difficult engineering problem. They are more difficult to recognize in the field, both because of the sinuosity of the fault trace and because of pervasive landsliding that tends to cover the trace. Furthermore, the exact locations of individual breaks are difficult to predict for a thrust fault; the upthrown block tends to be distorted and to break up pervasively on subsidiary fractures and faults, particularly in the narrow point of the wedge close to the surface trace of the fault (Fig. 21). In addition, the differentiation between an active and an inactive thrust fault may be difficult, inasmuch as fault scarps resulting from geologically recent movements are very rapidly obscured by landsliding and erosion.

Normal faults are generally intermediate between thrust and strike-slip faults in their ease of recognition and their potential engineering hazard. Scarps are generally better preserved than for thrust faults (Fig. 25). In contrast to thrust faults, it is the downthrown block of normal faults that tends to be broken up and subject to slumping and the formation of grabens bounded by subsidiary fractures. For normal faults the ground is not usually fractured to the same degree or to the same distance from the main fault as is typical of thrusts (Fig. 21).

'*Master*' *fault systems*

There are certain major active faults of great length which have been moving for millions of years and have displaced the ground on opposite sides by tens to hundreds of miles. These 'master' faults, particularly of the strike-slip type, may dominate the topography to such an extent that they can be easily recognized on aerial photographs by the lay person (Fig. 26). In general, displacements of these master faults are those which generate the very strongest world earthquakes.

The San Andreas Fault in California is a classic and probably the most intensely studied example. Geological evidence suggests that it has been moving with primarily horizontal displacement for at least 100 million years and that the total displacements caused by the repeated movements during this time may exceed 300 miles. As seen in Fig. 27 it runs nearly the full length of California and extends into northern Mexico. In some places it comprises a single fault zone but in southern California it comes into conflict with a great system of east–west structures causing a complex faulting pattern. Where the San Andreas Fault passes about 70 miles east of Los Angeles, it splits into many branches (Fig. 28). Epicentral locations of earthquakes leave no doubt that the zone as a whole extends into the northern end of the Gulf of California, but the fault frays out into a number of great branches and it is not clear which, if any, of the branches truly deserves the parent name.

The N. Anatolian fault system, which trends east–west across Turkey with a length of about 1000 km is another master (strike-slip) fault with characteristics and historic experiences which are very similar to those of the San Andreas Fault. At the west end, its location becomes indistinct and it probably frays out into many individual faults in a little understood pattern in western Turkey and northern Greece. The Boconó fault system (strike-slip) trending with a length of at least 500 km northeast–southwest across western Venezuela is another with similar feature (Figs 23 and 26). Many other such master faults are known to exist including the Alpine and Wellington Faults, New Zealand (Fig. 18) and the Philippine Fault which trends northwest through the entire Philippine archipelago for 1200 km (Fig. 22) and the Atacama Fault, which runs north–south for 1000 km in northern Chile (Fig. 29). Many of these faults extend beyond the land unknown distances into the oceans.

Although these master faults are easily identified by geologists accustomed to studying active faults, it is also true that master faults in various parts of the world have not been recognized during geological studies for nearby structures, primarily because the local geological literature made no mention of them and because the geologist studying the site was not orientated towards the study of active faults or did not examine the geology far enough from the dam site. In fact, many of the main known active faults in the world have only been called to the attention of the profession in the last 10–15 years, and it is believed that there are others yet to be recognized. Some of these major faults have not broken in historic time, but are nevertheless easily identified as active faults.

In portions of these master faults where the rupturing is generally confined to one main fault zone, such as the San Andreas Fault in northern California (Figs 2 and 27), there is no problem in locating and identifying the activity of the fault. It is unlikely that dams would be built across the fault in these locations, except for small reservoirs in special circumstances such as described earlier for the Palmdale Dam. The main problems of identification and evaluation occur along those portions of the system where the main fault branches out into many individual faults.

The most obvious master faults, such as those mentioned in the foregoing, are strike-slip faults because of the clear topographic expression. However, active thrust fault systems likewise may be master faults of great hazard, such as the frontal fault of the Himalayan range which has been the source of many damaging earthquakes. Although of similar tectonic significance, these are not as easily recognized, and are much more complicated. Some active faults have combined action with nearly equal horizontal and vertical movements (Fig. 30).

Crustal deformations, subsidiary faults and 'triggered' faulting

Large fault breaks and strong earthquakes are accompanied by deformations of the surrounding ground surface, and often by breaks on branch and subsidiary faults (Fig. 31). The deformed crustal area often extends a few miles or tens of miles from the fault. In major earthquakes on master faults, crustal deformation may occur over greater distances. In the great 1964 Alaska earthquake ($M=8{\cdot}5$), it is conservatively estimated that at elast 200 000 km^2 had appreciable vertical movement, about half going up and about half going down, with a maximum upward movement of about 12 m. These crustal deformations occur at about the same time as the fault break, and are part of the same action.

Breaks on branch and subsidiary faults generally occur in areas of greatest crustal deformation and are more commonly associated with breaks on thrust and normal faults than on strike-slip faults. The main breaks on thrust and normal faults usually have breaks on adjacent subsidiary faults, while for major breaks on strike-slip faults, there are sometimes essentially no branch or subsidiary fault breaks.

More than half of the historic fault events in North America have had a substantial amount of subsidiary faulting (Bonilla, 1970). For thrust faults most of the crustal deformations and breaks on subsidiary faults occur on the upthrown block, with often no great disturbance to the ground surface occurring on the downthrown block, for example the 9 February, 1971 break on the San Fernando Fault (thrust) in southern California ($M=6{\cdot}4$). Essentially no ground surface movement was measured on the downthrown block to the south of the main fault break, whereas there was considerable distortion and upheaval (more than 2 m) on the upthrown block. Many small faults broke on the upthrown block, mainly concentrated within one kilometre of the main break (Fig. 32), but one break with 150 mm displacement occurred at a distance of about 2·5 km from the main fault break.

For normal faults, crustal distortion and breaks on subsidiary branch faults is generally concentrated on the downthrown block (Fig. 33). This is well illustrated by the experience of the 1959 earthquake near Hebgen Dam, discussed earlier. As shown in Fig. 5 there was large crustal settlement and many subsidiary faults on the downthrown block (to south of main fault breaks) for at least 10 miles (16 km) from the fault break. On the upthrown block north of the main faults, there was no subsidiary faulting and the ground surface went up only a few centimetres.

In areas of severe crustal distortion there have been experiences where structures founded on bedrock have been tilted by crustal strains, although no fault displacements were seen. In the September 1954 earthquake near Orléansville, Algeria ($M=6{\cdot}75$), there was considerable vertical crustal deformation. Surveys showed that some rock tunnels were lifted by as much as 1·5 m. The Ponteba Dam, about 4 km from the epicentre, a 20 m high concrete gravity dam with a length of 80 m, was rotated vertically so that after the earthquake one end was 550 mm higher than the other (Thevenin, 1955). Apparently the dam remained tight on its bedrock foundation and the block of bedrock directly underneath tilted. To cause such sharp rotation of the bedrock, there must have been some movement on closely spaced minor faults. The dam itself was not severely damaged; the abutment blocks were cracked and spillway gates were twisted, could not be opened, and had to be removed by cutting them out.

A recent example of similar crustal distortion at a dam site occurred on the upthrown block during the break on the San Fernando (thrust) Fault, California, in February 1971. A new dam (Los Angeles Dam) is proposed to be built just upstream from the Lower San Fernando Dam, which has been badly damaged by the earthquake. Comparison of surveys on benchmarks before and after the earthquake showed that the bedrock comprising the left abutment of the proposed new dam was heaved up 0·55 m while the right abutment went up only 0·32 m, for a tilt of 0·23 m in a horizontal distance of the order of 400 m (Yerkes *et al.*, 1973).

Breaks on subsidiary faults have occurred at distances as great as 25 km from the main fault break; however, the further the distance from the main fault, the lesser the magnitude of the displacement. Bonilla (1970) has shown that displacements on branch and subsidiary faults located more than 3 km from the main fault break are generally less than 20% of the main fault displacement. Most breaks on branch and subsidiary faults have occurred along faults with previous histories of recent geological movement which could have been recognized as active faults.

Earthquakes do not, in general, seem to trigger one another around the world. There are some examples to show, however, that strong earthquakes originating on one fault have caused small displacements on other faults at considerable distances (other than on branch subsidiary faults). The 1968 Borego Mountain earthquake in California ($M=6{\cdot}4$) triggered small displacements (10–20 mm) on three other independent faults at distances of 45, 50 and 70 km from the causative fault (Allen *et al.*, 1970). The small 'triggered' displacements occurred on faults that had been recognized as being active from independent geological evidence.

It is not known how often small displacements may be triggered on independent faults by strong earthquakes on the causative faults, or how often small fault movements occur in zones of strong crustal deformation. The above experience makes it necessary, however, to suspect that small adjustments could occur on the myriad of small faults in seismic areas. Fairly wide occurrence might be difficult to recognize, since small movements on minor faults and shears would be difficult to differentiate from the many superficial cracks which develop on the ground surface during strong earthquakes from other causes, such as slumping and incipient landslides. From the standpoint of dam design, these possible widespread small fault displacements are mainly of interest for concrete dams, where even a centimetre or two of differential movement at a critical location can be unacceptable.

Amount of displacements and lengths of breaks

Individual fault breaks during single earthquakes have ranged in length from less than a kilometre to several hundred kilometres.

The displacements have ranged from only a few millimetres to a maximum of the order of 10 m. The largest fault displacement known during a single earthquake is about 11 m, which was the vertical displacement on the Chedrang fault reported by Oldham (1899) at the time of the 1897 Assam earthquake ($M=8{\cdot}7$). The largest corresponding horizontal displacement is 8·8 m, measured at the Bogdo Fault after the great 1957 Mongolian earthquake ($M=8{\cdot}3$) (Florensov and Solonenko, 1963). Still larger movements have been associated with regional deformation during large earthquakes, such as changes of elevation of more than 14 m at the 1899 Alaskan earthquake ($M=8{\cdot}5$) (Tarr and Martin, 1912), but the maximum observed displacements across discrete fault lines are believed to be those above. There is no reason to think that they will not be exceeded as more intensive observations are made in the future, although the fact that earthquakes appear to have a maximum magnitude of about 8·9 implies that there is also an upper limit to the amount of fault displacement that is credible during a single earthquake.

A number of investigators have compiled statistics regarding fault breaks in historic time. Two of the more comprehensive compilations are given by Ambraseys (1969) and Bonilla (1970). For both of these compilations the investigators went back to the original records or literature, written by the man who actually studied the fault break in the field, in order to avoid errors introduced by second or third-hand reporting. Also, they included only records that they considered reliable. These compilations were made independently and include many of the same faults. The Ambraseys compilation contains 63 records between 1847 and 1967. Both compilations include faults occurring in all parts of the world (Table 2).

Photo: R. E. Wallace, U.S. Geological Survey

Fig. 29. Atacama Fault near Antofagasta, Chile. No historic breaks are recorded on this fault but geologic evidence that it has offset all but the youngest alluvium indicates conclusively that it is active

Photo: V. Solonenko

Fig. 30. Scarp (arrows) of 1957 Gobi-Altai earthquake ($M=8{\cdot}3$) in Mongolia. Average fault break displacement in area of photograph was 5 m horizontal component and 5 m thrust component

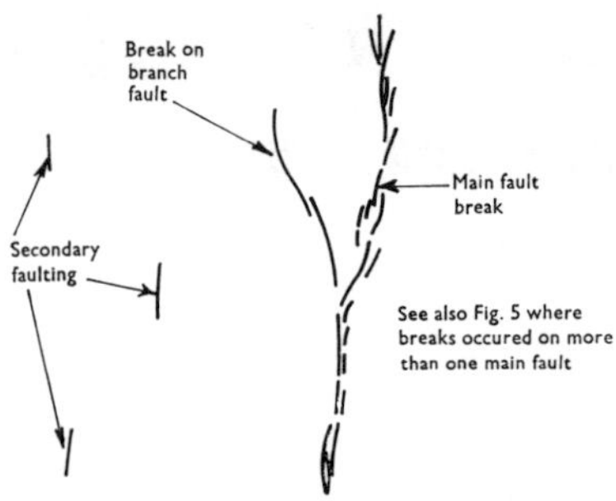

Fig. 31. Schematic relations among breaks on main, branch and secondary faults (after Bonilla, 1967)

Photo: C. R. Allen

Fig. 32. Thrust fault that broke during 1971 San Fernando, California earthquake (M=6·4). This is a secondary fault in the upthrown block, about 1 km from main fault break. Fault plane dips about 60° with the horizontal (black arrow in lower right hand corner)

Photo: N. Florensov

Fig. 33. Break on a secondary normal fault with a displacement of about 1·0 m during the 1957 Gobi-Altai earthquake, Mongolia (M=8·3). Several branch faults are broken on the downthrown block

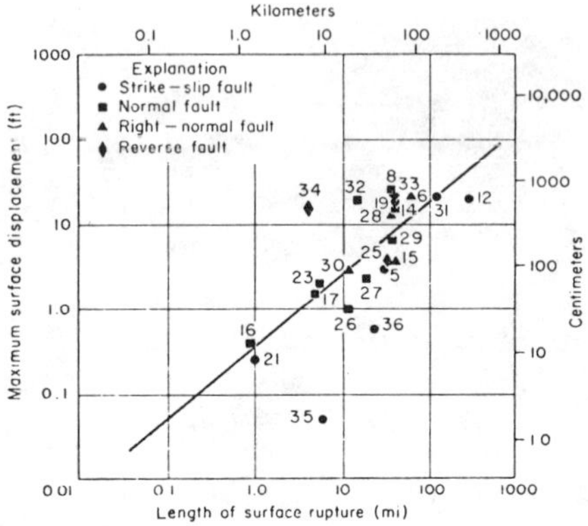

Fig. 34. Rough correlation between length of break and maximum relative displacement of ground on opposite sides (after Bonilla, 1970; numbers on chart refer to individual faults in the Bonilla tabulation)

Photo: C. R. Allen

Fig. 35. 700 km long master fault near Niihama, Shikoku, Japan. No major historic earthquakes or fault breaks have occurred along this fault for at least 1500 years, and there is no indication from instrumental measurements that the fault is active, leading to considerable controversy over its potential hazard. Obviously the fault has broken in recent geologic time as indicated not only by the abrupt relief of mountain front (reflecting vertical displacement) but also cutting of very young stream terraces, as show at the point of the large arrow

While there is some difference between the two compilations, it is seen that they show the same basic trend; that is, the smaller the fault displacement, the greater the number of observed fault breaks. Both the Ambraseys and Bonilla compilations contain a considerable total number of faults, for practical purposes including all the breaks for which reasonably reliable records are available.

Table 2 shows the maximum displacement measured on the main fault. The average displacement across the fault breaks was considerably less. In his study, Ambraseys (1969) made a rough attempt to study the distribution of the displacement along the break. For half the total number (27 of 54) of historic fault breaks included in his compilation, he found enough information given to make an estimate of the average displacement. The ratio of average to maximum displacement ranged roughly between 0·2 and 0·7 with an average of about 0·4.

From these figures it is seen that for the great majority of fault breaks, the maximum displacement was less than 5 or 6 m. The average displacement along the length was less than 50% of the maximum. These figures constitute one of the main sources of confidence that one can build safe embankment dams at sites with active faults. This is true because we are dealing with offset displacements generally of less than 3 m and there is a practical maximum value of the order of 7 to 8 m. These are values for which a conservative embankment dam can be designed to withstand, simply because the critical zones in the dam can practically be made much larger than the fault offsets.

There is a rough relationship in observed historic fault breaks between the length of the fault break and the amount of the movement, and both are related to the magnitude of the earthquake produced. The longer fault breaks have the greater displacements and generate the larger earthquakes. The plots showing these relationships, prepared by Bonilla (1970), are used as a guide to judgement for the geologist in predicting the probable maximum displacement on a given fault (Fig. 34).

The length of the fault break during a particular earthquake generally is only a fraction of the length of the fault as it can be traced on the ground. The length of historic breaks in southern California has been shown to range between 20 and 50% of the length of fault system on which the earthquake occurred (Albee and Smith, 1966). Although this is a very rough and tenuous correlation, it is employed by geologists to guide the estimate of displacements which might occur on a given fault under investigation.

Table 2. Magnitudes of historic displacements

Maximum displacement, m	Number of fault breaks	
	Bonilla	Ambraseys
1– 1·00	15	18
1·01– 2·00	16	13
2·01– 3·00	8	4
3·01– 4·00	8	7
4·01– 5·00	4	5
5·01– 6·00	4	2
6·01– 7·00	4	2
7·01– 8·00	1	1
8·01– 9·01	1	1
9·01–10·00	1	0
10·01–11·00	1	0
11·01–12·00	0	0
12·01–13·00	0	1
	63	54

Discrete fault displacements have now been recognized in association with earthquakes as small as magnitude 3·6, with displacements in the order of 1 cm (Brune and Allen, 1967), and more careful searches will undoubtedly uncover displacements associated with still smaller earthquakes in the future. Relatively continuous fault creep was first documented along the San Andreas fault in California by Steinbrugge and Zacher (1960), and it has now been recognized on a number of other faults including the North Anatolian Fault of Turkey (Ambraseys, 1970) (Fig. 20). Maximum reported slip rates are 1 to 2 cm per year. Creep may be uniform or spasmodic, and accelerated creep has been observed both before earthquakes (Allen and Smith, 1967) and after earthquakes (Smith and Wyss, 1968). It is a matter of some debate among geologists and geophysicists as to whether continuous creep on a fault is relieving strain that might otherwise build up to a large earthquake, or whether it is a manifestation of high strain that may be precursory to a large event. It is important to note, however, that all reported instances of creep are along fault traces that would be classified as active on the basis of other geological evidence.

Time relationships (periodicity or frequency of fault breaks)

Little helpful information is available to the dam engineer from historic records on this aspect of faulting. The time of record is too short.

On the master faults there can be little doubt that faulting has recurred thousands of times, and there probably is some cyclic average period. Geodetic measurements of crustal strain are providing general useful information in this regard. For example, in California it now appears that the land mass to the west of the San Andreas Fault is moving north with respect to the land east at a rate of about 3 cm per year. Assuming that all this strain is concentrated finally in periodic breaks on the San Andreas Fault, fault movements of 3 to 6 m could be anticipated at intervals of 100 to 200 years, which appears to be about the frequency being experienced.

For the breaks on lesser faults with a few exceptions no displacements have occurred in historic time and it is completely impossible to form any reliable estimate as to when the next break may occur (or when the previous prehistoric break occurred). There have been a few examples where geological studies of individual faults, utilizing carbon-14 dating techniques have shown that faults have moved several times displacing Holocene alluvium at intervals of the order of several hundred years. In several examples historic breaks have occurred on lesser faults which had been known to have broken previously in Holocene time (last 10 000 years). Also, there have been a few examples, other than on master faults, where historic fault breaks have occurred twice at the same location (as described earlier).

One can conclude that some currently active faults, including the master faults, are breaking at intervals of the order of 50 to 200 years. On the other hand it seems quite probable that some active faults may move at much longer intervals, such as thousands or tens of thousands of years. This conclusion is speculative, but otherwise it would be difficult to account for the discrepancy between the relatively great number of active faults which can be recognized on the ground by geological evidence and the relatively smaller number of historic fault breaks.

From the standpoint of time there is one conclusion of considerable importance from study of historic events. Experience shows clearly that there are long periods (of 50 years or more) along certain sections of known active master faults where there is no earthquake activity which can be measured even with sensitive instruments (Allen *et al.*, 1965). It is also known that in some areas very strong earthquakes have occurred at intervals of several hundred years, with periods of quiescence in between (Ambraseys, 1971). There are also faults with clear geological evidence of recent activity which have not broken or caused any earthquakes in many centuries of observation (see, particularly, Fig. 35). These observations indicate that neither the absence of instrumentally measured small earthquakes or the absence of felt earthquakes for several hundred years can in itself give assurance that a given fault is inactive.

A question of interest to the engineer is 'How long does it take for the fault break to occur at any given location?' Unfortunately there is little good evidence available on this subject, either from instrumental measurement or from direct observations. Obviously the break does not occur instantaneously or no buildings would remain standing (and there is no evidence that structures located very close to fault breaks are subjected to much greater acceleration forces than those located 1–2 km away). On the other hand, fault displacements must occur fairly abruptly, or high frequency seismic waves would not, therefore, be generated. Probably the principal fault displacement during large earthquakes takes place over a period of a few seconds to tens of seconds. In many cases only a part of the surface displacement occurs immediately during the earthquake and a second part occurs as a 'rapid creep' minutes, hours, days (or even months) after the earthquake.

GEOLOGICAL STUDIES OF FAULT ACTIVITY

General

Evaluation of potential fault activity is too large a subject for complete or extensive treatment here. Some main aspects are described with emphasis on recent practice.

For every dam where the potential activity of a fault in the foundation is being studied, generally the geologist is also studying the seismic activity of the area to form an opinion on the probable earthquake shaking to which the site may be subjected. For this latter purpose he must locate all the active faults in the area, even though they do not pass through the dam foundation. Large landslides during earthquakes along active faults are common, primarily due to the shattered condition of the rock along the faults. Hence, evaluation of possible damaging landslides also requires investigation of active faults. The same general study is made for all three purposes.

Most of the reliable evidence for evaluating fault activity comes from geological investigations. Important peripheral assistance in some cases is given by studies of locations of historic earthquakes and by instrumental measurements of small earthquakes and geodetic measurements of creep and crustal strain, but detailed geological surface evidence that there has or has not been fault displacement in recent geological time is the primary investigative tool.

Largely because of the increasing emphasis in recent years on the safety of nuclear power plants and large dams, there has been a great increase in the number of fault studies which have been made, and the effort spent on them. While no large technical breakthroughs have been made, accumulated knowledge concerning techniques for evaluating faults, and knowledge as to what can and cannot be done by extensive and detailed geological studies, is increasing rapidly. In earthquake regions, geologists are specializing and devoting their practice exclusively to the field of earthquake activity and seismic geology ('seismotectonics'), a subject which previously received the attention of only a relatively few scientists.

One important result of the recent concentration of study in this field has been to demonstrate that when a fault has been thoroughly and competently investigated, geologic evidence can usually be found suggesting whether it should be considered active or not. The Authors' experience indicates that it is the exceptional case for which no evidence can be found to render any judgement. This conclusion could only be reached after a considerable number of extensive fault investigations had been made, and would have have been necessarily anticipated on the basis of the knowledge available only a few years ago (e.g., before 1970). An active fault with recurrent displacements must have an appreciable influence on the surface geology and topography, and it would not be expected that all evidence could be completely obliterated by the weathering and erosional processes.

Nature of fault investigations

Geological fault evaluation is based on the assumptions that a fault is likely to break if it has broken in the recent geological past and is not likely to move if it can be shown that it has not moved in a very long time. The date of last displacement is estimated from stratigraphic and geomorphic relations. The time since last fault displacement must be less than the age of the youngest faulted rock (or soil) and greater than the age of the oldest unfaulted rock which lies across the fault. Estimates of the probable type, amount and location of movement are guided evidence from past geological movements.

In the geological study, the first problem is to determine whether or not there is a fault crossing the proposed dam site. This may or may not be easily established during preliminary investigations of the site geology. Once the existence of one or more faults is demonstrated, study of potential activity is commenced. In a few cases, there may be geological evidence near the site which can be seen in test pits, such as the example of the Cedar Springs Dam described earlier. More commonly, a much more extensive investigation is required in order to find sufficient evidence to render a judgement. Even in a circumstance such as that at Cedar Springs Dam, where observations in the test pits at the site showed the faults are active, a more extensive investigation is desirable to evaluate the length of the faults and their relationship to the regional geological framework in order to assess maximum fault displacement (and credible earthquake).

The investigations are usually carried out in two phases. The first phase is directed towards: obtaining information on the geology, regional tectonics and seismicity of the general region, perhaps within a radius of as much as 200–300 km; delineating major through-going structural trends that may be close enough to affect the site, assuming that the investigation is also being directed toward evaluation of the earthquake shaking potential; assembling information of all features detected from aerial photographs and other types of imagery which might represent active faults that are close enough to the dam to be of concern; planning and preparing for the detailed field study. A comprehensive review of pertinent published and unpublished literature is made during the first phase of the study. An aerial reconnaissance by the geologist in a low-flying aeroplane or helicopter is usually helpful. From the first phase it should be hopefully possible to trace suspected faults in both directions to considerable distances from the dam site, and to locate branches and related faults.

The second phase includes detailed geological mapping. Study is made of all locations where the faults can be recognized on the ground. Special aerial photographs may be obtained to assist the work. In addition to detailed surface geological studies, supplementary investigations are employed including test trenches and age dating and sometimes geophysical studies. The type of fault (strike-slip, thrust, or normal) is determined from general study of the geological features and surface evidence.

The most readily obtainable evidence for fault activity comes from surface manifestations of movement which have not yet been obliterated. Some of the more common surface features identifying active faults are

(*a*) scarps from previous displacements which are not obliterated by surface erosion or buried by younger deposition (Fig. 35)

(*b*) off-set streams and ridge lines
(*c*) closed depressions (sag ponds)
(*d*) faults which act as groundwater barriers and have a higher water table on one side, and sometimes lines of springs
(*e*) lineaments of vegetation resulting from different depths to water tables on opposite sides of a fault
(*f*) lines of landslides
(*g*) displaced alluvial terraces.

Most of these surface features, which are usually first identified on air photos, can also have other geological origins. For example, outcrops of hard rock beds can create linear ridges very similar in appearance to fault scarps, but the difference can be readily determined from field studies. Similarly, closed depressions and ponds are sometimes created along lineaments by limestone solution. Except for limestone solution, however, there are very few actions of nature except recurrently moving faults which create closed surface depressions more rapidly than they are filled in by deposition (Fig. 36). Closed depressions are very common on active strike-slip and normal faults, and occasionally are found along thrust faults.

Groundwater barriers in alluvium are a peculiar phenomenon of active faults. For reasons not well understood faults passing through alluvium commonly become less pervious than the adjacent alluvium by inclusion of soil fines. The resultant higher ground water table on one side of the fault commonly causes a different type of vegetation. Occasionally lines of trees grow along the line of the fault, perhaps because of the different soil along the fault as well as the higher ground water. Figure 37 shows an example (taken about 25 years after the last displacement) of an active fault trace in recent river alluvium identified both because it is a groundwater barrier and because a trace of scarp remains.

Studies needed over a large area

Evidence for the last time of movement on an active fault commonly exists only at certain places along its length. Hence, conditions existing at any given small area along the length of a fault, such as in the immediate vicinity of a dam site, will rarely provide conclusive evidence. Studies must be carried considerable distances beyond the dam, along the trends of the main faults of interest. The degree of confidence with which conclusions can be drawn on a given site depends greatly on the amount of effort spent in the studies, and the distance from the dam site to which the investigations are carried. Examination of photographs and field investigations of points of interest along the faults may be required to a distance of more than 100 km from the dam site.

A common problem is the fault which runs down the valley through the dam site, under the river alluvium. As the river alluvium may be removed each year at flood, all surface evidence of recent fault displacement is obliterated. The evaluation of such a fault can usually only be made at locations away from the dam where the fault leaves the river valley, which may be at some distance from the dam site.

Drainage divides, where the fault trace passes from one stream system to another, should be scrutinized carefully; geomorphic features of recent faulting are likely to be particularly well preserved at such localities (e.g., Fig. 23).

In past practice, many dams with foundation faults in earthquake regions have been designed with no detailed geological study having been carried out beyond the immediate vicinity of the dam site. This should no longer be considered acceptable practice at sites where there is any question about the fault activity.

Aerial photographs and 'low sun-angle' photography

Aerial photographs are an essential part of fault investigations; fault features can be seen on them which could not be found by ground studies. Faults appear as lineaments, reflecting various different physical conditions created by faulting (such as disruptions of the normal ground surface along the lines of previous breaks, different soils on opposite sides, different depths of water tables on opposite sides creating different vegetation, lines of landslides along the upthrown block of thrust faults, offset talus deposits, etc.). For some strike-slip master faults, the line of the fault can be easily and immediately identified (Figs 18, 22 and 29). At the other extreme, for some active faults, the surface features seen on the aerial photographs may be so small and relatively insignificant that they can only be recognized as a 'suspicious' feature by a man with considerable experience in aerial photograph fault studies.

In any event, it is emphasized that air photos are a critical tool for the geologist in the study and recognition of active faults. If for any reason, systematic aerial coverage cannot be made available, the geologist cannot normally make a competent evaluation of the problem, any more than the engineer can design the dam without exploration of the foundation.

Because there is a large element of experience and judgement in aerial photograph interpretation, and particularly for interpretation of recent fault activity, photographs of any kind should be studied independently by at least two experienced men on the team.

Until the last few years, ordinary black and white photographs have been used. Much of the entire world has been photographed by the United States Army Map Service at scales generally of 1:30 000 to 1:60 000. For many countries these photographs are available from local government agencies, or they can be obtained in Washington DC, with permission of the local government. These types of photographs are especially good for preliminary regional investigations and should be obtained wherever possible, because a large area is covered with a few photographs. For detailed studies, scales of 1:12 000 to 1:20 000 are preferable. For stereoscopic viewing, which is absolutely necessary, 60% overlap is standard practice. Ability to interpret the photograph is dependent to some degree on the quality of the photography but to a larger degree on the quality of printing. Specifically, a recently developed printing process called 'electronic dodging' is more economical but does not provide suitable photographs for geological interpretation. Orders for prints should specify maximum contrast and specifically prohibit the use of electronic dodging.

Ordinary aerial photographs have usually been taken during the middle of the day. Starting about 1967, a research programme was carried out jointly by the University of Nevada Department of Geology and one of the Authors (L. S. Cluff) in which new photographic techniques were developed for the special purpose of studying active faults. This technique, called 'low sun angle' aerial photography utilizes photographs taken at different times of the day (and sometimes in different parts of the year), depending on the fault trend and the position of the sun (Slemmons, 1969; Cluff and Slemmons, 1972). The technique is an extension of a widely used practice among geologists in which early morning and late afternoon field observations were long known to be often superior. The purpose is to enhance shadows from fault scarps and other subdued ground surface irregularities characteristic of active fault zones. The optimum condition obtains when the sun is shining nearly perpendicularly to the trend of the fault, in the direction from the upthrown side towards the downthrown side and when the sun angle is low enough for the scarps to cast a shadow but not so low as to put large parts of the whole area in shadow.

Figures 38 and 39 illustrate the effectiveness of low sun angle photography during studies of the major active (normal) fault which bounds the west edge of the Wasatch Mountains, at a

Photo: L. S. Cluff

Fig. 36. A typical sag pond in closed depression along the Boconó Fault in Venezuela, a 'master' (strike-slip) active fault. Note ancient scarps of breaks on secondary faults on hillside to left of main fault trace. Note also notch in ridge line in far distance, a common manifestation of highly active faults

Photo: C. R. Allen

Fig. 37. Major river valley eroded along an active strike-slip fault, typical situation at many dam sites: Soganli River Canyon, Turkey, following the North Anatolian Fault. The four white arrows indicate the remnant of the last fault break during the 1944 Bolu earthquake ($M=7{\cdot}6$). The photograph was taken 25 years after the last fault break. The fault is recognizable because of a subtle scarp which remains in the flood plain and the fact that the fault passing up through the river alluvium is evidently acting as a groundwater barrier. The last movement on the fault was predominantly strike-slip and about 1·5 m

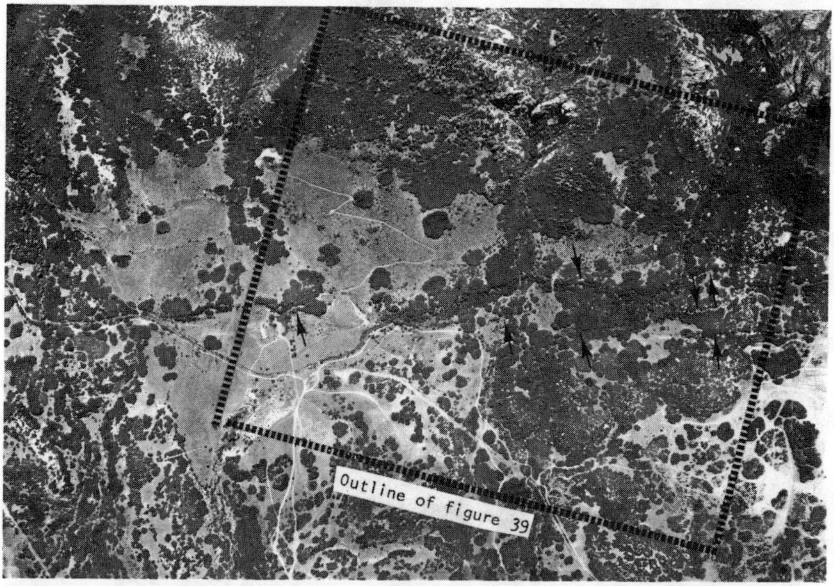

Fig. 38. Conventional air photograph showing a section of the main Wasatch Fault in central Utah, a 'master' normal fault which has not broken in historic time but is obviously active from much geologic evidence. The fault is shown by the black arrows (after Cluff and Slemmons, 1972)

Photo: G. E. Brogan

Fig. 40. Series of parallel active faults, Fish Lake Valley, California, about 500–1500 ft apart, showing the use of low-sun-angle photography. Some of these fault scarps are less than 50 cm high and would not be seen in photographs taken at midday

Fig. 39. A low-sun-angle photograph of the area included in the central square of Fig. 38, with the arrows in the same locations, showing how much more clearly the fault scarp can be seen (after Cluff and Slemmons, 1972)

site near Salt Lake City, Utah. Fig. 38 is a conventional vertical photograph of a section of the fault. Fig. 39 is a vertical photograph of the area outlined in Fig. 38. Fig. 38 was taken at approximately midday. Fig. 39 was taken at approximately 09.00 h, catching longer shadows. The example of Figs 38 and 39 is not extreme. In some cases, ordinary photographs may show almost no evidence, whereas the faults stand out clearly on low sun angle photographs (Fig. 40).

Low sun angle photography has proved very helpful and has been employed successfully during the last few years by the Authors for fault studies in many projects in the western United States, Alaska, Venezuela, Peru, Chile, Italy and Iran. The value of the procedure is so well demonstrated by these experiences that it is clear that the method should be used as a routine investigatory tool. In future studies all photographs taken for fault investigations should be at low sun angle. The most universally useful scale is about 1:12 000; 1:6000 and 1:3000 scales are good for detailed analyses.

Other types of photograph which are occasionally helpful include infrared and colour film, and images which are now available commercially from satellites (ERTS), high altitude aircraft (NASA), and sidelooking airborne radar (SLAR). The Authors have utilized all these extensively on recent projects and have general opinions as follows.

By far the most useful type of photograph is the black and white low sun angle vertical photograph which chould be the standard procedure for this work.

The infrared photograph occasionally gives a better indication of differences in groundwater level, or vegetation changes, on opposite sides of the fault, but rarely have infrared photographs provided much additional information that was not also available from black and white low sun angle ones. Low sun angle photography is not yet widely used in the profession. Some colleagues have suggested that infrared photographs are superior to black and white photographs of any kind. The Authors' experience shows the opposite. Comparisons of low sun angle black and white photographs with infrared photographs for several active faults have led to the conclusion that the low sun angle photographs are almost always superior for the purpose.

Sidelooking airborne radar (SLAR), which has only recently become commercially available, has a considerable advantage in that (*a*) it can be obtained in virtually all weather conditions; (*b*) the direction and depths of 'shadows' can be controlled by the flight patterns; (*c*) it reduces the contrast between different types of vegetation, thus representing the ground surface more clearly. Coverage by SLAR is considerably more expensive than ordinary photos. Even in the last year or two (1974–74), there have been large improvements in SLAR imagery; however, it is currently available only at a scale of 1:250 000, and enlargements are fuzzy and of limited usefulness. Hence, for current applications SLAR has value only for preliminary regional study. It is probable that recent military technical developments, not yet available commercially, will allow clear imagery at scale of 1:5000. Consequently, it can be expected that SLAR imagery will be more valuable for fault studies in future, particularly in areas of heavy vegetation (rain forests) where ordinary photographs may not be very helpful.

Colour photographs are superior to black and white photographs taken at midday and should be used wherever they are already available. Where colour photographs are not available and where photographs must be taken for the purpose of a given fault study, the Authors recommend low sun angle, black and white photographs, which are much superior to colour photographs taken at midday. Colour photography cannot be employed for low sun angle photography as it is necessary to take colour photographs near midday.

ERTS and NASA photographs can be very useful for preliminary regional studies, for identifying the problems which need more careful evaluation with low sun angle photography.

They are not imperative, but as they are available it is desirable to use them for preliminary studies, especially in areas where good regional geological maps are not available.

Photographs for study of faults do not have to be taken with the precision required for aerial photographs used for making topographic maps. For photographs intended only for geological evaluation, the Authors have found that it is much less costly and technically preferable to lease an aircraft with pilot, photographer and camera hourly and to have the photographs taken at the direction of the geologist who accompanies the flight. Usually a reconnaissance flight is made first without taking photographs. Subsequently a flight plan is drawn up and plotted on a map for the pilot to follow and the best time chosen for low sun angle photographs. During the flight the geologist directs all activities and he may change the order of flying to improve the photographs as guided by his observation of the shadow projection. After a short time the geologist learns to operate the common cameras and much of the work is done with no photographer. Although there has been some resistance to this practice among the air photo companies, it has now become the Authors' common practice. It gives much greater flexibility in taking photos at the proper sun angle and a large cost difference.

Test trenches

A main lesson learned from the recent concentration of study in this field is that test trenches across suspected fault lines are absolutely necessary parts of routine modern investigations. In the last few years, trenching has been used with increasing frequency. Commonly there may be no surface evidence in the vicinity of the test trench but examination of the geological profile in the test trench walls may clearly indicate whether the fault is active or inactive, as in the case of Cedar Springs dam.

The main purpose of the trenches is to determine whether surface deposits lying across the fault are displaced or not. The trenches also provide opportunities to obtain samples of carbonaceous material which can be radiometrically dated. It is not usually necessary to excavate very deep; trenches 2–4 m deep have commonly been adequate

Interpretation of evidence in the test pit walls requires some specialized experience and observation techniques. The trench walls must be cleaned carefully with hand labour using shovels to remove disturbed material. Brushes and scrapers of various kinds are used for final cleaning to allow examination of minute details.

It is not uncommon in the walls of test trenches for clear evidence of several recurrent fault displacements to be exposed, as indicated by the fact that deeper, older deposits are offset greater amounts by the fault than the younger surface deposits. For example, in a current investigation in Nicaragua, not considered unusual, evidence has been found in a number of test trenches that surface deposits less than 10 000 years old have been displaced 2–5 times by faulting at different times. Similar situations have been the subject of several careful scientific investigations. An example is the study by Clark *et al.* (1972) of the Coyote Creek Fault (southern California) in which analysis of evidence seen in the walls of test trenches indicates recurrent displacements of recent geological surface deposits at intervals of 150–200 years during the past 3000 years (Fig. 41).

The appearance of the fault gouge (clay-rich sheared material in the fault zone) as exposed in the test trenches may be important in forming an opinion regarding the erosion potential through the dam foundation along the fault. The apparent 'freshness' of the fault gouge cannot be used as reliable evidence of fault activity. Fault gouge below the depth of greatest surface weathering frequently has the pasty and unweathered appearance of freshly ground and crushed rock even in faults that have not moved for many millions of years.

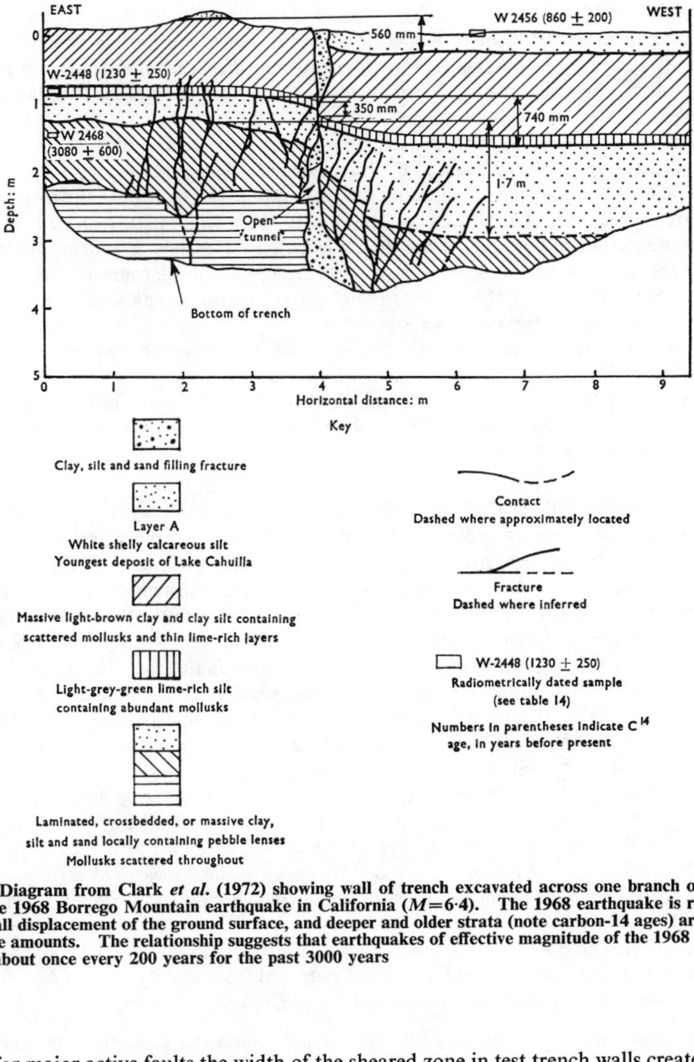

Fig. 41. Diagram from Clark *et al.* (1972) showing wall of trench excavated across one branch of the fault causing the 1968 Borrego Mountain earthquake in California ($M=6{\cdot}4$). The 1968 earthquake is represented by the small displacement of the ground surface, and deeper and older strata (note carbon-14 ages) are offset by progressive amounts. The relationship suggests that earthquakes of effective magnitude of the 1968 event have occurred about once every 200 years for the past 3000 years

Even for major active faults the width of the sheared zone in test trench walls created by the last fault movements through very recent surface deposits, may be very narrow, 10–20 mm or less.

The direction of slickensides and shearing striations in fault gouge can be misleading when used to predict the direction of fault movement. Near the ground surface there is tendency for the gouge to be squeezed upward in the direction of least resistance regardless of the direction of displacement of the fault.

Estimating fault displacement

The geologist can estimate the probable amount of displacement during the next break on a given active fault in several ways, such as from previous displacements as seen in test trench walls or as estimated from displaced surface features, or from the statistical data for historic fault breaks.

It is particularly important to trace the fault as far as possible and to estimate its probable length, in order to guide the estimate of the displacement which might occur. This estimate is finally made based on judgement guided by the geological evidence. Both the length and type of fault are considered. On strike-slip faults many tens of kilometres of fault break often occur with displacement not much more than one metre. However, a thrust fault with a ten kilometre break may have two to three metres of surface fault displacement. As an example, in the 1971 San Fernando, California earthquake, the length of break was only about 15 km and the maximum displacement was about 2·0 m.

Several methods are employed to estimate the displacement which may occur, all utilizing past experience with historic fault breaks. One procedure is simply to assume that the length of fault break may be an appreciable portion of the total length of the fault, such as 50%, and to estimate that the displacement will be similar to that on faults of a similar type with the same length of fault break (Fig. 34).

Seismological studies and epicentral locations

What is the importance of geophysical studies, as opposed to purely geological studies, in establishing the presence and nature of active faults in a dam site area? It is sometimes suggested, for example, that a dense network of seismographic stations might reveal the alignment of epicentres of small earthquakes, which may be too small to be 'felt', along a fault of unknown activity or one whose surface expression is concealed. Except in unusual circumstances, however, thorough geological studies will reveal an active fault more directly and far more cheaply than an extensive and time-consuming geophysical programme, particularly as faults that may be seismically active at depth but do not break the ground surface are not of concern. If future surface displacements are possible along a fault, there is almost always geological evidence of similar displacements in the recent geological past.

Another difficulty with geophysical investigations is that some very active faults are not marked by an abundance of small earthquakes, as is well demonstrated by certain segments of the San Andreas Fault that apparently are wholly devoid of minor seismicity between infrequent great earthquakes (Allen *et al.*, 1965). However, if small measured earthquakes could be shown to line up well along the trace of a fault being studied, it would provide strong evidence of activity.

For these reasons the Authors have not employed seismographs for the purpose and do not consider them to be a primary tool for study of potential fault activity. Circumstances where they might be considered desirable would include some of the following.

(*a*) A situation where it is very difficult to understand the geology from conventional studies, such as areas of exceedingly heavy vegetation, water cover, or river valleys covered everywhere with flood deposits.

(*b*) A project where two to four years of observation time can be allowed in order to obtain the results.

(*c*) A project where it is desired to install seismographic stations anyway for the purpose of studying the possible increase in seismic activity caused by filling the reservoir, which is rapidly becoming standard practice for all very large reservoirs.

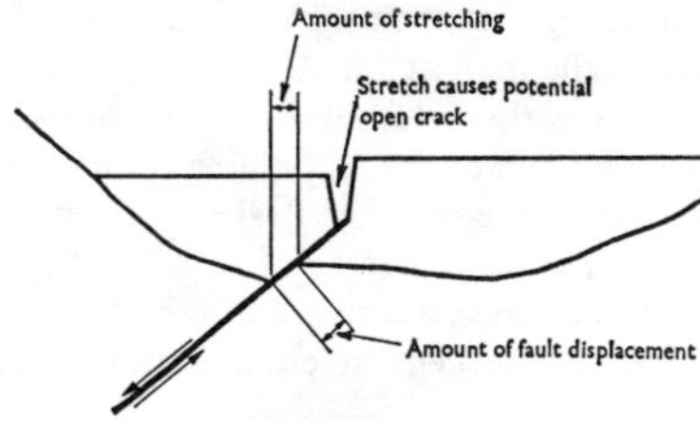

Fig. 42. Displacement of normal fault can stretch dam

In general studies of dam sites, frequently epicentres of instrumentally measured earthquakes are plotted on a map to show the general seismicity of the surrounding region. Often the epicentral locations are obtained from the records accumulated by a seismological laboratory or government agency. The locations of epicentres are computed from wave travel times and may be very rough estimates, especially when the instruments are located at considerable distances from the site. Hence, from the standpoint of fault activity studies, little reliance can usually be given to the fact that a typical epicentre map shows a general scatter of epicentres with no particular alignment along a fault. Also, unfortunately, there are many unreliable published assignments of earthquake magnitudes.

ACTIVE FAULT CONSIDERATIONS IN DAM DESIGN

Fault activity evaluation

After the geological studies are completed, the dam designer must finally decide whether he will assume that the fault may break under the dam or whether it will not. It is not possible to incorporate an intermediate decision, reflecting the geologist's uncertainty, into the design in any direct fashion. That is, if the geologist is only 50% confident that the fault is truly active, it does not make sense to design the dam for only 50% of the possible fault displacement. The decision must ultimately be one of 'go' or 'no go'.

The dam designer's decision is simple if the geologist is willing to make a flat statement of the activity or inactivity of the fault in question. However, a range of conditions may exist between these two extremes.

Relative activity of faults

In any given area, geologists can form opinions regarding the relative likelihood that various faults will break in the lifetime of a dam. The classification of relative activity has very great significance in preliminary planning and design studies to assist in the evaluation of alternate sites where the dam might be located, but such a relative classification may have little value in choosing the design details for a dam at a given site.

This conclusion is illustrated by a simplified example. Two alternative sites several kilometres apart are under consideration for a given dam, each site with a different fault through the foundation. At site A, convincing evidence exists that the fault has broken several times during the last 10 000 years, and thus is clearly an active fault in the sense that futher movement is sufficiently likely to be considered in the dam design. At site B the fault is overlain by unbroken, young surface deposits (indicating that the fault has not ruptured during the past 10 000 years), but there is abundant evidence of faulting prior to that time—within the past

35 000 years. This fault, too, probably must be considered active (by the AEC definition) but obviously not to the same degree as the fault at site A.

Other things being equal, site B is clearly preferable to site A; the likelihood of fault displacement during the life of a dam is far greater at site A, judging from its recent geological history. If, then, it is decided to build the dam at the safer site B, should the design be any different from that at site A, to take into account the lesser degree of activity of the fault? The answer is probably no, as both faults must be considered active, and therefore the dam must be designed to withstand the largest credible displacement at either locality, regardless of the greater likelihood of its occurrence at site A.

Faults of uncertain activity

After completion of his studies of a given fault, the geologist may conclude that there is insufficient evidence to make a firm judgement one way or the other regarding the activity of the fault. This may be either because insufficient exploration has been carried out or because no pertinent geological evidence was found to exist even after the most careful and exhaustive field investigation programme. For example, the geologist might find that mid-Pleistocene strata of 1 000 000 year age are broken by a fault, but that no datable younger formations are present to indicate whether the fault might have moved as recently as, say, 35 000 years and therefore be an active fault by his definition. Thus the fault is of uncertain activity. Nevertheless, even if the geologist is unable to make a firm statement as to the activity of the fault, he is obliged to make some judgement, with an appropriate disclaimer as to his degree of confidence in that judgement. Assuming the absence of diagnostic local geological evidence, such as in the preceding example, 'suggestive' evidence bearing on the fault's activity is almost always available from studies of the regional geology and geomorphology, the regional earthquake history, and the instrumentally recorded seismicity. Even if the geologist can do no more than estimate that, in his best judgement, the probability is 50% that a fault has broken within say, the past 35 000 years, this statement is far more meaningful to the dam designer than the simple (though honest) statement of 'I don't know.' The dam designer, in turn, must take into account the geologist's confidence in his conclusions, as well as exercising his own degree of conservatism, in reaching the final decision as to whether the dam should or should not be designed for a specified fault displacement.

In the Authors' opinion, major faults located in geological provinces where strong earthquakes have occurred (and thus are expected) should be viewed conservatively, by assuming for dam design purposes that they are potentially active unless convincing evidence exists to the contrary. This is believed to be the only reasonable and prudent assumption for critical structures such as dams, where literally thousands of lives may be at stake. Engineers have been surprised too often by unexpected engineering and geological phenomena associated with major earthquakes; if dam design is to err, clearly it should be on the side of conservatism.

Faults in non-seismic regions

In areas of the world commonly considered to be non-seismic, usually there have been few or no earthquakes in historic time, and there probably exists little evidence of fault displacements in recent geologic time. In these areas it is apparent that the likelihood of a fault displacement under a dam is very much smaller than in an area of known seismicity.

Nevertheless, the Authors believe that it is prudent and desirable to conduct reconnaissance studies to confirm the absence of active faults at and near any proposed site of a major dam in any part of the world. In the developing science of seismology it is being learned that there are fewer and fewer large areas of the world in which one can be wholly confident that strong natural earthquakes could not occur, and it is probable that some areas now considered seismically inactive are only areas with long quiescent periods between strong earthquakes. In many truly non-seismic areas evidence may be quickly found to demonstrate the absence of active faults, but certainly such evidence should be sought and documented. The larger the reservoir, the higher the dam, the less well known the regional geology, the greater is the need for such studies of possible fault activity even in such non-seismic areas.

A considerably lesser effort is justified in the study of possible fault activity than would be necessary for a similar site in a region of known strong earthquake activity. In the general case, the Authors would be satisfied to categorize a fault in such an area as being inactive if no suggestion of recent displacement were found from competent studies of good aerial photographs supplemented by field reconnaissance. This is in contrast to the situation in a known seismic area, where there should be a much greater tendency to consider a fault active until proved otherwise.

Reservoir induced earthquakes

Within the past few years, a number of instances have been cited in which the filling of large reservoirs has allegedly triggered earthquakes, even in some regions with little or no prior earthquake history. The Authors believe that there is now sufficient evidence to make it necessary to conclude that the reservoir filling may trigger strong earthquakes. This introduces a new dimension into seismic design considerations for major dams, particularly in areas normally considered free of earthquakes. Insofar as the subject of this Paper is concerned, several critical questions arise. If a dam for a major reservoir in a general area of earthquake activity is to be constructed across a regional fault of some length (e.g., 10–20 km), should the dam be designed for possible fault displacement even if the fault is demonstrably inactive as defined in this Paper (e.g., a fault covered by unbroken Pleistocene deposits)? Also, is it credible that a major fault in a non-seismic area could become active solely because of the filling of a reservoir on or near it? These are difficult questions to answer, partly because of the limited geological evidence available from some of the localities where reservoir-induced earthquakes have been demonstrated.

It might be argued that the surface faulting under these circumstances is possible inasmuch as (*a*) the reservoir induced earthquakes have been as large as $M = 6{\cdot}4$, which is large enough to be associated with appreciable fault displacement under normal circumstances; (*b*) the induced earthquakes have been shallow; (*c*) in several instances the induced earthquakes have been shown to have been on pre-existing faults; and (*d*) focal mechanism studies at several localities have indicated convincingly that the earthquakes represented tectonic strain release rather than a direct mechanical effect of the weight of the water. Thus the argument might be made that the reservoir induced earthquakes are in every sense normal tectonic earthquakes, even those in previously quiescent areas, and that surface faulting is just as likely to be associated with these shocks as with any others in the same magnitude range. Nevertheless, none of the larger reservoir induced earthquakes reported to date are known to have been associated with surface fault displacements, although some effects at the Koyna Dam (India) have been debated. Also, the reservoir induced earthquakes have been relatively rare events, even in areas of high natural seismicity. The Authors cannot make a definite statement that surface faulting should or should not be considered credible under these circumstances but believe the possibility cannot be completely dismissed at the present state of knowledge. Clearly this is an area where further intensive geological and geophysical studies are necessary. The possibility remains that if competent studies were made of the main reservoir induced earthquakes in the areas previously believed to be non-seismic, evidence would be found that nearby faults are active, which the Authors would not find surprising.

Embankment versus concrete dams

Where demonstrably active faults cross the foundation, concrete dams should not be considered. This is the strong opinion of the Authors and is believed to be the general opinion in the industry. Embankment dams can be designed with confidence to withstand safely the deformations (and damaging results thereof) imposed by fault movement, whereas any design of a concrete dam on an active fault is questionable.

If a foundation fault displaces, the dam sections on opposite sides will be offset by the same general amount and direction as the fault displacement. For an embankment dam the main potential hazard is from a concentrated leak developing suddenly along the rupture surface formed in the dam. The hazard can be eliminated by the provision of adequate internal zoning to control the leakage safely. Depending on the dimensions and the conservatism of the protective zones, an embankment dam can be theoretically made safe against any conceivable fault displacement.

All types of concrete dams are rigid and brittle. With few exceptions they are founded on fairly sound rock and depend for security on the bond between the concrete and the rock foundation and/or abutments. Any foundation fault displacement with vertical component must either lift the dam off the foundation, breaking its bond with the rock, or cause a break in the concrete. For a concrete gravity dam, fault movement can break the contact between the dam and the foundation, allowing full uplift pressure to act underneath, reducing the shearing resistance along the dam base and causing failure by sliding on the foundation of the general type which resulted in the disaster at the St. Francis Dam in southern California in 1928. For a concrete arch dam an abrupt displacement of two parts of the dam of the order of 0·25–0·50 m in almost any direction could cause complete sudden failure by several mechanisms. The concrete could be crushed or one end of the dam lifted off its abutment, destroying the continuity of the arch.

The Authors believe that concrete dams should also generally be avoided in earthquake regions at sites with faults for which the studies of the activity are inconclusive. The main reason is simply that the possibility of fault movement exists, no matter that the likelihood may be extremely small. The greater the consequences of the failure, the greater the desirability of arbitrarily excluding the concrete dam in the preliminary design stage.

The arbitrary choice of an embankment dam has other potentially important advantages at such sites. Frequently vital additional information about the geology is obtained from exposures during foundation excavations. Conceivably additional evidence for fault activity may be obtained. For the embankment dam, modifications of the design during construction to increase the conservatism are a relatively simple matter. For the concrete dam, however, it might be necessary to abandon the work and start over again with an embankment dam design. Without even considering fault activity, many dams have been changed during construction from concrete to embankment because of conditions exposed in the excavations, but not vice versa.

Dam sites located within a few kilometres of important thrust faults on the upthrown side, and within a few kilometres of important normal faults on the downthrown side, are particularly suspect for concrete dams. Areas surrounding such sites can be expected to be distorted during earthquakes originating on the main faults, and it is impossible to be confident that small displacements could not occur on almost any existing rock shear zone or fracture, or even possibly new breaks.

Magnitude and type of fault displacement

The dam designer needs an estimate of maximum probable magnitude and direction of the fault break which might occur in order to proportion the internal protective zones. The estimate should be made by the geologist on the basis of his study, and judgement guided by historic experience with similar faults. Some general observations can be made concerning these estimates.

First, the direction of fault displacement relative to the longitudinal dam axis can be as important as the magnitude of the displacement. Particularly, inclined normal faults striking parallel with the river whose displacement would probably tend to cause portions of the dam on opposite sides to be pulled apart could result in a wider initial crack than other types of fault movement (Fig. 42).

Displacements on thrust faults (and on some strike-slip faults) result in compression on the shearing surface. However, displacement on a vertical strike-slip fault which strikes at a small angle with the dam axis (Fig. 43) could also conceivably elongate the dam and cause potentially larger cracks than similar magnitudes of fault displacements in other directions.

Estimates of fault displacement magnitudes should not be dependent on any evaluation of 'relative activity' of the faults.

EMBANKMENT DAM DESIGN

Fundamental basis for confidence

The examples of the Coyote, Cedar Springs and Palmdale Dams demonstrate that many engineers are confident that conservatively designed embankment dams can safely withstand displacement of foundation faults. The Authors agree with this general conclusion.

This confidence has two main bases. First, there is considerable experience to show (records of the many fault breaks all over the world during the last 200 years) that the magnitude of maximum displacements on faults is generally less than 5–7 m, the average displacement along average fault breaks is 1 m or less and, in the extreme, displacements exceed 10 m only rarely and in exceptional world earthquakes on 'master' active faults. Also, while present ability to predict the maximum displacement which may occur on any given active fault has limitations, there is considerable knowledge of the subject and a reasonable estimate can be made. Second, we are confident that conservatively designed embankment dams will withstand without failure the worst conceivable damage imposed by foundation fault movements of these expected magnitudes, provided that there are materials available from which to construct thick, protective zones of cohesionless transition material.

Cracks cannot exist in cohesionless materials

Sands, gravels, sand-gravel mixtures and masses of hard rock fragments, when these materials are free of any appreciable admixtures of fine-grained soil (clays or silts), are cohesionless in the sense that each individual particle is not held to its neighbours. There is no cementation of any kind. Hence, a mass of these materials will not stand on an unsupported vertical face of any appreciable height and it is impossible for an open crack to exist in the mass. Any open crack which might tend to be formed momentarily by the fault displacement must immediately close by the action of the walls of the crack collapsing. Hence, internal zones of these materials act as 'crack stoppers' in dam embankments.

For any high embankment dam it is known from experience that cracks develop without earthquake or fault movement because of differential settlement (Sherard, 1973). Hence, cohesionless transition zones frequently are employed as crack stoppers in any event. The main difference in designing against fault displacement is that the zones may need to be made thicker.

For an open crack to exist in an embankment, it is necessary for the embankment material comprising the wall of the crack to have sufficient unconfined compressive strength to prevent the crack from closing. The mechanics are shown in a simplified fashion in Fig. 44. If we assume an open crack in the embankment with approximate depth h, the vertical stress P_v existing on the horizontal plane forming the top of the small shaded element a must be approximately the weight of the column of overburden with height h and density γ. As the crack is open and there is no horizontal stress on the side of element a ($P_h=0$). Hence, the crack can only remain open if the unconfined compressive strength U of embankment element a exceeds P_v. Otherwise element a will fail in compression and the crack must close. As shown in Fig. 41, the maximum depth to which a crack can be sustained open without collapsing is $h_{max}=U/\gamma$. For cohesionless zones, $U=0$, and no crack can be sustained. For typical compacted fine-grained soils commonly employed for the impervious cores of earth dams, the unconfined compressive strength is generally in the range between 20 and 40 t/m^2, and γ is of the order of 2 t/m^3. From these rough figures we can see that open cracks can be sustained in the cores of embankment dams to depths of the order of 10–20 m, below which the crack walls must collapse and close the crack, at least in the absence of internal fluid pressure in the crack. Hence, for high dams it is difficult to imagine that large open leakage channels could exist in the core at depths below the crest exceeding about 30 m.

It should be acknowledged that sand and gravel deposits in nature sometimes become partially cemented and in this form are observed to stand on fairly high vertical cliffs in river banks, and man-made excavations. The question has sometimes been raised as to whether cohesionless zones in dams might become cemented in some way over long periods of time. Cementation in natural sand and gravel deposits generally results from either a sufficient quantity of clayey fines to bind the mass together or cementation by chemical deposits from groundwater flow. In the extreme, the cemented deposits are partially transformed to soft formations (sandstones, conglomerates, etc.). For most clean, natural sands and gravels and crushed hard rocks, there is no reason to suspect that they will recement themselves when compacted in an embankment. The individual fragments are hard and inert and in time spans such as 100 to 200 years, under the pressures existing in a dam embankment, no changes in the material properties occur. This is true for the normal alluvial sands and gravels which are the hardest and most durable portions of their parent rocks, after being worked by nature for thousands of years. For some badly weathered and very old alluvial terrace deposits, and some quarried soft rocks, recementation might occur in an embankment, and these materials are avoided for use in the cohesionless transition zones of dams.

Large leaks can be safely controlled

With the cohesionless transition zones acting as crack stoppers, even if a large open crack develops in the impervious core the maximum leakage will be limited by the permeability of the transition material to a volume which can be controlled safely. The leakage is controlled

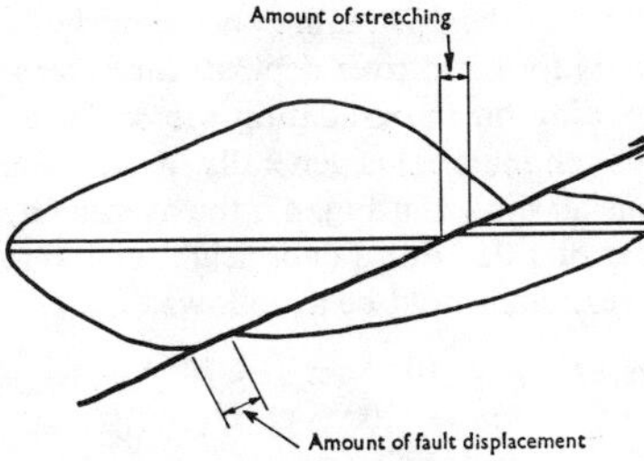

Fig. 43. Displacement of strike-slip fault can stretch dam

by providing a zone of considerably coarser and more pervious material (usually quarried rock fill or screened large cobbles) downstream from the downstream transition zone, so that the leakage must pass from the transition zone into the rockfill zone, and exit from the dam through the rockfill zone.

Since the permeability of the transition zone will always be at least an order of magnitude (and usually several orders of magnitude) less than that of the rockfill zone, the amount of leakage water will always fall well below the quantity which would tax the hydraulic capacity of the rockfill zone. Also, the relative gradation of the transition zone and the rockfill zones are made such that the individual particles of transition zone materials cannot penetrate the voids of the rockfill zone. Using these design principles, the maximum possible leakage quantity will emerge safely under complete control through the toe of the downstream rockfill zone.

There is no doubt concerning these conclusions. In addition to the theoretical considerations as outlined in general terms in the foregoing, there are many experiences in the industry in which large leaks have developed through dams with downstream rockfill zones with no danger or threat of failure. Many such examples could be described which make it completely sure that as long as the flow entering the downstream rockfill zone is controlled (choked) by the permeability of the finer transition material upstream, the flowing water entering the rockfill zone cannot cause threat of instability.

The 1964 failure of the Hell Hole Dam (California) is frequently quoted as an example of failure of a rockfill (Leps, 1973). The Hell Hole Dam failed during construction in a storm when the reservoir rose behind the partially completed dam. The water flowed through the main rockfill embankment section without control by an upstream transition zone which had not been constructed at the time of failure. If the upstream transition had been in place it is clear that no failure could have occurred.

Simplified illustrative example of fundamentally safe dam

The dam which has fundamental, unquestionable safety against abrupt rupturing caused by foundation fault displacement has two main elements. First, the cohesionless transition zone and second, the coarser rockfill zone downstream. Fig. 45 shows a simplified illustrative example of an embankment dam which consists only of these two elements, with an upstream zone of alluvial sand and gravel and a downstream zone of rockfill, with no impervious core. The leakage through such a dam would be fairly high relative to other dams, but it would be a completely safe dam.

As an example let us consider that the upstream zone comprises coarse, well-graded sand-gravel mixture, which is frequently found in river deposits (and from which many major dams have been constructed). Depending on the gradation the coefficient of permeability of embankment zones constructed of such material is generally in the range between 10^{-4} and 10^{-2} cm/s. With general proportions as shown in Fig. 45, the average hydraulic gradient through the upstream zone is of the order of 1·0. For a dam height of 100 m, the quantity of seepage through the upstream sand-gravel zone would be as follows:

$$\text{assume } k = 10^{-2}\ \text{cm/s} = 10^{-4}\ \text{m/s}$$
$$Q = kia = (10^{-4})\,(1{\cdot}0)\,(100)$$
$$= 10^{-2}\ \text{m}^3/\text{s/m}$$
$$Q = 10\ \text{litres/s/m of dam length}$$

It is apparent that such leakage quantities can easily be accommodated safely in a large downstream rockfill zone.

If the dam were ruptured by foundation fault displacement, a portion of the upstream alluvial material would be displaced and sheared and possibly loosened (made less dense) increasing the permeability locally and the quantity of leakage. However, the massive hydraulic capacity of the downstream rockfill zone is so overwhelming that no hazard would result.

The fact that large quantities of water can flow safely through rockfill zones is demonstrated by many recent experiences, and by laboratory and mathematical studies. Leps (1973) presented an excellent review and analysis of several experiences in which large flows have passed through rockfill embankments, including examples of hundreds of cubic metres per second. Dams have even been built in which small spillway floods are discharged directly into the downstream rockfill zones (Wilkins, 1963; Olivier, 1967; Curtis and Lawson, 1967). Also there have been several cases where floods backed up water behind rockfill dams during construction, before the impervious element was constructed, causing very large flows to pass through the rockfill without damage.[1]

Design principles

At any site where a fault may displace in the foundation of a dam, it must also be anticipated that strong earthquakes may occur and the dam must be designed to resist the strong shaking motion. Since one of the main ways in which an embankment dam may be threatened by earthquake shaking is a general slumping and spreading of the embankment with associated possible cracks through the core, some of the main design measures employed to increase the security of an embankment dam against earthquake shaking also increases the safety against fault offsets. For this reason a dam which is conservatively designed against strong earthquake shaking would probably also safely withstand moderate movements of foundation faults. Opinions concerning main aspects of earthquake design have been given elsewhere (Sherard, 1967). Only a few of the main points are re-emphasized here briefly.

It is not uncommon in areas of strong earthquakes that small faults are found for the first time when the dam is under construction, during excavation for the foundation. Although they may be obviously minor features, their unexpected discovery and the need for evaluating their potential activity can cause problems and delays. For this reason, at any site for an important dam or reservoir in a general earthquake region, consideration should be given to

[1] One of many examples: Dix River Dam, 85 metres high, concrete faced, rockfill dam in Kentucky. Major flood during construction (1925) before concrete facing constructed. Water reached about 18 m depth in reservoir, flowed through dam for six days with estimated maximum flow of 85 m^3/s without damage.

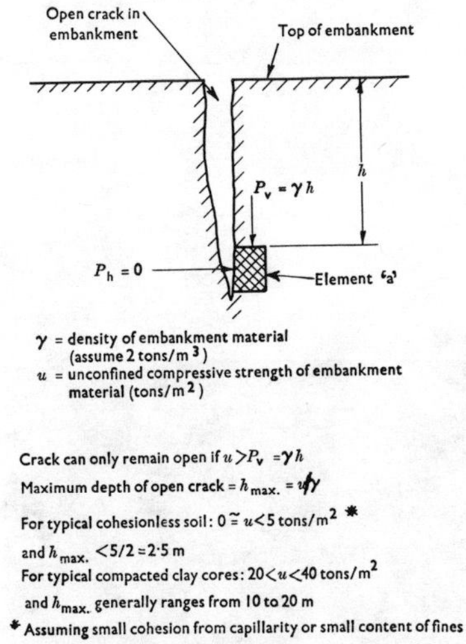

Fig. 44. Theoretical maximum depths of open cracks in embankments

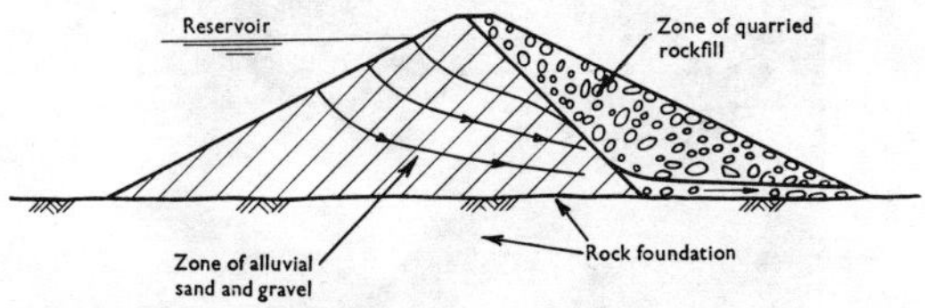

Fig. 45. Simplified illustrative example showing the two main elements of an embankment dam providing fundamental safety against foundation fault displacement

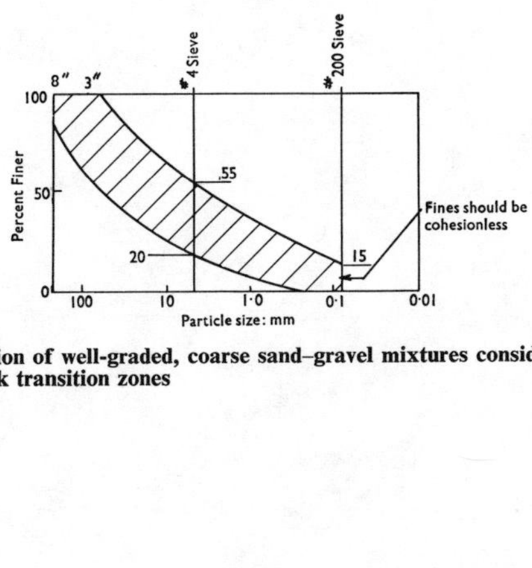

Fig. 46. Range of gradation of well-graded, coarse sand–gravel mixtures considered to be especially suitable without processing for thick transition zones

Fig. 47. Typical natural alluvial sand–gravel mixture meeting the gradation of Fig. 46. Thick zones of material of this general nature, used without processing, constitute good, self-healing and internally stable transitions

providing a design which would be safe under a small displacement on any previously unforeseen minor fault which may be exposed during construction.

The selection of the main design details governing the fundamental safety of the dam (such as the width of the top of the dam, the thickness of the cohesionless transition zones, etc.) must be made finally arbitrarily guided by the estimate of the probable magnitude of fault displacement. While it is impossible to give quantitative guidelines, it is clear that the degree of conservatism which is arbitrarily provided in the design of these vital details should be made larger for those dams located above populated areas, and for dams with especially large reservoirs.

The top of the dam is particularly vulnerable because it is the thinnest part and has less internal pressure tending to close cracks. For any given dam design, efforts to increase the fundamental safety should consider particularly the details of the top of the dam, which especially needs good zoning and conservative width and freeboard.

The risk involved from fault offset is dependent on the erosion resistance of the foundation. For foundations other than hard rock, and particularly foundations which do not have high resistance to erosion, special design details should be employed to make the water travel a long distance through the foundation. In the extreme case, if the foundation materials are highly erodible and the potential consequences of failure are great, very conservative design details are justified.

Cohesionless transitions must always be used. Satisfactory designs range from central core dams with transitions on both sides of the core to homogeneous dams with internal 'chimneys' consisting of bands of cohesionless transitions and drainage material. The thicker the transition zones, the greater the security against a given fault displacement. The cohesionless transition zones are generally manufactured by washing, screening and blending, to provide materials with uniform and reliable properties.

At sites where cohesionless natural sands and gravels are in large supply they can be used for the transitions without processing by making the widths of the zones larger to compensate for some lack of uniformity in the properties. Natural well-graded, sand–gravel mixtures, having grain size distributions generally as shown in Fig. 46, are particularly good materials for transition zones (Fig. 47). The fine portion should be cohesionless. In common with other cohesionless transition materials, this material will act as a crack stopper. It also has the advantage that, while it is relatively pervious compared to the dam core, its fundamental permeability is not high. Another desirable attribute is the fact that it is inherently stable against internal piping, being in itself a natural filter. Also, this material is generally coarse enough so that there is no danger that any appreciable quantity of the material could be washed into the voids of adjacent rockfill zones.

Sand-gravel river deposits, with gradation in the range of Fig. 46, are one of the most common materials available for building earth dams in many parts of the world. In these areas this sand and gravel may be the most economical material available and almost the whole dam may be built of it, in which case the most economical type dam is often the fundamentally safe dam. At some sites transition materials with the desirable gradation range can be obtained economically by processing (screening and blending) alluvial soils. Suitable material of such gradation can also be made by crushing or screening quarried rock. Usually hard rock crushed to a maximum size of 3 in. or 6 in. will have suitable gradation.

The properties of the impervious core material can be considered of lesser importance. A fundamentally safe dam can be made using a core material of any type. It is preferable to spend extra money improving the transition zones rather than for bringing in impervious core material of supposedly superior characteristics from long-haul distances. Other things being equal, preference should be given to cohesionless core materials over clayey materials, at least in the upper part of the dam. Impervious materials with fines of dispersive clay should be avoided (Sherard *et al.*, 1972).

Central earth core rockfill dams

The rockfill dam with central earth core and thick cohesionless soil transitions upstream and downstream, as exemplified by the design of the Cedar Springs Dam (Fig. 14), is a fundamentally safe structure. The basic type of dam is also frequently the most economical type which can be constructed at a given site, at least for high dams.

The use of the second sand transition zones at Cedar Springs Dam both upstream and downstream of the core (zone 2), certainly increases the fundamental safety. If a large leak developed through a crack in the clay core at the top of the dam, some of the sand from the upstream zone 2 would undoubtedly be washed into the crack, tending to seal it. Also, assuming that the sand comprising the zone 2 layer downstream of the core could not sustain a crack, the volume of leakage passing through the top of the dam would be limited by the permeability of this sand layer to a lower value than otherwise.

The symmetrical design, with transitions both upstream and downstream from the core, as shown in Fig. 14, has a potential advantage from the standpoint of making rapid and simple repairs if the dam were damaged by fault displacement. Because of the existence of the upstream and downstream transitions it would be possible to seal open fissures and cavities in the dam core by grouting the cracks. Cement and soil-cement grouts will not enter the transition zones. Hence, grout pumped into the core of the dam will remain in the core and fill the openings. This has been done successfully on a number of recent central-core dams where cracks and erosion channels have developed, caused by differential settlement. Examples are the recent experiences at the Yards Creek Upper Reservoir Dam in New Jersey and the Hyttejuvet Dam in Norway (Sherard, 1973).

ACKNOWLEDGEMENTS

A number of colleagues, both from the geology and engineering sides, reviewed the Paper in an earlier draft and provided suggestions for improvement. The Authors are grateful for this generous assistance and wish they could do more than simply acknowledge the reviewers, many of whom deserve far greater credit for their contributions to the Paper. Particular thanks are due to N. N. Ambraseys, Imperial College of Science and Technology, London; M. G. Bonilla, United States Geological Survey, Menlo Park, California; G. Post and P. Londe, Coyne and Bellier, Consulting Engineers, Paris; C. J. Cortright and E. W. Stroppini, California Department of Water Resources, Sacramento; O. Papageorgiou, Public Power Corporation, Athens; J. M. Wool, Los Angeles Department of Water and Power, California; J. M. Kellberg, Tennessee Valley Authority, Knoxville; G. E. Brogan, G. E. Hervert, M. K. Korringa, B. B. Gordon and D. Tocher, Woodward-Lundgren and Associates, Oakland, California.

REFERENCES

Albee, A. L. & Smith, J. L. (1966). Earthquake characteristics and fault activity in southern California. *Engineering Geology in Southern California*, Association of Engineering Geologists Special Publication, October 1962.

Allen, C. R. (1957). The San Andreas Fault. *Engineering and Science Monthly*, *California Institute of Technology*, May.

Allen, C. R., St. Amand, P., Richter, C. F. & Nordquist, J. M. (1965). Relationship between seismicity and geologic structure in the southern California region. *Bulletin, Seismological Society of America* **55**, 753.

Allen C. R. & Smith, S. W. (1966). Pre-earthquake and post-earthquake surficial displacements. *Bulletin, Seismological Society of America*, **56**, 961.

Allen, C. R., Wyss, M., Brune, J. N., Grantz, A. & Wallace, R. E. (1972). Displacements on the Imperial Superstition Hills, and San Andreas Faults triggered by the Borrego Mountains Earthquake. *US Geological Survey Prof.*, Paper 787, 87.

Ambraseys, N. N. (1969). Maximum intensity of ground movements caused by faulting. *Proceedings, Fourth World Conference on Earthquake Engineering, Chile*, **1**, 154.

Ambraseys, N. N. (1970). Some characteristics of the Anatolian fault zone. *Tectonophysics* **9**, 143.

Ambraseys, N. N. (1971). Value of historic records of earthquakes. *Nature* **232**, Aug. 6, 375.

Bonilla, M. G. (1967). Historic surface faulting in continental United States and adjacent parts of Mexico. *US Atomic Energy Commission Report* TID-24124 (available from the clearing house for federal scientific and technical information, National Bureau of Standards, Springfield, Virginia, 22151).

Bonilla, M. G. (1970). Surface faulting and related effects. Chap. 3 in *Earthquake Engineering* (R. L. Wiegel, Editor). Prentice-Hall, Inc., Englewood Cliffs, New Jersey.

Bonilla, M. G. & Buchanan, J. M. (1970). Interim report on worldwide historic surface faulting. *US Geological Survey open file report.*

Brown, R. D. & Vedder, J. G. (1967). Surface tectonic fractures along the San Andreas Fault. *US Geological Survey Prof.* Paper 579, 2–23.

Brune, J. N. & Allen, C. R. (1967). A low-stress-drop, low-magnitude earthquake with surface faulting. The Imperial, California, Earthquake of March 4, 1966. *Seismological Society of America Bulletin* **57**, 501–514.

California Department of Water Resources (1964). Investigation of Baldwin Hills Reservoir. *Report of Engineering Board of Inquiry.*

Carlifornia Department of Water Resources (1968). *Cedar Springs Dam specifications and bid documents*, No. 68-30.

Casagrande, A., Wilson, S. D. & Schwantes, E. D. (1972). The Baldwin Hills Reservoir failure in retrospect. *Proceedings of the Specialty Conference on the Performance of Earth and Earth Supported Structures*, ASCE **1**, Part 1, 551.

Clark, M. M., Grantz, A. & Rubin, M. (1972). Holocene activity of the Coyote Creek Fault as recorded in sediments of Lake Cahuilla. *Geological Survey Professional Paper* 787 U.S. Dept. of Interior, p. 112.

Castle, R. O. & Youd, T. L. (1972). Baldwin Hills Reservoir failure. *Proceedings, Specialty Conference on Performance of Earth and Earth-Supported Structures*, ASCE **3**, 91.

Cluff, L. S. & Plafker, G. (1966). Implications derived from a study of thrust faulting associated with the March 27, 1964, Alaskan earthquake on Montague Island (abstract). *Association of Engineering Geologists National Meeting*, October, 1966.

Cluff, L. S. & Slemmons, D. B. (1972). Wasatch Fault Zone—features defined by low-sun-angle photography. *Utah Geological Society Publication* 1, 6–1.

Cluff, L. S. & Steinbrugge, K. V. (1966). Hayward Fault slippage in the Irvington-Niles districts of Fremont, California. *Bull. Seismol. Soc. Am.* **56**, No. 2, 257.

Cortright, C. J. (1970). Re-evaluation and reconstruction of California dams. *Jnl Power Div.* ASCE, Jan., 55.

Curtis, R. P. & Lawson, J. D. (1967). Flow over and through rockfill banks. *Jnl Hydraulics Div.*, ASCE, September, 1.

Florensov, N. A. & Solonenko, V. P. (1963). The Gobi-Altai earthquake. *Akademiĭ Nauk. SSSR*, 424 p.

Hadley, J. B. (1964). Landslides and related phenomena accompanying the Hebgen Lake earthquake of August 7, 1959. *US Geological Survey*, Paper 435, 107.

Jansen, R. B. *et al.* (1967). Earth movement at Baldwin Hills Reservoir. *Jnl Soil Mech. Fdns Div.*, ASCE, 551.

Lawson, A. C. (1908). The California earthquake of April 18, 1906. Report of the state earthquake investigation commission. *Carnegie Institute, Washington* Pub. 87.

Leps, T. M. (1972). Analysis of failure of Baldwin Hills Reservoir. *Proceedings of the specialty conference on the performance of earth and earth-supported structures*, ASCE **1**, Part 1, p. 507.

Leps, T. M. (1973). *Flow through rockfill. Chapter on embankment dam engineering*, 87. R. C. Hirschfeld & S. J. Poulos, Editors. New York: John Wiley and Sons.

Louderback, G. D. (1950). Faults and engineering geology. *Geological Society of America.* Berkley, 125.

Louderback, G. D. (1937). Characteristics of active faults in the central coast ranges of California with applications to the safety of dams. *Bull. Seismol. Soc. Am.*, Jan., 1.

Morante, E. M. & Allen, C. R. (1973). Displacement on the Philippine Fault during the Ragay Gulf earthquake of 17 March 1973. (Abstract) *Geological Society of American Abstracts with Programs* **5**, 744.

Myers, W. B. & Hamilton, W. (1964). Deformation accompanying the Lake Hebgen earthquake. *US Geological Survey*, Paper 435, 55.

Nason, R. D. (1972). Creep in Anatolia. *Geotimes* **17**, No. 9, 30.

Oldham, R. D. (1899). Report on the great earthquake of 12 June 1897. *Geological Survey India Mem.* **29**.

Olivier, H. (1967). Through and overflow rockfill dams—new design techniques. *Proc. Inst. Civ. Engrs*, paper 7012.

Press, F., Ben-Menahem, A. & Toksoz, M. N. (1961). Experimental determination of earthquake fault and rupture velocity. *Jnl Geophysics Research* **66**, 3471–3485.

Reitz, R. (1971). Horizontal and vertical displacement of Pacoima Dam. *Los Angeles County flood control district, Survey Division*, November.

Richter, C. F. (1958). *Elementary seismology*. W. F. Freeman and Company, San Francisco and London.

Sewell, J. S. (1907). The effects of the earthquake and fire on buildings and engineering structures. *Bulletin, US Geological Survey*, No. 324.

Sherard, J. L. (1959). A report on the damage to Hebgen Dam in the West Yellowstone earthquake. *Engineering News Record*, 17 August.

Sherard, J. L. (1967). Earthquake considerations in earth dam design. *Jnl Soil Mech. Fdns Engng Div.*, ASCE, 377.

Sherard, J. L. (1973). Embankment dam cracking. Chapter on embankment dam engineering 271., R. C. Hirschfeld & S. J. Poulos, Editors. New York: John Wiley and Sons.

Sherard, J. L., Woodward, R. J., Gizienski, S. F. & Clevenger, W. A. (1963). Earth and earth-rock dams. New York: John Wiley and Sons.

Sherard, J. L., Decker, R. S. & Ryker, N. L. (1972). Piping of earth dams of dispersive clay. *Proceedings, Specialty Conference on the Performance of Earth and Earth-supported Structures*, ASCE **1**, 589.

Slemmons, D. B. (1969). New methods of studying regional seismicity and surface faulting. *Transactions, American Geophysical Union* **50**, 397.

Slemmons, D. B., Steinbrugge, K. V., Tocher, D., Oakeshott, G. B. & Gianella, V. P. (1959). Wonder, Nevada, earthquake of 1903. *Bull. Seismol. Soc. Am.* **49**, 251.

Smith, S. W. & Wyss, M. (1968). Displacement on the San Andreas Fault subsequent to the 1966 Parkfield earthquake. *Bull. Seismol. Soc. Am.* **58**, 1955–1973.

Stroppini, E. W., Babbitt, D. H. & Struckmeyer, H. E. (1972). Foundation treatment for embankment dams on rock. *Jnl, Soil Mech. Fdns Div.*, ASCE, 1073.

Steinbrugge, K. V. & Zacher, E. G. (1960). Creep on the San Andreas Fault—fault creep and property damage. *Bull. Seismol. Soc. Am.* **50**, 389–396.

Tarr, R. S. & Martin, L. (1912). The earthquake at Yakutat Bay, Alaska, in September, 1899, with a preface by G. K. Gilbert. *US Geological Survey Prof.* Paper 69.

Talley, H. C., Jr. & Cloud, W. K. (1962). *United States earthquakes, 1960*. Washington: US Coast and Geodetic Survey.

Terzaghi, K. & Lacroix, Y. (1964). Mission Dam, an earth and rockfill dam on a highly compressible foundation. *Géotechnique* **14**, No. 1, 14.

Thevenin, J. (1955). The effects of the September 9, 1954 earthquake on the hydraulic works in the Orleansville area. *Terres et Eaux*, No. 24 (in French).

Tibbetts, F. H. (1936). Earthquake proof earth dams. *Engineering News-Record*, July 2.

US Atomic Energy Commission (1973). Seismic and geologic siting criteria for nuclear power plants. *US Federal Register* **38**, No. 218, 23 Nov.

USCOLD (1967). *General Paper No. 7, Ninth Congress on Large Dams, Istanbul* IV, 760.

USCOLD (1970). *General Paper No. 7, Tenth Congress on Large Dams, Montréal* IV, 807.

USCOLD (1971). Terzaghi Dam. *Newsletter of the US Committee on Large Dams*, January, 14.

USCOLD (1973). *General Paper No. 1, Eleventh Congress on Large Dams, Madrid* V, 366.

Wilkins, J. K. (1963). The stability of overtopped rockfill dams. *Proceedings, Fourth Australia–New Zealand Conference on Soil Mech. and Fdn Engng*, 141.

Witkind, I. J. *et al.* (1962). Geologic features of the earthquake at Hebgen Lake, Montana. *Bull. Seismol. Soc. Am.* **52**, 163.

Yerkes, R. F., Youd, T. L. & Alt, J. N. (1973). Deformation of the Van Norman Reservoir area, northern San Fernando Valley, California. *Fifth World Congress on Earthquake Engineering, Rome*, Paper No. 299.

Introduction to Paper 8

Pinhole Test For Identifying Dispersive Soils

by J.L. Sherard, L.P. Dunnigan, R.S. Decker, and E.F. Steele
Journal of the Geotechnical Engineering Division,
Vol 102, No.GT1, January 1976, pp 69-85.

Prepared by Lorn P. Dunnigan[3], M. ASCE

Dr. Sherard became interested in dispersive clay erosion while reviewing reports of small dam failures in Australia. These failures were attributed to dispersive clay erosion. In about 1970, while consulting on a U.S. Department of Agriculture Soil Conservation (SCS) dam, Jim learned that the SCS had experiences similar to those described in the Australian literature. Because of Dr. Sherard's interest in this phenomenon, the SCS contracted with him to make an independent review of SCS dam failures where dispersive clay was believed to be a contributing cause.

During his study of SCS dams it became apparent that a better method of identifying dispersive clay was necessary. The need for a new identification method provided the first opportunity for me to work with Dr. Sherard. We initiated a research project which was rather unique for the following reasons: First, we had a repository of soil samples on which field performance, relative to dispersive clay erosion, was known. Secondly, the research was completed without special funding or without a fee charged by either party. This was accomplished because of a common interest in a geotechnical problem. Our goal in this effort was to develop a test that matched, as closely as possible, the action of dispersive clay erosion in an earth embankment.

[3]Former Head, Soil Mechanics Laboratory, Soil Conservation Service, U.S. Department of Agriculture, Lincoln, NE.

JOURNAL OF THE GEOTECHNICAL ENGINEERING DIVISION

Pinhole Test for Identifying Dispersive Soils

By James L. Sherard,[1] F. ASCE, Lorn P. Dunnigan,[2] Rey S. Decker,[3] Members, ASCE, and Edgar F. Steele[4]

Introduction

A new laboratory test has been developed for direct measurement of the dispersibility (colloidal erodibility) of compacted fine-grained soils in which water is caused to flow through a small hole punched in a specimen. The water running through samples of dispersive clay carries a cloudy colored suspension of colloidal particles, whereas the water running through erosion resistant clays is crystal clear. The main purpose of the test is identification and improved understanding of dispersive high sodium highly erodible fine-grained soils, previously evaluated by indirect testing methods. A considerable experimental effort was carried out before methods of testing were found which gave reliably reproducible results. Since 1973 the writers have been using the test in practice, mainly for studies of earth dams. It has also been used by other engineers in more than a dozen different laboratories in various parts of the world, with no reported large problems of setting up the apparatus, making the tests or interpreting the results.

In the last few years it has become more clearly understood that there are in nature certain clays (dispersive clays) that are highly erodible. The discovery of the existence of these clays has explained many heretofore difficult-to-explain failures and troubles with embankment dams. The principal difference between dispersive clays and ordinary erosion-resistant clays is the nature of the cations in the pore water. Dispersive clays have a preponderance of sodium, whereas ordinary clays have a preponderance of calcium and magnesium cations in the pore water.

The basic nature of dispersive clays was fairly well understood by soil scientists

Note.—Discussion open until June 1, 1976. To extend the closing date one month, a written request must be filed with the Editor of Technical Publications, ASCE. This paper is part of the copyrighted Journal of the Geotechnical Engineering Division, Proceedings of the American Society of Civil Engineers, Vol. 102, No. GT1, January, 1976. Manuscript was submitted for review for possible publication on April 1, 1975.

[1] Consulting Engr., Berkeley, Calif.

[2] Head, Soil Mechanics Lab., Soil Conservation Service, Lincoln, Neb.

[3] Sr. Soils Engr., Hoskins, Western, Sonderegger, Consulting Engrs., Lincoln, Neb.

[4] Civ. Engr., Soil Mechanics Lab., Soil Conservation Service, Lincoln, Neb.

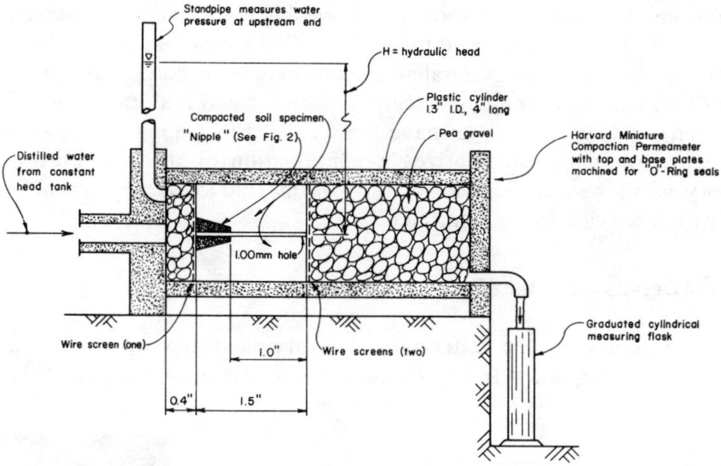

FIG. 1.—Pinhole Test Apparatus (1 in. = 25.4 mm)

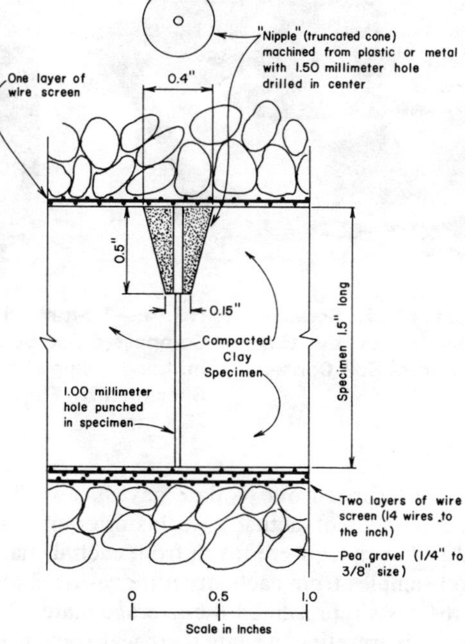

FIG. 2.—Section through Pinhole Test Specimen (1 in. = 25.4 mm)

and agricultural engineers of the U.S. Soil Conservation Service more than 40 yr ago (2,4), but the importance of the subject in civil engineering practice was not recognized widely until the early 1960's when many failures of small clay dams in Australia were shown to have been caused by dispersive clays (1). A 1970–1971 study sponsored by the U.S. Soil Conservation Service showed conclusively that the main Australian conclusions were correct and that similar failures of clay dams in the United States, which could not readily be explained by the then existing knowledge, have been caused by piping of dispersive clay (3). The present paper summarizes a continuation of this investigation. The laboratory work was carried out in the soil mechanics laboratory of the Soil Conservation Service in Lincoln, Neb.

Reason for Developing Pinhole Test

At the present time the criteria for identifying dispersive clays are based on tests on samples taken from eroded areas of damaged and failed dams (3).

FIG. 3.—Pinhole Test Mold, Screens, Nipple, and Needle (Photo by U.S. Department of Agriculture Soil Conservation Service)

FIG. 4.—1.0-mm Hole Punched in Compacted Soil Specimen Through 0.5-in. Long "Nipple" (Pushed into Top of Specimen by Finger Pressure) (1 in. = 25.4 mm)

While the investigations carried out to date have been fairly extensive, there were only a limited number of actual case histories of failures and damage. Also, while several soil samples were taken from each damaged dam, the tested properties of the soil samples from each structure varied, and it was not possible to know which of the tests represented the average material governing the dam performance. Thus, in interpreting the data there was some need for extrapolation and some of the criteria were not as strongly supported by the performance experience as others.

In an effort to gain more experience, without the necessity of having actual records of the behavior of dams with leaks in them, a test was devised in

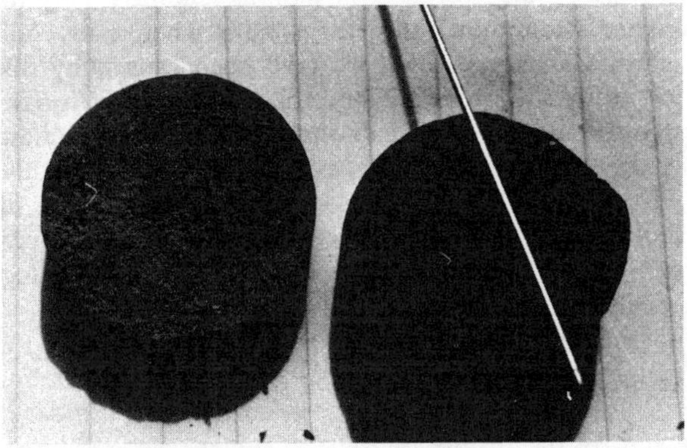

FIG. 5.—Appearance of Typical Clay Specimens after Test Is Completed [Sample on Left Show Typical Eroded Hole in Dispersive Clay after 5-Min Flow at 2-in. Head; Sample on Right Shows Hole with No Discernible Erosion (Compare with Needle Size) in Sample of Nondispersive Clay Tested with 40-in. Head; When Latter Sample is Broken along Pinhole Line, Needle Fits Exactly Cast of Hole Showing That There Has Been No Measurable Erosion] (1 in. = 25.4 mm)

FIG. 6.—Comparison of Cloudy Colloidal Appearance of Water Flowing through Sample of Dispersive Clay (Left Cylinder) with Perfectly Clear Water Obtained in Test on Typical Erosion-Resistant Clay (Right Cylinder) (When Allowed to Sit, Cloudy Water Remains Same Color for Days, after First Several Hours When Coarser Particles Settle Out; Clear Water Is Judged by Looking At it through Side and from Top through Vertical Column of Water, as Shown in Sample Data Sheets of Figs. 8 and 9)

the laboratory simulating the action of a leak in a clay dam. Starting in 1973 various details were studied for several months leading finally to the adoption of a testing apparatus as shown in Figs. 1-6 and a standard testing procedure as described in detail in Appendix I.

Specifically, the test was devised for the purpose of identifying dispersive clays. It is not intended to be used as a quantitative test for measuring rates of erosion as a function of the velocity of flowing water.

General Test Description

As seen in Fig. 1, distilled water is caused to flow through a 1.0-mm (0.039 in.) diam hole punched in a 1.0-in. (25.4-mm) long specimen of clay. Initially the water is caused to flow under a hydraulic head of 2 in. (50 mm) of water. The principal differentiation between dispersive clay and ordinary erosion-resistant clay is given by the results of the test under 2 in. (50 mm) of head. For the dispersive clay the flow emerging from the specimen is visibly colored with a colloidal cloud and does not clear with time. Within 10 min the hole enlarges to about 3 mm (0.12 in.), or larger, and the test is completed.

For the ordinary erosion-resistant clay the flow emerging is completely clear, or becomes completely clear in a few seconds, and the hole does not erode.

TABLE 1.—Categories of Test Results

Classification of individual test results (1)	Classification of soil (2)
D1 and D2	Dispersive soils: fail rapidly under 2-in. (50-mm) head
ND4 and ND3	Intermediate soils: erode slowly under 2-in. (50-mm) or 7-in. (180-mm) head
ND2 and ND1	Nondispersive soil: no colloidal erosion under 15-in. (380-mm) or 40-in. (1,020-mm) head

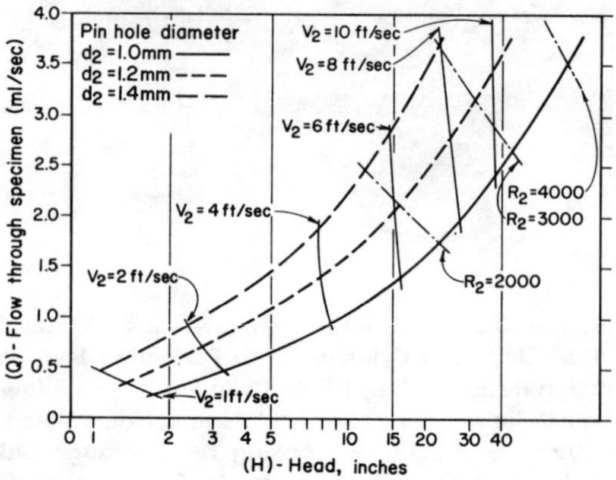

FIG. 7.—Calculated Flow Quantities (from Conventional Computations of Flow through Pipe) (1 in. = 25.4 mm; 1 fps = 0.305 m/s)

At the end of 5 min the rate of flow is unchanged. The hydraulic head is raised in progressive steps, for 5 min each, to 7.0 in., 15.0 in., and 40.0 in. (180 mm, 380 mm and 1,020 mm). At each progressively higher gradient the rate of flow is measured and the color of the water, if any, observed.

For highly erosion-resistant clays, there is never any visible color or cloud in the water and the rate of flow is constant under each flow condition. For the typical clay of high erosion resistance, the hole in the clay specimen does not increase in size, as shown by dismantling the specimen and comparing it with the needle used to punch the hole. Figs. 5 and 6 show the typical final appearance of soil specimens and of the waters which flowed through the specimens during the test, for dispersive and nondispersive clays.

As described in Appendix I, the test result is evaluated from the appearance of the water, the rate of flow, and final size of the hole in the specimen.

The test is highly reproducible and the results of each individual test can be categorized easily into one of the six categories shown in Table 1 with little room for differences in personal judgment between different technicians. The test is simple and can be made in any soil mechanics laboratory. While used primarily for recompacted specimens, the test can also be used easily with undisturbed samples, cut to the size of the cylinder and pushed in.

Pipe Flow Computations

Fig. 7 shows computed quantities of flow through the hole in the nipple and soil specimen using conventional pipe flow calculations, assuming that the head is lost only in causing flow through the hole:

$$\text{Head} = \frac{V_1^2}{2g}\left(0.5 + \frac{f_1\, l_1}{d_1}\right) + \frac{V_2^2}{2g}\left(K_c + \frac{f_2 l_2}{d_2} + 1.0\right) \quad \text{(1)}$$

$$\text{assuming laminar flow:} \quad f = \frac{64}{\mathsf{R}} \frac{\dfrac{64}{Vd}}{\nu} \quad \text{(2)}$$

in which V_1 and V_2 are velocities; f_1 and f_2 are friction coefficients; l_1 and l_2 are lengths; d_1 and d_2 are diameters for the holes in the nipple and soil specimen, respectively; R = Reynold's Number; ν = the kinematic viscosity of water; and K_c is the contraction coefficient for head loss between the hole in the nipple and soil sample.

The flow quantities in Fig. 7 are computed assuming laminar flow, although the Reynold's numbers exceed 2,000 for the higher heads, as shown. Similar calculations made assuming turbulent flow are not presented here because it was found that the flow quantities as measured in the laboratory on the nonerodible samples approximated the computed quantities in Fig. 7 surprisingly well. Evidently for the small diameter pinhole and the gradually increased velocities used in the test, the flow remains laminar even at Reynold's numbers up to 4,000.

As seen in Fig. 7 the computed velocity of flow through the hole in the the compacted soil specimen is between 1 fps–2 fps (0.305 m/s–0.6 m/s) when the head is 2 in. (50 mm) and about 10 fps (3.05 m/s) under 40 in. (1,020 mm) of head.

Description of Test Results

A main conclusion is that essentially all of the high sodium clays, previously designated as dispersive and highly erodible from the studies of failed and damaged

PIN HOLE TEST DATA

Pin Hole Test No. Example

Sample No. ____

Compaction Characteristics good

Water Content near optimum

Distilled water added: ✓ Yes or ___ No

Curing time: none

Date: ____

Page: ____

Specimen after test:

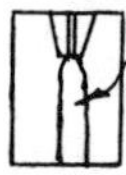

Flow started on 1st trial.

Clock Time	Head	Flow Rate ml	Flow Rate sec	Color from Side: Dark	Color from Side: Slight to Medium	Color from Side: Barely Visible	Color from Side: Completely Clear	Completely Clear From Top	Particles Falling: None	Particles Falling: Few	Particles Falling: Heavy	Remarks
3:05	2"	10	25	✓					✓			highly colloidal
		10	10	✓					✓			flow
		25	18	✓					✓			
		25	18	✓						✓		" "
		25	17	✓						✓		
		25	16	✓						✓		
		25	15	✓						✓		" "
3:08		25	15	✓						✓		Stop test-
												Obvious Failure
												(D1)

FIG. 8.—Typical Data for Dispersive Clay D1 (1 in. = 25.4 mm)

dams (3), erode rapidly in the pinhole test. Conversely, essentially all the low sodium clays, previously designated as "ordinary, erosion resistant clays," do not erode in the pinhole test. There are a few exceptions of low-sodium clays which erode rapidly. These exceptions are apparently not very common. Some of these are not recognizable as dispersive clays when tested using the other available tests for clay dispersion (3).

Fig. 8 presents the basic data from a typical test on highly dispersive clay which shows that colored water came through the specimen under the 2-in. (50-mm) head at an initial rate of 10 ml/25 sec (0.4 ml/s). The discharge from the specimen continued dark colored with increasing rate until the test was completed in about 4 min, at which time the rate of flow had increased to approx 1.7 ml/s. On dismantling the specimen it was found that the hole had eroded to approx 2.5 times the initial diameter.

PIN HOLE TEST DATA

Pin Hole Test No. *Example*

Sample No. ______

Compaction Characteristics *soft sample*

Water Content *19.3%*

Distilled water added: ____ Yes or *✓* No *None.*

Curing time: *Compacted at natural water content*

Date: ______

Page: ______

Specimen after test:

Flow started on *1st* trial.

Clock Time	Head	Flow Rate ml	Flow Rate sec	Color from Side: Dark	Slight to Medium	Barely Visible	Completely Clear	Completely Clear From Top	Particles Falling: None	Few	Heavy	Remarks
9:33	2"	10	22				✓	✓	✓			Sparkling Clear
		10	20					✓	✓			
		=										
		10	19					✓	✓			
9:38		10	19					✓	✓			
9:38	7"	25	20					✓	✓			
		25	20					✓	✓			
		=										
9:42		25	20					✓	✓			" "
9:43		25	20					✓	✓			
9:43	15"	50	27					✓	✓			
		50	27					✓	✓			
9:45		50	26					✓	✓			
9:47		50	27					✓	✓			" "
		50	26.5					✓	✓			
9:48	40"	50	15					✓	✓			
		50	15					✓	✓			
		=										
9:50		50	16					✓	✓			
		50	15.5					✓	✓			" "
		=										
9:52		50	15					✓	✓			
		50	16					✓	✓			
9:53		50	16					✓	✓			End Test (ND1)

FIG. 9.—Typical Data for Nondispersive Clay ND1 (1 in. = 25.4 mm)

Fig. 9 shows typical results for a highly erosion-resistant clay. The hydraulic head was progressively raised from 2 in.-40 in. (50 mm-1,020 mm). At each head, the flow through the specimen was constant, increasing from about 0.45 ml/s under 2 in. (50 mm) of head to about 3.3 ml/s under 40 in. (1,020 mm). At all times the water flowing through the specimen was perfectly clear. At the end of the test, the apparatus was dismantled and the hole in the specimen was observed to be the same size as originally punched with the needle.

For highly dispersive clays the color in the water is usually completely colloidal, although occasionally larger particles are carried along and are seen to be falling through the water in the cylindrical flask. For some nondispersive and moderately erosion-resistant soils, the water may be completely clear as seen from both the side and the top of the measuring flask, but some slight erosion occurs in the form of barely visible particles being carried along. Probably these visible particles consist of large flocs (aggregates) of clay particles. Sometimes a few such small particles are seen falling through water in the measuring flask which is otherwise crystal clear, indicating that the erosion conditions are sufficiently great to detach clay flocs from the wall of the hole, but there is no colloidal erosion. Usually when such action occurs, the hole erodes more or less uniformly along its length. Occasionally a concentration of erosion occurs at the downstream end, giving a small bell-shaped cavity.

Using the testing methods described in Appendix I, the test results are generally reliably reproducible for highly dispersive and erodible soils (D1 and D2) and

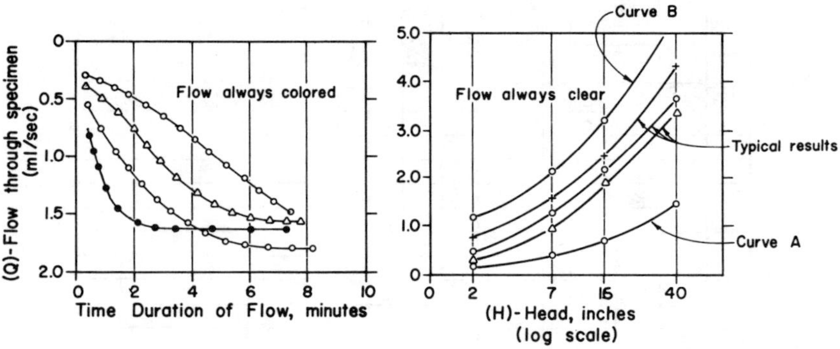

FIG. 10.—Typical Test Results

also for highly erosion-resistant soils (ND1 and ND2). For soils of intermediate erosion resistance, the test results are reproducible with some moderate scatter; for example, the results of a series of repetitive tests performed on specimens of the same soil prepared in the same manner might vary: ND3, ND4, ND3, ND4, etc.

Fig. 10(*a*) shows typical test results for a dispersive clay.

Fig. 10(*b*) shows typical test results on specimens of nondispersive erosion resistant soils (ND1 and ND2). In most tests the measured flow coming through the specimens approximate the values calculated for hole diameters in the range between 1.00 mm and 1.20 mm (Fig. 7). Although the hole was formed in the sample with 1.00-mm diam hypodermic needle, it was inevitable that the hole in the soil sample was slightly larger than the needle simply because the needle cannot be pushed in and pulled out without making a hole slightly larger than the needle (Fig. 4).

While most test results fall within a fairly narrow band when plotted as shown in Fig. 10(*b*), occasional tests on the nondispersive soils give considerably higher or lower measured flow quantities. Curve A in Fig. 10(*b*), shows a test result where the hole diameter evidently was reduced by swelling or other action. Curve B shows a test result where the hole diameter was considerably greater, possibly caused by lateral movement of the needle as the hole was being punched. In both the tests of Curves A and B, the flow passing through the soil sample remained completely clear and did not increase with time, and the materials were obviously erosion resistant.

For the soils which give greatest resistance to erosion in the pinhole test (ND1) there is no observable erosion under 40 in. (1,020 mm) of head. In some of the tests the water coming through the specimen may have a trace of color for a few seconds, evidently caused by the flow washing some loosely held fragments from the hole wall before equilibrium is reached, after which the water is perfectly clear. More commonly, for the specimens giving high erosion resistance (ND1), the water runs completely clear during the entire test.

For those soils that give intermediate reaction to the pinhole test (ND3 and ND4) the flowing water has a slight color, indicating that slow colloidal erosion is occurring. For some of these intermediate soils, the water passing through the specimen under 2 in. (50 mm) of head runs completely clear and the color is observed to appear only at the higher velocities; for others, the water is observed to have a slight color even under 2 in. (50 mm) of head. Soils of this type can be considered intermediate between dispersive clays and nondispersive clays. However, the erosion is enormously slower than that experienced by the typical high sodium dispersive clay.

Fortunately it appears from the writers' studies that these intermediate soils comprise only a small fraction of the total number of fine-grained soils in nature. The large majority of soils tested, other than the soils which fail rapidly in the pinhole test, indicated high erosion resistance in the pinhole test (ND1 and ND2), with no clay particles going into colloidal suspension in the water.

Influence of Compaction Water Content and Density

Almost all of the writers' pinhole tests have been made on specimens compacted at a water content near the plastic limit, and no extensive study has been made of the possible influence of the compaction water content on the results of pinhole tests for various soil types. This is a subject which needs further study. The writers have some opinions and observations from the work carried out to date.

It has been established conclusively that high sodium dispersive clays will erode rapidly in the pinhole test regardless of the water content or density at which the specimens are compacted.

For the great majority of erosion-resistant soils (ND1 and ND2), it is clear that the pinhole test results are not changed by moderate differences in the compaction water content. For most of the tests in this study, the compaction water content was adjusted to a value judged visually to be near the plastic limit or "good for compaction." The result was that many tests were performed with the same soil on specimens compacted over a fairly wide range of water content without causing erratic results.

For the smaller group of soils in the intermediate category of erosion resistance (ND3 and ND4), possibly moderate differences in the compaction water content can have an important influence on the pinhole test results.

Investigations Leading to Adoption of Details for Standard Test

Working over a period of about 18 months, more than 1,200 individual pinhole tests have been performed on about 200 different soils. The samples used in this testing program included samples from routine testing for design purposes. Also tested were many of the samples from failed and damaged dams included in the previous investigation (3) and more recently acquired samples from areas where clays have eroded or resisted erosion in nature. The procedures for making the tests and interpreting the results were gradually modified based on the results of the previous tests, striving towards a test procedure that provided reproducible results and differentiated between known dispersive and nondispersive soils.

In the early stage of the work various sizes of pinholes were tried, ranging from a minimum of 0.016 in.-0.10 in. (0.406 mm-2.54 mm). The 1.0 mm size was chosen for the standard procedure for the following main reasons:

1. Smaller holes gave about the same results but they frequently became plugged, whereas only rarely does the 1.0-mm hole become plugged.
2. Larger holes above 1.0 mm could be used and give generally similar results, but they require a larger quantity of water and have no compensating advantage.
3. The 1.0-mm hole is about the size of a medium sand particle, so that it is much larger than most of the grains in the clayey soils being tested. It is a convenient hole size to punch reliably and the rate of flow through it is convenient to measure and supply.

Many tests were made before the 0.5-in. (12.7-mm) long nipple was adopted (Figs. 1 and 2) with no great difference in the average results. Experiments showed that the nipple decreased the frequency that the hole in the sample was plugged, since it deters collapse of the hole. It also serves as a convenient template for punching the hole (Fig. 4).

The hydraulic heads employed [2 in.-40 in. (50 mm-1,020 mm)] were chosen finally as being convenient for use in the laboratory and because they generate velocities of flow [1 fps-10 fps (0.305 m/s-3.05 m/s)] which approximate the initial velocity that might be expected in leaks within embankment dams and other structures. In the experimental stage, hydraulic heads as low as 0.5 in. (12.7 mm) were used, with the conclusion that there was no value to using a head lower than 2 in. (50 mm) in the standard test since the dispersive clays eroded in the test equally easily at 0.5 in. (12.7 mm) as at 2.0 in. (50 mm). Experiments were also made at heads of the order of 80 in.-100 in. (2,030 mm-2,540 mm) with the conclusion that there was no general value in the routine test to using more than 40 in. (1,020 mm).

The investigation showed clearly that it is necessary to protect samples at their natural water content and not allow them to air dry before testing. The drying and rewetting apparently causes a temporary (and perhaps partially permanent for some soils) condition of disequilibrium in the interaction forces between clay particles, which gives variable results in the pinhole tests, particularly if the dried soil has been pulverized by mechanical action before water is added. For some soils, drying, pulverizing, and rewetting does not change the pinhole test results. For other soils it has been found that the sample preparation can be a critical aspect of the testing procedure, as presented in Appendix II.

Use of River Water in Pinhole Test

Colloidal clay flocculation theory suggests that: (1) Clays are more likely to disperse in pure water than salty water; and (2) small amounts of salt in the water percolating through the specimen could have a large influence on the test results. The standard pinhole test procedure using distilled water measures the erodibility of the clay in pure water, which is considered a basic property of the material. A few tests have been made with waters having small quantities of dissolved salts (such as 0.1 meq/l-1.0 meq/l of $CaCl_2$ and some natural river waters). In these tests, some of the dispersive clays eroded in exactly the same fasion as with distilled water, but for others the dispersive clay did not erode in the pinhole test with the slightly salty water running through the hole.

Significance of Test Results on Nondispersive Clay

For the typical nondispersive low sodium clay there is no visible erosion in the pinhole test. Tests have not been performed to establish beyond all question that water coming through the specimen in these tests, which appears to be crystal clear, does not carry rare occasional colloidal particles accompanied by a very slow rate of erosion. However, a few tests have been made with more than 300 hr of total flow—with no measurable increase in flow quantity. Studies of this subject are continuing.

Results of these tests show promise that compacted embankments of some nondispersive clays may withstand at least a small diameter leak having a velocity of 10 fps (3.05 m/s) with no erosion whatsoever. This pinhole test experience combined with the knowledge that concentrated leaks have occurred through some clay dams with unchanged rate of flow for decades makes it seem probable that the clay particles are held together with such tenacity that not a single one is dislodged from the walls of the leakage channel. If this tentative conclusion is finally confirmed, it would have important practical implications with regard to the need for filters for these nondispersive clays.

Summary and Conclusions

The paper describes a newly developed laboratory test for direct measurement of the erodibility of fine-grained soil, using a flow of water passing through a small hole in a specimen at velocities of 1 fps-10 fps (0.305 m/s-3.05 m/s). A main purpose of the test is reliable identification of dispersive clays, which have recently been shown to be responsible for many troubles and failures of earth dams and serious erosion to other earth structures.

The developmental work needed to assure that the test gives reproducible results and reliably identifies the highly erodible dispersive soils is completed. It is unlikely that further basic investigations will show that there are any important improvements to be obtained by modifying the test procedure described in Appendix I.

The discovery that water flowing through specimens of nondispersive clay at 10 fps (3.05 m/s) causes no visible erosion in the test shows promise for future improvements in present filter criteria for these clays, with potentially very important cost savings.

The large majority of several hundred samples of fine-grained soils tested with the pinhole test in the last several years are shown to be either nondispersive (ND1 or ND2) or dispersive (D1 or ND2) or dispersive (D1 or D2). Only a relatively small percentage gave intermediate test results (ND3 and ND4).

Acknowledgments

The writers are grateful to the Soil Conservation Service for permission to publish the results of the work. Thanks are due to the personnel of the SCS Soil Mechanics Laboratory in Lincoln, Neb., for assistance in performing the tests beyond the call of duty. For review of the manuscript and many useful comments we are indebted to David C. Ralston, Edward B. Perry, and Oliver H. Raine.

Appendix I.—Pinhole Test Laboratory Procedure

The general procedures that must be followed are:

1. Soil samples must be preserved at the natural water content by shipping and storing before testing in air tight bags or containers (see Appendix II).
2. If the material contains coarse sand or gravel particles, these are removed by pushing the material through the No. 10 sieve (2 mm).
3. The natural water content is measured and the water content is brought to near the plastic limit by adding the required water (or by gradually drying, if too wet). All water added should be distilled water.
4. The test specimen and apparatus are shown in Figs. 1 and 2.
5. The 1.5-in. (38-mm) long clay specimen is compacted in the cylinder on top of pea gravel and wire screen in five layers with 16 tamps each layer using a 15-lb (6.8-kg) spring with Harvard miniature compaction test tamper (which gives about 95% of Standard Proctor Compaction Effort).
6. The plastic nipple is pushed into the top of the specimen with finger pressure and the hole is punched through the nipple with a 1.0-mm diam stiff steel wire (hypodermic needle) (Fig. 4).
7. After the apparatus is assembled, water is percolated through the hole in the sample under heads of 2 in., 7 in., 15 in. and 40 in. (50 mm, 180 mm, 380 mm, and 1,020 mm) for time periods of 5 min–10 min at each head. The quantity of flow is measured with a stop watch and cylindrical measuring flasks (10 ml, 25 ml, and 50 ml) continuously and the time recorded, as shown in the typical data sheets (Figs. 8–9). The color of water is observed by looking both through the side of the flask and vertically through the column of fluid in the flask.
8. At the end of the successive flow tests, the apparatus is dismantled and the soil specimen is extruded from the cylinder and broken open to examine the size of the hole, which is measured approximately by comparison with the needle used for punching the hole.

At 2-in. (50-mm) head:

9. The test is started with head of 2 in. (50 mm). (If no flow occurs, stop the test, dismantle top of apparatus and repunch the hole, or seal the first hole and make a new one. This occurs rarely.)

10. The principal differentiation between dispersive and nondispersive soils is given by the test results under 2 in. (50 mm) of head. If the flow under 2 in. (50 mm) of head is visibly cloudy and does not become clearer with time, the specimen is failing in the fashion typical of dispersive clays. The main indicator of failure is the colloidal color of the water. Most dispersive clays erode rapidly under 2 in. (50 mm) of head with strong color in the water coming through the specimen. For dispersive clays usually the flow continuously increases and reaches a maximum value limited by the hydraulic capacity of the equipment in 2 min–5 min of flow {about 1.5 ml/s–2.0 ml/s, using the nipple and Harvard miniature compaction mold permeameter [Figs. 8 and 10(*a*)]}. Continue running the test for a total time of 10 min. Unless the color of the flow clears substantially, the test is then completed. For the typical dispersive clay the hole will normally be increased to about three needle diameters after 10 min of flow. Classify as highly dispersive (D1).

11. If the flow at 2 in. (50 mm) of head has a slight but easily visible color as seen from the side of the flask at the end of 5 min, continue test for total of 10 min. If flow continues colored, stop the test. If the rate of flow at the end of 10 min has not exceeded 0.8 ml/s, and the hole diameter does not exceed 1.5 needle diameters, classify as intermediate (ND4). If the flow at the end of 10 min exceeds 1.0 ml/s, and the hole diameter exceeds 2 needle diameters, classify as dispersive (D2). If the test is stopped at the end of 10 min and the results classified as ND4 and D2, the test should be repeated with a new specimen to see what happens by raising the head to 7 in. (180 mm).

12. If the flow under 2 in. (50 mm) of head is clear (or has only a very slight trace of color as seen from the side of the flask) at the end of 5 min, raise the head to 7 in. (180 mm) and continue the test {rate of flow is usually 0.3 ml/s–0.6 ml/s [Figure 10(*b*)]}.

At 7-in. (180-mm) Head:

13. At 7-in. (180-mm) head if the water continues to flow clear, or has only a trace of color as seen from the side of the flask, after 5 min raise the head to 15 in. (380 mm) and continue the test {rate of flow is usually less than 1.8 ml/s [Figure 10(*b*)]}.

14. If the water has color and rate of flow increases rapidly, stop the test. Classify as intermediate (ND3). The flow at end of test will generally exceed 2.5 ml/s, and the hole size will be larger than 2 needle diameters.

At 15-in. (380-mm) Head:

15. If the flow is completely clear as seen from the top of the measuring flask, raise the head to 40 in. (1,020 mm) after 5 min {rate of flow is usually less than 2.5 ml/s [Figure 10(*b*)]}.

16. If the flow has a slight color or exceeds 3.5 ml/s, stop test and classify nondispersive (ND3).

At 40-in. (1,020-mm) Head:

17. If the flow continues completely clear at rate usually less than 4.0 ml/s, classify as nondispersive (ND1) [Figs. 9 and 10(*b*)]. There should be no noticeable erosion of the hole in the sample at the end of the test.

18. If the flow has a bare trace of color or exceeds 5.0 ml/s, classify as nondispersive (ND2).

19. The criteria of steps 10–15 are summarized in Table 2.

Appendix II.—Problems with Testing Air-Dried Samples

Near the midpoint of the program of experimentation, a considerable perplexity arose when it was discovered that the test procedures then employed did not

TABLE 2.—Summary of Criteria for Evaluating Results

Classification (Table 1) (1)	Head, in inches (2)	Test time for given head, in minutes (3)	Visual final flow through specimen, in milliliters per second (4)	Color of flow at end of test (cloudy or color) (5)	Hole size after test (needle diameter) (6)
D1	2	5	>1.5	Very distinct	2x
D2	2	10	>1.0	Distinct to slight	2x
ND4	2	10	<0.8	Slight but easily visible	1.5x
ND3	7–15	5	>2.5	Slight but easily visible	2x
ND2	40	5	>3.5	Clear or barely visible	2x
ND1	40	5	<5.0	Crystal clear	No erosion

Note: 1 in. = 25.4 mm.

TABLE 3.—Results of Typical Pinhole Tests Showing Influence of Curing in Compacted State on Samples That Were Air Dried, Pulverized, and Rewetted before Compaction

Time of curing in compacted state, in days (1)	Soil Sample Number			
	S-32 (2)	S-118 (3)	S-110 (4)	S-96 (5)
1	ND1	D2	D2	D2
2	ND1	ND2	D2	ND4
5	ND1	ND1	D2	D2
12	ND1	ND1	D2	ND3
19	ND1	ND1	ND4	ND2
29	ND1	ND1	ND1	ND1
39	ND1	ND1	ND1	ND2
44	ND1	ND1	ND1	ND1
96		ND1		ND1

give reproducible results for a certain group of clays. These were clays which were believed to be nondispersive, as determined by the tests of the sodium content in the pore water and others. For identical specimens of some of these soils, the test results ranged from highly erosive (D1) to completely erosion resistant (ND1), when tests were performed at different times with essentially the same technique. The tests giving variable results were on samples which had been air dried during storage in the laboratory and had to have considerable water added before compaction in the pinhole test cylinder.

Much effort was spent trying to determine what was causing the variation in test results. The main factors studied were: (1) Small variations in the compaction water content; (2) the time of curing after adding the water to the dry soil and before compaction; (3) various compaction energies; (4) the influence of small differences in the distilled water percolated through the hole (waters obtained from the laboratory still at different times, which had different pH values); and (5) various methods of creating a constant-sized hole at the downstream end, such as provision of a piece of metal foil at the end through which the percolation hole was punched. (This latter procedure was shown to be of no value in the testing procedure.)

All these studies proved fruitless. Tests made in which all the main individual factors were held constant continued to give unpredictable and variable results until more or less by accident the solution was found. It was learned that if these soils were first compacted at water content near the plastic limit and then allowed to cure in the compacted state for 24 hr, or more depending on the type of soil, before testing, the results of all subsequent pinhole tests were then entirely reproducible. Also, no troubles of this kind have been experienced when the samples taken in the field at the natural density and maintained at the natural water content until tested. This rather surprising large influence of drying and pulverizing the dried soil before rewetting was proved conclusively by many tests.

Experiments were made in which a considerable number of low sodium nondispersive clays were air dried, pulverized, rewetted to a water content near the plastic limit, and then compacted in a standard Proctor mold. The compacted soil cylinder was then pushed out of the mold and cured in a plastic bag to prevent drying. For each soil, pinhole tests were made on the material after it had been allowed to cure in the compacted state for various lengths of time. Typical results from these experiments are shown in Table 3. For most nondispersive soils the results shown for samples S-32 and S-118 are typical i.e., either there was no influence (S-32) or only 1 day or 2 days of curing in the compacted state sufficed to cause the material to give a nondispersive reaction in the pinhole test (S-118). For a few samples, however, weeks of curing in the compacted state were required before the pinhole test results changed from dispersive to nondispersive [Samples S-110 and S-96 (Table 2)].

For these air-dried soils where it was demonstrated necessary to allow the samples to cure in the compacted state, the difference caused by curing in the compacted state could be visually seen during the pinhole test. For samples which had been cured in the compacted state, when the apparatus is dismantled at the end of the test the plastic nipple is pulled out of the specimen without any clay sticking to it. The nipple comes out perfectly free of soil and does not have to be washed before making the next test. For these previously dried and rewetted soils not cured in the compacted state, when the nipple is pulled out of the specimen at the end of the test, the adjacent soil clings to it tenaciously and must be washed off before being used in the next test. This visual observation gave additional evidence that a substantial change in the interaction forces between the clay particles occurs for these particular soils during a period of curing in the compacted state.

For these soils for which air drying influences the results, allowing the material to cure in a plastic bag after adding water to the sample but without compaction generally does not achieve the same results, somewhat to the writers' surprise. Evidently in order to achieve equilibrium in the clay interaction forces, it is necessary that the clay particles be forced together by compaction before the process of interacting commences.

For these specimens of nondispersive soil which eroded in the pinhole test after being dried, pulverized, and rewetted, the results of the pinhole test are in every way identical to the results of tests on dispersive clays. The colored water flowing through the sample is apparently a truly colloidal suspension, showing that for these soils the mechanical action of drying and rewetting temporarily causes repulsive forces between the clay particles which act in the same way as the permanent interparticle repulsive forces within the high sodium dispersive clay. After curing in the compacted state, however, they are apparently transformed again to the nondispersive erosion-resistant state.

There are certain nondispersive soils which are always completely erosion resistant (ND1) in the pinhole tests regardless of method of sample preparation. Many pinhole tests were made on some of these soils over the extreme range of water content from dust dry to water content which was so high that the material had the consistency of soft butter and had to be placed in the pinhole apparatus with light finger pressure. Nevertheless the sample was always shown to be completely erosion resistant (ND1), even at heads greater than 40 in. (1,020 mm). These preliminary observations give promise that future research will demonstrate that there are large differences in the fundamental erosion resistance among the common nondispersive clays. At the present time, the writers do not know what fraction of the total number of nondispersive soils tested might have continued to demonstrate erosion resistance had they been compacted at the extreme conditions of water content and density described herein.

Appendix III.—References

1. Aitchison, G. D., and Wood, C. C., "Some Interactions of Compaction, Permeability, and Post-Construction Deflocculation Affecting the Probability of Piping Failures in Small Dams," *Proceedings*, 6th International Conference on Soil Mechanics and Foundation Engineering, Montreal, Canada, International Society of Soil Mechanics and Foundation Engineering, Vol. II, 1965, p. 442.
2. Richards, L. A., *Diagnosis and Improvement of Saline and Alkali Soils, U.S. Dept. of Agriculture Handbook No. 60,* U.S. Government Printing Office, Washington, D.C., 1954.
3. Sherard, J. L., Decker, R. S., and Ryker, N. L., "Piping in Earth Dams of Dispersive Clay," *Proceedings*, ASCE Specialty Conference on the Performance of Earth and Earth-Supported Structures, Vol. 1, 1972, pp. 589–626.
4. Volk, G. M., "Method of Determination of the Degree of Dispersion of the Clay Fraction of Soils," *Proceedings*, Soil Science Society of America, Vol. 2, 1937, p. 561.

Introduction to Paper 9

Identification and Nature of Dispersive Soils

by J.L. Sherard, L.P. Dunnigan, and R.S. Decker
Journal of the Geotechnical Engineering Division,
Vol 102, No. GT4, April 1976, pp 287-301.

Prepared by Lorn P. Dunnigan[4], M. ASCE

In 1970, the U.S. Department of Agriculture Soil Conservation Service (SCS) contracted with Dr. Sherard to study piping failures and erosion damage from rain in clay dams in Oklahoma and Mississippi. As part of this study, Dr. Sherard sampled many SCS dams. He also sampled other dams in problem areas such as Wister Dam. During these field reviews he collected soil samples from both damaged and undamaged dams. For each sample location he would carefully document field observations. Because of his strong interest in dispersive clay, he continued collecting samples for about two years from projects he was working on outside the United States.

It was my good fortune to work with Dr. Sherard throughout this period as we jointly studied dispersive clays. We would meet at the SCS Soil Mechanics Laboratory in Lincoln, Nebraska, a couple of times a year. Each of these meetings would last for about ten days. During this time our efforts concentrated on testing. After each of these visits we studied and reviewed the test results and planned the next series of tests.

Dr. Sherard always displayed a great enthusiasm for research and had tremendous ability in generating and analyzing data.

---

[4]Former Head, Soil Mechanics Laboratory, Soil Conservation Service, U.S. Department of Agriculture, Lincoln, NE.

JOURNAL OF THE GEOTECHNICAL ENGINEERING DIVISION

IDENTIFICATION AND NATURE OF DISPERSIVE SOILS

By James L. Sherard,[1] F. ASCE, Lorn P. Dunnigan,[2] and Rey S. Decker,[3] Members, ASCE

INTRODUCTION

Certain fine-grained soils in nature are highly erodible (dispersive soils). These soils have a higher content of dissolved sodium in the pore water than ordinary soils. They erode by a process in which the individual colloidal clay particles go into suspension in practically still water, whereas the erosion process for ordinary clays is quite different, requiring considerable velocity in the eroding water.

Existence of dispersive clays was first generally recognized by engineers after pioneer investigations of eroded clay dams in Australia in the early 1960's (1,2,7,8,9,10,12). More than 20 yr earlier, soil scientists knew that sodium clays were responsible for much trouble with erosion to structures and agricultural land, and they were describing them as "dispersive" soils in the soil science literature in the mid-1930's (6,11,21). Since 1970, studies identified widespread dispersive clay erosion damage in the United States and provided rough criteria for identifying dispersive clays (14,15). In 1973, a new laboratory test (pinhole test) was developed which measures colloidal erosion directly by passing water through a hole in a compacted soil specimen (17). This paper presents the results of 1972-1975 studies which led to improved criteria for identifying dispersive clay. The effort included laboratory tests on a wide range of soils, and additional studies of case histories of erosion damage in nature.

FOUR LABORATORY TESTS (DESCRIPTIONS AND DEFINITIONS)

Four laboratory tests for identifying dispersive clays have been used by the

Note.—Discussion open until September 1, 1976. To extend the closing date one month, a written request must be filed with the Editor of Technical Publications, ASCE. This paper is part of the copyrighted Journal of the Geotechnical Engineering Division, Proceedings of the American Society of Civil Engineers, Vol. 102, No. GT4, April, 1976. Manuscript was submitted for review for possible publication on July 17, 1975.

[1] Consulting Engr., Berkeley, Calif.

[2] Head, Soil Mech. Lab., Soil Conservation Service, Lincoln, Neb.

[3] Sr. Soils Engr., Hoskins, Western, Sonderegger, Consulting Engr., Lincoln, Neb.

writers. Since the tests are not generally familiar to the engineer, a brief review follows.

Pinhole Test (17).—Distilled water flows through a 1.0-mm diam hole in a compacted specimen. For dispersive clays, the water becomes colored and the hole rapidly erodes. For nondispersive clay the water is clear and there is no erosion.

Soluble Salts in Pore Water (19).—This is a standard test of the agricultural soil scientist, used by the U.S. Department of Agriculture (USDA) and others, described briefly: (1) Soil is mixed with distilled water to a consistency near the Atterberg liquid limit; (2) a pore-water sample ("saturation extract") is sucked out by vacuum using a filter; and (3) the saturation extract is tested to determine the quantities of the four main metallic cations in solution (calcium, magnesium, sodium, and potassium) in milliequivalents per liter. By definition in this paper, the "total dissolved salts" (TDS) equals the total of these four cations, and "percent sodium" equals the quantity of sodium divided by the TDS, all in milliequivalents per liter.

These tests have been made for engineering projects using one of two procedures: (1) The soil sample is sent to an agricultural soil laboratory (many government and private laboratories all over the world are following the same procedures as the USDA; and (2) a pore-water sample (saturation extract) is extracted from the soil in a soil mechanics laboratory, a simple procedure described in Ref. 14, and tested in a commercial water chemical laboratory.

SCS Laboratory Dispersion Test (Sometimes Called Double Hydrometer Test).—The test was developed by Volk (21) and has been widely used by the U.S. Soil Conservation Service since 1940, although it was never described in the general engineering literature. The particle size distribution is first measured using the standard hydrometer test, in which the sample is dispersed in the hydrometer bath with strong mechanical agitation and a chemical dispersant (Curve 1, Fig. 1). A second hydrometer test is made without strong mechanical agitation and without a chemical dispersant (Curve 2, Fig. 1). Curve 2 shows less colloidal particles than Curve 1 and the difference is a measure of the tendency of the clay to disperse naturally. By definition, "percent dispersion" is the ratio of clay size particles (0.005 mm) in the two tests (Fig. 1). The test procedure is described in detail in an Appendix of Ref. 14.

Crumb Test.—A small soil crumb is immersed in water and the dispersion is observed directly, as described later in this paper.

Laboratory Investigation

All four tests were performed on each of 195 samples of fine-grained soils of widely varying properties and origins. Eighty of these were dispersive, with samples taken from many geographic areas including failed and damaged dams in several parts of the United States and abroad. Because of the large number of samples from widespread sites, the dispersive clays are probably representative of a wide range of such soils in nature.

The nondispersive soils were routine samples passing through the SCS Central Soil Mechanics Laboratory (Lincoln, Neb.), plus some samples from dams that had resisted leaks. Also, an effort was made to cover the range of fine-grained soils commonly used in earth dams.

In pinhole tests most soils either do or do not erode. In this study the few soils giving intermediate results were excluded.

Pinhole Test Results Related to Pore-Water Salts

A strong correlation was found between pinhole test results and pore-water sodium content. The two dotted lines in Fig. 2 were criteria proposed for the

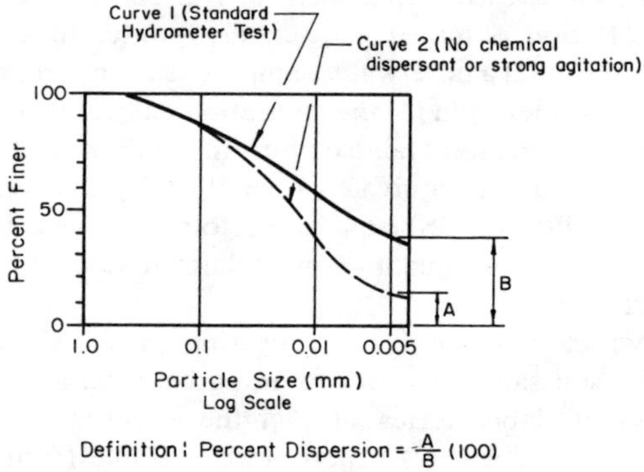

FIG. 1.—SCS Dispersion Test

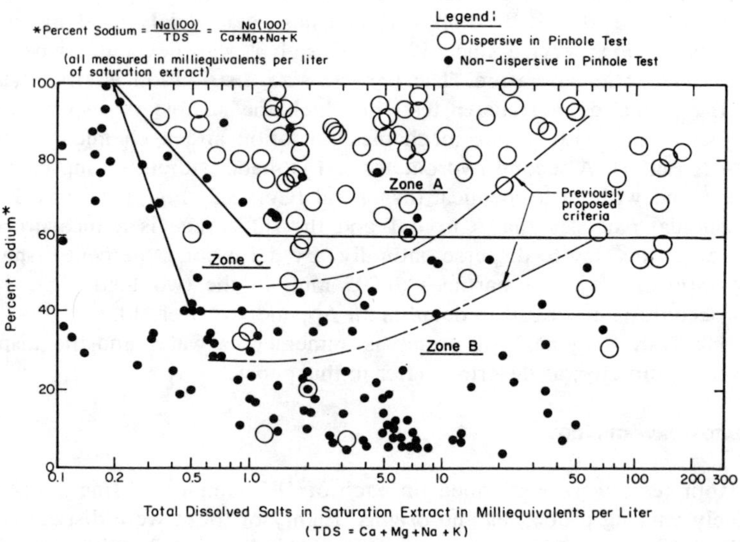

FIG. 2.—Relationship of Pore-Water Salts and Dispersion of Compacted Samples Measured in Laboratory with Pinhole Test

transition between dispersive and nondispersive soils from an earlier study of erosion of damaged dams (14). These earlier criteria fit the present data reasonably well in the center part of the plot, but there are significant differences at greater and lesser salt contents.

Few of the soils taken from eroded dams in the previous study had TDS above 30 meq/1, so that the earlier criteria were not strongly supported by the then existing experience for such soils. In 1972, the failure of the La Escondida Dam in Mexico (4) by dispersive piping of clay with high TDS showed that the earlier criteria needed modification. In this study many tests show that soils with high pore water TDS and 50% sodium, or more, are dispersive, confirming that the dotted curves in Fig. 2 are too high. Most soils with low TDS (less than 1.0 meq/1) were nondispersive in the pinhole test, even when the sodium content was well above 60%. This tendency had been seen in the previous study without sufficient data for a conclusion.

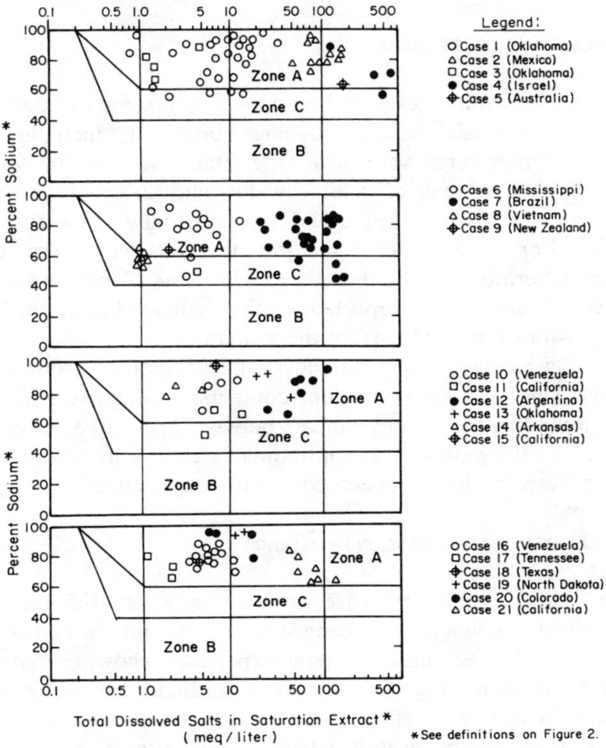

FIG. 3.—Pore-Water Salts in Soils from Dispersive Piping Case Histories

Six soils with sodium content less than 40% were dispersive in the pinhole test. These exceptions were investigated with many tests. No explanation has been found for these exceptions; they did not have a compensating high potassium content. These six exceptions were soils being routinely tested and were not among samples taken from failed dams or erosion areas.

The soils tested comprise a wide range of different clay minerals in various mixtures. Flocculation tests of clay-water suspensions show that the boundary between dispersed (deflocculated) and flocculated states varies considerably among different clays. Thus, it is not unexpected that the transition between dispersive and nondispersive soils in Fig. 2 is wide. It is surprising, in fact, that pinhole tests for the many diverse clays tested give such a strong correlation with the pore-water salts.

A few preliminary experiments were made in which the pore-water salts of a given clay were modified. In these, a dispersive clay was mixed with a water soluble flocculant (alum), gradually increasing the pore-water TDS and decreasing the percentage sodium. Each treated sample was processed to a water content near the plastic limit, compacted, and pinhole tested. The pore-water TDS and percentage sodium were measured after mixing with the alum. In these experiments the clays changed from dispersive to nondispersive at a boundary near 60% sodium.

Erosion Experience—Case Histories

Fig. 3 presents laboratory tests on fine-grained soil samples taken from cases in nature where colloidal erosion damage occurred, including two main categories: (1) Dam failures soon after the first reservoir filling, usually by piping of an initial small leak of colored water; and (2) severe erosion damage by rainfall in embankments and natural deposits, usually in the form of tunnels or deep gullies. For 13 of the case histories the soils have been tested in the SCS Nebraska laboratories. For the other eight, tests of pore-water salts were provided by colleagues or were reported previously in the literature. Descriptions of these case histories, available from the writers.

As seen in Fig. 3, essentially all the samples obtained from eroded clays in nature have high pore-water sodium contents. Comparison of Figs. 2 and 3 leaves little doubt that the upper boundary between dispersive and nondispersive soils as defined by the pinhole test approximates closely the condition in nature above which severe erosion has been commonly experienced.

Case Histories Demonstrating Erosion Resistance

The writers have only a few case histories in which fine-grained soils in nature were reliably known to have been subjected to and resisted strong erosive action. There is, however, much general experience showing that natural soil deposits and dams consisting of low sodium (ordinary erosion resistant) soils are not eroded in the same spectacular fashion in tunnels and deep gullies. In all cases studied by the writers where rainfall runoff caused tunnels and deep gully erosion in earth dams and natural slopes, the material was high sodium dispersive clay. Also, in nearly all recent investigations of piping failures in dams, the soils were found to be dispersive.

The large difference in behavior of the two materials results from the fact that for the nondispersive soil there is a definite threshhold velocity below which flowing water causes no erosion. The individual clay particles cling to each other and can only be removed by eroding water flowing with a certain erosive energy. For the dispersive clay, there is no threshhold velocity; the colloidal clay particles go into suspension even in quiet water.

Internal erosion damage (tunneling) starts with a small initial leak through a soil mass, such as rainwater flowing in a drying crack in a natural soil formation or a small concentrated flow through a fissure in a clay dam. Because of the gradients involved and the common widths of the cracks, the initial velocity of such leaks in soil masses is not high, usually a few centimeters a second, and rarely exceeding a few feet per second. In the dispersive clay the water becomes muddy regardless of the velocity and progressive erosion of the clay particles proceeds. For the nondispersive clay, no individual colloidal clay particles go into suspension and the velocity is too low to cause flocs of clay particles to be pulled from the soil mass. Therefore, in the nondispersive clay, the leakage water either continues to flow perfectly clear, or the leak is gradually closed off by swelling or by softening and collapse of the walls of the leakage channel.

Engineers familiar with the behavior of many small homogeneous dams will agree that it is not uncommon to find small leaks of clear water on the downstream slope at or above the toe. Where such leaks occur through an embankment of fine-grained relatively impervious soil, the leakage water is traveling through some kind of open flow channel which, although very narrow (such as a fraction of a millimeter), is large compared to the average size of pores in a fine-grained soil. Even though the seepage is apparently completely clear, the natural question is raised, "Could the leak be eroding very slowly by carrying suspended clay particles in such small concentration that they are not visible?" Many dams have sustained such small leaks for decades without any increase in flow, showing that no erosion is taking place; i.e., the soil must have a threshhold velocity. Experience with the pinhole test confirms definitely that the typical nondispersive soil has a threshhold velocity exceeding several feet per second, below which no erosion occurs. Some tests have been run with water flowing continuously through the hole in the sample at 10 fps for weeks with no measurable increase in hole size.

Influence of Pore-Water Salts

The case histories and pinhole tests confirm that the main factor governing dispersibility is the relative content of pore-water sodium, and a general relationship, valid for most soils, has now been established with considerable confidence as shown in Figs. 2 and 3.

Zone A.—Much experience shows that damaged and failed dams all over the world have been constructed of these dispersive soils. Almost all soils are dispersive in the pinhole test.

Zone B.—The great majority of these soils are nondispersive. These are the soils generally considered "ordinary erosion resistant clays," but include silts of low plasticity (ML), also nondispersive. A small percentage of exceptional soils in Zone B erode in the pinhole test in exactly the same fashion as soils of Zone A, and some of these can be identified only by the pinhole test.

Zone C.—Soils in this group may range from dispersive to nondispersive. This group contains a few soils which give intermediate reaction in pinhole tests, with apparently colloidal erosion but at a very slow rate compared with soils of Zone A.

These relationships are only valid for relatively pure eroding waters (such as TDS less than 0.5 meq/1 or about 300 ppm). Ths is true because they are based on results of pinhole tests using distilled water and case histories of erosion in nature caused mainly by rainwater and relatively pure reservoir waters.

Colloidal flocculation theory shows that clays are more likely to disperse in pure water than in salty water. Thus, if the 195 soils included in this study had been tested with the pinhole test using an eroding water other than distilled water, such as a typical reservoir water with several milliequivalents per liter of dissolved salts of mixed cations, it is likely that a lesser number would have been designated dispersive, and the boundaries shown in Fig. 2 would have been at high sodium percentages.

While clearly dispersive and erodible from the civil engineering standpoint, soils with pore-water salts falling in Zone A which have pore-water TDS greater than about 40 meq/l are commonly flocculated in their natural state (because even high sodium soils are flocculated at very high pore-water TDS). These soils, called "saline" and "saline-alkali" by the agricultural soil scientist (13), are often flocculated and fairly pervious and tillable as long as an excess of water is not applied to the land. When leached by downward percolation of excess water, the soil disperses and the ground becomes impervious and unfavorable for agriculture. In dispersion tests the flocculated state of these soils is never seen because the relatively large amounts of water used in the laboratory tests dilute the pore-water salts, i.e. the soil is transformed from its natural flocculated state to the dispersed state by the water used in the test. For engineering purposes, the temporary flocculated state is not of interest, because when subjected to the erosive action of a concentrated leak, such as through an earth dam, the flocculated structure disappears rapidly as the pore-water salt is diluted.

Soils with Total Pore Water Salts Less Than 1.0 meq/l.—Soil chemists reviewing an earlier draft of this paper raised a question about the proposed criteria for soils with pore-water TDS less than 1.0 meq/l and high sodium content, citing theoretical considerations indicating dispersion would be expected. Subsequently, increased attention was given to these, with many pinhole tests, some using dilute solutions of various salts in the eroding water.

The result was to confirm that very few soils with less than 1.0 meq/l are dispersive. The writers have no doubt about this conclusion, from other evidence as well: (1) Fig. 3 shows that practically none of the samples from the erosion case histories had pore water TDS much less than 1.0 meq/l; and (2) in experiments with dilute sodium clay-water suspensions, the measured viscosity (shear strength) of the suspension is often considerably higher for TDS values below 1.0 meq/l than it is at and just above 1.0 meq/l (e.g., see Ref. 20, pp. 98–99).

This is an important point because of the relatively large number of soils with TDS below 1.0 meq/l which apparently exist in nature.

Comparison with Earlier Australian Criteria.—The pore-water percent sodium can be expressed also in terms of the Sodium Absorption Ratio (SAR), a term used by soil scientists. By neglecting the dissolved pore-water potassium, which was present in negligible quantities in all 195 samples tested, every point in a plot such as Fig. 2 has a single SAR value, as shown by the contours of Fig. 4.

The Australian investigators postulated that two main factors governed soil dispersion: (1) The exchangeable sodium percentage of the soil (which is usually related closely to the pore-water SAR); and (2) the TDS and SAR of the eroding (percolating) water. The pore-water TDS was not considered a factor. For pure eroding water (TDS much less than 0.5 meq/l), the Australian criteria (1,7,9,10) show that all soils are dispersive if the pore-water SAR exceeds 1 or 2; i.e., all soils in Fig. 4 with SAR above the contour of SAR = 1 (or 2) would be dispersive. Therefore, the writers' results, summarized in Fig. 4, are in reasonable agreement with the Australian criteria for soils with pore-water TDS approximately from 0.5-3 meq/l, but not for soils with higher and lower pore-water TDS.

This discrepancy raises the question, "Could a dam of clay with moderately high pore-water SAR (5-10), with less than 60% pore-water sodium, which is nondispersive in the pinhole test, be transformed gradually to the dispersive state by long-time seepage of low-salt reservoir water?" While more study of this question is desirable, there are a number of reasons for concluding that such action is probably impossible:

1. If such action could occur, it would be expected that we would have records of piping failures of homogeneous clay dams occuring after long periods

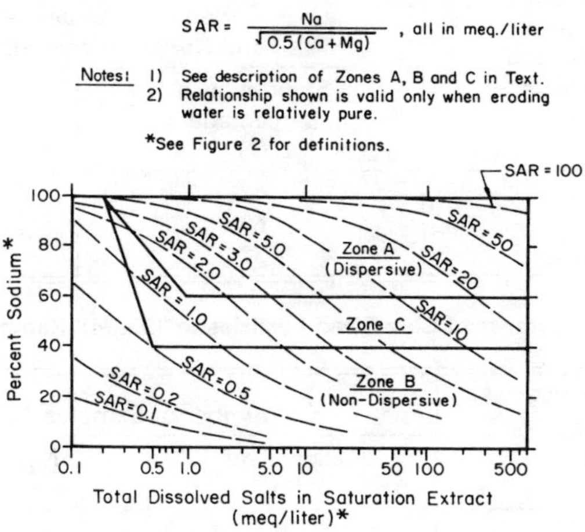

FIG. 4.—Relationship Between Dispersibility (Susceptibility to Colloidal Erosion) and Dissolved Pore-Water Salts Based on Pinhole Tests and Experience with Erosion in Nature

of successful operation, but such records are practically nonexistent. In all records known to the writers, dispersive clay piping has occurred either on the first reservoir filling or, less frequently, after raising the reservoir level to a higher level than previously.

2. In pinhole tests the compacted sample either disperses immediately or it does not. In tests on hundreds of soils giving nondispersive reaction, there was no evidence of change with time as the water flowing through the pinhole and pores of adjacent soil dilutes the pore-water salts, even when the test is continued for long periods of time (hundreds of hours). If dilution of the pore water could cause a change from the nondispersed to the dispersed state, it would be expected that at least a few soils which gave a nondispersive reaction at the beginning of the pinhole test would start to disperse toward the end of the test.

3. As the pore-water salts are diluted by seepage, there is no reason to believe that the percentage sodium in the pore water would increase. On the contrary, a considerable number of tests performed by the writers show that as the pore-water salt is diluted, the percentage sodium usually decreases as additional calcium cations are generated by solution of gypsum and other partially soluble calcium compounds or it remains constant.

Occasionally, rainfall erosion tunnels on dam slopes of dispersive clay appear only after several years. The delay is believed to be explained readily by the time required for sufficient rain to fall to cause the erosion.

TABLE 1.—Test Results Showing Dispersive and Nondispersive States

Pore-water salt concentration of test sample, in milliequivalents per liter (1)	Approximate Sodium Content Defining Boundary Between Dispersive and Nondispersive States	
	Sodium, as a percentage (2)	SAR (3)
5	57	2.7
10	60	4.2
100	58	13.0

TABLE 2.—Contents of Clay-Sized Particles in 195 Soil Samples Tested

Percentage clay (finer than 0.005 mm) (1)	Number of Samples Tested	
	Dispersive[a] (2)	Nondispersive[a] (3)
0–5	0	0
6–10	0	1
11–15	4	5
16–20	16	13
21–30	29	36
31–40	13	22
41–50	8	19
More than 50	10	19
Totals	80	115

[a] As determined by pinhole tests.

Other Current Investigations.—Valuable clay erosion research has been carried out at the University of California, at Davis, Calif. using a cylindrical soil specimen eroded by a rotating water bath. In one set of tests, a single clay was treated to bring it to three levels of pore-water TDS (5 meq/l, 10 meq/l, and 100 meq/l) and at each level several specimens were made at varying sodium contents, using distilled water as the eroding fluid (3). A certain viscous "critical" shear stress (rotating velocity) was shown to be needed to initiate erosion in the low sodium samples. For samples with higher sodium, this critical shear stress was zero (erosion without velocity), showing the soil to be in the dispersive state. These test results (3) can be interpreted to define the boundary between dispersive and nondispersive states as shown in Table 1.

Thus the water bath erosion test, quite different in all details from the pinhole test, defined a boundary between dispersive and nondispersive states for the soil tested which is identical to the upper curve (60% sodium) of Fig. 4.

Content of Clay-Sized Particles

The questions have been raised: "Do dispersive clays have higher contents of clay-sized particles than ordinary clays?", and "What minimum content of clay-sized particles must a clay have in order to cause trouble by dispersive clay erosion?" As seen in Table 2, there was no significant difference in the clay contents of the dispersive and nondispersive soils. Essentially none of the samples (only 1 out of 195) had less than 10% clay fraction and only a small percentage had less than 15% clay fraction. These data suggest that soils with less than 10% clay particles may not have enough colloids to support dispersive piping.

Atterberg Limits

In the previous study (14), no useful correlation was found between the soil plasticity and the incidence of damage to dams. As seen in Fig. 5, the Atterberg Limits for the 80 dispersive and 115 nondispersive soils both cover the general range of ordinary clays (CL and CH), confirming that the Atterberg Limits for a given clay provide no clue as to whether it may be dispersive or not. Tough CH clays with high liquid limit and plasticity index may be dispersive, and clayey silts (CL-ML and ML) may be nondispersive in the pinhole test.

Despite these results, the writers are cautious about concluding that a homogeneous dam built of tough CH dispersive clay might fail by piping as easily as a similar dam of dispersive clay of low plasticity, all other conditions being equal. Piping failures in clay dams usually commence with a slow-moving initial concentrated leak, such as in a very narrow crack (15). As the crack erodes by dispersive piping, the clay forming the crack walls simultaneously tends to swell and close the leakage channel. Therefore, in every such situation there is a contest between the rate of erosion and the rate of swelling. Possibly future investigations will show that certain swelling clays are rarely, if ever, damaged by tunnel erosion, even when they are in the dispersive state. Also, in the pinhole test no measurement is made of the rate at which erosion occurs in various dispersive soils. Possibly some clays classified as dispersive actually erode much less rapidly than others, making it improbable that erosion will proceed at a faster rate than swelling.

Although it is clear that dispersive clays cannot be differentiated from nondispersive clays by their Atterberg Limits, the tests on the 195 samples show some correlations: (1) Among soils with intermediate pore-water sodium (Zone C), those with plasticity index less than 10 are more likely to erode in the pinhole test than those of higher plasticity index; (2) among the low sodium soils (Zone B) the few exceptions which eroded in the pinhole test generally had plasticity index less than 10; and (3) for the exceptional soils with pore-water salts in the category of dispersive soils (Zone A) which resist erosion in the pinhole test, the Atterberg Limits are not significantly different.

Fig. 5 shows a suggestion of one possible basic difference between the two main groups of soils related to the Atterberg Limits, i.e., the group of nondispersive soils includes 12 samples with liquid limit greater than 40 classified as silts of medium to high plasticity in the USC System (MH and ML-MH). In contrast, none of the 80 dispersive soils were silts of these properties. This could be coincidence, but the writers suspect that it is not. Three of the MH soils (Samples S-196, S-197 and S-213) were taken from two sites which had demonstrated high resistance to erosion (Fig. 5). Also, pinhole tests have been made recently

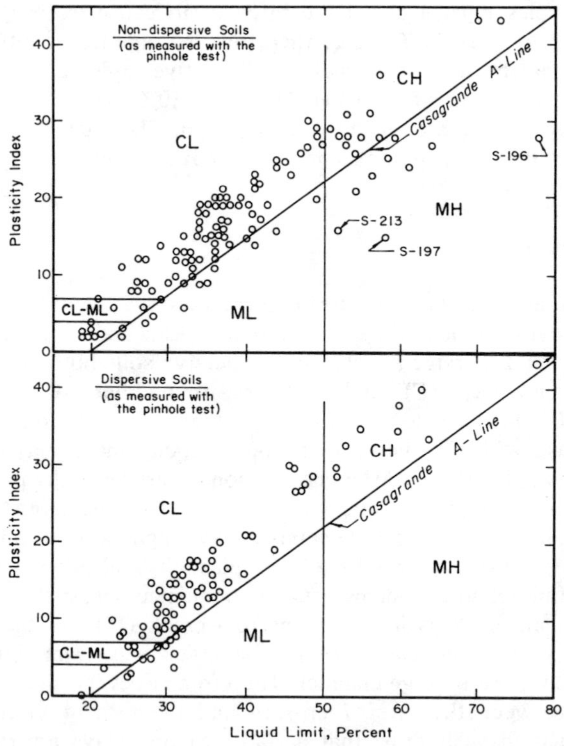

FIG. 5.—Relationship Between Dispersibility and Atterberg Limits

on many residual soils from tropical weathering of basalts, diorities, and gneisses in several countries, frequently MH and ML-MH soils, all showing non-dispersion in the pinhole test. Previous investigators have suggested that all dispersive soils may contain a moderate content of active clay minerals (montmorillonite, etc.), which soils are generally CL or CH. Possibly future investigations will demonstrate that most MH soils are nondispersive.

SCS Dispersion Test

There was a good agreement among results of pinhole tests, SCS dispersion tests, and tests of pore-water salts (Fig. 6): (1) For soils showing more than 50% dispersion in the SCS dispersion test, the great majority (54 out of 61 samples) were also dispersive in the pinhole test; (2) in a small percentage of cases, the SCS dispersion test apparently fails to identify dispersive clay, but for the samples with less than 30% dispersion, 10 out of 104 were dispersive in the pinhole test, including six with more than 60% sodium; and (3) soils with 30%-50% dispersion in the SCS dispersion test may be either dispersive or nondispersive in the pinhole test.

In summary, the present study shows that the SCS dispersion test is a generally reliable test. It is believed unlikely it would fail to recognize most dispersive

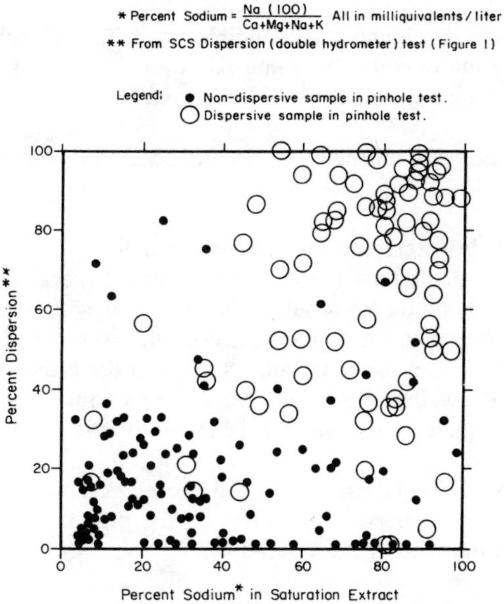

FIG. 6.—Relationship Among Pinhole, SCS Dispersion (Double Hydrometer) and Pore-Water Salinity Tests

soils in nature if tests were made on many samples. The pinhole test is the preferred physical test because it is a direct measure of dispersion of the compacted soil.

Crumb Test

After study of dam failures, Rallings (12) concluded that crumb tests (5) identified dispersive soils as well as any other test then available and that a dilute NaOH solution (0.001N = 1.0 meq/1) worked better than distilled water.

The writers have been routinely making crumb tests since 1971, using both distilled water and a 0.001 Normal NaOH solution for each sample, as follows:

Procedure.—A crumb of soil preserved at the natural water content (about 1/4 in.-3/8 in.) is dropped into a beaker of water (about 150 ml). The tendency for the clay particles to go into colloidal suspension is observed after 5-10 min of immersion, using the following interpretation guide:

Grade 1. No Reaction: Crumb may slake and run out on bottom of the beaker in flat pile but no sign of cloudy water caused by colloids in suspension.

Grade 2. Slight Reaction: Bare hint of cloud in water at the surface of crumb. (If the cloud is easily visible, use Group 3).

Grade 3. Moderate Reaction: Easily recognizable cloud of colloids in suspension. Usually spreading out in thin streaks on bottom of beaker.

Grade 4. Strong Reaction: Colloidal cloud covers nearly whole bottom of beaker, usually in a very thin skin. In extreme cases all the water in the beaker becomes cloudy.

Crumb test results are highly reproducible by different technicians working independently. For many soils, the same reaction is obtained regardless of the water content or previous handling of the crumb, but for some soils drying influences the results.

Primarily from the tests on the 195 soils included in this study, the following conclusions are justified:

1. Using distilled water, the crumb test is a very good indicator, but in one direction only; i.e., if the crumb test indicates dispersion (Reaction Grade 3 or 4), it is overwhelmingly probable that the soil will be dispersive in the pinhole test. Unfortunately, the converse is not true. About 40% of the dispersive soils had nondispersive reactions (Grades 1 or 2) in the crumb test.

2. Using distilled water, practically all samples found to be nondispersive in the pinhole test gave nondispersive reaction in the crumb test (Grades 1 or 2).

3. Using the NaOH solution, no useful correlation was found; about 40% of the soils giving nondispersive reactions in the pinhole test gave dispersive reactions in the crumb test (Grades 3 and 4).

Lime Treated Dispersive Clay

Pinhole tests have been made on dozens of different dispersive clays treated with small amounts of $Ca(OH)_2$ (1%–4% by weight). Without exception the material was transformed to a nondispersive state by the lime treatment.

Summary and Conclusions

Many different fine-grained soils of diverse origins were tested with the new pinhole test, as well as with three other laboratory tests previously used for evaluating soil dispersion. Case histories of erosion in nature were studied.

Dispersive clays cannot be recognized by the identification tests currently employed in civil engineering practice.

The amount of dissolved sodium relative to other salts in the pore water is the main factor determining whether a clay is dispersive or not. The great majority of soils in nature will be dispersive or nondispersive if the pore-water salts are within Zones A and B, respectively, as defined in Fig. 4. This relationship has now been established with considerable experience.

For evaluation of a given soil, both pinhole tests and tests of the pore water salts should be made. For soils in Zone C and for exceptional soils in Zones A and B, only the pinhole test can reliably identify the state of dispersion.

The SCS dispersion test is also a good indicator, though somewhat less definitive than the pinhole test. The crumb test using distilled water is a useful indicator only in one direction. If the crumb test indicates dispersion, the soil is probably dispersive; however, many dispersive clays do not react to the crumb test.

Acknowledgments

The writers are grateful to the Soil Conservation Service for permission to publish the results of the work. D. C. Ralston, S. H. Nickel, and O. H. Raine reviewed the manuscript and provided many useful suggestions. Special thanks are due E. B. Perry for continuing valued assistance during preparation of this paper.

Appendix.—References

1. Aitchison, M. E., Ingles, O. G., and Wood, C. C., "Post-Construction Deflocculation as a Contributory Factor in Failure of Earth Dams," *Proceedings*, 4th Australia-New Zealand Conference on Soil Mechanics and Foundation Engineering, Montreal, Canada, 1963, pp. 275–279.
2. Aitchison, M. E., and Wood, C. C., "Some Interactions of Compaction, Permeability, and Post-Construction Deflocculation Affecting the Probability of Failure of Small Dams," *Proceedings*, 7th International Conference on Soil Mechanics and Foundation Engineering, Vol. 11, Montreal, Canada, 1965, pp. 442–446.
3. Arulanandan, K., Loganathan, P., and Krone, R. B., "Pore and Eroding Fluid Influences on Surface Erosion of Soil," *Journal of the Geotechnical Division*, ASCE, Vol. 100, No. GT1, Proc. Paper 11078, January, 1975, pp. 55–66.
4. Benassini, A. and Casales, V., discussion of "Piping in Earth Dams of Dispersive Clay," by J. L. Sherand, N. L. Ryker, and R. S. Decker, *Proceedings*, Speciality Conference on the Performance of Earth and Earth-Supported Structures, ASCE, Vol. 3, 1972, pp. 135–141.
5. Emerson, W. W., "A Classification of Soil Aggregates Based on Their Coherence in Water," *Australian Journal of Soil Research*, Vol. 2, 1967, pp. 211–217.
6. Fletcher, J. E., and Carroll, P. H., "Some Properties of Soils Associated with Piping in Southeastern Arizona," *Proceedings*, Soil Science Society of America, 1948, pp. 545–547.
7. Ingles, O. G., and Metcalf, *Soil Stabilization*, Butterworths, Sydney, Australia, 1972.
8. Ingles, O. G., discussion of "Piping in Earth Dams of Dispersive Clay," by J. L. Sherard, N. L. Ryker and R. S. Decker, *Proceedings* Speciality Conference on the Performance of Earth & Earth-Supported Structures, ASCE, Vol. 3, 1972, pp. 135–141.
9. Ingles, O. G., and Aitchison, G. D., "Soil-Water Disequilibrium as a Cause of Subsidence in Natural Soils and Earth Embankments," Colloquium on Land Subsidence, International Association of Scientific Hydrology-United Nations Economic, Social and Cultural Organization, Tokyo, Japan, Vol. 11, 1969, pp. 342–352.
10. Ingles, O. G., Lang, J. G., and Richards, B. G., "Pre-Equilibrium Observations on the Reconstructed Flagstaff Gully Dam," *Proceedings*, Symposium on Earth and Rockfill Dams, Indian National Society of Soil Mechanics and Foundation Engineering, Vol. 1, Meeruit, India, 1968, pp. 162–170.
11. Lutz, J. F., "The Physico-Chemical Properties of Soils Affecting Erosion," *Research Bulletin 212*, Agricultural Experiment Station, University of Missouri, Columbia, Mo., July, 1934.
12. Rallings, R. A., "An Investigation of the Causes of Failure of Farm Dams in the Brigalow Belt of Central Queensland," *Bulletin 10*, Water Research Foundation of Australia, 1966.

13. Richards, L. A. ed., "Diagnosis and Improvement of Saline and Alkali Soils," *Handbook #60*, U.S. Department of Agriculture, U.S. Government Printing Office, Washington, D.C., Feb., 1954.
14. Sherard, J. L., Ryker, N. L., and Decker, R. S., "Piping in Earth Dams of Dispersive Clay," *Proceedings*, Specialty Conference on the Performance of Earth and Earth-Supported Structures, ASCE, Vol. 1, 1972, pp. 589-626.
15. Sherard, J. L., Ryker, N. L., and Decker, R. S., "Hydraulic Fracturing in Low Dams of Dispersive Clay," *Proceedings*, Specialty Conference on the Performance of Earth and Earth-Supported Structures, ASCE, Vol. 1, 1972, pp. 653-689.
16. Sherard, J. L., Ryker, N. L., and Decker, R. S., Closure to "Piping in Earth Dams of Dispersive Clay," *Proceedings*, Specialty Conference on the Performance of Earth and Earth-Supported Structures, ASCE, Vol. 3, 1972, pp. 150-161.
17. Sherard, J. L., et al., "Pinhole Test for Identifying Dispersive Soils," *Journal of the Geotechnical Division*, ASCE, Vol. 102, No. GT1, Proc. Paper 11846, Jan., 1976, pp. 69-85.
18. Sherard, J. L., discussion of "Pore and Eroding Fluid Influences on Surface Erosion of Soil," by Kandish Arulandan, Paripornonanclan Logenathan, and Ray B. Crone, *Journal of the Geotechnical Engineering Division*, ASCE, Vol. 101, No. GT12, Proc. Paper 11743, Dec., 1975, pp. 1299-1300.
19. "Soil Survey Laboratory Methods and Procedures for Collecting Soil Samples," Soil Conservation Service, *Soil Survey Investigation Report No. 1*, Department of Agriculture, Washington, D.C., 1967.
20. van Olphen, H., *Clay Colloid Chemistry*, John Wiley and Sons Inc., New York, N.Y., 1963.
21. Volk, G. M., "Method of Determination of the Degree of Dispersion of the Clay Fraction of Soils," *Proceedings*, Soil Science Society of America, 1937, pp. 432-445.

Introduction to Paper 10

Some Engineering Problems with Dispersive Clays

by J.L. Sherard, J.L. and R.S. Decker
Proceedings, Symposium on Dispersive Clays, Related Piping, Erosion in Geotechnical Projects, American Society for Testing and Materials, Special Technical Publication 623, 1977, American Society for Testing and Materials, Philadelphia, pp 3-12

Prepared by Lorn P. Dunnigan, M. ASCE

Please see Introduction to Paper 9.

J. L. Sherard,[1] *L. P. Dunnigan,*[2] *and R. S. Decker*[3]

Some Engineering Problems with Dispersive Clays

REFERENCE: Sherard, J. L., Dunnigan, L. P., and Decker, R. S., "**Some Engineering Problems with Dispersive Clays,**" *Dispersive Clays, Related Piping, and Erosion in Geotechnical Projects, ASTM STP 623,* J. L. Sherard and R. S. Decker, Eds., American Society for Testing and Materials, 1977, pp. 3–12.

ABSTRACT: Laboratory tests show that filters containing some fine sand sizes (such as ASTM Specification for Concrete Aggregates (C 33-74a) concrete sand) will generally control piping and seal leaks through dispersive clay. The authors believe that it is reasonble practice to build major embankment dams with cores of dispersive clay with appropriate design details. Existing dams of dispersive clay which have given no trouble are not usually considered unsatisfactory structures. Piping of dispersive clay in dam foundations is believed generally unlikely below the depth where the clay is continuously saturated in nature. Some observations are given on the geologic origin, surface topography, and appearance (color) of deposits of dispersive soil. Damage to canals is described.

KEY WORDS: clays, dispersions, earth dams, filters, piping, soil erosion

The belated widespread recognition that certain clays exist in nature in their dispersive state is an important and fundamental contribution to geotechnical engineering, and particularly to the art of building embankment dams. These dispersive clays which cannot be differentiated from ordinary erosion-resistant clays by the routine civil engineering tests erode rapidly in slow-moving (or even quiet) water by individual colloidal clay particles going into suspension. This recognition that there is such an enormous difference in fine-grained soils with similar index properties allows new insight into the behavior of dams subjected to concentrated leaks and readily explains certain past exceptional experiences of failures and serious piping damage in dams under conditions shown to be

[1] Consulting engineer, Devonshire, Bermuda.
[2] Head, Soil Mechanics Laboratory, Soil Conservation Service, Lincoln, Nebr. 68508.
[3] Senior soil engineer, Hoskins-Western-Sonderegger, Lincoln, Nebr. 68501.

similar to those of the great majority of similar dams designed and constructed by accepted standards.

In the last several years, the writers have been cooperating in studies of the nature of dispersive clay, including laboratory investigations and evaluation of damaged structures. This paper summarizes opinions from these efforts on some engineering problems relating to dispersive soils.

Filters for Dispersive Clay

A considerable number of laboratory filter tests have now been made with many different dispersive clays in which water is caused to flow through a hole (¼ to ½ in. diameter) in a compacted clay specimen (6 to 9 in. long) and discharged into a medium sand filter (such as ASTM Specification for Concrete Aggregates (C 33-74a) concrete sand), with hydraulic head of about 3 ft of water. In all tests, the results were similar, showing the filter was effective with details as follows.

1. At the beginning of the test, colloidal colored water comes directly through the sand filter, with the quantity limited by the permeability of the sand filter (such as 10 ml/s).
2. The rate of flow gradually decreases, reaching a low value (commonly less than 1.0 ml/s) after 10 or 15 min, and is finally sealed off completely or the discharged water continues to flow at a constant low rate, becoming completely clear.
3. On dismantling the specimen at the end of the test, it is found that the hole in the compacted clay specimen is filled partially or completely with soft mud, that the face of the sand filter at the downstream end of the hole is plastered with a thin skin of fines, and there is no visible sign of contamination of the sand filter, which is apparently as clean as it was at the beginning of the test, indicating that all colloidal clay particles carried into the filter in suspension in the leakage water were carried along with the water out of the filter.

Similar tests on highly dispersive clay have been made recently in Brazil as well as tests using larger compacted clay specimens (about 2 by 2 by 2 ft) with a ¼-in. slot for the leakage channel, with similar results.[4]

These experiments provide confidence that sand filters will seal and safely control leaks in dispersive clay. The sealing action is different from that normally expected from a conventional filter, which prevents all fine particles of the protected soil from entering the filter. In sealing and filtering a leak in dispersive clay, the filter cannot prevent the colloidal clay particles in suspension from passing through, but the silt-sized particles carried along with the flow cannot enter the sand filter and are retained in the leakage channel upstream of the filter, gradually sealing off the leak.

[4] Bordeaux, G. and Imaizumi, H., this publication, pp. 13–24.

Filter tests of more conventional type were also made in which no hole was made through the dispersive clay specimen. Tests were made with extreme conditions such as a 1.0-in.-thick compacted clay specimen in a permeameter backed up with pea gravel subjected to seepage with pressure head of 100 in. (gradient of 100) for more than three years. In all such tests, nothing happens. The dispersive clay is so impervious that it is essentially watertight. In order to initiate erosion and to test the filter, it is necessary to start a concentrated leak through the clay.

Additional evidence supporting the conclusion that sand filters will control leaks in dispersive clay is given by prototype dam experience, which shows that the considerable number of dispersive clay failures investigated in the last few years have all been through homogeneous dams in which the initial leak emerged in free air on the downstream slope, without passing through a sand filter.

From these experiences, which are consistent with theory, it is reasonable to conclude that sand filters that contain at least a small percentage of fine sand sizes (finer than the #40 sieve) will reliably protect dispersive clay against piping of concentrated leaks. For any given project, however, where the use of dispersive clay is being contemplated for an important dam, laboratory tests of the specific clay with the sand filter proposed for use should be considered.

Use of Dispersive Clay in Important Dams

In civil engineering practice, the general recognition of dispersive clay is so recent that there have not been many major dams (retaining important reservoirs) built since engineers have had the tools to evaluate the problem. Hence, at the present time, there is no large fund of precedent to guide a designer when he finds at a proposed site that the soil deposit which would previously have been considered the most economical core material is dispersive.

The writers believe that the present knowledge about dispersive clay does not require any important change in current practice for building embankment dams. Recognizing the problem should lead to safer dams at about the same cost if it is necessary to use dispersive clay in the dam, and should provide the opportunity for considering economies if highly nondispersive and erosion-resistant soils are available. Evidence for support of this main conclusion is given in the following paragraphs.

Experience during several decades with literally thousands of embankment dams built according to current accepted practice shows very few failures or problems from piping, and almost all of these have been attributable to some condition not foreseen by the designer (such as a lapse in the construction quality control or a geologic accident in the foundation not discovered in the explorations). The only important exceptions to this satisfactory experience have been failures of homogeneous dams where leaks emerged on the downstream slope without passing through filters, many of them relatively small and economically constructed projects, including farm pond dams built with minimal engineering attention. Hence, long before there was any widespread understanding of dispersive clay, the profession already had sound reasons for concluding that current design practice results in completely safe embankment dams if the foundation was adequately explored and evaluated.

It is inevitable that, among the thousands of dams with wholly satisfactory experience, many have impervious cores of dispersive clay merely because of the widespread existence of dispersive clays in nature. Hence, this experience indicates that current practice for building embankment dams, including filters and careful construction control, will provide safe dams whether the clay core is dispersive or not.

The final factor supporting the main conclusion is the fact that sand filters with adequate quantities of fine sand sizes are shown by laboratory tests to control piping and seal leaks in dispersive clay, even under very severe conditions.

The designer is now better equipped to build dams with higher fundamental safety from rare accidents of all kinds because of the improved understanding of dispersive clay. An example which will probably be encountered fairly frequently is a site at which there are two soil deposits available from which the dam core could be built, both at about the same estimated unit cost, one dispersive and one nondispersive. In such a case, the new understanding will allow the engineer to choose the nondispersive soil, with its intrinsically superior resistance to erosion of the concentrated leaks (which occur infrequently from various actions), and to rely less on the action of filters. Before the designer had the tools to identify the difference between the two soil deposits, he might have chosen the dispersive clay for the core (or left the choice to the contractor), making the filters the first line of defense.

Where the only economical source of impervious soil available is dispersive, and where the earth dam is more economical than alternate dams, it is considered reasonable and appropriate practice for the designer to use it, just as he would have done a few years ago when he would not have known it was dispersive. The resulting dam will probably be intrinsically superior against rare accidents than it would have been before the recognition of the problem, because the engineer will pay more attention to the defensive design details.

Sand Chimneys in Homogeneous Dams

One of the always present design problems for a homogeneous dam of low to moderate size is: "Is a vertical sand chimney drain needed or can seepage be safety controlled with a horizontal blanket drain under the downstream slope or a toe drain?"

Whatever other factors must be considered for answering this question at a given site (likelihood of cracking from foundation settlement or earthquake, importance of reservoir, consequences of failure, etc.), it is now clear that one of the most important factors is, "Is the embankment material dispersive clay?" Experience shows that homogeneous dams which have failed have been generally constructed of dispersive clay, and recent research has shown that low dams of nondispersive clay should withstand small concentrated leaks safely.

If no supply of clean and pervious sand of a type commonly used for a chimney drain is available, a zone of cohesionless silty sand or sandy silt can be considered as a line of defense against dispersive piping in a homogeneous dam. Such a chimney may not be pervious enough to act as a drain, but it would be expected to prevent piping. The typical dispersive clay piping failure in the homogeneous dam has started with a very small initial leak which eroded in a few hours to a tunnel of substantial diameter.[5] If this small initial leak (which probably started in a narrow crack in the clay dam) had to pass through a chimney of cohesionless silty sand on its way through the dam, piping should be prevented because silt-sized particles from the wall of the leakage channel upstream of the chimney could not enter the sand zone and the leak would be sealed and finally closed by swelling.

Existing Dams of Dispersive Clay

There are many hundreds of small homogeneous dams with no trouble existing in the world which consist of dispersive clay. This can be concluded from the knowledge that dispersive clays exist in widespread geographic areas and that these clays are the kind that heretofore would have been normally considered as good clays for earth dams. More specifically, in certain regions where it is known from investigations by the U.S. Soil Conservation Service that many soils are dispersive clay (such as in southeastern Oklahoma and north-central Mississippi), many dozens of clay dams have been holding water perfectly well for years.

For a dam to fail by dispersive clay piping, there must be a concentrated leak, such as in a crack from any source or leakage channel left open by a construction deficiency. There is much experimental and theoretical evidence supporting the conclusion that failures are not initiated by the general seepage of water through the pores of a well-compacted clay mass. Hence, once a dam of dispersive clay is retaining a reservoir and no leaks

[5] Sherard, J. L., Decker, R. S., and Ryker, N. L., *Proceedings,* Specialty Conference on Performance of Earth and Earth-Supported Structures, American Society of Civil Engineers, Vol. I, 1972, pp. 653–689.

develop, no failure occurs subsequently. There is no reason to suspect that a concentrated leak will develop later under ordinary conditions of reservoir operation.

On several recent occasions, engineering reviews have been made of existing homogeneous dams (without filters) of dispersive clay, some retaining important reservoirs. The tendency has been to conclude that there is no reason to consider the dams unsatisfactory, reasoning that, since there was no leak through the dam for several years, there is no reason to predict one will develop under normal conditions. However, failure could result if a leak developed later under unusual conditions, such as cracking from earthquake shaking or drying cracking following a prolonged period of low reservoir. In a situation where it is desired to increase the security of an existing dam of dispersive clay, a blanket of sand filter on the downstream slope covered with a weighting berm has been considered the most practical approach.

Dispersive Piping in Natural Formations Comprising Dam Foundations and Abutments and Reservoir Rim Ridges

Several times in the last few years, sites have been studied where only dispersive clay is available for the impervious section of the dam, and the abutments and foundation are also of dispersive clay of sufficient thickness that a seepage cutoff is not possible. The question rises, "If we design a safe dam with good sand filters, how do we prevent dispersive piping in the natural clay underneath the dam?"

This is a difficult question at the present state of the art; however, the writers suspect it may be reasonable to conclude that the risk of piping tunnels extending deep into saturated formations of dispersive clay (below the depth of drying cracks or potential collapse on saturation) is probably negligible and does not justify extensive protective design measures.

There are a few records of low dams in arid regions where piping tunnels passed through natural soil foundations or abutments a few feet below the bottom of the dam embankment,[6] probably from leaks originating in drying or settlement cracks. However, almost all the recent troubles with dams leading to the present interest in dispersive clays have been tunnels in the embankment itself, and no cases are known of deep tunnels in the foundation passing below the phreatic surface.

In addition to this experience, a main reason for suspecting deep foundation erosion tunnels are improbable in saturated formations of dispersive clay is the fact that the ground water dissolved salts are already in equilibrium with the salts on the exchange complex of the clay, so that, if any cracks exist that can support concentrated leaks, the first flow of

[6] Decker, R. S. and Dunnigan, L. P., this publication, pp. 94–109.

water after filling the reservoir will be this ground water which is probably not capable of causing erosion. This should generally ensure that the leak will remain small in volume and the new water gradually seeping from the reservoir will have time to pick up the salts from the pore water, causing it to become nonerosive. Also, the weight of the dam increases the compressive stresses in the foundation, tending to close cracks in the already impervious clay formation.

Consideration in such a case should always be given to providing seepage barriers extending through the upper zone of drying cracks and unsaturated clay, and removing all material likely to settle abruptly when saturated from the reservoir seepage.

Geologic Origin and Topography

The various origins of dispersive clay are not well known. Most of them encountered to date by the writers have been alluvial clays in the form of flood-plain deposits, slope wash, lake bed deposits, and loess deposits. In some areas, it has been found that clay stone and shales, laid down as marine deposits, have the same pore water salts as dispersive clay, and their residual soils are dispersive.

With only one exception known to the writers, all tested fine-grained soils known to be derived from *in situ* weathering of igneous and metamorphic rocks have been nondispersive as well as all soils derived from limestones.

Dispersive clays have been red, brown, grey (some nearly white), yellow, and all transitions among these. No black-colored soils, with obviously high organic content, have tested dispersive.

Many deposits of dispersive clay studied are situated in flat flood plains or gentle rolling topography with low relief and smooth, relatively flat slopes (such as the Oklahoma plains and the low bluff hills of north-central Mississippi). Usually in areas where there are no manmade excavations or embankments in the area, there is little or no evidence from the surface appearance that the soil is highly erodible. The ground surface is covered with a thin layer of protective silty sand from which the dispersive clay particles have been removed, or has a protective layer of topsoil and vegetation.

Where steep topography and dispersive clay coexist, the peculiar ground surface topography is easily recognized by deep rapidly forming channels and tunnels. It seems likely that understanding of dispersive clay will provide a tool for geologists to explain many geomorphic features. For example, it seems clear that steep high mountains would be very rapidly (in geologic time) eroded away if they weather to dispersive soil.[7]

[7] Villegas, F., this publication, pp. 408–418.

Earlier investigations, both by soil scientists and engineers of the Soil Conservation Service[6] indicated that dispersive soils in nature, causing trouble with agriculture and hydraulic structures, were found most commonly in arid and semiarid regions and commonly had alkaline pore water (pH greater than about 8.5). Studies of the last few years for engineering projects (including studies of damaged dams) have found many dispersive soils in humid areas, and many have pH well on the acid side. In recent years, deposits of dispersive clay have been found (by laboratory test and erosion patterns) in so many parts of the world that it must be suspected they could be found anywhere.

General Filter Criteria for Clay

The recognition that certain clays are subject to rapid colloidal erosion reopens the whole question of filters for clays, long suspected to be one of the aspects of earth dam design about which there was still a serious question.

Recent laboratory tests have shown that, for many nondispersive clays, concentrated flows of water through small-holes compacted specimens, with velocities of 5 to 10 ft/s, can be maintained for hundreds of hours with no erosion, even though no filter whatsoever is used in the test.[8] These laboratory results are compatible with many experiences in nature where concentrated leaks have existed through homogeneous clay dams for decades at constant volume, which can only be explained if there is absolutely no erosion occurring: if even a few rare particles were being eroded from the walls of the leakage channel, the leakage volume would increase.

The question naturally arises, "Do not these observations show that, for typical nondispersive clays, filters with fine sand sizes are not needed?" The answer is probably "yes"; however, before contemplating such a radical departure from past practice, the earth dam designer must pause awhile to examine the problem from all viewpoints.

Accepted filter criteria for clays have been established from early tests on soil specimens, including both dispersive and nondispersive clays, although the differences were not then known. Hence, it is not surprising that the attitude in the profession toward present criteria for filters for clay has been uncertain.

Difficulty in Predicting Hazard from Rainfall Erosion

One observation from nature that is still not easily explained in the fact that, while most dispersive clays erode rapidly in forms of tunnels and

[8] Sherard, J. L. et al, "Pinhole Test for Identifying Dispersive Soils," *Journal of the Geotechnical Division,* American Society of Civil Engineers, Jan. 1976.

deep gullies from rainfall in walls of excavations and on slopes of embankments, some do not. Some clays shown to be dispersive by laboratory tests do not suffer severe erosion from rainfall. Some of the dams in Oklahoma that failed rapidly on the first reservoir filling have developed essentially no surface erosion after being repaired and put into service for many years, even though storms with concentrated heavy rainfall are common and other similar dams nearby are badly eroded. Early studies seemed to indicate that dispersive clays with total dissolved salts in the pore water exceeding about 15 milliequivalents per litre would not develop rainfall erosion tunnels. This correlation still appears fairly good but more recently exceptionally severe rainfall erosion tunnels developed rapidly in clays with much higher pore water salts. This difference in behavior may be explained in the future by relative differences in the rate of swelling to close cracks and at the rate at which colloidal particles go into suspension from the face of dispersive clay mass exposed to water. Also, it is suspected that the nature of the rainfall could be a factor. Sudden heavy rainstorms directly following periods of hot, dry weather would be expected to be much more likely to erode tunnels starting in drying cracks than gentle steady rains that would give the ground time to soak up the water and close drying cracks before they can erode.

Unlined and Lined Canals

Many excavated channels for flood control and irrigation, and also natural stream channels that traverse dispersive soil areas, are badly damaged by severe tunnel and gully erosion of the banks and degradation of the bottom.[9] It is also found, however, that some sections of canals in soils shown to be dispersive by laboratory tests are not more severely eroded than those in nondispersive soils. Hence in designing canals where dispersive clays are recognized from laboratory tests, the engineer must look to erosion patterns in the area, if any, to help guide the decision as to whether extensive erosion protection measures may be warranted.

Several examples have been studied where dispersive clay piping has badly damaged or completely destroyed concrete canal linings. Rain runoff on the ground at the top of the canal bank percolates downward into drying cracks and then flows into the canal through construction joints in the lining (when canal is empty). Erosion tunnels develop that undermine the lining. In an extreme case, after a few years, the concrete-lined canal is transformed to an earth canal about twice as wide and much deeper with only disconnected remnants of concrete slabs lying around.

It seems clear that concrete-lined canals and dispersive clay are not compatible. It can be speculated with some confidence that there have

[9] Ryker, N. L., this publication, pp. 370–389.

been many such failures that have not been described in the literature or diagnosed as dispersive clay piping.

Summary

Laboratory tests show that filters containing some fine sand sizes (such as ASTM Specification C 33-74a concrete sand) will generally control piping and seal leaks through dispersive clay. The authors believe that it is reasonable practice to build major embankment dams with cores of dispersive clay with appropriate design details. Existing dams of dispersive clay which have given no trouble are not usually considered unsatisfactory structures. Piping of dispersive clay in dam foundations is believed generally unlikely below the depth where the clay is continuously saturated in nature. Some observations are given on the geologic origin, surface topography, and appearance (color) of deposits of dispersive soil. Concrete-lined canals have been destroyed by dispersive clay erosion.

Introduction to Paper 11

Sinkholes in Dams of Coarse, Broadly Graded Soils

by J.L. Sherard
Transactions, Thirteenth International Congress on Large Dams, Vol 2, 1979, pp 25-34.

Prepared by Jean Lafleur[5], M. ASCE

"Sinkholes..." is a landmark in Dr. Sherard's sequence of communications; at this point in his career, his interest shifted from dispersive clays to granular filters, a subject he explored with great interest until his premature death. The paper discussed the misfunctioning of some earth dams having in common a core made of broadly graded material; the analysis was challenging, thought-provoking and it eventually lead for the following decade, to a thorough reexamination of universally accepted filter criteria. Internal stability of particles, selective erosion, and hydraulic fracturing were evoked and given their full sense in the presentation. The shape of the gradation curves of these broadly graded materials was also questioned. The possibility that their own finer particles might be entrained by the flowing water and leaving open sinkholes, was advocated in convincing terms.

These ideas were to be seriously considered at the end of the seventies since at that time, the James Bay project in northern Quebec was just being undertaken; the impervious cores of more than 200 dams and dykes were to be constituted of broadly graded cohesionless tills. In fact, the gradation curves of these tills fell right in the range of those of the cores of the damaged dams reported by Sherard. Filters had therefore to play an adequate line of defense against internal erosion. The evaluation of their compatibility was therefore the essence of the research program initiated at Ecole Polytechnique on filters for broadly graded cohesionless soils. During the realization of this program, Sherard's well organized memoranda and correspondence were always a privileged vehicle for exchanging debatable ideas and current research results.

[5]Professor of Civil Engineering, Ecole Polytechnique, University of Montreal, Montreal, Quebec, Canada.

COMMISSION INTERNATIONAL
DES GRANDS BARRAGES

Treizième Congrès
des Grands Barrages
New Delhi, 1979

Q. 49
R. 2

SINKHOLES IN DAMS OF COARSE, BROADLY GRADED SOILS (*)

James L. SHERARD
Consulting Engineer
Bermuda

U.S.A.

I. INTRODUCTION

In the last 20 years there have been a number of incidents in which sinkholes (craters) have appeared on the crests and slopes of embankment dams. Almost all have occurred in dams comprised of remarkably similar coarse, broadly graded soils frequently of glicial, but sometimes of alluvial, origin. The troubles have developed on dams designed according to accepted current standards.

These troubles can be explained by the fact that the broadly graded coarse soils are internally unstable, in the sense that the fine portion is not compatible with the coarser particles from the standpoint of filter requirements. The sinkhole is caused by erosion of a concentrated leak causing the finer soil particles to migrate out of the compacted soil mass, exiting through cracks in the foundation rock, or through filters which are too coarse to retain the selectively eroded fines.

These experiences support the conclusion that fine to medium sand filters should be considered for important dams with thin cores of such materials. Also consideration should be given to sealing cracks in rock foundations under dam shells of these materials, in addition to the treatment of rock foundation surfaces under the impervious core.

(*) *Cratères dans les barrages en terre grossière à large fuseau granulométrique.*

II. REVIEW OF EXPERIENCE

a) Coarse soils of glacial origin.

Internal erosion causing collapse and sinkholes on the surface has been described in the literature as occuring at the following dams: Balderhead [10]; Hyttejuvet [12]; Yards Creek Upper Reservoir [8]; Viddalsvatn [11]; Messaure [1]; Seitevare [1]; Bastusel [1]; and Churchill Falls Dykes (3.7). In all these dams the erosion and sinkholes developed in embankment material of glacial origin of very similar properties, being comprised of a widely graded mixture ranging from coarse gravels and cobbles down to silt and clay sized particles.

A characteristic of these coarse glacial soils is that they are similar in particle size distribution in many parts of the world. Nearly all the materials used in all these dams had grain-size distributions falling within the range shown in Figure 1. Usually the grain size distribution plots as a nearly straight line on the commonly used semi-logarithmic chart, and the curves for individual samples plot as straight lines or as curves with only slight curvature within the range shown in Figure 1. The fines in these soils are either non-plastic or have low to medium plasticity.

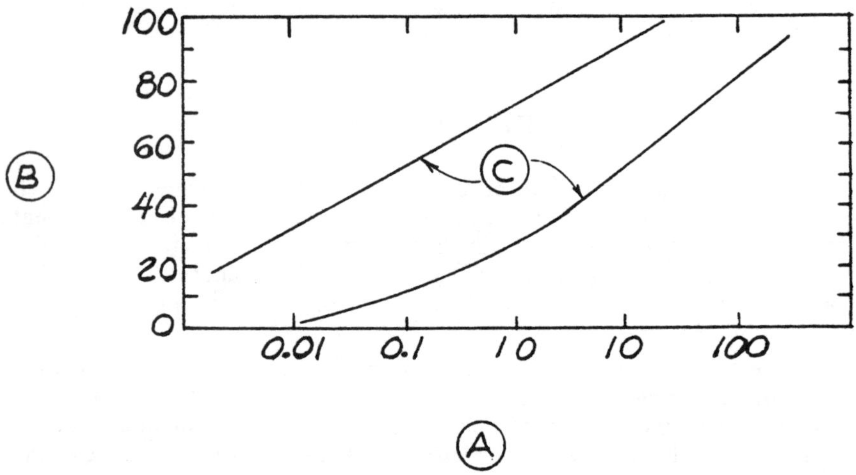

Fig. 1

(A) Particle Size, millimeters.

(B) Percent Finer, by weight.

(C) Range of Particle Size Distribution of Coarse, Broadly Graded Soils in Damaged Dams.

(*A*) *Dimension des particules, en millimètres.*

(*B*) *Finesse en pourcentage du poids.*

(*C*) *Largeur du fuseau granulométrique pour les matériaux grossiers à large fuseau dans les barrages endommagés.*

The problems with these dams, already described in various degress of detail in the references cited, are all similar with the following main common elements:

1. The glacial soil causing the trouble was generally used as the central core of a dam with rock fill or gravel fill shells.

2. Soon after the first reservoir filling a substantial leak of colored water, carrying soil fines, appeared abruptly at the downstream toe. The amount of fines carried in the water ranged from a relatively small quantity (water barely colored), as at Balderhead Dam, to large (dark colored) as at Yards Creek Upper Reservoir Dam.

3. The turbid leakage was usually followed by one or more sinkholes (crater like depressions) often several meters in diameter and several meters deep, appearing on the dam crest or on the upper elevations of the upstream or downstream slope. The sinkholes sometimes appeared within a few weeks of the turbid leakage but sometimes after a considerable time, such as several years. Sometimes sinkholes appeared after repairs were made by grouting. Sketches or photographs of some of the sinkholes are given in references [2], [3], [8] and [10].

4. Analyses of these experiences generally led to the conclusion that a concentrated leak developed through the core which caused a type of internal erosion in which the soil fines are eroded selectively and carried out of the core, leaving the coarse sand and gravel particles behind to act as a pervious drain. The volume of fine material eroded was larger than the volume of the void spaces between the coarser soil particles causing progressive collapse of the material above the initial leakage channel, which action finally reaches the dam surface as manifested by the sinkhole or crater. In most of the damaged dams a considerable length of the dam was involved. At the long low dykes at the Churchill Falls Project, seven different areas were affected.

5. No complete failures occurred and usually there was no realistic threat of a disaster, but in some situations a serious situation could have developed if the conditions had been somewhat more severe. Repairs were usually made successfully by grouting, though some recurrences of the trouble developed. At Balderhead an extensive slurry wall cutoff was put down through the core and one of the Churchill Falls Dykes was excavated and rebuilt after lowering the reservoir.

6. Since the erosion takes place inside the dam where it cannot be directly observed, in some cases there was some question as to where the piping of fines took place directly through the filter downstream of the core or whether the leakage water by-passed the filter and went into a crack in the rock foundation. Both actions have occurred. At the Balderhead Dam and Matahina Dam (below), it was demonstrated that the eroded fines entered the filter, which was too coarse. At one of the Churchill Falls Dykes it was shown that the eroded fines were carried into cracks in the foundation rock.

b) Coarse soils of other geologic origin.

A similar experience (with a very large crater on the downstream edge of the crest) occurred at the Matahina Dam, New Zealand [5, 8], where the core is comprised of a weathered and broken sandstone excavated as a soil, which material had grain-size distribution almost identical with the glacial tills, Figure 1.

At the 100-mater high Hill's Creek Dam, Oregon, with a core of coarse alluvium with a gradation very similar to that of Figure 1, definite evidence was found that fines had been eroded out of the core leaving "open work gravel" behind, but the leak was small and grouted off before sufficient material was removed to form a sinkhole. The investigators concluded "unfortunately the question of why the seepage developed so rapidly after six years of normal performance remains largely unresolved" [6].

The writer knows of several other very similar experiences not yet described in the engineering literature.

III. INTERNAL INSTABILITY OF COARSE SOILS WITH GRADATION OF FIGURE 1

The experiences cited above leave little doubt that the trouble is related to the fundamental properties of the soil through which the sinkhole develops. There are too many nearly identical experiences in dams of essentially the same material (and very few experiences of sinkholes in dams of other materials) to conclude otherwise. The erosion in these dams, which propagates itself upward tens of meters above the initial leakage point, possibly can only develop in a soil in which the fines are washed out leaving a substantial quantity of coarse material behind.

The typical coarse broadly graded soil of type shown in Figure 1 has been commonly considered a reasonably well-graded material with "self-healing" properties when used as dam impervious core. These soils, however, are not well-graded in the sense that a concrete aggregate and many deposits of river sand and gravel are well-graded. The well-graded materials have a particle size distribution which is sharply concave upward when plotted on the conventional semi-logarithmic chart, indicating just the correct quantity of each particle size to fill the voids of the progressively larger particles. The straight line particle size distribution of the typical coarse soils which have been associated with the troubles summarized here in (Fig. 1) shows that the volume of fine particles is greater than the volume of the voids of the coarse sand and gravel fraction, and the coarser particles therefore are floating in a matrix of the fines.

Internal erosion stability, or the resistance of the fines to migration out of the soil mass, can be approached analytically, by assuming that the material is divided into two parts, such as by sieving, and examining the particle size distribution of the two parts with respect to accepted filter criteria. This is a tecnhique the writer first saw proposed by Professor Victor De Mello [4], for evaluating gap-graded natural soils for use in earthdams. Figure 2 shows the particle size distribution of a typical glacial till and the particle size distribution of the two materials which would be obtained if the till were sieved into two parts, dividing it at 1.0 millimeter size. As seen in Figure 2, the D_{85F} size of the finer fraction is 0.37 millimeters and the D_{15C} size of the coarser fraction is 3.9 millimeters with a ratio of $D_{15C}/D_{85F} = 10.5$. When similar calculations for the same glacial soil (Fig. 2) are made comparing the D_{85}/D_{15} ratio for finer and coarser portions of the soil when divided at other sizes the results are as follows (for soil of Fig. 2):

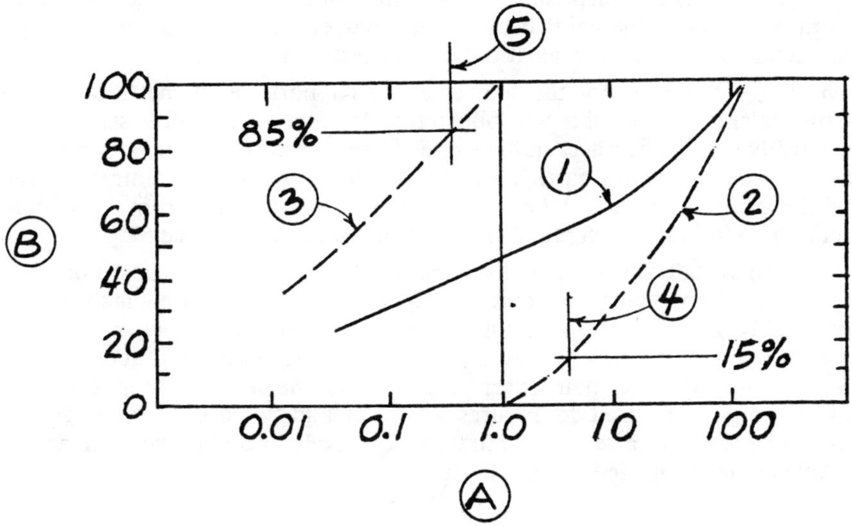

Fig. 2

(A) Particle Size, millimeters.
(B) Percent Finer, by weight.
(1) Typical Broadly Graded Unstable Soil.
(2) Particle Size Distribution of Coarse, Fraction of Typical Soil (1).
(3) Particle Size Distribution of Finer Fraction of Typical Soil (1).
(4) $D_{15C} = 3.9$ mm.
(5) $D_{85F} = 0.37$ mm.
$D_{15C}/D_{85F} = 3.9/0.37 = 10.5$.

(A) Dimension des particules, en millimètres.
(B) Finesse en pourcentage du poids.
(1) Sol type instable et à large fuseau granulométrique.
(2) Courbe granulométrique de la partie la plus grossière du sol type.
(3) Courbe granulométrique de la partie la plus fine du sol type.
(4) $D_{15C} = 3.9$ mm.
(5) $D_{85F} = 0.37$ mm.
$D_{15C}/D_{85F} = 3.9/0.37 = 10.5$.

Size at Which Soil is Separated (mm)	D_{85F} (Finer Fraction) (mm)	D_{15C} (Coarser Fraction)	Ratio D_{15C}/D_{85F}
10.0	3.5	18	5.1
5.0	1.6	9.6	6.0
1.0	0.37	3.9	10.5
0.5	0.18	2.0	11.1
0.1	0.046	0.54	11.7

These results show that the finer and coarser fractions of such a soil are not compatible from the standpoint of accepted filter criteria which require that the D_{15}/D_{185} ratio be less than 4 or 5. Hence, it would be expected that if two independent soils with the particle size distribution of the coarser and finer fractions (Fig. 2) were placed adjacent to each other in a dam, the finer material could be caused to enter the voids of the coarser material under the action of the seeping water. Conversely, with such a high ratio of D_{15}/D_{85}, it is not surprising that the fine portion of the till can be caused to migrate out of the voids between the coarser particles explaining readily the action which has damaged the dams cited.

The D_{15C}/D_{85F} ratio for the coarser and finer portions of the till obtained from calculations of the type illustrated in Figure 2 is very sensitive to relatively small differences in the slope of the particle size distribution curve. Similar materials with particle size curves falling within the central portion of the range of Figure 1 have D_{15C}/D_{85F} ratios varying from 5 to more than 20, with only moderate differences in the slope of the curve.

Similar calculations made on the great majority of the common soils in nature from which the cores of embankment dams are made almost invariably give D_{15C}/D_{85F} ratios in the range from 2 to 4 when they are assumed to be divided into two parts at any size exceeding about 0.1 or 0.2 millimeters. The main exceptions are the coarse broadly graded soils of the type illustrated in Figures 1 and 2 and some coarse gap-graded alluvial soils comprised of a mixture of gravels and silts, without appreciable contents of sand size particles.

IV. FILTERS FOR COARSE SOILS WITH GRADATION OF FIGURE 2

Using any of the current accepted filter criteria for a soil of the particle sized distribution of Figure 2, it would be considered acceptable to utilize a filter comprised of a well graded sand-gravel mixture with 5-15 % by weight passing the N° 200 sieve, an average size (D_{50}) of about 10 millimeters and maximum size up to 100 or 150 millimeters. Such a filter is not uncommon in dams with coarse core materials, especially in areas where there are no economical supplies of finer filter material easily available. A coarse filter of these properties is *theoretically* perfectly satisfactory for a soil of gradation of Figure 2; however, practically it is not possible to construct such a coarse filter without appreciable particle segregation.

This segregation assures that there will be many locations at the boundary between the protected material and the filter, where the filter will consist only of gravel sized particles. At these segregated locations the fines from a soil such as that of Figure 2 can enter the filter if a concentrated leak developes, starting the action which leads to the development of sinkholes.

For these reasons the writer believe that is is appropriate to consider using a sand filter downstream of the core of soils such as Figure 2 which filter would satisfactorily prevent movement of the silt sized particles in the core material, especially for important dams with thin impervious cores and thin filter zones.

In this regard one very interesting precedent is published with regard to the filter design of the Cow Green Dam in England. Following the bad experience with erosion and sinkholes in the coarse glacial core of the Balderhead Dam in England [10], for which a filter without appreciable fine to medium sand was used, the subsequent design of the Cow Green Dam with similar core material provided a medium sand filter with more than 10 % of fine sand sizes, after special filter tests were made in the design stage to show that the filter selected would retain the finest particles which could be eroded out of the core material by the particular water contemplated for storage in the reservoir [9].

It should be emphasized that coarse broadly graded soils (especially of glacial origin) have been used widely for the impervious cores of many completely successful embankment dams, including high, thin core rockfill dams and all types of zoned dams as well as homogeneous dams, with and without chimney drains, and also some of the largest dams in the world without fine sand filters.

The accidents summarized in this paper have occurred only on a small fraction of the total number of dams using such soils. Probably in the average case the erosion started from very small concentrated leaks in narrow channels caused by hydraulic fracturing, as discussed in more detail in the writer's description of the problems at the Yard's Creek Dam, and others [8].

Most of the dams with sinkhole damage had relatively thin cores, conducive to stress transfer and hydraulic fracturing. Unless a concentrated leak develops, it is believed essentially impossible for the fines to be washed out of the soil, regardless of the filter coarseness and hydraulic gradient; when seeping through the voids of the well compacted coarse soil the percolating water travels too slowly and with too little energy to move even the finest and most easily dislodged particle out of the core material.

It appears that the erosion trouble in dams with cores of coarse, broadly graded soils develops only when an unfavorable combination of the following conditions exists: thin core, usually vertical; downstream filter of coarse sand and gravels, with little or no fine sand sizes (usually not a processed material except screening to remove cobbles); steep or jagged rock foundation not adequately sealed; and very rapid reservoir filling. For the average dam with core of glacial till, either no leak develops or leaks which do develop do not get out of control before they are sealed by the action of the filter and the squeezing shut of the leakage channel through the core.

For any given existing dam with these coarse soils for core which has performed satisfactorily for several years without large leakage, there is no reason to suspect there will be trouble in the future. This opinion is based on the facts that: [1] all of the bad experiences have developed during the first few years of reservoir operation; [2] all theory and experience confirms that leaks caused by hydraulic fracturing develop soon after the first reservoir filling as the final settlement of the embankment is occurring, after which there is no new action to cause a change in the dam behavior; and [3] there are a great many dams of soils of type of Figures 1 and 2 with all types of filters and foundations treatment and completely successful behavior.

V. SEALING CRACKS IN ROCK FOUNDATIONS UNDER SHELLS OF COARSE SOILS WITH GRADATION OF FIGURE 1

The experience of several of the dams damaged by sinkholes left no doubt that the fines were washed directly into rock cracks in the foundation under the dam shells. Leakage water travelling through rock cracks probably percolated into the embankment from a crack with relatively high pressure head and then back into another open crack with lesser pressure head. Such conditions can be imagined to occur in any rock foundation, where different rock cracks can have greater or lesser direct connection with the reservoir head and greater or lesser pressure in the percolating water.

There has been a recent (1975) case (not previously described in this paper) in which an important dam with thin upstream sloping clay core and upstream and downstream shells of coarse gap-graded gravel-silt mixture was damaged with details briefly as follows:

1. The foundation was horizontally bedded sandstone with thin shale beds. The sandstone had large relief cracks, common to such formations, in the valley walls.

2. The rock surface under the clay core was well treated and grouted, but less attention was paid to the treatment of rock surface under the coarse gravel shells, where some rock cracks were left untreated, reasoning that there was no substantial tendency for the gravel shell material to enter the rock cracks.

3. On first reservoir filling, about 300 liters/second of highly turbid leakage appeared at the toe, followed by a sinkhole on the downstream slope at the top of the dam (about 8 m deep and 4 m average diameter), at a point where the dam was about 45 m high.

4. Later investigations with borings through the dam showed that directly below the sinkhole a more or less cylindrical column of the downstream shell had been transformed from a well — compacted silt-gravel-cobble mixture to loose pervious gravels and cobbles. with most of the fines removed.

5. Satisfactory repairs were made by extensive grouting of the dam and foundation.

6. It is believed that the leak causing the erosion leading to the sinkhole emerged from one of the sandstone beds under the downstream dam shell, penetrated the dam and then exited in a lower sandstone bed which had less pressure head, carrying with it the fine fraction of the shell material.

7. Making calculations of the type shown in Figure 2 for the typical shell material, the D_{15C}/D_{85F} ratio was in the range 17-25.

The experiences show the desirability of sealing foundation rock cracks under shells of unstable, coarse, broadly graded soils, at least for a sufficient distance upstream and downstream from the dam axis such that sinkholes cannot develop in the central section of the dam. Probably sinkholes in the downstream shell are more hazardous that in the upstream shell if the dam has a central core, because the sinkhole could remove lateral support for the core allowing it to move locally downstream under the water pressure.

REFERENCES

[1] C. E. BERNELL. — 1976, "Control of Leakage Through Dams Founded on Glacial Till Deposits", 12th International Congress on Large Dams (Mexico), Vol. 11, pp. 937-950.

[2] C. E. BERNELL. — 1973, Discussion of Question 42, 11th International Congress on Large Dams (Madrid), Vol. V, pp. 462-465.

[3] R. D. BOIVIN and R. N. SEEMEL. — 1973, "The Churchill Falls Power Development — Construction of the Dykes", Paper presented at Canadian National Committee on Large Dams Annual Meeting, Quebec City, Quebec.

[4] V. F. DEMELLO. — 1975, "Some Lessons from Unsuspected, Real and Fictitious Problems in Earth Dam Engineering in Brazil", 6th Regional Conference on Soil Mechanics and Foundation Engineering, Durbin, South Africa, Vol. 11, p. 304.

[5] G. H. H. GALLOWAY. — 1967, "Installed Instruments In the Matahina Dam", 5th Australian-New Zealand Conference on Soil Mechanics and Foundation Engineering (1 Vol.), pp. 147-152.

[6] J. D. JENKINS and D. E. BANKSFIER. — 1972, "Hills Creek Dam Seepage Correction", Proceedings, Earth and Earth-Supported Structures, American Society of Civil Engineers, Vol. 1, Part 1, pp. 723-733.

[7] R. N. SEEMEL and C. N. COLWELL. — 1976, "Drainage Provisions and Leakage Investigations of Churchill Falls Dams and Dykes", 12th International Congress on Large Dams (Mexico), Vol. 11, pp. 107-127.

[8] J. L. SHERARD. — 1973, "Embankment Dam Cracking", Embankment Dam Engineering (Casagrande Memorial Volume), John Wiley and Sons, New York, pp. 272-353.

[9] P. R. VAUGHAN, H. T. LOVENBURY and P. HORSWILL. — 1975, "The Design, Construction and Performance of the Cow Green Embankment Dam", *Geotechnique*, London, September, pp. 555-580.

[10] R. R. VAUGHAN et al. — 1979, "Cracking and Erosion of Balderhead Dam and the Remedial Works for "Its Repair"", 10th International Congress on Large Dams (Montreal), Vol. 1, p. 73.

[11] H. VESTAD. — 1976, "Viddalsvate Dam. A History of Leakages and Investigations", 12th International Congress on Large Dams (Mexico), Vol. 11, pp. 369-389.

[12] D. M. WOOD, B. KJAERNSLI and K. HOEG. — 1976, "Thoughts Concers ning the Unusual Behavior of Hyttejuvet Dam", 12th International Congress on Large Dams (Mexico), Vol. 11, pp. 391-414.

SUMMARY

A number of important dams in the last 20 years have been damaged by internal erosion causing sinkholes (craters) to develop near or on the crest. Almost all of these experiences have occurred in dams of coarse, broadly graded soils containing particle sizes ranging from gravels and cobbles to silt and clay, frequently of glacial, but sometimes of alluvial, origin. Fines are usually cohesionless or of low to medium plasticity.

The damage has resulted form a type of internal erosion in which the fines are caused to migrate out of the voids by a concentrated leak, leaving the coarse sand, gravels and cobbles behind. The leaks carry the fine particles through coarse filters or into cracks in the rock foundation. The volume of the fines carried away exceeds the normal void volume of the remaining coarse particles, causing progressive collapse of the embankment material above the initial leakage channel, and finally sinkholes at the dam surface.

The typical coarse soils in which the trouble has developed (Fig. 1) are internally unstable in the sens that the finer fraction is too fine to be compatible with the coarse fraction from the standpoint of accepted filter criteria.

Many completely successful dams have been constructed of the same coarse unstable soils. The trouble develops only under the unusual circumstances where a concentrated leak gets started through the impervious dam section and where the leakage water can carry away the fines (through coarse filters into rock shells or into unsealed cracks in rock foundations). Existing filter criteria for these broadly graded unstable soils need reconsideration. Use of a sand filter which will retain the silt sized fraction is conservative.

RÉSUMÉ

Pendant les vingt dernières années plusieurs barrages importants ont été endommagés par une érosion interne. Cette érosion a causé des cratères au niveau du couronnement. Presque tous les dommages ont eu lieu dans les barrages faits de terre grossière et à large fuseau granulométrique contenant des particules allant du limon et argile au gravier et grosses pierres. Les gros matériaux sont d'origine glaciaire et alluviale. Les boues manquent de cohésion ou sont de basse à moyenne plasticité.

La cause des dommages est une érosion interne où les particules fines sous l'effet d'une fuite concentrée quittent les vides. Les sables à grain grossier, les graviers et les galets restent à leur place. La fuite transporte les particules fines à travers un filtre grossier où dans des crevasses de la fondation rocheuse. La quantité de particules fines transportées dépasse le volume des vides des particules grossières restantes. Par conséquent, le corps du barrage s'écroule progressivement au-dessus du canal de fuite initial, et des cratères apparaissent à la surface du barrage.

Les terres grossières typiques où les troubles se développent (Fig. 1) ont une instabilité interne, en ce que la fraction fine n'est pas compatible avec la fraction grossière; et le phénomène ne répond pas aux critères des filtres.

Plusieurs barrages qui ont été un succès ont déjà été construits avec les mêmes terres grossières et instables. Le dommage apparaît seulement dans les cas exceptionnels quand une fuite concentrée commence dans la section imperméable, et quand les matériaux fins peuvent être transportés par la fuite hydraulique. Les critères existants pour les filtres à terres instables de classification large doivent être réexaminés. L'utilisation de filtres qui peuvent retenir des éléments de la taille du limon est une technique sûre.

Introduction to Paper 12

Piezometers in Earth Dam Impervious Sections

by J.L. Sherard
Recent Developments in Geotechnical Engineering For Hydro Projects, Kulhawy, F.H. ed., American Society of Civil Engineers, 1981, pp 125-165.

Prepared by R.W. Beene[6], M. ASCE and
Clifford LeRoy McAnear[7], F. ASCE

James L. Sherard was an astute observer of phenomena associated with the performance of earth dams. A hallmark of Jim Sherard's work was his intense interest in finding out why a problem had occurred, and in devising a practical solution. After the 1961 slide at Waco Dam, which was discussed in the 1967 Conference on Stability and Performance of Slopes and Embankments, Jim visited the site and discussed the features of the problem with us. One of the most remarkable features was the unusually high pore pressures that we had measured in the clay shale foundation. Fearing there would not be a sufficient funding for investigating the slide (to his standards), Jim wrote Arthur Casagrande, who was a consultant on the project, and urged that he insist on adequate time and funds to determine why the situation existed. We solved the problem sufficiently within a couple of years to complete the dam successfully, but a detailed technical analysis of the development of the pore pressures was not accomplished until 1981. By this time Jim had finally found enough time and data to complete his own study of piezometers, upon which this paper was based. During the period of preparation of his paper, Dr. Sherard visited and corresponded with us concerning on-going research related to development of pore pressures and their measurements. His insights and encouragement were greatly valued.

[6]Geotechnical Consultant, Fairfax, VA.

[7]Project Manager, Louis Berger & Associates, Vicksburg, MS.

PIEZOMETERS IN EARTH DAM IMPERVIOUS SECTIONS

James L. Sherard*, F. ASCE

ABSTRACT

Double tube hydraulic piezometers are satisfactory instruments but there is a trend away from using them, toward diaphragm piezometers which are easier to use. Both vibrating wire and the better types of pneumatic piezometers are good instruments for use in unsaturated impervious embankments, but the vibrating wire piezometer has substantial advantages. Some models of pneumatic piezometers have not given uniformly good service. When used with coarse porous tips (low air entry value) vibrating wire and pneumatic piezometers measure the pressure in the pore air when it is significantly different than the pore water pressure. It is reasonable current practice to use vibrating wire and pneumatic piezometers with coarse porous tips (low air entry value) for routine instrumentation. For measuring pore water pressures (as opposed to pore air pressure), using high air entry value ceramic tips, the vibrating wire piezometer is superior to the pneumatic piezometer. Vibrating wire piezometers can measure sub-atmospheric pore water pressure but there isn't much experience as yet.

1. INTRODUCTION

This paper presents experience and opinions regarding remote reading piezometers, with tips placed during construction in the impervious sections of earth dams and tubes or cables leading to measuring points outside the dam (that is, piezometers other than standpipe piezometers with a vertical riser pipe). These are piezometers used for routine measurement of pore pressures both during construction of the dam and operation of the reservoir. While much of the subject matter applies also to other applications of these piezometers, emphasis is on the special problems of placing them in the impervious, unsaturated core of a central core dam and in the body of a homogeneous type dam, where the pore water contains air bubbles.

For some of the main points discussed there is no widespread agreement in the profession, with different engineers holding considerably different opinions. Because of this range of opinion in current practice, published discussions of this paper in the Journal of the Geotechnical Engineering Division, ASCE, are particularly solicited.

In preparation of the paper the writer gathered opinions and experience from practice from a considerable number of colleagues (See ACKNOWLEDGEMENTS, Section 15).

*Consulting Engineer, San Diego, California

2. THREE PIEZOMETER TYPES IN COMMON USE

There are three basic types used commonly as routine instrumentation:

1) Double Tube Hydraulic Piezometer with water filled tubes (two tubes per piezometer) carried out of the dam to a point where measurement is made of the water pressure at the end of the tubes.

2) Vibrating Wire Piezometer with pressure acting on a diaphragm connected to a tensioned steel wire with electric cables carried out of the dam to a point where measurement is made of the frequency of vibration of the vibrating wire, which is directly related to the pressure on the diaphragm.

3) Pneumatic Piezometers with pressure acting on a diaphragm with tubes carried out of the dam to a point where measurement is made by circulating pressurized gas (usually nitrogen) through the tubes and determining the pressure needed to balance the pressure acting on the diaphragm.

Although a few other types of remote sensing piezometers have been used on various projects (such as diaphragm piezometers operated with water or oil, and other types of electric piezometers using resistance strain gages), the frequency of application has been negligible compared with the above three types.

In world practice there is currently a considerable range of opinion among dam designers regarding the relative merits of these three piezometer types and the interpretation of their experience records. Each of the three types is currently favored by different groups of engineers. A few examples described in Sections 3 to 5 illustrate the diversity of current practice and opinions.

3. CURRENT USE OF DOUBLE TUBE HYDRAULIC PIEZOMETERS

Among the three common types, the double tube hydraulic piezometer has been in use the longest. The earliest widespread use was by the U.S. Bureau of Reclamation in the Western USA (recently renamed "Water and Power Resources Service"), with first installations in 1939 at the Fresno Dam, and continued use on their projects until about 1978 with moderate improvements in the details of the piezometer tip and tubing. A significant improvement was the change from the coarse porous tip to a "high air entry value" (sometimes called "high bubbling pressure") ceramic tip with very fine pores (Section 8), about 1960. This reduced problems with air bubbles in the tubes and allowed better measurement of low pressures and pressures below atmospheric pressure. Starting in 1978, hydraulic piezometers were abandoned by the U.S.B.R. in favor of pneumatic piezometers, primarily because they are simpler to use. This practice is not yet reflected in their widely read manuals, which still show the previously used hydraulic piezometers (W.P.R.S. 1974).

The Bureau of Reclamation practice with double tube hydraulic piezometers was widely followed in the years 1950 - 1965 by many other engineers in the USA and abroad, frequently with troublesome experiences causing the practice to be dropped when the other types of piezometers became available. Hydraulic piezometers have been favored by engineers in England, influenced largely by geotechnical engineering research at Imperial College which have been widely discussed in the literature. Hydraulic piezometers have been used on projects in various parts of the world by English engineers and others under the influence of this English practice.

One example is the 80 m. high, 10 million m^3 Pournari Dam in Greece, completed in 1980, where the designers used double tube hydraulic (Imperial College Type) piezometers in the central clay core, while using pneumatic piezometers in the upstream pervious gravel shell. The details of hydraulic piezometers for the core were largely based on English practice, using high air entry value ceramic porous tips and tubing consisting of an inner nylon tube with an exterior jacket of polyethylene (in an effort to keep air from entering the tubes from the surrounding embankment). Nevertheless, considerable trouble developed with air bubbles in the tubes as the dam was being constructed. Pneumatic piezometers were chosen for the upstream pervious shell because the lines could be brought to a measuring point on the crest, without crossing the central impervious core (which cannot be done with double tube hydraulic piezometers). Hydraulic piezometers were favored over pneumatic piezometers in the impervious clay core because it was believed that they would measure the water pressure (as opposed to air pressure) and be able to measure pressure below atmospheric pressure. Also, there had been some bad experiences with vibrating wire piezometers in earlier installations in Greece.

Another recent example is the 180 m. high, central core, rockfill Dartmouth Dam in Australia completed in 1979, in which the majority of the piezometers in the impervious core are hydraulic piezometers of the Imperial College type, with details similar to those at the Pournari Dam. In addition a lesser number of vibrating wire piezometers were installed "to indicate total pore air and pore water pressure during construction when this may have departed from pore water pressure" (Cummings, 1979).

While the above examples illustrate a continuing preference among some engineers for the hydraulic piezometer, the general trend in recent years has been away from hydraulic piezometers, because of the need for a relatively elaborate terminal measuring house, difficulties in measurements caused by air bubbles in the tubes, and the need for considerable skill and care in keeping the system operating. Before about 1960 there were no alternatives widely available. Starting about 1960, instrument manufacturers offered the two types of diaphragm piezometer (vibrating wire and pneumatic) which eliminated the water filled tubes. There were early questions about the suitability of diaphragm piezometers in the unsaturated impervious sections of dams, because of the difference between the pressures in the water and air phases of the pore fluid. While there is still some debate on this point, installations in which hydraulic and diaphragm piezometers were placed side by side have given general confidence that there is no significant difference in the measured pressures (Section 7).

Double tube hydraulic piezometers are very satisfactory instruments when properly designed and maintained. With high bubbling pressure tips little air should enter the system from the tips. With a double jacket of nylon and polyethylene for the tubes research shows that little air will enter the hydraulic lines (Penman and Kennard, 1981). The use of mercury manometers for pressure measurement can eliminate the problems of long time measurements and maintenance with Bourdon gages. Especially where well trained instrumentation staff is available at the construction site, reliable and accurate measurements can be made with double tube hydraulic piezometers. However, less skill and care is needed to maintain the newer diaphragm piezometers which are much simpler.

4. CURRENT USE OF VIBRATING WIRE PIEZOMETERS

Developed and first used in Europe, vibrating wire piezometers have been used with increasing frequency in recent years. Early applications were in the 40 m. high Rosshaupten Dam in Germany, completed in 1953 (Trieber, 1958) and the 48 m. high Lokvarka Dam in Yugoslavia, completed in 1956 (Nonveiller, 1957). In both cases the piezometers were reported as giving reliable and reproducible measurements. At the Lokvarka Dam, 20 of the originally installed vibrating wire piezometers were functioning well 14 years after construction, whereas hydraulic piezometers also used in the core never worked satisfactorily (Nonveiller, 1969).

Another early major application was in the Serre-Ponçon Dam, a 125 m. high central core dam with gravel shells constructed in France during 1957 - 1960, where hydraulic piezometers were also used. The excellent performance at the Serre-Ponçon Dam led the engineer (Coyne and Bellier) to utilize vibrating wire piezometers in the impervious cores of all their subsequent major dams including Saint-Cassien (1965) and Mont Cenis (1967) in France, Ait Aadel Dam (1970) in Morocco, and others. The experience has been uniformly good; the vibrating wire piezometers in Serre-Ponçon Dam, now 20 years old, are functioning well (Douillet and Ledeuil, 1979). The French Government Electric Company (EDF) has been using Telemac vibrating wire extensometers and piezometers for 30 years and 20 years, respectively. They report that about 90% of the more than 3000 extensometers are still functioning and that the piezometers are highly reliable.

During the 1960's vibrating wire piezometers were installed in the impervious section of a number of major dams in Brazil and there were several unsatisfactory experiences in which a large percentage of the instruments were rendered inoperative by lightning. Similar troubles occurred from lightning damage during this period in other countries. Most of the troubles from lightning damage occurred during construction before the cables were covered by embankment material, and in many cases, the cables were not those supplied by the instrument manufacturer.

Subsequently, manufacturers provided better shielded cables, grounding and over-voltage protection and there have been no recent problems from lightning to the writer's knowledge (Section 9). These troubles from lightning were well known, causing vibrating wire piezometers to be temporarily in disfavor. In some countries, such as Brazil, where vibrating wire piezometers were widely used previously, some engineers are currently favoring pneumatic piezometers. For example, at the 100 m. high, 60 million m^3 embankment dam for the Tucuruí Hydroelectric Project, currently under construction in Brazil, pneumatic piezometers are used inside the impervious section by an engineer who had used vibrating wire piezometers on previous large dams. At the same time other Brazilian engineers are installing vibrating wire piezometers in recent and currently constructed dams.

Since 1970 vibrating wire piezometers have been used with increased frequency in various parts of the world and many dam specialists believe that they are the best instrument available. One example is the engineering staff of Hydro-Quebec, which first used vibrating wire piezometers along with hydraulic piezometers in the 400 foot high Outardes 4 Dam (completed, 1968), and have been using vibrating wire piezometers routinely in the earth cores of subsequent large dams. Another example illustrative of this current trend is the experience at the large Guri Dam (120 m. high, 60 million m^3 embankment) under construction (1980) in Venezuela, where pneumatic piezometers were originally specified. During the early part of construction the engineers working with the instruments became dissatisfied with the length of time required to make readings, and with the care and skill required to make reliably reproducible measurements, and have concluded that the Telemac vibrating wire piezometer is a much better instrument, which allows instantaneous readings easily reproducible by technicians without previous training.

The trend toward the use of vibrating wire piezoemeters has not extended to the USA. Most of the U.S. instrument manufacturers have concentrated on pneumatic piezometers. The writer does not know of any dam in the USA with vibrating wire piezometers in the impervious core, though a few have been used in dam foundations and many have been used in other geotechnical applications. To the writer's knowledge there is only one U.S. manufacturer of vibrating wire piezometers at the present time (IRAD Gage, Inc. Lebanon, New Hampshire).

5. CURRENT USE OF PNEUMATIC PIEZOMETERS

Starting about 1960 pneumatic piezometers became available and were introduced into practice. The early development and use of pneumatic piezometers was largely in the USA and by U.S. engineers working outside the USA.

An early application was the 280 foot (88 m.) high Briones Dam in California, constructed 1961 - 1964, in which both USBR double tube hydraulic and pneumatic piezometers were installed (Anton and Dayton, 1972). As with most subsequent applications the performance was generally satisfactory. In the first years of operation, both types of piezometers gave about the same measured values. The owner reports (1980) that all the pneumatic piezometers are still functioning about 18 years after installation, whereas all the USBR type hydraulic piezometers became plugged and ceased functioning at the end of about 10 to 12 years of operation.

By the mid-1960's pneumatic piezometers were being widely used inside impervious sections of many dams, including some of the largest projects in the USA and around the world. In an early application outside the

USA pneumatic piezometers were placed in the impervious section of the Xavantes Dam in Brazil in 1964, with pairs of double tube hydraulic, vibrating wire and pneumatic piezometers to compare readings, with satisfactory agreement.

Some instruments require considerably longer time to make readings than others. At the Punchina Dam presently (1980) under construction in Colombia, a 70 m. high, 5 million m^3 homogeneous dam of fine-grained soil, 140 pneumatic piezometers were specified. At the beginning of the job about 50 piezometers of one manufacturer were installed. This type commonly required 10 to 20 minutes to make a reading on each piezometer, and considerable care and skill was needed to obtain reproducible results. Also the field staff had difficulties making air tight connections to the piezometer and splices in the tubing when necessary, and with malfunctioning of the readout box. For the subsequent pneumatic piezometers, instruments from another manufacturer were obtained which allowed readings to be made in 5 to 10 minutes per piezometer and which had details that eliminated the other problems.

The various pneumatic piezometers available have significantly different designs, materials and performance. It is unfortunate that some of the unsatisfactory experiences with pneumatic piezometers, which create a generally negative opinion in the profession, have been with instruments having less desirable details than the average type on the market. A review and evaluation of the main basic types of pneumatic piezometers is given by Dunnicliff (1980 and 1981).

At the present time the pneumatic piezometer is in general use by many engineers for measuring pore pressures inside impervious sections of dams, as well as in foundations and pervious zones of dams. Sometimes pneumatic piezometers are used in conjunction with adjacent hydraulic or vibrating wire (or standpipe) piezometers to provide a check, but on many large dams they are the only piezometers used in the impervious core.

6. RELATIVE ADVANTAGES OF DIAPHRAGM (VIBRATING WIRE AND PNEUMATIC) AND DOUBLE TUBE HYDRAULIC PIEZOMETERS

While there are differences of opinion among different experienced engineers on the importance of the problems with hydraulic piezometers, there is consistent agreement that diaphragm piezometers are simpler to install and require less skill and care to operate and maintain.

For hydraulic piezometers it is necessary to locate the tubes and the terminal measuring house in such a way that the water pressure at any point in the hydraulic tubes is not much lower than 0.5 kg/cm^2 below atmospheric pressure, which restricts the layout of the instrumentation system. For diaphragm piezometers there are no restrictions on the relative elevations of tips and the lines, and the measuring point can be located anywhere, a major advantage over the double tube hydraulic piezometer.

An intrinsic advantage of the diaphragm piezometer is that it has practically no influence on the pore pressures in the embankment. For the double tube hydraulic piezometer it is inevitable that some water from the piezometer tip penetrates the embankment, increasing the water content in the immediately adjacent soil. This occurs mainly during the periods of time that water is circulated to remove air bubbles from the system. Especially in a clayey embankment with initially high capillary suction, the local conditions at the piezometer tip can be changed appreciably, causing the pressures measured to be higher than they would be otherwise, at least for some length of time. It is difficult to judge how important this action may be in some situations.

Another important advantage of the diaphram piezometer is the relatively simple measuring terminal needed. For both pneumatic and vibrating wire piezometers the measuring points of several dozen piezometers can be brought into a relatively small space (often simply an outdoor panel supported by posts without an enclosing house) and measurements are made by means of a portable "readout" box. For the hydraulic piezometer a substantial enclosed house is needed, heated against freezing, with relatively elaborate equipment including facilities for circulating deaired water through the tubes and measuring gages for each piezometer. Because of the time needed to build the terminal house for hydraulic piezometers and assemble the needed equipment, it is common that no measurements are obtained in the early part of construction. For the diaphragm piezometers measurements can be made immediately after installing the piezometer in the embankment, while the tubing or cable is still on temporary reels.

The purchase cost of a hydraulic piezometer and connecting tubes is less than that of a vibrating wire piezometer and cables. However, the total cost of a group of hydraulic piezometers, including the cost of the terminal house and the extra cost of operation is of the same order as the cost of installation and operation of a similar number of vibrating wire piezometers.

Some proponents of the hydraulic piezometer cite evidence that it has a longer life than diaphragm piezometers. The experience as accumulated by this writer does not support the general conclusion that hydraulic piezometers have a consistent advantage of greater useful life. Where the two types are installed in the same dam and maintained by the same group, the experience is as likely to be that the hydraulic piezometers cease to function first, as typified in the experience at the Briones Dam (Section 5) and at the Lokvarta Dam (Section 4). There are many experiences where piezometers of all three types have continued to function satisfactorily since installation for 10 to 15 years and more, and some cases where all three types of piezometers ceased to function soon after installation. The likely useful life of any of the three types of piezometer is more related to the conservatism of the design and construction details of the specific piezometer, connecting lines and installation care, and to the conditions in which it is placed (differential settlement of the embankment through which the lines are placed, corrosive environment, etc.), than it is related to fundamental differences among the three types of piezometers. Any one of the three types of piezometers can be made and installed with sufficiently conservative details to provide an adequately long functioning life. The main need for piezometers in the impervious sections of embankment dams is to provide measurements during the first reservoir filling and the first several years of operation. The liklihood that long term measurements will be of any significant value on the typical project is negligible.

7. DIFFERENTRATING BETWEEN AIR AND WATER PRESSURES

There has been much discussion in the profession on the difference between the pressure existing in the air and water in the pores of compacted impervious embankments, and the related problems of measuring pore pressures with piezometers in earth dams. A considerable volume of literature has been published describing research on the subject, including theoretical treatment, laboratory tests and measurements in embankments (Hilf, 1956, Vaughan 1973, Gibbs and Coffee 1969 and Bishop 1969). There still remains some question about the maximum difference between air and water pressures that could exist in impervious sections of dams under certain circumstances, and how the differences should be evaluated. However, it is believed that the following are generally valid:

1) When a layer of impervious embankment material is placed and compacted, the soil voids are partially filled with air. Immediately after compaction the pressure in the pore air is near atmospheric pressure and the pressure in the pore water is below atmospheric pressure because of the capillary suction that exists in the small soil pores. Hence, immediately after compaction the difference between air and water pressures equals the capillary suction created by the water menisci in the soil, and this is the maximum difference which can exist at any time.

2) The capillary suction is a function of the fineness of the soil and is only of significant magnitude in clays and coarser soils with an appreciable content of clay fines. The finer and more plastic the clay and the lower the compaction water content, the higher the capillary suction and the greater the difference between air and water pressures in the soil voids during the construction of the dam.

3) The initial capillary suction in compacted clay embankments, just after compaction and before loading in the dam, is commonly in the range of -20 to -40 psi (14 to 28 m. of water head below atmospheric pressure). This initial capillary suction can be easily measured in the laboratory using special techniques developed by Hilf (1956) in which the sample is placed on a saturated high air entry porous ceramic and back pressured with air in such a way that water in the measuring lines does not cavitate (U.S.B.R., 1966 and W.P.R.S., 1974). Figure 1 shows measured initial capillary suction in a typical sandy clay of medium plasticity compacted at different water contents and densities, with the capillary suction measured immediately after compaction.

4) As the embankment is constructed above the point under consideration, the soil compresses and the volume occupied by the air decreases. As the embankment compresses, the degree of saturation increases, causing a decrease in the capillary suction, and the pressure in the air increases, but the difference between the air and water pressures decreases. For most dams, there is not enough compression during construction to compress all the air, and the embankment remains unsaturated (though nearly completely saturated in the lower part of high dams). As long as the embankment is unsaturated and air fills some of the soil voids,

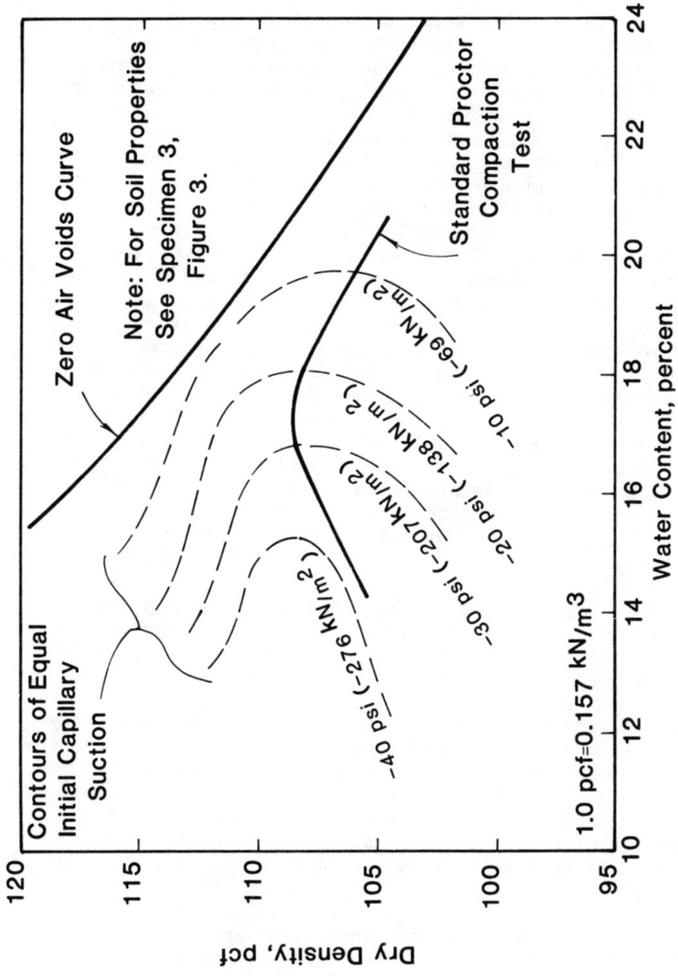

Summary of Measurements of Initial Capillary Suction in a Clayey Soil Compacted at Various Water Contents and Densities (After U.S.B.R. 1966)

Figure 1

there are menisci in the voids and capillary forces. This is true also when the water pressure is above atmospheric pressure. The difference between the water and air pressures is always equal to the capillary suction in the soil, with the air pressure being always the larger (Hilf, 1956).

5) Both the pore air and pore water pressures in compacted clay specimens can be measured in the laboratory while the specimen is subjected to undrained compression, simulating the dam construction condition. The decrease in capillary suction which occurs as the specimen is compressed can also be measured in other tests. Figure 2 shows representative results of such tests on a compacted specimen of clayey soil. As seen in Figure 2, the initial capillary suction was below -20 psi (-138 kN/m^2) and the capillary suction became smaller in magnitude (approached zero gage pressure) as the specimen was compressed. The measured air pressure started out at atmospheric pressure and gradually increased as air bubbles were compressed. The measured pore water pressure is always less than the air pressure by an amount equal to the magnitude of the capillary suction; the only reason there is a difference between the pore air and water pressures is the existence of the water menisci and the capillary tension. Since the capillary suction decreases as the specimen is compressed the difference between the pore air and pore water pressures also decreases as the embankment material compresses.

6) The pore air pressure which develops in a laboratory specimen during undrained compression can be calculated from the classic laws governing compression and solution of air (Hilf, 1956 and W.P.R.S., 1974). The pore air pressures measured in the laboratory tests, giving results as shown in Figure 2, generally approximate calculated pressures very well (U.S.B.R., 1966).

7) In a dam, diaphragm piezometers with coarse porous tips (with low air entry value) measure the pressure in the pore air, whenever there is a significant difference between the air and water pressure. This is discussed in more detail in Section 11.

Figure 3 shows results of laboratory tests on three compacted clay specimens subjected to undrained compression, with simultaneous measurement of pore air and pore water pressures (on different ends of the specimens). The results are for two clayey fine-grained soils compacted with water content a little on the dry side of optimum water content. The laboratory test results for the three specimens shown in Figure 3 are for the same type of test as for the specimen of Figure 2 but plotted in a different way.

The results (Figure 3) show that when these clay specimens have compressed to the point where the pore water pressure reaches atmospheric pressure, the air pressure ranges approximately between 4 and 6 m. of water head. As the sample is compressed further the differences continue to decrease. When the water pressure has reached 20 m. of head, the difference is in the range of 0.5 to 1.5 m., or about 5%.

The data for Figure 3 are for clay specimens compacted moderately dry of optimum water content. For specimens of the same soils compacted near optimum the results would plot to the right of the curve for

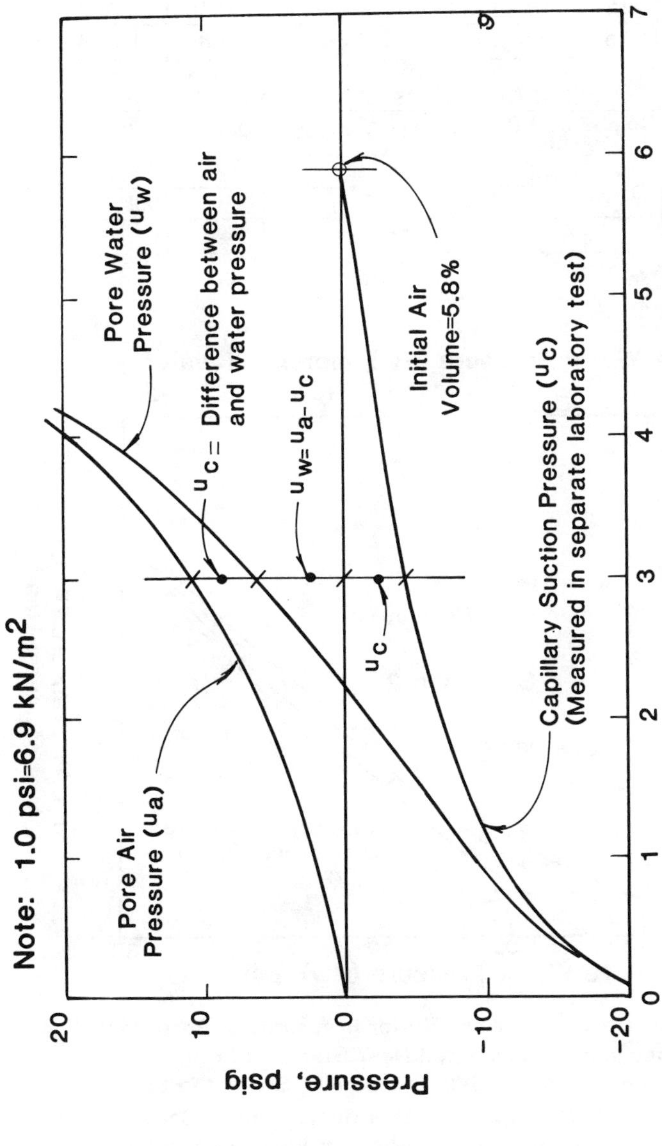

Relationship Among Pore Air Pressure, Pore Water Pressure and Capillary Suction During Undrained Compression of a Clay Embankment Material Compacted on the Dry Side of Optimum (After U.S.B.R. 1966)

Figure 2

Soil Properties

	LL	PI	Percent Finer		Standard Proctor	
			0.002mm	0.080mm	Max.Dens.	OMC
Specimens 1& 2	32	19	34	92	110.1pcf	16.2%
Specimen 3	40	24	26	64	108.6	17.4

Initial Conditions of Specimens

	Compacted Dry Density	Water Content	Degree of Saturation
Specimen 1	110.6pcf	15.3%	80.4
Specimen 2	107.5	14.5	70.2
Specimen 3	109.8	15.2	76.1

1.0 pcf=0.157 kN/m^3

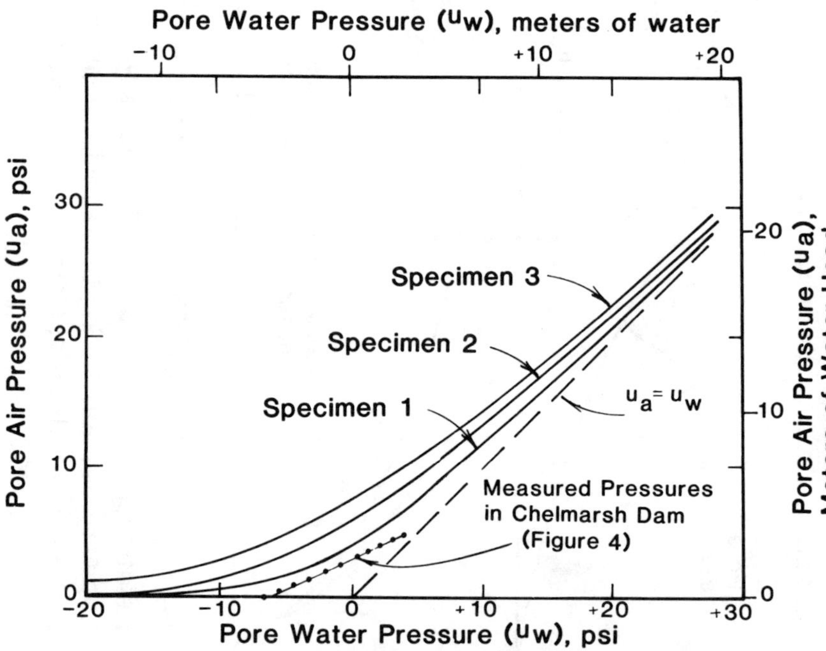

Relationship Between Air and Water Pressures in Pore Voids of Clay Specimens Compacted Near Standard Proctor Maximum Density on Dry Side of Optimum Subjected to Undrained Compression in the Laboratory. Values Shown are Computed and Generally Verified by Measurements. (From U.S.B.R. 1966)

Figure 3

Specimen 1, with lower differences between air and water pressures.

During construction of a dam, some of the air in a given compacted layer escapes upward out of the construction surface as the material compresses under additional load and the action of traffic on the construction surface, whereas under laboratory conditions all the air is retained in the specimen during the test. Hence, it is likely that this difference between air and water pressure in a clay dam during construction will be considerably less than the difference obtained in laboratory test of the type which provided the results of Figures 2 and 3.

The maximum difference between the air and water pressure exists during the early part of construction when the pore water pressure is at its lowest value. There have been only a few careful measurements of this difference made with instruments in dams, generally for research purposes. The comparative measurements have been made generally in situations where the pore water pressure was relatively low.

Figure 4 shows one example reported by Penman (1979), which is believed probably representative of the order of magnitude of differences which exist in the initial stages in many clayey embankment dams. As seen in Figure 4, in one location in the Chelmarsh Dam, a double tube hydraulic piezometer with (high air entry tip) and a vibrating wire piezometer, with coarse tip which measures the air pressure (Section 11), were placed side by side, and readings of both were commenced immediately after installation. Frequent measurements were made with both piezometers during the period of time that 14 m. of embankment were placed above the piezometers. The initial water pressure measurement was about 6.5 m. of water head below atmospheric pressure and the initial air pressure was approximately atmospheric pressure. Both measured pressures increased as the embankment compressed under the increasing weight of overlying material, with the water pressure increasing more rapidly than the air pressure. By the time that there was 11 m. of fill weight over the piezometers, the water pressure had increased to atmospheric pressure and the air pressure was about 3 m. of water pressure. As the fill height was increased to 14 m. the difference between the air and water pressure had decreased to less than 1.5 m. of water head. For higher loads it is apparent that the difference would continue to decrease and would be small compared to the water pressure.

When diaphragm piezometers began to be used widely starting about 1960, a number of dams were instrumented with side-by-side piezometers of different types to compare the measurements obtained, particularly for the "during construction" condition where the difference in the pore air and water pressures is greatest. These experiences showed generally that there was no significant difference in the pressures measured by different types of piezometers, some of which were clearly measuring air pressure.

An instructive example of these experiences was that of the Três Marias Dam, a 60 m. high homogenous clay dam constructed in Brazil during 1958 - 1961 (Areas, 1963). In several representative locations three types of piezometers were installed and read during the construction: vibrating wire, open standpipe and double tube hydraulic piezometers.

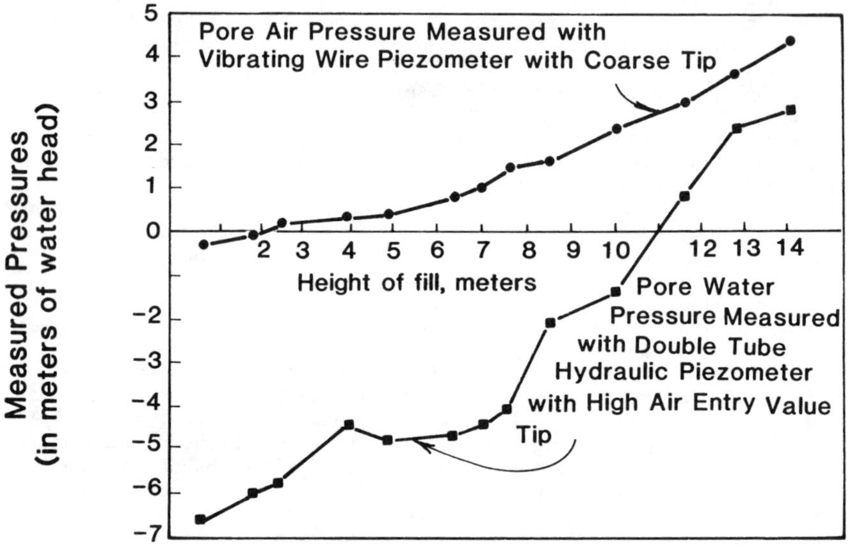

Measurements of Pore Air and Pore Water Pressure in Side-by-Side Piezometers in the Impervious Section of Chelmarsh Dam During the Time that 14 meters of Embankment Were Placed Above the Piezometer Location. (After Penman, 1979)

Figure 4

The Três Marias Dam was constructed before there was any general practice of using fine ceramic tips with high air entry value, and both the vibrating wire and double tube hydraulic piezometers were installed with coarse porous tips of the type then in common use. Figure 5 shows measurements obtained from a set of piezometers located near the bottom of the dam directly under the crest. The embankment material was an alluvial clay of medium plasticity compacted near Standard Proctor optimum water content, with average properties as shown in Figure 5. The measurements showed that after the measured pore pressure exceeded about 5 m. of water head all piezometers gave essentially the same results, with maximum difference at any given time not exceeding about 1.5 m. of water head.

The open standpipe piezometer (Figure 5) measured the pressure in the pore water. The double tube hydraulic piezometer probably measured pore water pressure, though sometimes hydraulic piezometers can measure higher pressure if air which enters the tip is not removed by circulating water through the double tube system. The vibrating wire piezometer was probably measuring pressure of the air in the voids (Section 11).

If it had been possible to measure the pressure in the pore water reliably in the range below atmospheric pressure the measured values would probably have been similar to those represented by the dashed Curve A, Figure 5. Hence, the water pressure probably became equal to atmospheric pressure when the weight of the overlying embankment reached about 4.8 kg/cm^2 (470 kN/m^2), at which time the measured air pressure was probably about +1.0 m. of water head as shown. It appears from these measurements that after the water pressure reached atmospheric pressure the difference between air and water pressure was somewhat less than in the measurements shown in Figure 4, and considerably less than the range of values from the laboratory tests shown in Figure 3. The Três Marias Dam was built of a fine-grained clay with high initial capillary suctions, probably of the order of those shown in Figure 1.

The Três Marias Dam measurements confirm that in embankments of fine-grained clayey soils which have relatively large potential for differences between air and water pressures, the differences are not significant during construction after the pore water pressure exceeds atmospheric pressure by about 5 m. of water head. The Três Marias embankment material is a clay of alluvial origin with properties similar to those used in many dams, and there is no reason to believe that the measurements are not representative of the pore pressures which develop commonly in embankment dams of clayey soil.

The magnitude of the initial capillary suction, which exists in the embankment immediately after compaction, is a function primarily of the content and plasticity of the soil fines. The values of capillary suction (and differences between air and water pressure), discussed above and shown in Figures 1 - 5, are for typical fine-grained clays of medium plasticity. For clays of high plasticity the capillary suction is higher. While preparing this paper the writer was not able to find much data on measured capillary suctions in fine-grained compacted soils of low plasticity. For one fine-grained, slightly clayey silt, of loessial origin (ML), with $LL \simeq 25$ and $PI \simeq 5$ the measured initial capillary suction of specimens compacted near optimum water content was about 7 psi

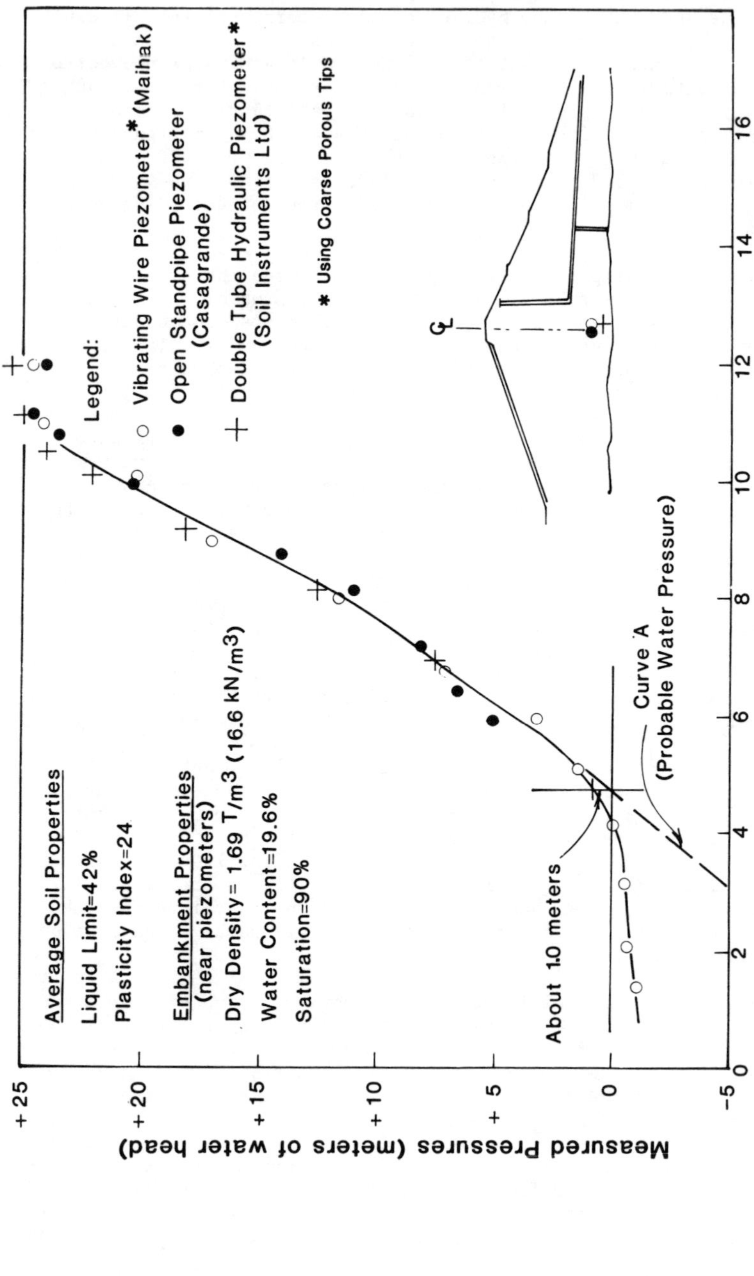

Comparison of Measurements with Side-by-Side Piezometers During Construction of Clay (Tres Marias Dam). After Areas, 1963.

Figure 5

(48 kN/m^2). For embankments of sand and gravel soils with cohesionless fines, materials frequently used for the impervious sections of embankment dams, the pore sizes are much larger than for embankments of clayey soils and the capillary suctions are negligible.

The maximum differences between air and water pressure are those which exist before the reservoir is filled. When water seeps into the impervious section of the embankment from the reservoir, the degree of saturation increases and the capillary suction in the soil decreases. Consequently, the difference between air and water pressure must decrease as seepage water penetrates the dam.

In impervious dam sections which are influenced by seepage from the reservoir, at the steady state condition, the differences between the pressures in the water and air phases of the pore fluid are generally negligible. This conclusion is supported by measurements with diaphragm piezometers equipped with coarse porous tips compared with the expected values from seepage computations (flow nets from the La Place Equation); that is, piezometers near the upstream face of the core show nearly full reservoir pressure as expected, piezometers near the central portion of the core show intermediate values of the approximate magnitude expected, etc. Also, measurements of side by side diaphragm piezometers with and without high air entry tips give the same values within the scatter of reading of both types. For example, in several dams in France, where the impervious material was of low to medium plasticity, compacted with water content near or slightly above optimum, measurements with vibrating wire piezometers with both coarse tips and fine ceramic (high air entry value) tips gave comparable values (Post and Picaut, 1969, and Cour and Vincent, 1973).

In summary, the writer believes that it is reasonable to draw the following conclusions from the experience presently available regarding the difference between air and water pressures in impervious dam sections:

1) For embankments consisting of soils with cohensionless fines the difference is always negligible.

2) For embankments consisting of soils with clayey fines the difference is greatest, but is not likely to exceed 2 or 3 m. of water head when the pore water pressure exceeds about 5 m. of water head and for higher pore water pressures the difference rapidly decreases.

3) The difference is always highest during construction and will later decrease in any portion of the dam influenced by seepage from the reservoir.

4) The differences during construction can be studied in advance by laboratory tests. The differences measured in the laboratory will probably be always higher than those which develop in the dam, because some of the air escapes from the compacted embankment during the dam construction, whereas in the compacted laboratory specimen all the air is retained in the undrained compression test.

5) There are very few situations where the difference between measured pore air pressure and true pore water pressure could have any significant influence on the evaluation of the behavior of the dam.

8. THE USE OF HIGH AIR ENTRY TIPS (HIGH "BUBBLING PRESSURE")

High air entry tips consist simply of porous ceramic with very fine pores of more or less uniform size. These ceramics are utilized as industrial filters in various chemical manufacturing processes. They were first used in geotechnical engineering in laboratory tests for measuring water pressures in unsaturated soils. For piezometers they were first used in double tube hydraulic piezometers in England (Bishop et al, 1960) primarily for the purpose of preventing air bubbles in the adjacent soil from entering the piezometer tip.

Later, high air entry tips were used for diaphragm piezometers, primarily for the purpose of keeping the tip and the space between the tip and the diaphragm saturated, so that the piezometer would measure water pressure rather than air pressure during construction of the dam. High air entry tips were first used on vibrating wire piezometers experiementally in the early 1960's and have been used fairly commonly by some engineers since 1970-75. Since about 1975 there have been a few installations of pneumatic piezometers with high air entry tips, though this has been less common.

When a saturated piece of high air entry ceramic has water on one side at atmospheric pressure and is subjected to increasing air pressure on the other side, the air will not penetrate the pores until the air pressure exceeds the capillary tension in the largest continuous pores, above which pressure the air will blow the water out of the pores. This threshold pressure is called the "bubbling pressure" or air entry value. The bubbling pressure of a given fine ceramic is easily measured in the laboratory as the pressure at which bubbles are seen emerging on the water side of the ceramic in such a test. The principal manufacturer of high air entry ceramics in the USA is the Coors Porcelain Company, Golden, Colorado, which supplies a series of different fine porous ceramics with bubbling pressures ranging generally from about 15 to 90 psi (100 to 600 kN/m^2). Most piezometers have been equipped with tips having bubbling pressures of the order of 15 to 30 psi (100 to 200 kN/m^2). Representative properties for three such ceramics of varying pore sizes were measured by Gibbs, et al (1960):

Bubbling Pressure psi (kN/m^2)	Coefficient Of Permeability (cm/sec)	Pore Diameter (mm)
4.4 (29)	3×10^{-5}	0.0095
23.3 (160)	5×10^{-6}	0.0018
39.0 (269)	3×10^{-7}	0.0011

The Coors Porcelain Company reports that the measured coefficient of permeability of their fine porous ceramic with bubbling pressure of 80 psi (500 kN/m^2) is approximately 1 to 3×10^{-7} cm/sec.

The pore diameters of the fine ceramics, as listed in the table above, are not commonly measured, but are computed from the measured bubbling pressure by the equation for water pressure in a capillary tube, $u = 4Ts/d$, where Ts is the surface tension of water and d is the pore diameter. The computed pore diameters for high air entry ceramics with bubbling pressures ranging from 15 to 90 psi (100 to 600 kN/m^2) are in the range from about 0.002 to less than 0.0005 millimeters.

In making tests of the bubbling pressure of the type described above for their fine ceramics, the Coors Porcelain Company staff determine two threshold pressures: (1) the air pressure at which the first isolated bubble penetrates the ceramic, indicating that the air is getting through the largest continuous pore and (2) the higher air pressure at which air bubbles emerge uniformly from the entire surface of the ceramic specimen being tested, indicating that the air is flowing through the pores of average diameter. The methods used to manufacture these fine porous ceramics have been devised to obtain a product in which there is not a large difference between the largest and average sized pore. The Coors Porcelain Company reports that for their ceramic with bubbling pressure of 80 psi (550 kN/m^2), which is quoted for the average pore size, the bubbling pressure for the maximum pore size is of the order of 72 to 75 psi (480 to 520 kN/m^2).

These fine porous ceramics can be formed by the manufacturer in any reasonable shape and dimensions that may be desired for piezometers.

There is no doubt that the use of fine ceramic (high air entry) tips is highly desirable for double tube hydraulic piezometers. They have the important advantage of keeping air bubbles out of the piezometer and there is no significant disadvantage. They reduce the scattering of measured pressures and permit more reliable measurements when the pore water pressure is near or just below atmospheric pressure. At the present time it is standard general practice to use high air entry ceramic tips for double tube hydraulic piezometers.

There are two principal reasons for using high air entry tips with diaphragm piezometers: (1) to allow measurement of the pressure in the water phase of the soil voids (instead of air pressure) and (2) to allow measurement of sub-atmospheric (capillary) water pressure in clayey embankments during construction (with vibrating wire piezometers, (section 10).

Sub-atmospheric pressures cannot be measured with pneumatic piezometers of the kinds in current use because measurement is made by pushing with a circulating gas on one side of the diaphragm to balance the soil pore pressure acting on the other side; when the soil pore pressure acting on the diaphragm is below atmospheric pressure no measurement is obtained.

Many engineers using vibrating wire and pneumatic piezometers have not been concerned about the difference between air and water pressures and have not believed it important to try to measure the initial capillary water pressures, and have used the ordinary coarse grained porous stones commonly supplied as routine equipment by the manufacturer. The writer believes that this can still be considered reasonable practice for routine instrumentation. When the pore water pressure is high enough to be significant, the air pressure is essentially the same magnitude.

In the future it is probable that the high air entry tips will be used more commonly for embankments of clayey soils, because there is no significant extra cost and a better fundamental understanding of the initial pore pressure development can be obtained. For this purpose the vibrating wire piezometer has considerable advantages over the pneumatic piezometer, as discussed in Section 12.

When coarse grained porous tips (with low air entry value) are used for diaphragm piezometers, there is no value in attempting to saturate the piezometers before installing them in the dam, as the water will run out of the piezometer by gravity flow and in clayey embankments will be sucked into the adjacent soil by the capillary action a short time after the piezometer is buried in the dam. For these installations the piezometers will measure the air pressure in the soil voids during the dam construction when the air pressure is significantly greater than the water pressure (Section 11).

If high air entry tips are to be used it is necessary to saturate the fine porous ceramic completely to remove all the air. While some efforts have been made to do this by sucking water in with vacuum the most satisfactory and common procedure is to submerge the tip in boiling water. There is some debate about the length of boiling time needed to drive out all the air from the small pores of the fine ceramic, and periods of two to six hours are commonly used. The finer the pores and the higher the bubbling pressure, the longer the time needed. Since there is a question about the needed time and how it can be done simply, it seems sensible to keep the tips in boiling water for an arbitrarily long time, such as 12 or 24 hours. When boiled such lengths of time current experience shows that it can be assumed the tips are saturated without making tests for confirmation.

For the last few years some piezometers manufacturers have supplied the high air entry stones already saturated at the factory, and shipped them in separate water filled containers. Even so, some engineers have boiled them in the construction laboratory, which is believed a good practice.

For pneumatic and vibrating wire piezometers it is necessary also to saturate completely the interior cavity between the fine ceramic tip and the diaphragm. The only practical way to do this is to attach the previously saturated tip to the piezometer working in a bath of de-aired water.

For some current kinds of pneumatic piezometers the porous tips are not detachable and it is not possible easily to boil the tip separately and saturate the interior cavity by reattaching the tip to the piezometer under water. In the last few years, some manufacturers have supplied a fully saturated piezometer (tip and interior cavity) shipped to the site in a liquid filled bag, sometimes including antifreeze. This practice has the disadvantage that there is no way for the engineer at the site to confirm that the porous ceramic tip and interior cavity are properly saturated. For installations using diaphragm piezometers with high air entry tips, it is preferable practice to saturate the porous tip and interior cavity at the construction site.

There is no widespread discussion or agreement in the profession as to what fineness (bubbling pressure) of high air entry ceramic tip is needed or desirable under various circumstances. At the present time the design engineer generally accepts the type of fine ceramic stone offered by the manufacturer, usually with bubbling pressure of the order of 15 to 30 psi (100 to 200 kN/m^2). In the future it is desirable that finer stones with higher bubbling pressure be used, and the following ideas are believed reasonable:

1) For double tube hydraulic piezometers, ceramic tips with the highest bubbling pressure commercially available are desirable. The higher the bubbling pressure and the thicker the fine ceramic the less the air that will get into the tips, both air bubbles and diffused air (Section 10). Tips with high bubbling pressures are more impervious but this is not an important disadvantage. Since pore water pressures in the impervious sections of embankment dams change relatively slowly, there is plenty of time for water to flow through even a high bubbling pressure tip when the pressure in the dam changes to reach a new equilibrium. Since fine ceramics with bubbling pressures of 80 to 90 psi (550 to 620 kN/m^2) are commercially available they should be currently used.

2) The saturated vibrating wire piezometer is a "no flow" piezometer in the sense that essentially no flow of water in or out of the tip is needed when the pressure in the pore water changes, and the diaphragm does not move when a pressure measurement is made. Hence, the permeability of the tip is not an important factor, and the higher the bubbling pressure (and thicker the ceramic tip) the better, in order to minimize flow of air into the piezometer cavity by diffusion (Section 10). Ceramic tips of 80 to 90 psi (550 to 620 kN/m^2) should be used, and perhaps even higher in the future.

3) For the saturated pneumatic piezometer, the process of making the measurement causes a slight deflection of the diaphragm, requiring a small flow of water through the porous tip. Because of this, the permeability of the tip could have an influence on the measurements. If the tip were very impervious the resistance to flow through the tip could cause the piezometer to read an erroneously high pressure. Also it can be imagined that the permeability of the fine ceramic tips might decrease with time. Hence, there is some reservation about the advisability of using a ceramic tip with very high bubbling pressure, unless it can be shown that the low permeability of the tip does not influence the reliability of the measurement. This depends on the volume of water displaced by the deflection of

the diaphragm during the pressure measurement and the surface area of the tip through which water can flow in and out of the cavity, details that vary among the various available pneumatic piezometers. Nevertheless the bubbling pressure of the tip should exceed the initial capillary suction in the compacted embankment in order to prevent the water in the piezometer cavity and pores of the ceramic tip from being sucked into the adjacent soil, after which the cavity would fill with air at pressure near the air pressure in the soil air voids. For these reasons it is desirable to determine the capillary tension (suction) which will exist in the embankment directly after compaction, by laboratory tests, and to provide porous tips with bubbling pressures which exceed this value, but are not much higher. This general approach was used recently for the McPhee Dam by the U.S. Water and Power Resources Service, where ceramic tips with different bubbling pressures were specified for pneumatic piezometers embedded in different portions of the dam where different initial capillary suctions were expected (W.P.R.S., 1980).

9. RELATIVE ADVANTAGES OF VIBRATING WIRE AND PNEUMATIC PIEZOMETERS

As discussed above both types of piezometers are in wide and general use by different groups of engineers. Generally European engineers are using the vibrating wire piezometers and U.S. engineers are using the pneumatic piezometers. In other areas either or both are being used with varying success, and opinions are being developed on the relative merits of the two types. The writer believes that the following would be generally agreed upon by engineers with experience with both types:

1) Installation Problems - From the standpoint of ease of installation there is no significant difference. Both types are simply installed and both can be read soon after installation while the tubes and cables are still on reels and before a permanent terminal is established.

2) Terminal Measuring Points - For both piezometer types very simple measuring points (terminal posts) are used and they can be placed at any convenient location with no advantage for one over the other.

3) Cold Weather - Neither is influenced greatly by freezing weather.

4) Time Required for Making Measurements - From standpoint of time and need for skill to obtain reliable and reproducible readings the vibrating wire piezometer is superior. For dams with relatively short lengths of measuring tubes, pneumatic piezometers of good design can be read reliably in time periods of 1 to 2 minutes. However, for dams of greater height and piezometer tube length, time required for making a reading for some types of pneumatic piezometers has been frequently in the range of 10 to 30 minutes per piezometer with the longer times required for longer tubes. For the vibrating wire piezometer the readings are made essentially instantaneously, regardless of the length of the cables. Also there is less "human factor" involved in reading the vibrating wire piezometer. The most skilled and conscientious engineer and the least trained technician

will generally obtain the same reading with the vibrating wire piezometer, whereas for the pneumatic piezometer with long tubes there are many experiences to indicate that considerable skill and care may be required to obtain reliably reproducible measurements.

This difference in the time required to make measurements and need for care and skill for pneumatic piezometers are principal reasons that have led many engineers to favor the vibrating wire piezometer, as typified by the current experience at the Guri Dam, cited in Section 4 above. There is an important difference among the various types of pneumatic piezometers commercially available with respect to the time and skill needed to make measurements. A common current practice with some of them is to make the initial gas filling of the input tube at higher rates, before reverting to the slower rate of gas feed needed to make the measurement. Using this technique it is now possible to make the readings in 3 to 5 minutes per piezometer, even for long tubing, such as 300 to 500 m..

Another aspect of this problem is that pneumatic piezometers which perform well at first sometimes get water in the tubes, probably from leakage through the tube walls, or condensation from the gas used to make the measurement. This requires that the water be removed by circulating dry gas, further increasing the time needed for measurements. There is no similar problem with vibrating wire piezometers. Different piezometers with different types of tubes have had different amounts of trouble of this type.

5) Rupture of Tubes and Cables - From the standpoint of relative resistance of the cables of vibrating wire piezometers and tubes of pneumatic piezometers to damage resulting from strains within the embankment, stretching and differential settlement between zones, there is not much difference but the electric cable is probably tougher in general and more likely to survive under similar imposed conditions.

6) Cost - The pneumatic piezometer is less costly. Both the piezometer and the electric cables of the vibrating wire piezometer are more expensive. This should not be a deciding factor, however, because the total relative cost of instrumentation is too small to justify making economies in the overall project cost by choosing instruments of minimum cost.

7) Applications with High Air Entry Ceramic Tips to Allow Measurement of Pore Water Pressure (As Opposed to Pore Air Pressure) - As discussed in more detail in Section 12, when used with fine ceramic tips to measure pore water pressure, the vibrating wire piezometer has a substantial advantage over the pneumatic piezometer because the initial saturation and airtightness of the piezometer can be tested before installation.

8) Longevity - From the information available to him the writer has not been able to form an opinion that either type has an intrinsic advantage from the standpoint of the length of useful life.

9) Lightning Damage to Vibrating Wire Piezometers - The potential for lightning damage is difficult to evaluate with complete confidence.

The writer believes it is reasonable to conclude that the technical problems are now solved. The major manufacturers are familiar with the past troubles. Some of them have included protective mechanisms in their current instruments (shielded cables, grounding and over-voltage arrestors), evidently using practices common in the telephone industry. Others provide the lightning protection only on request by users. There have been no recurrences of the problem in recent years while increasing numbers of dams have been instrumented with vibrating wire piezometers by engineers who are familiar with the lightning damage troubles in the past. While some areas are more prone to lightning storms then others it seems desirable that all installations be equipped with protection against lightning damage. This subject is discussed in detail by Dunnicliff (1981).

10) Zero Drift of Vibrating Wire Piezometers - The question of "zero drift" of the instrument with time, such as might be caused by creep in the tensioned wire, the diaphragm or supports, has been raised by some engineers as a possible important disadvantage. Most engineers using vibrating wire piezometers have not been concerned and have not given much consideration to the possibility. The writer has not been able to find any experiences where measurements of vibrating wire piezometers have indicated that significant zero drift, or a change in the calibration, has occurred with time. There are several main reasons for believing that zero drift is not a likely problem:

a) There are thousands of vibrating wire instruments in use for many years all over the world, for strain gages, tiltmeters, and other applications, as well as piezometers. From this mass of experience there are only rare isolated experiences reporting zero drift. (See, for example, O'Rourke and Cording, 1974). The great majority of experiences indicate that vibrating wire instruments of all kinds have been fundamentally stable.

b) Evaluation of zero drift in a given instrument is very simple. It is only necessary to place an instrument in operation under constant conditions in an experimental installation and make continuing measurements with time. If it is going to drift, the drift will be seen in the first few days or weeks. All the major manufacturers have an incentive to devise details which will assure that the zero position is stable.

c) Piezometers are located inside embankments in a tranquil and constant environment, not subjected to movements or shocks. Hence, piezometers are less likely to be disturbed by outside factors than vibrating wire instruments for other purposes.

d) Many long time measurements of vibrating wire piezometers placed in areas of constant known water pressure, such as in dams with essentially constant reservoir elevation and in dam foundations near the downstream toe, have given constant readings for long periods of time, such as the nearly 20 years of measurement in the foundation of the Serre-Ponçon Dam (Combelles, 1976).

e) There is no theoretical reason for zero drift in a vibrating instrument. While any given instrument could be defective with a poor connection which could slip or creep, for instruments without such defects there is no fundamental reason to suspect changes at constant condition. As long as the main elements are designed with low working stresses relative to the yield point, high strength metals do not creep.

Scott and Kilgour (1967) present an excellent summary of the history of development and experience with vibrating wire piezometers, and results of laboratory tests on the Maihak piezometer, concluding that it is a very good instrument. Measurements are presented for a vibrating wire piezometer placed near the upstream face of the impervious core of the Wildwood Dam (Canada) showing that it measured the reservoir level very closely with no zero drift over the first two years of reservoir operation.

10. MEASURING SUB-ATMOSPHERIC PORE WATER PRESSURES IN FINE-GRAINED CLAYEY EMBANKMENTS WITH VIBRATING WIRE PIEZOMETERS

The practice of using vibrating wire piezometers to measure water pressures in clayey embankments with initial sub-atmospheric pore water pressures is only a few years old. There have not been a great number of measurements as yet and few of these have been presented in the engineering literature.

Measurements in different dams of clayey soil have shown different patterns reflecting differences in behavior of different clays, but probably also differences in the details of the different piezometers used. Some of the measurements are not easily explained. This section presents some representative measurements and a discussion of probable problems.

The pressure acting on the diaphragm of a vibrating wire piezometer is determined by measuring the length of a pretensioned steel wire attached to the diaphragm. When the diaphragm is subjected to increasing pressures it deflects slightly inward, decreasing the wire length. When the diaphragm is subjected to decreasing pressure (such as by vacuum) it deflects slightly outward, increasing the wire length, and the change is measured in the same way. Hence, from the standpoint of the mechanism of the instrument the vibrating wire piezometer measures pressures below atmospheric pressure in the same way, and same reliability and precision, as it measures pressures above atmospheric pressure.

When a diaphragm piezometer with high air entry ceramic tip and interior cavity saturated with deaired water is placed in a clayey embankment having sub-atmospheric (capillary) pore water pressure, the water inside the piezometer cavity is in direct contact with the soil pore water at the exterior face of the tip. The water pressure in the cavity, (between the ceramic tip and the piezometer diaphragm) which was at atmospheric pressure when placed in the embankment, decreases rapidly, since it is sucked by the soil pore water. If the capillary water pressure in the adjacent soil is not lower than about -0.8 atmospheres (about 8 m. of water head below atmospheric pressure), no flow of water out of the piezometer cavity will occur, the cavity will remain saturated and the pressure in the cavity will rapidly become equal to the soil pore water pressure.

In the relatively common situation where the initial soil pore water pressure is below -1.0 atmospheres, the piezometer cavity could remain full only if the water in the cavity could sustain a tension (below -1.0 atmosphere). Completely deaired water is capable of sustaining high tensile stresses in carefully controlled laboratory experiments, without the water column parting. However, in a piezometer cavity it is likely that remnants of dissolved air will come out of solution forming small bubbles which will start to expand when the water pressure approaches -1.0 atmospheres. This is the general experience in measuring capillary suction in soil engineering laboratories and is also confirmed by the fact that the recorded measurements in piezometers in clay dams never are lower than 8 or 9 m. of water pressure below atmospheric pressure.

It is likely, therefore, that when the soil pore water pressure is much lower than -1.0 atmospheres, a remnant of dissolved air in the water in the cavity will come out of solution and start to expand and the water will be sucked out of the cavity of a diaphragm piezometer into the adjacent soil. Since the high air entry value ceramic tip prevents air bubbles from the adjacent soil from entering the piezometer, when the water is sucked out a partial vacuum is left behind in the cavity.

The water leaving the piezometer will increase the water content of the unsaturated soil directly adjacent by a small amount, at least temporarily, and decrease the capillary suction. Since the volume of water leaving the piezometer is insignificant compared to the volume of water in the soil mass within a short distance from the piezometer, probably equilibrium will be reached quickly with essentially no change in the soil pore water pressure immediately adjacent to the tip.

Hence, when the soil pore water pressure is well below -1.0 atmospheres all the free water in the piezometer cavity will be sucked out into the adjacent soil. The high air value ceramic tip will remain saturated, with menisci developed on the interior face, as long as the difference in pressure between the interior of the cavity and the soil pore water pressure does not exceed the bubbling pressure of the ceramic tip.

Up to this point, as described in the above paragraphs, we can be fairly confident of the action which takes place in a clay dam with high initial capillary suction. Probably within a short time (a day or two, or perhaps a few hours) after the piezometer is placed in the dam embankment, at least a temporary equilibrium condition is reached in which there is a partial vacuum in the piezometer cavity and the ceramic tip remains saturated with capillary menisci developed on the inside face to resist the suction of the pore water outside. This condition has been often measured with vibrating wire piezometers; it is relatively common that within a few hours (or the next day) after installation of the piezometer in the embankment, the measured pressure is in the range of 6 to 9 m. of water pressure below atmospheric pressure, and continuing measurements over the next period of time give exactly the same pressure.

In this initial condition of equilibrium the air pressure in the soil voids near the piezometer tip is probably near atmospheric pressure, considerably higher than the partial vacuum existing in the piezometer cavity on the other side of the ceramic tip. While the fine pores in the ceramic tip prevent air in the form of bubbles from travelling through the tip, it is possible that a significant quantity of air, dissolved in the water, could travel through the water filled pores of the ceramic tip by the general process of diffusion and enter the piezometer cavity, gradually increasing the pressure acting on the diaphragm. (Scott and Kilgour, 1967 and Bishop et al, 1964).

The rate at which air can enter the piezometer cavity by diffusion through the pores of a fine ceramic tip is not known, and is probably dependent greatly on the bubbling pressure and thickness of the ceramic tip and on the permeability and nature and content of air in the adjacent soil. For a given small volume of air which may travel through the ceramic tip by diffusion, the influence on the pressure measured by the piezometer will also be influenced by the volume of the piezometer cavity (Vaughan, 1973). If appreciable air does enter the cavity of diaphragm piezometers by diffusion, the piezometer will measure a pressure which is somewhat higher than the pore water pressure and probably somewhat lower than the pore air pressure. At the present time it is not known whether the influence of air diffusion is significant or not, or if it is significant under some circumstances whether it could be rendered unimportant by different piezometer details. This is a subject for future study. Measurements in dams show a few cases where air diffusion could not have been important, and others where it could have occurred.

Starting about 1970 a few engineers commenced installing vibrating wire piezometers, with saturated high air entry tips and cavities, in impervious sections of dams. Some were placed in homogeneous dams or central cores consisting of very fine plastic clays which should have had high initial capillary suction. Figure 6 shows typical patterns of measurements during construction in four such very fine-grained, highly impervious clay dams constructed since 1975. All were constructed with conventional methods. In each case the measurement shown is representative of the readings of several piezometers with similar measurements. These define several quite different categories of measured pressures:

(a) For Dam 1, the initial piezometer measurement was about 8 m. of water head below atmospheric pressure. During construction of the dam above the piezometer elevation the piezometers were read frequently showing that the pressure gradually decreased to about -9 m. of water head at mid-construction, and then gradually increased again to -8 m.. The period of time for construction after installation of the piezometer was about 12 months. This is an experience where it seems probable that air diffusion into the piezometer had no significant influence on the measurement. Since the partial vacuum in the piezometer cavity at the end of construction was about the same as it was when the piezometer was first placed in the embankment, the experience of Dam 1 indicates that the measurement could not have been

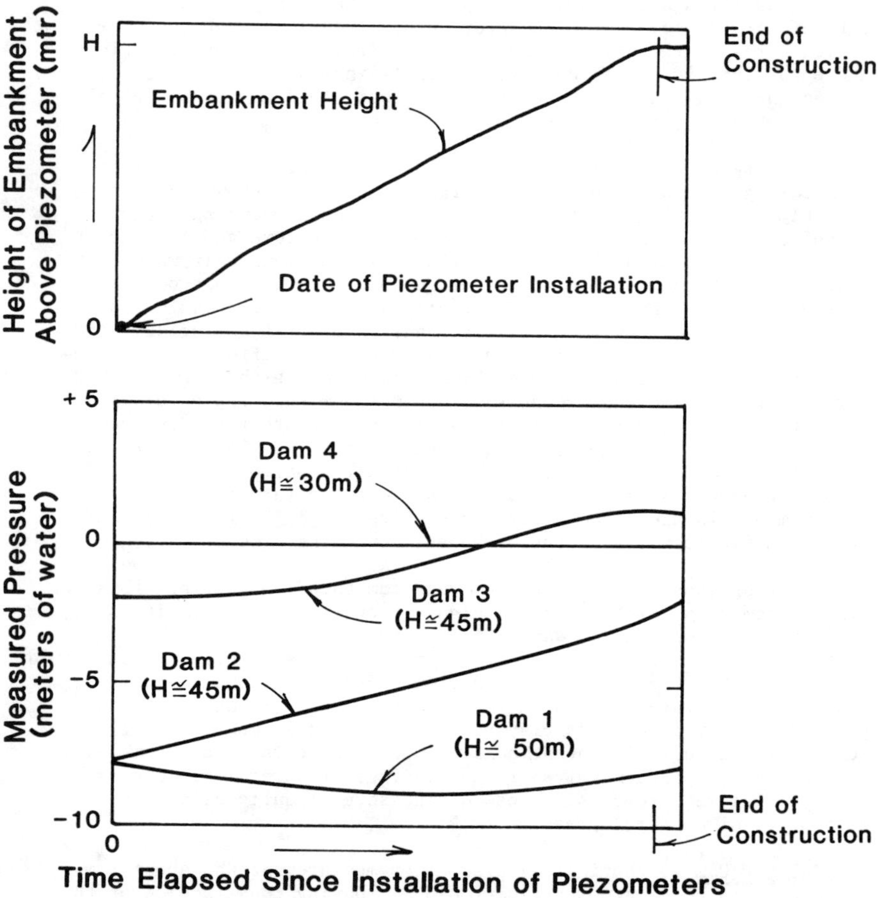

Various Patterns of Measurements of Sub-Atmospheric Pore Water Pressure During Construction in Embankments of Fine-Grained Clayey Soil with Fully Saturated Vibrating Wire Piezometers with High Bubbling Pressure Tips

Figure 6

significantly influenced by diffusion of air into the piezometers. Probably the interior cavities of the piezometers were empty during the entire construction period and the partial vacuum was maintained successfully. Later when the reservoir filled and the pore water pressure increased, water from the soil probably flowed back into the cavity.

(b) In Dam 2 the initial measurement was also about -8 m. of water head, but the pressure increased gradually during construction, reaching -2 m. at completion. The increase in pressure could be due wholly to the compression of the embankment under load, which decreased the capillarity in the adjacent soil, and the piezometers may have read the pore water pressure correctly. Possibly some of the pressure increase measured during construction was due to diffusion of air into the piezometer cavity.

(c) In Dam 3 the initial pressure measured was about -2 m. of water head. This remained more or less constant for several months and then increased gradually, reaching about +2 m. at the end of construction. The clay of this dam was very fine and of high plasticity. Probably the initial true pore water pressure was considerably lower than the value measured, and the piezometers were not functioning properly to measure sub-atmospheric pore water pressure, for some unknown reason. Conceivably there was not a satisfactory air seal between the ceramic tip and the piezometer body, or the capillarity of the soil was greater than the bubbling pressure of the ceramic tip, allowing the water to be sucked out of the pores of the tip, letting air flow back in. Another hypothesis was that there might have been small defects (cracks) in the ceramic tips, which allowed air to enter the piezometer cavity. (As discussed in Section 12, the air tightness of the connections and possible defects in the ceramic tips can be detected by simple tests before installing the piezometer in the dam). In any event it seems highly likely that air had penetrated the piezometers in Dam 3, making it act like a piezometer with coarse porous tip, and the pressure measured was the pore air pressure (Section 11).

(d) In Dam 4, Figure 6, all the piezometers gave a measurement essentially equal to zero (atmospheric pressure) during the entire construction. Clearly they were not measuring water pressure in the clay dam during the construction and it is difficult to explain how they could measure neither air or water pressure. Among the possible explanations it seemed probable that the piezometers were placed in the dam in such a way that the high air entry tips were not in close contact with the compacted embankment. The piezometers were of the type (Maihak) with a cylindrical disc porous tip on the end of the piezometer. They were installed by first forming a hole in the embankment with the approximate shape of the piezometer body and then pushing the piezometer into the hole. It is speculated that the piezometers were not pushed into the hole with sufficient force, and that a small air space was left in the bottom opposite the ceramic tip. This possibility would explain the constant zero readings during construction. The piezometers would not measure water pressure unless the stone was in good contact with the soil and they wouldn't measure air pressure because the air pressure was too low to cause any appreciable movement of the water menisci formed on the face of the saturated high air entry value ceramic tip. As in the case of Dams 1-3 there was every indication from the later performance that the piezometers were functioning properly. They reacted as expected to the reservoir filling. Especially several piezometers in the lowest elevation of the downstream slope of the homogeneous clay dam (below a tailwater level) showed tail water level when the level was stationary, and followed fluctuations very well.

There have been a number of experiences (besides that of Dam 4, above) to indicate that, when attempting to measure sub-atmospheric pressures with vibrating wire piezometers, it is very important to obtain a good contact between ceramic tip and the compacted embankment. Evidently even a small thickness of poorly compacted soil with larger air content at the face of ceramic tip can prevent good communication between the soil pore water and the water in the pores of the ceramic tip (Section 12).

At one project, Geonor vibrating wire piezometers (which have the high air entry value ceramic on the sides of the cylindrical piezometer body) were installed by two methods: (1) pushing the piezometer into a preformed hole of slightly smaller diameter, which is a common standard procedure, and (2) making the preformed hole a little larger than the diameter of the piezometer and embedding the piezometer surrounding by thin skin of a saturated paste of the clay mixed to a water content near the liquid limit, thereby obtaining a sure seal between the ceramic tip and the soft paste and a sure contact between the paste and the pore water in the adjacent compacted embankment. Piezometers installed by both methods gave similar results but those using the soft paste measured the initial capillary pressure immediately, whereas the others needed some time after installation to registar the capillary suction. This thin paste of soil was also used on the tip for the installation of Maihak vibrating wire piezometers at the Wildwood Dam (Scott and Kilgour, 1967). In the future the writer believes that it is desirable to use the saturated paste method of installation. The necessity of making a good seal also applies to the use of high air entry ceramic tips with pneumatic piezometers.

At the Wildwood Dam in Canada (Scott and Kilgour, 1967) Maihak vibrating wire piezometers were used with fine ceramic tips having a measured bubbling pressure of 28 psi (193 kN/m^2) in an embankment with measured initial capillary suction of 20 psi (138 kN/m^2). The initial piezometer measurements were never lower than -1.5 psi (-10 kN/m^2) below atmospheric pressure. It was speculated that the inability of the piezometers to measure lower pressures may have been due to entry of air into the cavity by diffusion. It could also have been due to entry of air through air leaks between the fine ceramic stone and the piezometer body.

In clay dams vibrating wire piezometers should measure the capillary suction as soon as they are installed. If the piezometers do not read substantial sub-atmospheric pressure within a day or two after placement in the embankment, consideration should be given to digging them out and finding a method of placement such that they do measure the initial capillary pressure, or find out why they are not measuring it.

In summary, from experiences such as that of Dam 1, Figure 6, there is every reason to believe that vibrating wire piezometers can measure initial sub-atmospheric pressures in compacted clay embankments correctly using adequate details and that the problem of air diffusion can be made unimportant. However, there are current problems and much to be learned about the influence of the details before it can be done reliably. Field or laboratory research could probably solve some of the problems relatively easily, embedding piezometers in compacted clays of different types under constant conditions of high capillarity, and studying the ability of the piezometer to sustain the vacuum conditions, the question of possible air diffusion through high air entry ceramic tips of various kinds, whether the water is actually sucked out of the piezometer cavity as postulated above, etc.

Probably future studies will show that using certain improved details or certain combinations of the details presently available, vibrating wire piezometers will always be able to measure reliably water pressures at least down to pressures approaching absolute zero. At the present state of knowledge the engineer who wishes to measure sub-atmospheric pressures should pay considerable attention to the details of the piezometers and installation.

Air diffusion into the piezometer cavity will cause the piezometer to read a pressure which is lower than the pore air pressure and higher than the pore water pressure. For example, in the case of Dam 2, Figure 6, it is possible that at the end of construction the true pore water pressure is still at -8 m. of water head or lower and the piezometer is measuring -2 m. of water head because some air penetrated the ceramic tip and increased the air pressure acting on the diaphragm. In this circumstance, when the pore water pressure subsequently rises, water will flow into the tip after the pore water pressure rises above the air pressure in the piezometer cavity. After this point, as the pore water pressure continues to rise, the piezometer will measure the true pore water pressure, even though the piezometer cavity continues to be partially filled with air, since the air and water inside the cavity will have the same pressure, which will be equal to the pressure in the water entering the piezometer.

For a dam in which it is planned to measure sub-atmospheric pressure in a fine, clayey embankment section, it is desirable to study in the laboratory the initial suction pressures which exist in the typical material compacted at different water contents (Figure 1). The ceramic tip must have a bubbling pressure exceeding the initial soil capillary suction pressure. If it does not, the soil suction will not only pull the water out of the piezometer cavity as described above, it will also suck the water out of the pores of the ceramic tip, allowing air from the soil to flow through the ceramic tip. When the pores are not saturated, even the finest high air entry ceramics will allow easy penetration of air. An alternative approach is to use a very fine ceramic tip, such as one with bubbling pressure of 80 psi (550 kN/m^2), which is probably higher than the initial suction pressure in even the finest clay embankment.

11. DIAPHRAGM PIEZOMETERS WITH COARSE POROUS TIPS MEASURE APPROXIMATELY THE PRESSURE OF THE AIR IN THE PORE VOIDS DURING CONSTRUCTION

As discussed in Section 7, as long as there is air in the soil voids, which is the common situation in impervious sections of embankment dams, there are capillary menisci in the soil voids providing the boundaries between the water and air in the soil pores. During construction of a compacted clayey embankment, the relationship among pore air pressure, pore water pressure and the capillary pressure is generally as shown in Figure 2, assuming that the initial air pressure equals atmospheric pressure. As seen in Figure 2 the capillary suction (u_c) continues to exist until all the air volume is compressed, and the pore air pressure is always greater than the pore water pressure by an amount equal to the capillary suction, even when the pore water pressure is above atmospheric pressure.

A diaphragm piezometer with ordinary coarse porous tip (low air entry value) installed in a clay embankment initially has air in the voids of the porous tip and in the interior cavity, at atmospheric pressure. As the embankment is constructed above the piezometers, compressing the adjacent compacted soil, the air pressure in the soil pores increases. The pressure differential between air in the soil voids and the air in the piezometer causes a flow of air toward the piezometer which will continue until the pore air pressure and piezometer pressure are equal.

The length of time needed for the air pressure in the piezometer to come to equilibrium depends on the volume of the air space in the voids of the porous tips and of the interior cavity of the piezometer, as well as the air permeability of the adjacent soil and the amount of air in the soil voids. If the air space in the piezometer is relatively small it would be expected that the piezometer would measure fairly rapidly the soil air pressure. If the air space in the piezometer is relatively large, and especially if a sand pocket is provided around the piezometer, a considerable volume of air must be squeezed from the adjacent embankment before the piezometer measurement will reach equilibrium, and the equilibrium pressure could be slightly lower than the true air pressure in the soil voids.

All efforts to measure air pressures in soil voids have the problem that it is necessary to cause the air in the pores to flow from the soil voids out into an instrument cavity where the pressure can be measured. When the air flows out of the soil into the measuring instrument it must expand, reducing the air pressure both in the instrument and in the soil just adjacent.

Measurements of air pressures in soil voids made at the ends of compacted specimens in the laboratory are made by a technique in which a very small space is provided for the air squeezed out of the specimen to accumulate (U.S.B.R., 1966 and Bishop, 1960). Since the air space in the laboratory test apparatus is small compared to the volume of air in the voids of the adjacent soil, the measurements obtained are believed to be accurate. Since the relative volume of air in the voids of an embankment dam surrounding a piezometer tip, compared to the volume of space within the piezometer, is greater than the relative conditions in the laboratory test, probably the small flow of air from the embankment toward the piezometer is not sufficient to have an important influence on the measured air pressure.

In two laboratory experiments, vibrating wire piezometers with coarse porous tips (low air entry value) were embedded in compacted clay specimens which were subjected to undrained compression with measurements of both pore air and pore water pressures using laboratory techniques as well as measurements of the vibrating wire piezometer (Bishop, Kennard and Vaughan, 1964 and Scott and Kilgour, 1967). In both experiments the pressure measured by the vibrating wire piezometer was essentially equal to the pore air pressure measured by the laboratory techniques and completely different from the measured pore water pressure. Both of these experiments used a Maihak vibrating wire piezometer, which instrument has a very small interior cavity space between the diaphragm and the inside face of the porous tip. In both of these laboratory experiments, the relationship between measured air and water pressures were very similar to that shown in Figure 4, with measured air pressures of about 2 to 3 psi (1.5 to 2 m. of water head) existing when the measured water was equal to atmospheric pressure.

As seen in Figure 5, the initial measurements in Tres Marias Dam, with the vibrating wire piezometers with coarse porous tip, showed a pressure below atmospheric pressure. Other vibrating wire piezometers in the same dam showed considerably lower initial measurements, in the range of 4 to 5 m. of water head below atmospheric pressure. Similar sub-atmospheric pressures have been recorded at other dams with vibrating wire piezometers equipped with coarse porous tips. There is no doubt that these piezometers are correctly measuring sub-atmospheric air pressure; the only question is whether they are measuring sub-atmospheric pressure because the flow of the air from the soil pores into the piezometer cavity changed the conditions in the fill, or because the pressure in the soil air voids is sub-atmospheric. There are several reasons for believing that the piezometers are measuring correctly sub-atmospheric pressure in the air voids.

One reason is that it would not be surprising if the initial pressure in the air voids of a compacted clay were slightly below atmospheric pressure. When the material is first placed and spread on the embankment surface, the air in the soil must be at atmospheric pressure, because it is in direct contact with the air. When compacted with a roller, and subsequently travelled over repetitively with wheels of heavy trucks, with tire pressure of the order of 80 psi (550 kN/m^2), the soil is temporarily loaded to high pressure, causing temporary high pressures in the air bubbles, probably driving out some of the air to the surface of the compacted layer. When the truck wheel load is removed the embankment material rebounds, causing the air bubbles to expand again. By the repetitive action it is easily possible that sufficient air could be pumped out of the soil mass such that the final average initial air pressure is below atmospheric pressure.

Another reason is that initial sub-atmospheric air pressures are also sometimes measured in laboratory tests on compacted specimens (where the same air pumping action could occur due to the repeated hammer blows used to compact the specimen). More commonly the initial air pressure measured in laboratory tests is equal to atmospheric pressure. In laboratory tests the same apparatus and techniques are used when initial measurements on different specimens give atmospheric and sub-atmospheric pressures; hence, the influence of the measuring technique should be the same. Similarly when several vibrating piezometers of the same kind were placed in different parts of the same clay dam, such as at the Três Marias Dam (Areas, 1963), the initial measurements, ranged from 1 to 5 m. of water pressure below atmospheric pressure. These piezometers read reliably to a precision of the order of 0.05 m. of water head, and were undoubtedly measuring different sub-atmospheric air pressures, which actually existed.

From the evidence currently available the writer believes that the following conclusions are valid for diaphragm piezometers with coarse porous tips (low bubbling pressure) placed in impervious dam sections:

1) If the piezometer is initially saturated the water inside the piezometer will rapidly flow out leaving the interior cavity air filled.

2) In a clay dam with high initial capillarity the piezometer cavity will remain air filled during the entire dam construction, even if the pore water pressures rise above atmospheric pressure, and the piezometer will measure a pressure approximately equal to (or slightly below) the air pressure in the soil voids, but never less than the pore water pressure.

3) In a dam section of low plasticity soil, with low initial capillary suction, the air and water pressure are about the same and the piezometer will measure this pressure.

4) In clay dam sections the initial pressure measured may often be several meters of water pressure below atmospheric pressure, and this is probably a true measurement of the pressure in the soil air voids.

On a few projects in recent years diaphragm piezometers have been installed in unsaturated embankment sections using high air entry value tips, because they were available from the manufacturer, but without saturating the ceramic or the piezometer interior cavity. These piezometers will act in the same way as they would if they had been provided with low air entry value tips.

12. VIBRATING WIRE PIEZOMETERS ARE SUPERIOR TO PNEUMATIC PIEZOMETERS FOR USE WITH HIGH AIR ENTRY TIPS TO MEASURE WATER PRESSURE (AS OPPOSED TO AIR PRESSURE)

Since the cost and effort of providing a fine ceramic tip (high air entry value) and saturating the piezometer is not great, the question arises: "Why not equip all diaphragm type piezometers used in impervious dam sections with high air entry tips?" If they could always be made to function as intended, the writer would agree that it would probably be good general practice. However, there are two troublesome technical problems: (1) the problem of assuring complete saturation and air tightness of the piezometer, and (2) the problem of assuring that the water in the pores of the fine ceramic tip is in good contact with the soil pore water.

If diaphragm piezometers with fine ceramic tips are going to measure water pressure satisfactorily in clay dams it is essential that the piezometer is saturated and that assembly is completely air tight. If there is an air passage in the connection between the ceramic tip and the piezometer body, or if the ceramic tip is not completely saturated (or has a defect, such as a crack), air bubbles will enter the piezometer soon after installation in the dam, and the piezometer will subsequently act in the same way as it would with a coarse porous tip.

For the vibrating wire piezometer the methods used to saturate the ceramic tip and the interior cavity, and the airtightness of the seal between the ceramic tip and the piezometer body, can be easily tested at the dam site before installing the piezometer in the embankment. After removing the finally assembled, saturated piezometer from the water bath it can be allowed to sit in the air with periodic measurements of pressure being made while evaporation of the water in the pores of the fine ceramic tip occurs. The evaporation causes the creation of capillary menisci in the face of the ceramic tip, rapidly reducing the water pressure in the cavity. If the evaporation causes the measured water pressure to reduce rapidly to sub-atmospheric pressures, the piezometer is fully saturated and the connections are airtight. If the measured pressures are not rapidly reduced to sub-atmospheric pressures, then either air is entering the cavity through the ceramic tip or the seals, or an air bubble in the cavity is expanding. In this test the development of the sub-atmospheric pressures by evaporation can be stopped at any time by splashing water on the face of the tip, or submerging it in water again. Tests of this kind cannot be made for pneumatic piezometers. This aspect of the vibrating wire piezometer constitutes a substantial intrinsic advantage of the vibrating wire piezometer over the pneumatic piezometer for use in situations where it is desired to use high air entry ceramic tips in an effort to measure pressure in the pore water (as opposed to air pressure).

As discussed in Section 10, the ability of the vibrating wire piezometer to measure sub-atmospheric pressure also allows verification of the adequacy contact between the ceramic tip and the embankment soil. When placed in a clayey embankment, the effectiveness of the contact can be tested immediately by measuring the initial capillary suction. This is not possible for the pneumatic piezometer.

For these two main reasons, and also because the vibrating wire piezometer can measure sub-atmospheric pore water pressures, the writer believes that vibrating wire piezometers should be used whenever an effort is being made to measure pore water pressure with fine ceramic tips. Pneumatic piezometers are best adapted for use with coarse porous tips (low air entry value).

When using diaphragm piezometers with fine ceramic tips to measure water pressure (as opposed to air pressure) it is imperative that a good contact be obtained between the water in the pores of the fine ceramic tip and the undisturbed compacted embankment material. Experience, such as the example of Dam 4, Section 10, shows clearly that even a small thickness of poorly compacted embankment material adjacent to the tip can prevent the water in the tip from communicating with the soil pore water, in which case the piezometer may measure neither the water or air pressure until the water pressure becomes fairly high after the reservoir fills.

For cylindrical piezometers with hollow cylindrical fine ceramic tips comprising a portion of the side wall, good practice is to make a hole in the embankment, with a mandrel or a drill, with diameter slightly smaller than the piezometer body and force the piezometer into the hold. The technique of applying a thin coat of saturated soil paste to the face of the fine ceramic, mixed to a consistency near the liquid limit, is probably desirable to aid in assuring continuity between the water in the pores of the fine ceramic and the pore water of the undistrubed compacted embankment. For piezometers with the fine ceramic tip located on the end (instead of the side) of the piezometer, special care is needed to make a good contact. Piezometers with the ceramic tip set back in a recess from the exterior surface of the piezometer are not desirable for this purpose.

For diaphragm piezometers with fine ceramic tips the good contact between the tip and undisturbed compacted soil is much more important than for double tube hydraulic piezometers. For double tube hydraulic piezometers some water from the piezometer tip flows out into the embankment during the initial deairing of the lines, providing a source of water to establish contact with the soil pore water, even if a perfect contact is not made between the tip and the soil. With the diaphragm piezometer no water will flow out of the tip if a satisfactory contact is not obtained and the saturated ceramic tip can be isolated from the soil pore water by a thin air boundary.

13. SAND POCKET AROUND PIEZOMETER TIP NOT DESIRABLE IN IMPERVIOUS EMBANKMENT SECTIONS

All three of the basic type of piezometers being discussed are essentially completely "no-flow" piezometers. Changes in water pressures in the embankment adjacent to the tip are immediately transmitted to the sensing element with no flow of water required in the adjacent embankment, except for the negligible flow which may be necessary to compress air bubbles collected in the tip area. Hence, for these piezometers it is not necessary to provide a sand pocket to increase the sensing area (decrease the response time). Also even coarse porous stone tips are fine enough to prevent soil particles from entering the piezometer tip.

In current U.S. practice pneumatic piezometers are often placed in a pocket of sand, or in a burlap bag filled with sand. This is sometimes recommended by piezometer manufacturers and is occasionally done also for vibrating wire piezometers. If the sand pocket is originally saturated by pouring water into it when the piezometer is placed, the water will run out of it in a short time leaving the sand pocket with air-filled voids. When used at the tip of a diaphragm piezometer equipped with coarse porous tip, the sand pocket will have little or no influence on the pressure measurement in embankments of cohesionless soils, but it has no value and is not desirable. In clayey embankments where the air pressure is higher than the pore water pressure, and the piezometer is measuring air pressure, the pocket of sand may cause the piezometer to read a slightly lower pressure. This is true because the air bubbles squeezed out of the adjacent soil expand in the voids of the air pocket (Section 11).

When a diaphragm piezometer is used with a fine ceramic tip (high air entry value) for the purpose of measuring pore water pressure (as opposed to pore air pressure), a sand pocket should never be used, as it is necessary for the ceramic tip to be in direct contact with the compacted embankment (Section 12).

In summary, these sand pockets should never be used for any of the three main types of piezometers discussed in this paper when used in the impervious sections of embankment dams.

14. SUMMARY OF MAIN POINTS

1. Double tube hydraulic piezometers are good instruments, but they require considerably more skill and care in their maintenance than the diaphragm piezometers (pneumatic and vibrating wire).

2. In current practice there is a trend away from the use of double tube hydraulic piezometers toward the newer pneumatic and vibrating wire piezometers, which are easier to use.

3. Vibrating wire and the better types of pneumatic piezometers are both satisfactory instruments, but the vibrating wire piezometer has substantial advantages and will probably be used more widely in the future.

4. There have been basic improvements in details of pneumatic piezometers in the last few years, but there are still important differences in the details and performance of different models. Some models have not given uniformly good service.

5. When used with coarse porous tips, both pneumatic and vibrating wire piezometers measure the pore air pressure (as opposed to the pore water pressure) when there is a significant difference.

6. In impervious embankments of most soils with cohesionless fines the difference between pore air and pore water pressure is usually negligible.

7. In impervious embankments with clayey fines, which have appreciable initial capillary suction, the air pressure is always greater than the water pressure. The difference is generally not great when the water pressure is above atmospheric pressure. There are very few situations where the difference between measured air pressure and true water pressure could have a significant influence on the evaluation of the behavior of a dam.

8. When vibrating wire or pneumatic piezometers are used for routine instrumentation in the impervious sections of embankment dams, it is considered reasonable current practice to use coarse porous tips (low air entry value).

9. There is an increasing trend toward the use of diaphragm piezometers with fine ceramic tips (high bubbling pressure), with saturated tips and interior cavity, to measure pore water pressure (as opposed to air pressure). For this application the vibrating wire piezometer has important advantages over the pneumatic piezometer because the initial saturation and air tightness of the piezometer can be tested before installation.

10. For diaphragm piezometers used with fine ceramic tips to measure pore water pressure, the tip and piezometer should be saturated at the site. For this application it is essential that a good contact be obtained between the tip and the undisturbed compacted embankment material. Even a small thickness of loose, poorly compacted soil at the face of the fine ceramic tip can prevent the piezometer from functioning. This good contact is much more important than it is for the tips of double tube hydraulic piezometers. For cylindrical piezometers with the fine ceramic tip on the side walls, good practice is to form a cylindrical hole in the embankment with diameter slightly smaller than the piezometer body, and force the piezometer into the hole. Smearing the ceramic tip with a thin coat of the saturated soil paste, mixed to a consistency near the liquid limit, is probably desirable to aid in assuring continuity between the water in the pores of the fine ceramic tip and the pore water of the undisturbed embankment material.

11. For double tube hydraulic and vibrating wire piezometers fine ceramic tips with bubbling pressures higher than used commonly at present should be used in the future.

12. There are experiences with vibrating wire piezometers with saturated fine ceramic tips (high bubbling pressure) which show they can measure sub-atmospheric pore water pressure in clay dams for many months using appropriate details, but there have been unsuccessful attempts and practice must be considered somewhat experimental at the present time. Probably it will be found that with good seals between the fine ceramic tip and the piezometer body, and with thick ceramic tips with bubbling pressure of 80 psi (550 kN/m^2) or higher, vibrating wire piezometers will consistently measure sub-atmospheric pressures with no significant air entering the piezometer by diffusion.

13. Sand pockets around the tips should never be used for any of the three main types of remote reading piezometers placed in impervious sections of embankment dams.

14. In some clay dams the initial measured pore air pressure, soon after compaction has been 1 to 7 psi (7 to 40 kN/m^2) below atmospheric pressure.

15 ACKNOWLEDGEMENTS

This paper could not have been completed without the liberal assistance of many colleagues. The greatest debt is owed Mr. John Dunnicliff, Geotechnical Instrumentation Consultant, Lexington, Massachusetts, who is also contributing the lead paper to this symposium. Mr. Dunnicliff gave so much of his time and knowledge to assist the writer's effort that he should be more appropriately listed as an author rather than a main contributor. The writer also gratefully acknowledges the assistance provided by the following colleagues: Kilian De Fries (Consulting Engineer, Caracas); Hamilton Oliveira (HIDROSERVICE, Sao Paulo); Joao Silveira (PROMON, Sao Paulo); George Post and Pierre Londe (Coyne and Bellier, Paris); Roberto Celeri and George Melios (CESP, Sao Paulo); Oscar Dscal (Hydro-Quebec, Montreal); Serge Hsu (THEMAG, Sao Paulo); Orestis Papageorgiou and George Pavlakis (Public Power Corporation, Athens); Carroll Coffey, Duncan Tharp and Luther Davidson (Water and Power Resources Service, Denver); Fabio Villegas (INTEGRAL, Medellin, Colombia); Alinor Figeirdo (THEMAG, Tucurui, Brazil); David Dayton (East Bay Municipal Utility District, Oakland, California); George Kruse (California Dept. of Water Resource, Sacramento); Hyun-Ha Lee (BC Hydro, Vancouver); David Kleiner (Harza Engineering Company, Chicago). Professor Iraj Noorany of San Diego State University reviewed a late draft of the manuscript and provided many helpful suggestions.

REFERENCES

Anton, W. F. and Dayton, D. J. (1972), "Performance of Briones Dam", Specialty Conference on Performance of Earth and Earth Supported Structures, ASCE, 1972, Vol. 1, pp. 833-866.

Areas, O. M. (1963), "Piezometros em Três Marias" 2nd Pan American Conference on Soil Mechanics and Foundation Engineering, Brazil, 1963, Vol II, pp. 413-440.

Bishop, A. W. (1969), "Pore Pressure Measurements in the Field and in the Laboratory", 7th International Conference on Soil Mechanics and Foundation Engineering, Mexico, Vol. 3, pp. 427-444.

Bishop, A. W., Kennard, M. F. and Penman, A. D. M. (1960), "Pore Pressure Observations at Selset Dam" Pore Pressure and Suction in Soils, Butterworths (London) 1960, pp. 91-102.

Bishop, A. W., Kennard, M. F. and Vaughan, P. R. (1964), "Developments in the Measurement of Pore Pressure in Earth Dams" 8th International Congress on Large Dams, Edinburgh, Question 29, Report 4, pp. 47-70.

Combelles, P. (1976), "Measurements of Piezometric Pressures and Leakage by E.D.F.", 12th International Congress on Large Dams, Mexico Vol. V, pp. 289-295.

Cour, F. and Vincent, A. (1973), "Determination de la Perméabilité du Noyan d'un Barrage a' partir des Mesures In Situ des Pressions Interstitielles" Geotechnique (London), September 1973, pp. 383-405.

Douillet, G. and Ledeuil, E., (1979), "L'Auscultation du Barrage de Serre-Ponçon Depuis Sa Naissance et Sur 20 Ans D'Exploitation", Revue Française de Geotechnique, No. 10, November 1979, pp. 21-38.

Cummins, P. J. (1979), "Instrumentation of Dartmouth Dam", Bulletin, Australian National Committee on Large Dams, April 1979, pp. 20-28.

Dunnicliff, J. (1980), Notes for U. S. Department of Transportation, Federal Highway Administration Training Course on Geotechnical Instrumentation.

Dunnicliff, J. (1981), "Long Term Performance of Embankment Dam Instrumentation", Paper submitted to the Symposium on "Instrumentation Reliability and Long-Term Performance Monitoring of Embankment Dams" ASCE, New York, May 1981.

Gibbs, H. J., Hilf, J. W., Holtz, W. G. and Walker, F. C. (1960), "Shear Strength of Cohesive Soils" Research Conference on Shear Strength of Cohesive Soils, ASCE, Boulder, Colorado, June 1960, pp. 33-162.

Gibbs, H. G. and Coffey, C. T., (1969), "Techniques of Pore Pressure Measurement and Shear Testing of Soil", 7th International Conference on Soil Mechanics and Foundation Engineering, Mexico, Vol. I, pp. 151-157.

Hilf, J. W. (1956), "An Investigation of the Pore Water Pressure in Compacted Cohesive Soils", Technical Memorandum No. 654, U. S. Bureau of Reclamation, Denver, Colorado.

Nonveiller, E. (1957), "Pore Pressure in the Lokvarta Dam" 4th International Conference on Soil Mechanics and Foundation Engineering, London, Vol. II, pp. 341-347.

Nonveiller, E. (1969), "Discussion on Long Term Behavior of Vibrating Wire Piezometers" 7th International Conference on Soil Mechanics and Foundation Engineering, (Mexico), Vol. 3, pp. 431-2.

O'Rourke, T. D. and Cording, E. J. (1974), "Measurement of Strut Loads By Means of Vibrating Wire Strain Gages" ASTM Special Technical Publication No. 584 (Performance Monitoring for Geotechnical Construction), pp. 58-77.

Penman, A. D. M., (1979), "Construction Pore Pressures in Two Earth Dams", Conference on Clay Fills, The Institution of Civil Engineers, London, 1979, pp. 177-188.

Penman, A. D. M. and Kennard, M. F. (1981), "Long-Term Monitoring of Embankment Dams in Brittain", Paper submitted to "Symposium on Instrumentation Reliability and Long-Term Performance Monitoring of Embankment Dams" ASCE Annual Convention, May 1981, New York.

Post, G. and Picaut, J. (1969), Discussion at Specialty Session on Pore Pressure Measurements in Field and Laboratory, 7th International Conference on Soil Mechanics and Foundation Engineering, Mexico, Vol. 3, p. 431.

Trieber, F. (1958), "Compaction Methods for the Construction of Rosshaupten Dam, Their Effectiveness and the Behavior of the Core", 6th International Congress on Large Dams, New York, Vol. 3, pp. 123-137.

U.S.B.R. (1966), "Measurements of Negative Pore Pressures of Unsaturated Soil Shear and Pore Pressure Research", Report EM-738, U. S. Bureau of Reclamation, Denver, Colorado, October 11, 1966.

Vaughan, P. R. (1973), "The Measurements of Pore Pressures with Piezometers", Symposium on Field Instrumentation in Geotechnical Engineering" Butterworths, London, pp. 411-422.

W.P.R.S. (1974), "Earth Manual" Second Edition, Water and Power Resources Service, U.S. Government Printing Office. (Formerly U.S. Bureau of Reclamation).

W.P.R.S. (1980), "Specifications, McPhee Dam, Dolores Project, Colorado", Specification No. DC-7420, U.S. Water and Power Resources Service, Denver, Colorado.

Introduction to Paper 13

Basic Properties of Sand and Gravel Filters

by James L. Sherard, Lorn P. Dunnigan, and James R. Talbot
ASCE Journal of Geotechnical Engineering, Vol. 110, No. 6, June 1984

Prepared by James R. Talbot

This paper was the result of the first phase of a three phase study of sand and gravel filters for earth dams started in late 1979 by the Soil Conservation Service (SCS). Jim Sherard came to work with SCS on this study because of his intense interest in filters and their ability to prevent concentrated leaks in embankment dams. Such leaks often develop from cracks associated with stress release, hydraulic fracturing, and related phenomena.

It was invigorating and professionally rewarding to work with Jim on these filter studies. Jim was a very deep thinker and had an inquiring mind. He would take many notes and then study them over after work hours. We would talk over findings at dinner and meet every morning to review the work of the previous day. This led to decisions regarding changes in the testing procedure or new tests to be made. The laboratory personnel had to be very flexible. Jim wanted material and tests prepared as quickly as possible and was often like a child trying to open a Christmas package; he was anxious and excited to get the results. The laboratory staff called him "The White Tornado" because he was like a whirlwind going through the laboratory checking on the simultaneous operations of preparing equipment, material and specimens and performing tests.

Jim was a prolific writer. After leaving the laboratory from one of our intensive 15 to 20 day testing periods, he would always write a hundred page or more "Memorandum" on the plane going home or within the next few days to document what had happened and his interpretations or thinking. Sometimes, these would come to us in a handwritten form from a foreign country as he traveled on to these assignments after our study session. He was truly remarkable in this regard. Our file of his notes and memoranda is a valuable resource both for today and for the future. This paper and the others on related subjects are truly classic and of great value to the profession. It was an honor to know Jim, to have the privilege of working with him and to be a part of this work and these papers.

Basic Properties of Sand and Gravel Filters

By James L. Sherard,[1] F. ASCE, Lorn P. Dunnigan,[2] and James R. Talbot,[3] Members, ASCE

Abstract: Laboratory experiments have shown that uniform filters will catch particles with diameter of about $0.11 D_{15}$ when the particles are carried in suspension in seeping water. Smaller particles will be carried through the filter pores and larger particles will not enter. The main current filter criterion, $D_{15}/d_{85} \leq 5$, is shown to be conservative, but not excessively. Its use should be continued. Filter criteria using the ratios D_{50}/d_{50} and D_{15}/d_{15} are not supported by experiments or theory. Filters of angular particles of crushed rock are as satisfactory as those of rounded alluvial particles. It is not necessary for the filter particle size distribution to have a shape generally similar to that of the base particle size distribution.

Introduction

An investigation of filters was carried out during 1981–82 at the Soil Mechanics Laboratory, Midwest National Technical Center, Soil Conservation Service, U.S. Department of Agriculture, Lincoln, Nebraska. The investigation was directed at improving understanding of the fundamental properties and behavior of filters.

This paper mainly relates to experiments with relatively uniformly graded filters having D_{15} of about 1.0–10 mm. A companion paper (6) gives results of a study of filters for fine-grained silts and clays and filters with smaller D_{15}.

Filter Tests with Uniform Sand Base

Fig. 1 shows details and dimensions of the test apparatus. Very uniform base (protected) soils were used (Fig. 2); these were obtained by screening between the nearest two sizes of commercial sieves. Filters were mostly uniform sands and gravels consisting of subrounded to subangular particles of alluvial origin (Fig. 3).

The filters were well compacted while in a saturated condition on a vibrating table with a 10 kg surcharge weight. Filter densities were not measured but probably ranged from 80–100% relative density. The uniform base sand was placed dry in three lightly tamped layers. The "side material" (Fig. 1) was placed between the filter and the cylinder wall to eliminate large voids (pores) at the interface that were larger than the pore channels inside the filter. General test procedures were as follows:

1. A valve to the city water system (about 4 kg/cm^2 pressure) was

---

[1]Consulting Engr., San Diego, Calif.

[2]Head, Soil Mechanics Lab., National Technical Center, Soil Conservation Service, U.S.D.A., Lincoln, Neb.

[3]National Soil Engr., Soil Conservation Service, U.S.D.A., Washington, D.C.

Note.—Discussion open until November 1, 1984. Separate discussions should be submitted for the individual papers in this symposium. To extend the closing date one month, a written request must be filed with the ASCE Manager of Technical and Professional Publications. The manuscript for this paper was submitted for review and possible publication on May 4, 1983. This paper is part of the *Journal of Geotechnical Engineering*, Vol. 110, No. 6, June, 1984. ©ASCE, ISSN 0733-9410/84/0006-0684/$01.00. Paper No. 18933.

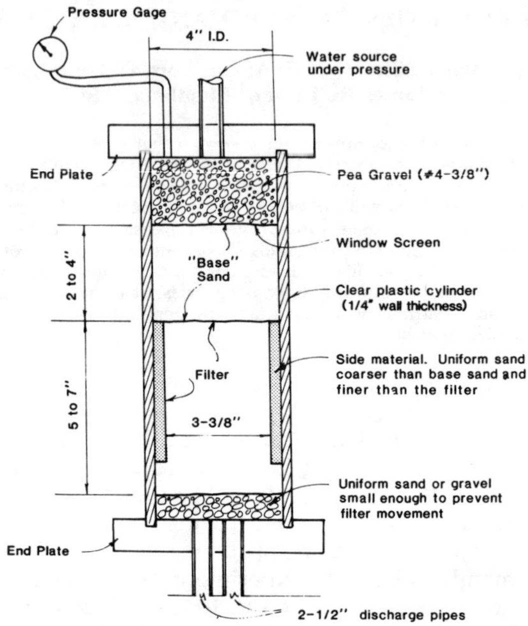

FIG. 1.—Filter Test Apparatus Details (Schematic. No Scale. Note: 1.0 in. = 25.4 mm)

opened rapidly, feeding water to the top of the apparatus.

2. After the valve was opened, water penetrated downward emerging through the pipes at the bottom a second or two later (not measured). The water and any base sand passing the filter was caught in buckets.

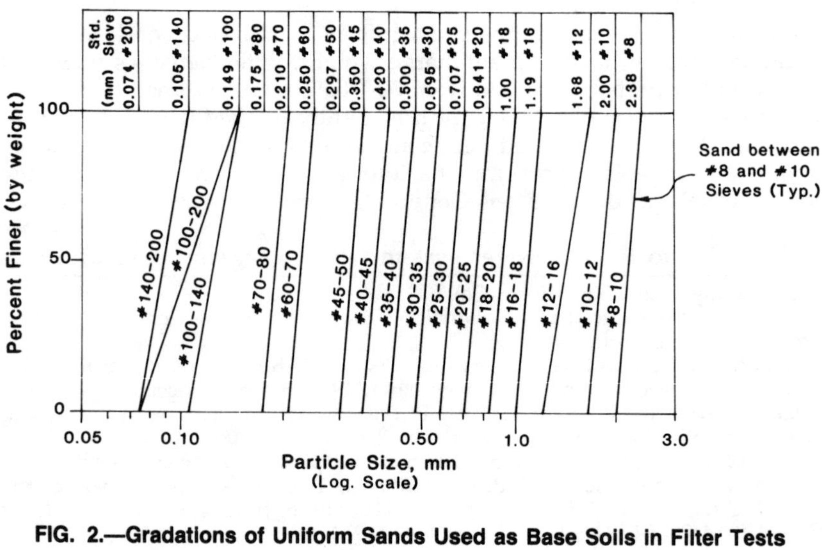

FIG. 2.—Gradations of Uniform Sands Used as Base Soils in Filter Tests

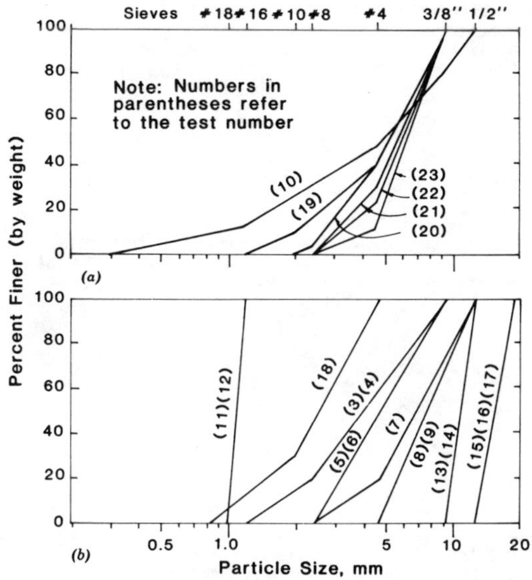

FIG. 3(*a*) and 3(*b*).—Gradations of Filters Used in Test Program

3. The amount of base sand in the first bucket was measured generally after 30–60 sec of flow. This ranged from zero (or a few grains) to more than 100 g. Generally, tests were continued with maximum pressure for 5–10 min. Experience showed that if little or no base sand was collected in the first 60 sec, little or no further base was collected with long-time flow. But if any significant quantity of base was collected in the first 60 sec, it continued to penetrate the filter at a more or less constant rate; such a result was considered a failure.

4. For those tests with little or no sand passing through the filter the entire apparatus, with water still flowing through it, was moved to a shaking table and held firmly by hand. The shaking table was then turned on at maximum amplitude and energy for 60 sec. This produced relatively strong shaking. The quantity of base material passing through the filter during the shaking period ranged from zero to 100 g or more.

5. Test results were judged as follows:

(a) *Successful.*—No significant quantity of base material got through the filter during either the water flow or vibration periods. The thickness and appearance of the base material were unchanged during the test.

(b) *Failure.*—A significant quantity of base material passed through the filter in the first 60 sec of flow, and continued at about the same rate. If the test was run for a long time, nearly all base material passed through the filter.

(c) *Borderline.*—No significant quantity of base material passed through the filter under flow of water alone, but a large quantity of base passed through during the vibration.

Tests were usually stopped as soon as failure was evident. The test

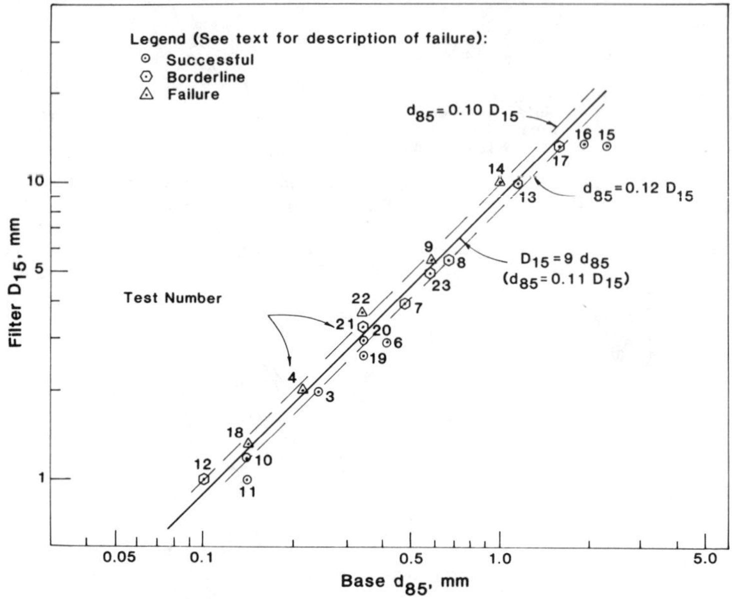

FIG. 4.—Relationship between D_{15} and d_{85} in SCS Filter Tests

results (Fig. 4) show that there was a very narrow boundary between filter failure and success, well defined by $D_{15}/d_{85} = 9$, or $d_{85} = 0.11D_{15}$. Uniform sand particles with d_{85} smaller than $0.10D_{15}$ always passed through the filter voids, and bases with d_{85} larger than $0.12D_{15}$ were always retained by the filter.

As seen in Fig. 4, the influence of heavy vibration of the test results was not very great. Therefore, it does not seem necessary to use more conservative criteria for filters that may be subjected to vibrations.

Filter Pore Sizes Observed Visually Using Molten Wax

Several attempts were made to study the nature of flow channels (continuous pores) through gravels by filling the voids with molten wax. Only a relatively small effort was devoted to this activity, but the results were instructive. After some experimentation, the following procedure was used:

1. A gravel specimen graded between 3/8 and 1.0 in. (9.5–25.4 mm), D_{15} = 11 mm, was placed in a cylindrical cardboard container, 15 cm in diam and 13 cm long, and vibrated to a dense state. No side material was used.
2. Hot molten wax was poured into the voids, filling them completely, after which the specimen was cooled in a freezer.
3. After the wax was cold, it had the consistency of hard, tough rubber. The cardboard container was then removed and the sample was cut into several pieces with a chisel. On several faces it was possible to pry

out gravel particles, leaving a network of wax with the exact configuration of the pore channels.

From study of this wax "skeleton," the following observations were made:

1. On a seepage path about 25–50 mm long (about 2–5 times the D_{15} size), all the flow channels had about the same maximum and minimum dimensions (measured normal to direction of flow). There were not some large flow channels and some small channels. All the flow channels were highly irregular in cross section, and the dimensions were repeated many times over a length of seepage path of about $10D_{15}$.
2. A rough attempt was made to measure the minimum flow channel dimension which would be expected to catch soil particles carried in suspension in the water seeping through the voids. This was generally of the order of 1–2 mm, or 1/11 to 2/11 = $0.09D_{15}$ to $0.18D_{15}$. While this was not an accurate measurement it was generally consistent with the results of the experiments summarized in Fig. 4.
3. The median "equivalent average" diameter of the flow channels was difficult to measure but was about 3–4 mm (or $0.25D_{15}$–$0.35D_{15}$), and there were no places along the seepage path where the maximum dimension was much more than 6 mm. Thus, the range of dimensions of the cross section of the continuous flow channels was approximately between $0.1D_{15}$ and $0.6D_{15}$.

At the cylinder wall interface the average pore channel was much larger than the typical interior pore channel.

Relation of Results to Earlier Research

Lund, 1949.—After completing the experiments summarized in Fig. 4, we studied a 1949 unpublished thesis describing similar tests (4) made using uniform sand and gravel filters with D_{15} in the range from about 1.0–15 mm. The main test details were as follows: (1) Downward flowing tests using a constant head tank; (2) cylindrical clear plastic permeameters, 3 and 8 in. (76 and 203 mm) in diam; (3) tailwater maintained above the base-filter interface, with the filter always saturated; (4) most tests performed with very uniform sand bases similar to those we used, but some with more broadly graded sand bases; and (5) permeability of filter was determined by measuring pressure gradient (with piezometers). The test results were grouped into three categories:

1. *Stable.*—No visible movement of the base into the filter and no change in filter permeability, even when apparatus was "gently tapped."
2. *Unstable.*—Significant drop in filter permeability when the apparatus was gently tapped, usually accompanied by visible penetration of base into filter.
3. *Completely Unstable.*—The whole base, or greater part of it, washed into the filter.

As seen in Fig. 5, the results were essentially identical to ours. All the

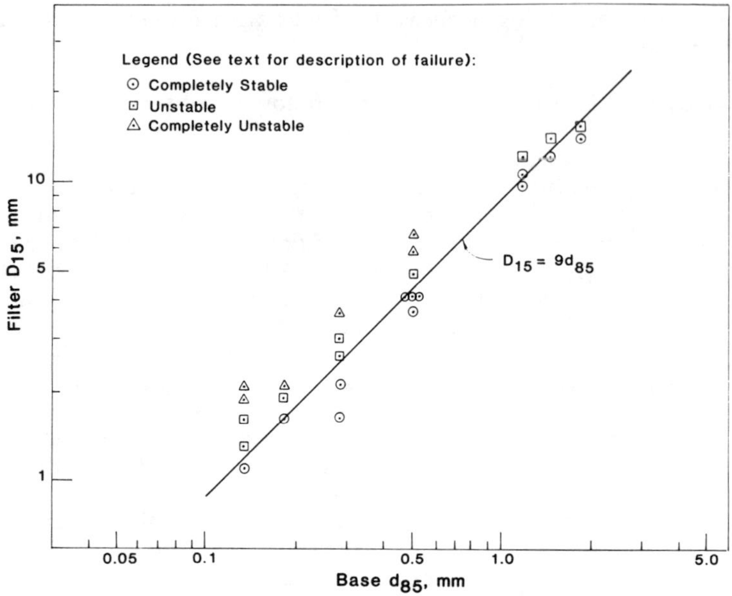

FIG. 5.—Relationship between D_{15} and d_{85} in Tests by Lund (4)

"stable" tests plot on or below the failure-success boundary, $d_{85} = 0.11 D_{15}$, and all the "unstable" tests plot above this boundary except for the single test with $d_{85} = 1.9$ mm, where the boundary is about $d_{85} = 0.12 D_{15}$.

Bertram, 1940.—In the early classic experiments of Bertram (1), similar tests were made using very uniform fine sand for the base, with d_{85} from 0.10 to 0.16 mm, and uniform filters. The failure-success boundary found in these early tests generally ranged $d_{85} = 0.10 D_{15}$ to $0.11 D_{15}$, consistent with our results.

U.S. Bureau of Reclamation, 1955.—In another filter research program carried out in the USBR central soil mechanics laboratory in Denver, Colo., some appreciably different results were obtained (9). The following main details were used: (1) Downward flowing tests using water fed from a constant head tank, with head raised in increments of 5 ft (1.6 m); (2) cylindrical clear plastic permeameters 8 in. (20.3 cm) in diam; (3) filters 8 in. (20.3 cm) long and base specimens 3 to 8 in. (7.6 to 20.3 cm) long; (4) no "side material" to control large voids at the filter-cylinder interface; and (5) total seepage through the test, as limited by permeability of the base specimen measured at each increment of head.

In the main research program, six base soils were tested using a number of filters for each (Series A, B, C, A_1, B_1 and C_1). For each base soil, at least one of the filters used was judged a failure, and for the 6 base soils a total of 7 filter-base combinations were judged to have been failures. Table 1 summarizes the results.

Test Series A tested a sandy silt base with $d_{85} = 0.12$ mm with a series of very uniform, medium to coarse sand filters. As seen in Table 1, filter 3 was judged successful and filters 4 and 5 were considered failures. The test with filter 4 had a D_{15}/d_{85} ratio of $0.66/0.12 = 5.5$, considerably

TABLE 1.—Summary of USBR Filter Tests (9)

Test series (1)	Filter (2)	D_{15}, in millimeters (3)	D_{15}/d_{85} (4)	C_u (5)	Result (6)
A (base soil was sandy silt with d_{85} = 0.12 mm, C_u = 7)	5	1.30	10.8	1.4	Failure
	4	0.66	5.5	1.4	Failure
	3	0.43	3.6	1.2	Success
B (base soil was very uniform fine sand, d_{85} = 0.135 mm, C_u = 1.58)	3	1.30	9.3	1.4	Failure
	2	0.64	4.6	1.4	Success
C (base soil was uniform, medium sand, d_{85} = 0.54 mm, C_u = 1.41)	3	5.3	9.8	1.4	Failure
	2	5.0	9.3	1.2	Success
	1	1.3	2.4	1.4	Success
A_1 (base soil was sandy silt, d_{85} = 0.135 mm, C_u = 7)	1	0.84	6.2	5.6	Failure
	2	0.37	2.7	12.2	Success
B_1 (same base as Series A_1)	1	0.66	4.9	18.5	Failure
	4	0.36	2.7	18.3	Success
C_1 (base soil was silty sand, d_{85} = 1.6 mm, C_u = 25)	1	1.6	1.0	30	Success
	2	6.6	4.1	6.0	Failure

different from our results, shown in Fig. 4. In addition to the tests summarized in Fig. 4, we ran many other tests with base soils similar to the base of the USBR Test Series A, all showing it was impossible for the sandy silt of Test Series A to have failed with filter 4 (6).

In the USBR program a test was judged "visible failure" when some movement of the base soil into the filter could be seen. This visible failure was observed only after the water pressure reached 15 ft (4.5 m) of head or higher (not seen at lower pressures). In the seven tests in which "visible failure" occurred (Table 1), the test was continued by increasing the water pressure in increments up to the maximum pressure, 30 ft (9.4 m). Fig. 6 gives the measured seepage quantities through the specimens of the seven failed tests, showing that the seepage quantity after "visible failure" continued to be about proportional to the increase in pressure. These measurements indicate conclusively that the base specimens remained intact, and there was no failure. If the base-filter combinations had failed and the pressure was then increased, holes would have eroded through the base specimens, and the seepage quantities would have increased greatly.

We made two laboratory filter tests to check the USBR results, using a silty sand base soil graded exactly the same as that of Test Series C_1 (Table 1). These were made in a 10-in. (212-mm) diam permeameter, with 4.0 kg/cm^2 water pressure and a "side material" as shown in Fig. 1, using filters with D_{15} = 9.5 and 12.4 mm giving D_{15}/d_{85} ratios of 5.7 and 7.6, respectively. In both tests the filter was successful, confirming our belief that the USBR test with filter 2 (Table 1), with D_{15}/d_{85} = 4 did not fail.

Based on the preceding observations and considerable recent experience with behavior of laboratory filter tests performed with a wide range of procedures, we have no doubt that all 7 of the USBR "failed" tests

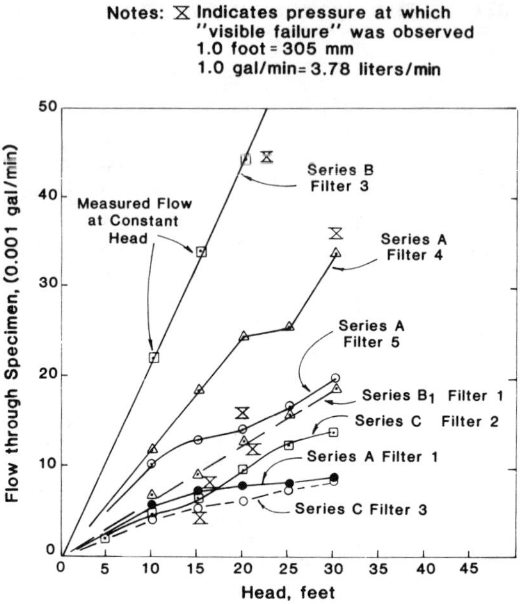

FIG. 6.—Flow Measurements Made During USBR Filter Tests and Judged "Visible Failure" (Table 1)

were actually successful. The "visible failure" observed was a small movement of base fines into the filter needed to allow base particles to enter and become trapped in the filter voids. When the pressure in such a test is raised high enough, such as 15 ft (4.5 m) of water head, the base sample deforms slightly. The deformation allows a small concentrated leak to develop, usually along the base-cylinder wall interface. This small leak erodes a small quantity of the base and carries it to the filter face where the eroded material is caught and seals the leak (if the filter is successful). This sealing action against the clear cylinder wall is what was thought to be a visible failure. We have seen such leaks suddenly develop in many filter tests in which the pressure is raised in increments (6). This action is more common in tests without side material such as that shown in Fig. 1.

Besides this observation of "visible failure," the USBR investigators used small measured differences in the permeability of the base specimens as another criterion for judging filter desirability. The permeability differences were those measured in successful tests with filters of different coarseness. This procedure is not considered reasonable because the measured permeability differences are small and not related to the basic filter action.

In summary, we believe that all the USBR tests (Table 1) should have been considered as successful, though some were near the failure boundary. The interpretation of the USBR research program results led to the recommendation to abandon the D_{15} size, as the main measure of the properties of the filter, in favor of D_{50} (9). As explained later in this paper, the D_{50} size is not a good measure of the filter behavior. This recommendation to abandon the D_{15} size was not followed in subsequent USBR published filter criteria recommendations for dams (8). Nevertheless, the USBR study (9) has been influential. The recommended use of the D_{50} size has been perpetuated in other general summaries of filter experience (10) and reference books (7).

Filter Permeability

In an earlier study, permeability tests were made on 15 different sands and gravels of the type commonly used for filters and drains in dams (11), with D_{15} in the range from about 0.1–10 mm (Fig. 7). For each material, tests were made on three different specimens, each compacted to about 70% relative density using identical procedures as far as possible. The purpose was to study the range of test results obtained for the same material. Tests were also made on three different samples of each material placed in the loosest possible state (zero relative density). Therefore, six specimens were tested with each of the 15 sands and gravels, for a total of 90 tests, a relatively large testing effort. All tests were constant head tests, with measurements made over a range of heads for each specimen. From the results summarized in Table 2, the following conclusions can be drawn:

1. When three independent tests are made in the same laboratory using the same procedures, the range in measured k is generally about

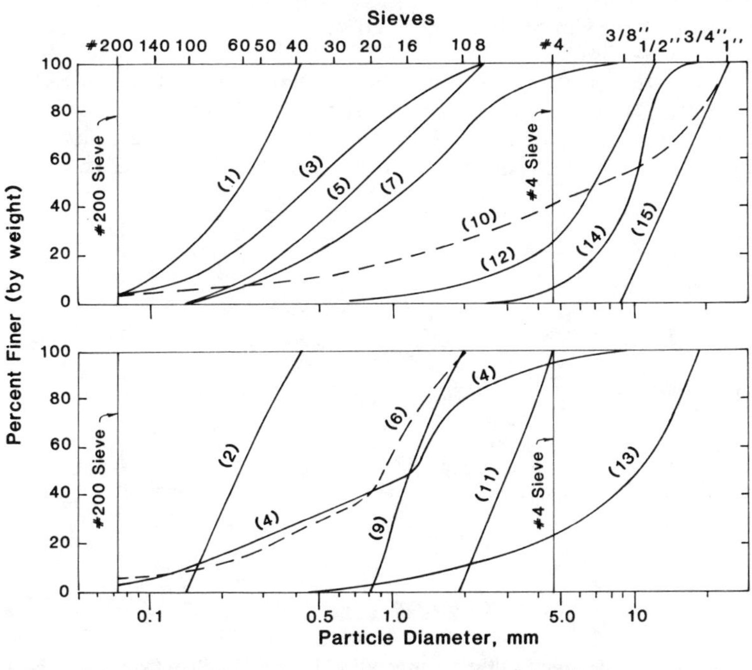

FIG. 7.—Sands and Gravels Used in Permeability Testing Program (Table 2)

TABLE 2.—Results of SCS Permeability Tests (11)

Specimen (1)	D_{15}, in millimeters (2)	$C_u = D_{60}/D_{10}$ (3)	COEFFICIENT OF PERMEABILITY, k, IN CENTIMETERS PER SECOND: Dense State, Median (4)	Dense State, Range[a] (5)	Loose State, Median (6)	Loose State, Range[a] (7)	Median k_{loose}/k_{dense} (8)
1	0.11	2.7	6.0×10^{-3}	$5.6\text{–}6.7 \times 10^{-3}$	1.8×10^{-2}	$1.4\text{–}2.1 \times 10^{-2}$	3.0
2	0.16	1.7	1.8×10^{-2}	$1.6\text{–}2.0 \times 10^{-2}$	2.5×10^{-2}	$2.0\text{–}2.7 \times 10^{-2}$	1.4
3	0.17	3.9	1.1×10^{-2}	$0.6\text{–}1.6 \times 10^{-2}$	1.9×10^{-2}	$1.6\text{–}2.3 \times 10^{-2}$	1.7
4	0.19	10	8.8×10^{-3}	$6.0\text{–}12.7 \times 10^{-3}$	1.2×10^{-1}	$1.1\text{–}1.6 \times 10^{-1}$	13.6
5	0.29	3.6	2.2×10^{-2}	$2.0\text{–}2.5 \times 10^{-2}$	4.9×10^{-2}	$3.5\text{–}5.6 \times 10^{-2}$	2.2
6	0.27	7.4	2.6×10^{-2}	$2.3\text{–}3.1 \times 10^{-2}$	3.0×10^{-2}	$2.6\text{–}3.2 \times 10^{-2}$	1.2
7	0.35	5.5	3.9×10^{-2}	$3.5\text{–}4.4 \times 10^{-2}$	5.1×10^{-2}	$4.2\text{–}6.1 \times 10^{-2}$	1.3
8	0.35	14.4	9.5×10^{-2}	$6.7\text{–}12.0 \times 10^{-2}$	0.24	0.21–0.28	2.5
9	0.9	1.4	0.25	0.21–0.28	0.56	0.49–0.67	2.2
10	0.7	37	0.14	0.07–0.25	0.78	0.65–1.4	5.6
11	2.1	1.4	2.6	2.2–3.2	3.8	2.9–4.3	1.5
12	3.0	3.1	4.8	4.3–5.1	5.3	5.1–5.6	1.1
13	2.6	7.1	1.8	0.7–2.1	5.6	4.8–6.8	3.1
14	5.7	2.0	8.7	8.1–9.6	12.2	11.5–12.7	1.4
15	10.1	1.7	13.9	12.6–15.0	15.2	14.0–16.0	1.1

[a]Range of three different tests with the same testing procedures.

Note: For specimens 11–15, the seepage was turbulent with decreasing values of $K = v/i$ with increasing gradient i. Values are for $i = 0.10$.

±10% to ±20% of the median. For the more broadly graded specimens the range was somewhat greater.

2. The median measured, k, was always greater for loose specimens than for dense specimens by a ratio generally in the range from 1.1–3.0, but reaching 13.7 maximum for specimen 4.

Fig. 8 shows the relation between the median value of measured k and D_{15} for the dense specimens (from Table 2), showing a good correlation with an average of about $k = 0.35\ (D_{15})^2$ (in which k is in centimeters per second and D_{15} in millimeters). Almost all the median k val-

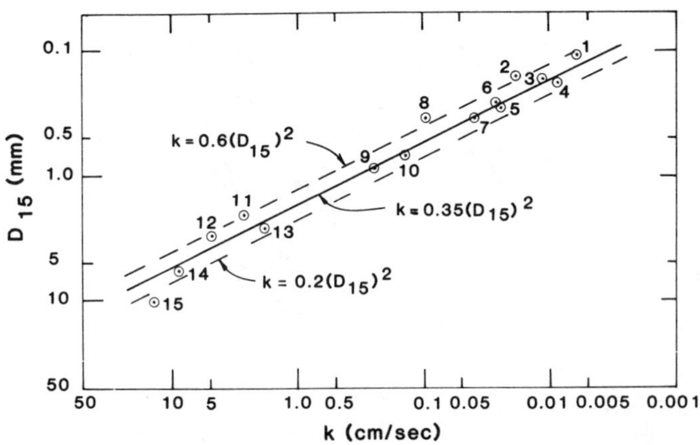

FIG. 8.—Relationship between Filter Permeability, k, and D_{15} Size [Note: See Particle Size Distributions (Fig. 7) and k Values (Table 2)]

ues fell between $k = 0.2\ (D_{15})^2$ and $0.6\ (D_{15})^2$. Similar plots show about equally good correlations between k and both D_{10} and D_{20}. Increasingly poorer correlation was obtained for plots of k versus D_{25} and coarser sizes. These relationships demonstrate: (1) That it is the sizes of the smaller particles of a filter that govern the sizes of pore channels through which the water flows; and (2) that the D_{50} size is not a good measure of the pore channel sizes.

The fact that k is directly related to the square of D_{15} shows that the velocity of water seeping through the pores of a filter follows the same physical law as laminar flow through a small pipe (capillary tube) of constant diameter. From conventional pipe flow theory, assuming water viscosity at 20° C, the velocity of laminar flow through a small pipe with diameter D is about

$$v = 31 D^2 i \qquad (1)$$

in which v is in centimeters per second; and D is in millimeters. The experimental average result of the permeability tests on dense sands and gravels (Fig. 8) can be written

$$k = 0.35 (D_{15})^2; \quad v = ki = 0.35 (D_{15})^2 i \qquad (2)$$

In order to compare Eqs. 1 and 2, it is necessary to make two adjustments:

1. Water seeping through the pores of a filter for a straight line distance, L, travels along an irregular path with a greater length L'. The average ratio, L'/L (called "tortuosity"), has been determined experimentally to be about 1.5 (3). Thus, the actual gradient forcing water through the pores of a sand or gravel filter is about 1.0/1.5 or 67% of the gradient as measured in a conventional permeability test.
2. Using the common definition of Darcy's Law for seepage through sands, $v = ki$, the velocity is defined as $v = Q/A$, in which A = the entire cross-sectional area of the sand. Thus, the actual velocity, v_a, of the water in the voids is greater by the ratio $v_a = V/n$, in which n = the sand porosity.

Making these two adjustments, and assuming the porosity of a dense sand equals 0.3, and combining Eqs. 1 and 2

$$31 D^2 i = 0.35 (D_{15})^2 i \frac{0.67}{0.3},$$

giving $D^2 = 0.025 (D_{15})^2$ and $D = 0.16 D_{15}$ .... (3)

From these calculations it can be seen that water under a given gradient travels at the same average velocity through pore channels in a dense sand or gravel filter as it would through a pipe with constant diameter, D, about $0.16 D_{15}$. The results of this simple calculation agree roughly with the more detailed development of "equivalent hydraulic diameter" presented by other investigators (2,14).

When water flows through a conduit that has an irregular cross-sectional area, such as the pore channel in a sand and gravel, the head loss is governed largely by the portions of the flow path with smallest diameter, because head loss is inversely related to the fourth power of the diameter. Thus, the diameter of an "equivalent" small pipe, computed earlier as $0.16D_{15}$, approaches the diameter of the smaller pores in the sand or gravel, and it corresponds very well with the experimental results of the filter tests (Fig. 4), which show that particles with diameter $0.12D_{15}$ will always be caught in the filter pores.

Broadly Graded Filters

The preceding results have been derived from tests on relatively uniform filters. For more broadly graded filters, these results are usually conservative since the permeability is often less than $k = 0.35D_{15}^2$ and minimum pore channel diameter is less than $0.1D_{15}$, but the difference generally is not great. A well-graded filter with C_u of 20 and a given D_{15} will catch particles that are roughly 1/2 the size of a uniform filter with the same D_{15} size. This is reviewed in more detail in a companion paper (6).

Well-Graded (Nonuniform) Base Soils

In addition to the tests with uniform sand bases, we have performed many other filter tests with conservative testing procedures (high water pressures and holes in the base samples to create initial concentrated leaks), using a large number of relatively widely graded sandy silts and sandy clays as base soils (with d_{85} in the range of 0.1–0.6 mm). For all these tests, the failure boundary was always at a D_{15}/d_{85} ratio in excess of 9, showing that the d_{85} size is a reasonable measure of the base size for this purpose for sandy silts and clays (6). Tests by others for well-graded (nonuniform) sand bases show similar results. Table 3 summarizes filter tests made by Lund (1949) as part of the research program described earlier on three relatively broadly graded sand bases, with d_{85} in the range of 1.2–1.9 mm using uniform gravel filters (4).

The preceding results confirm that, for the well-graded base soils tested (sandy silts and clays and well-graded pervious sands with d_{85} of 1.2–1.9 mm), a filter sufficiently fine to catch the d_{85} size will also catch the finer base particles. Thus, it can be concluded that, for the soils tested, the oldest and most widely used filter criterion, $D_{15}/d_{85} \leq 4$ or 5, is conservative. It can be considered to have a safety factor of about 2, since the experimental failure-success boundary is about $D_{15}/d_{85} = 9$.

TABLE 3.—Results of Filter Tests on Well-Graded Sand Bases [After Lund (4)]

Base (1)	d_{85}, in millimeters (2)	D_{15}, in millimeters (3)	D_{15}/d_{85} (4)	Test result (5)
Base 5	1.2	11.0	9.2	Stable
Base 5	1.2	12.3	10.2	Unstable
Base 6	1.5	12.3	8.2	Stable
Base 6	1.5	14.5	9.7	Unstable
Base 7	1.9	14.5	7.6	Stable
Base 7	1.9	16.0	8.4	Unstable

The experimental results support the general validity of the main currently used filter criterion, $D_{15}/d_{85} < 5$. It uses the most appropriate properties of the filter and base to describe their interaction (i.e., D_{15} and d_{85}), and it has a substantial built-in conservatism. Performance of filters in dams meeting this criterion has almost always been satisfactory, and filters satisfying the criterion are generally not costly. Thus, a main conclusion of this study is that it is appropriate to continue using $D_{15}/d_{85} \leq 5$ as the principal filter acceptance criterion.

This conclusion applies generally to filters with D_{15} larger than about 1.0 mm. For finer filters, generally used for protecting fine-grained silts and clays, research reported in a companion paper (6) shows that the failure-success boundary changes significantly.

The preceding conclusions apply to base soils that are internally stable. For these stable base soils, the filter prevents entry of the coarse particles which accumulate in a thin skin on the filter face and prevent entry of the finer base particles. For certain gap-graded and unstable, coarse, broadly graded base soils, usually graded from clay sizes to gravels with d_{85} larger than 2 mm, the base soil fines may be able to enter the filter voids even if the coarser particles cannot. Filter criteria for these internally unstable base soils need to be applied differently, using procedures outside the scope of this study.

Filter Criteria Employing D_{50}, d_{15}, and d_{50}

In a sand or gravel filter, the sizes of the internal pore channels are governed by the finer particles, D_{15}. The average particle size of the sand or gravel, D_{50}, is clearly not a satisfactory measure of the minimum pore sizes. For this reason, filter criteria in using D_{50}, such as $D_{50}/d_{50} \leq 25$ (10,13) and $12 < D_{50}/d_{50} < 58$ (9), are not founded on a satisfactory theoretical or experimental base. Use of the D_{50}/d_{50} ratio is sometimes defended on the grounds that it will prevent excessive segregation during construction. This argument, however, is not well founded; the base d_{50} size has practically no relationship to the maximum tolerable filter segregation. Filter segregation is an independent problem. It should be limited by restrictions on the range, not the average, of particle sizes in the filter.

Another fairly widely published filter criterion employs the d_{15} size of the base soil, $D_{15}/d_{15} < 40$ (9,10). This criterion has even less support from theory or experimental results than criteria employing D_{50}/d_{50}. The d_{15} size has no significant influence on the properties of the needed filter. In our experiments with impervious silt and clay base soils, the D_{15}/d_{15} ratio in successful tests commonly exceeded 1,000 (6).

Based on the preceding review, it is the personal opinion of the writers that the current filter criteria employing D_{50}/d_{50} and the criterion $D_{15}/d_{15} < 40$ (or 25) should be abandoned.

Requirement that Filter and Base Soil Gradation Curves Have Similar Shape, and Segregation Problem

One of the oldest general filter criteria is that the filter particle size distribution curve should be approximately the same shape as that of the base soil. This criterion is still widely quoted in the literature. For some time, however, designers have recognized that it is not generally necessary or desirable. Our research supports this opinion.

Uniform sands (with C_u of 2–5) are excellent filters for fine-grained clays that are broadly graded over wide range of particle sizes, $C_u > 50$, even though the gradation curves are not at all similar in shape (6). Uniformly graded sands (or gravels) with appropriate D_{15} size are always satisfactory filters regardless of the shape of the base soil particle size distribution curve. On the other extreme, well-graded gravelly sand can be an excellent filter for a very uniform silt or fine uniform sand.

In certain circumstances a broadly graded filter has two main potential advantages over an equivalent uniform filter: (1) It may cost less; and (2) it may allow the use of a single filter band instead of a multiple band.

The main technical reason for limiting the maximum range of filter particle size is to minimize segregation during construction. Some segregation of broadly graded filters is inevitable. The coarser gravel particles, with little or no sand filling the voids, accumulate in pockets or streaks, frequently at the base-filter interface. The dimensions of these segregated streaks or pockets increase rapidly with the size of the larger filter particles, such as the D_{90} size. For such a segregated filter to function, there must first be sufficient migration of base soil into the filter to fill the open voids in the segregated zone at the filter-base interface. Such migration is undesirable.

In current design practice there are no widely agreed upon quantitative criteria for determining the maximum desirable broadness of filter grading to limit segregation. Some existing guides limit C_u of the filter to a maximum value, such as 20, for this purpose. But this is not very satisfactory, since the particles larger than D_{60}, such as the D_{90} size, commonly govern the magnitude of the problem. Also, many coarse sandy gravels with C_u about 20 are difficult to place in a dam without undesirable segregation of the larger gravels. On a job where reasonable construction procedures are used, experience is the best judge of the maximum desirable range of gradation to limit segregation. It is beyond the scope of this paper to treat the subject fully.

Crushed Rock Filters with Angular Particles

For four of the materials shown in Fig. 7, sand specimens 3 and 7 and gravel specimens 12 and 14, duplicate specimens with exactly the same particle size distribution were made of subrounded alluvial particles and of angular particles of crushed limestone. Median results of three permeability tests on these duplicate specimens (compacted to 70% relative density) are shown in Table 4.

Permeability of the crushed rock specimens was less than that of the alluvial samples, except for specimen 14 where there was negligible difference. In addition, we have a considerable number of results of other permeability tests on crushed angular sands and gravels from various kinds of hard rock (granite, basalt, gneiss, sandstone, etc.). All these tests show that $k = 0.35 D_{15}^2$ is a reasonable average relationship for these materials in a dense state.

Since the permeability of the crushed materials is not consistently greater

TABLE 4.—Median Results of 3 Permeability Tests on Duplicate Specimens Compacted to 70% Relative Density

Specimen (Fig. 7) (1)	Coefficient of Permeability k, in Centimeters Per Second	
	Alluvial particles (2)	Crushed rock (3)
3	0.011	0.0035
7	0.039	0.026
12	4.5	2.0
14	8.8	9.5

than that of the alluvial materials with equivalent D_{15}, it is reasonable to assume that the pore channels are not larger and that the same criteria of filter acceptability can be used. This conclusion conflicts with the results of the USBR research program (9), probably because of the test result interpretations previously described.

Summary and Conclusions

The sizes of pore channels, which govern the permeability, are determined by the sizes of the finer filter particles and are well represented quantitatively by the D_{15} size. The conclusions summarized here are based on tests and observations of sands and gravels that are compacted to a dense state without appreciable visible segregation and do not have any significant quantity of material finer than No. 200 sieve. The conclusions apply primarily to uniform filters but are not greatly different for well-graded filters, at least up to $C_u = 10$.

The cross-sectional area of a continuous pore channel is irregular and changes rapidly and repetitively from maximum to minimum size. The channel's linear dimensions normal to the direction of flow vary over the approximate general range from $0.1–0.6D_{15}$. By the time seeping water has traveled in the pores a relatively short distance, such as about $5D_{15}$, it has already passed through all the combinations and permutations of pore channel sizes and shapes. The concept that some pore flow channels of significant length have a larger or smaller minimum size than the average is not correct.

Particles smaller than about $0.10D_{15}$ that are carried in water suspension will generally pass through the voids and out of the filter. Particles larger than about $0.12D_{15}$ will be retained near the filter face. Thus the filter acts like a laboratory sieve with openings about $0.11D_{15}$ wide.

The coefficient of permeability of dense filters is generally in the range between $K = 0.2D_{15}^2$ and $0.6D_{15}^2$, with an average of about $k = 0.35D_{15}^2$, in which k is in centimeters per second and D_{15} is in millimeters.

The oldest and most widely used of the existing filter criteria, $D_{15}/d_{85} \leq 5$, is shown to be conservative and to employ the appropriate characteristics of the filter and base (D_{15} and d_{85}). A main conclusion of this study is that, for filters with D_{15} larger than about 1.0 mm, the ratio $D_{15}/d_{85} \leq 5$ should be continued as the main criterion for judging filter acceptability. For finer filters used for protecting silts and clays, different criteria can be used (6).

Filter criteria that limit the D_{50}/d_{50} and D_{15}/d_{15} ratios are not founded on a sound theoretical or experimental basis and should be abandoned. Some published results of laboratory filter programs include evaluations of test results that have led to unsupportable conclusions and recommendations especially with regard to the criteria employing D_{50}/d_{50} and D_{15}/d_{15}.

It is not necessary that the particle size distribution curve of the filter be generally similar in shape to that of the base soil.

Angular particles of crushed rock are as satisfactory as rounded alluvial particles for filters and can be designed using the same criteria. The same criteria can be used for filters which may be subjected to vibrations.

Appendix I.—References

1. Bertram, G. E., "Experimental Investigation of Protective Filters," *Soil Mechanics Series No. 7,* Graduate School of Engineering, Harvard University, Cambridge, Mass., 1940.
2. Busch, K. F., and Luckner, L., "Geohydraulik," VEB Deutscher Verlag fur Grundstoffindustrie, Leipsig, 1972, 442 pp.
3. Carman, P. C., *Flow of Gases Through Porous Media,* Butterworth's Scientific Publications, London, 1956.
4. Lund, A., "An Experimental Study of Graded Filters," thesis presented to the University of London, at London, U.K., in 1949, in partial fulfillment of the requirements for the degree of Master of Science.
5. Sherard, J. L., "Sinkholes in Dams of Coarse Broadly Graded Soils," 13th ICOLD Congress, India, Vol. II, 1979, pp. 25–35.
6. Sherard, J. L., Dunnigan, L. P., and Talbot, J. R., "Filters for Silts and Clays," *Journal of Geotechnical Engineering,* ASCE, Vol. 110, No. GT6, June, 1984, pp. 701–718.
7. Terzaghi, K., and Peck, R. B., *Soil Mechanics in Engineering Practice,* 2nd ed., John Wiley and Sons, Inc., New York, N.Y., 1967.
8. U.S. Bureau of Reclamation, *Design of Small Dams,* U.S. Printing Office, Washington, D.C., 1973, pp. 234–236.
9. U.S. Bureau of Reclamation, "The Use of Laboratory Tests to Develop Design Criteria for Protective Filters," *Earth Laboratory Report No. EM-425, June 20, 1955* (also published in Proceedings, ASTM, Vol. 55, 1955, p. 1183).
10. U.S. Army Engineers, "Filter Experiments and Design Criteria," *Technical Memorandum No. 3-360,* Waterways Experiment Station, 1953.
11. U.S. Dept. of Agriculture, "Permeability of Selected Clean Sands and Gravels," *Soil Mechanics Note No. 9,* Soil Conservation Service, Dec., 1978.
12. Vaughan, P. R., and Soares, H. F., "Design of Filters for Clay Cores of Dams," *Journal of the Geotechnical Engineering Division,* ASCE, Vol. 108, No. GT1, Jan., 1982.
13. Wilson, S. D., and Marsal, R. J., *Current Trends in the Design and Construction of Embankment Dams,* ASCE, New York, N.Y., 1979.
14. Wittmann, L., "The Process of Soil Filtration—Its Physics and Approach in Engineering Practice," 7th European Conference on Soil Mechanics and Foundation Engineering, Brighton, U.K., 1979, pp. 303–310.

Appendix II.—Notation

The following symbols are used in this paper:

C_u = coefficient of uniformity = D_{60}/D_{10} (filter) and d_{60}/d_{10} (base);
D = diameter of small pipe of constant diameter;

D_{15} = particle size in filter for which 15% by weight of particles are smaller; similarly for D_{60}, D_{50}, etc.;
d_{85} = particle size in base soil for which 85% by weight of soil particles are smaller; similarly for d_{50}, d_{15}, etc.;
i = hydraulic gradient, in centimeters per centimeter;
k = coefficient of permeability, in centimeters per second;
n = porosity = volume of voids/total volume;
v = average seepage velocity, $v = ki$; and
v_a = seepage velocity through filter pores = v/n.

Introduction to Paper 14

Filters for Silts and Clays

by James L. Sherard, Lorn P. Dunnigan, and James R. Talbot
ASCE Journal of Geotechnical Engineering, Vol. 110, No. 6, June 1984

Prepared by James R. Talbot

In late 1979, the Soil Conservation Service (SCS) started a study of sand and gravel filters for earth dams. The main purpose of our study was to determine the proper grading criteria for filters to protect clay soils. The study had been underway for only a short time when I received a phone call from Jim Sherard. Without knowing about our study, Jim called because he had encountered some sink hole development and other evidence of filter failure in dams having clay cores as well as cores of broadly or widely graded soils. Upon hearing about our study, Jim expressed his desire to work with us on it. Jim had worked with SCS before on studies of dispersive clays.

SCS entered into an agreement with Jim to work together on the study with no exchange of funds. We supplied our laboratory facility, staff, and materials for testing, and he provided expertise and guidance. We were fortunate to have been able to work with Jim on these studies and to benefit from his vast experience, his knowledge, and his enthusiasm for finding answers to the many perplexing questions dealing with earth dam design and construction. We started by trying to define the basic properties of sand and gravel that provides filtering functions and then progressed on to a study of filters for silts and clays, and then to broadly graded soils.

One of the best writers I have ever known, Jim had a special ability to convey technical information in written word. It was a privilege and a learning experience to write a paper with him. This award winning paper (ASCE Norman Medal) and the companion papers on filters and hydraulic fracturing demonstrate his great ability.

Filters for Silts and Clays

By James L. Sherard,[1] F. ASCE, Lorn P. Dunnigan,[2] and James R. Talbot,[3] Members, ASCE

Abstract: An investigation was made of the filters needed in dams for fine-grained clays and silts. The "critical" downstream filter in a central core dam should be capable of controlling and sealing a concentrated leak through the core, and should also be stable in conventional laboratory filter tests under a relatively high gradient, such as 1,000. Two different types of laboratory tests were developed to simulate the action of a critical filter (slot and slurry tests). Both gave identical and reproducible results. For fine-grained clays a sand filter with D_{15} of 0.5 mm is conservative. For sandy clays and silts the filter criterion $D_{15}/d_{85} \leq 5$ is conservative and reasonable. The Atterberg limits of a clay have no significant influence on the needed critical filter. For nondispersive and dispersive clays having similar particle size distribution the needed critical filters are the same. For "noncritical" filters, such as filters upstream of a clay core, quantitative filter criteria are not necessary.

Introduction

An investigation of filters was carried out during 1981–82 at the Soil Mechanics Laboratory, Midwest National Technical Center, Soil Conservation Service, U.S. Department of Agriculture, Lincoln, Nebraska. The main object was to improve understanding of filters needed to protect clays and silts commonly used as the impervious sections of embankment dams. In current practice there is no widely agreed upon quantitative filter design criteria for fine clays.

In this paper only the "filter action" of a filter is considered, i.e., the properties of the filter needed to prevent any significant penetration of the base soil particles into the filter. It does not treat the problem of the permeability needed for a single-band filter that functions as both filter and drain, or the need for additional drainage zones.

A parallel study directed at improving basic understanding of sand and gravel filters, including the sizes of pore channels which catch particles carried in suspension in water flowing into the filter, is described in a companion paper (8).

"Critical" Filters Needed to Seal Concentrated Leaks Through Impervious Dam Sections

Small concentrated leaks occasionally develop through the impervious sections of embankment dams designed and constructed according to

[1]Consulting Engr., San Diego, Calif.

[2]Head, Soil Mechanics Lab., National Technical Center, Soil Conservation Service, U.S.D.A., Lincoln, Neb.

[3]National Soil Engr., Soil Conservation Service, U.S.D.A., Washington, D.C.

Note.—Discussion open until November 1, 1984. Separate discussions should be submitted for the individual papers in the symposium. To extend the closing date one month, a written request must be filed with the ASCE Manager of Technical and Professional Publications. The manuscript for this paper was submitted for review and possible publication on May 4, 1983. This paper is part of the *Journal of Geotechnical Engineering*, Vol. 110, No. 6, June, 1984. ©ASCE, ISSN 0733-9410/84/0006-0701/$01.00. Paper No. 18934.

good modern practice. These leaks may be caused by: (1) Development of open cracks from differential settlement; (2) flow at contacts with rock foundations or concrete structures; (3) construction deficiencies; and (4) hydraulic fracturing. The evidence is in the following main categories:

1. Concentrated leaks have emerged soon after the first reservoir filling on the downstream slope of small, well-constructed homogeneous dams without internal drains. Such leaks are usually at points of maximum differential settlement. They sometimes lead to breaching (3,7).

2. In some recently constructed central-core dams, piezometers at the downstream face of the core have registered nearly full reservoir pressure. When borings are made in the impervious core to a certain depth, water in the casing rises rapidly to the pressure head of nearby piezometers. This rise clearly shows that a concentrated leak was present (experiences not yet published).

3. In several instances, concentrated leaks have suddenly emerged at the downstream toe of modern central-core dams; these leaks have usually occurred during or soon after the first reservoir filling (5). In some of these instances, extensive investigation concluded that hydraulic fracturing was the only likely cause of the leak (1,3,10,11).

4. Before filling the reservoir, open cracks have been observed that extended below the full reservoir elevation.

5. Calculations by the finite element method confirm that, for well constructed dams, the stress in the core is frequently low enough to allow hydraulic fracturing even without unusual differential settlement (2,4).

6. Small concentrated leaks can occur along the inevitable small open cracks at rock foundation contacts. At Yard's Creek Upper Reservoir Dam, New Jersey, these leaks were observed in a test pit sunk through the core to the rock foundation (3).

We believe there is already sufficient evidence from dam behavior, supported by theory, to require the designer to assume that small concentrated leaks can develop through the impervious section of most embankment dams, even those without exceptional differential settlement.

Fig. 1 shows what happens when a concentrated leak discharges into the downstream filter. If the leak carries eroded particles of core material that seal the filter face, the water pressure in the leakage channel abruptly rises and approaches the reservoir head. This creates a high seepage gradient through a short distance in the core to the adjacent portion of the filter, shown by path *Y* in Fig. 1.

We believe that this situation probably develops at least temporarily in many dams, especially in high, thin-core dams and in dams with steep rock abutments. If the downstream filter successfully controls the leak, the core material along the walls of the leakage channel probably swells or softens and collapses. The pore-water pressure in the core then usually reverts to a nearly normal seepage pattern.

From this reasoning, we conclude that a satisfactory downstream filter should not only be capable of sealing an initial concentrated leak, but also be stable in a laboratory test simulating the seepage conditions of

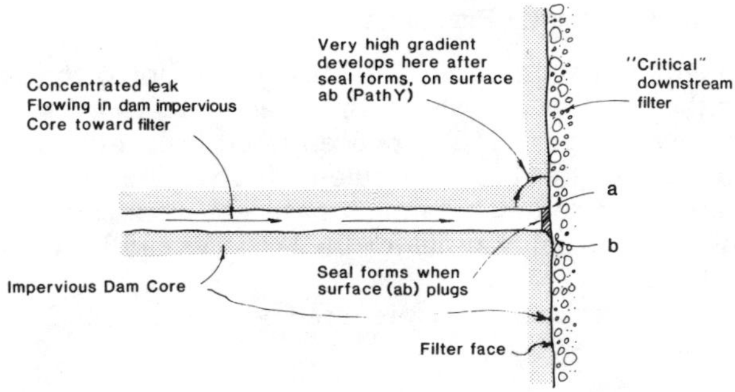

FIG. 1.—Sketch Showing Concentrated Leak Through Dam Core Discharging into Downstream Filter (No Scale)

Path *y*, Fig. 1—i.e., with high water pressure (near reservoir head) acting over a short length of compacted core material.

Conventional Filter Tests Using High Water Pressure

As a first effort, a number of conventional filter tests (without an initial hole in the base specimen) were made with compacted sand and sandy gravel filters using relatively thin (30–60-mm thick) base specimens of clay and silt. The specimens were compacted near Standard Proctor Optimum water content. In these tests, the water pressure acting across the base specimen was gradually increased to a maximum of about 6 kg/cm^2, giving a hydraulic gradient of about 1,000–2,000.

At relatively low pressures, generally below 1.0 kg/cm^2, no filter failures occurred, even for very coarse filter tests lasting many weeks. The small quantity of water seeping from the base sample into the filter had very little energy, and there was no tendency for the fine clay or silt base material to enter the filter pores.

During the tests the water pressure was increased in increments of 0.5 kg/cm^2 until a concentrated leak of colored water developed, usually when the pressure was above about 1.5 kg/cm^2. These leaks appeared suddenly, and visual observations showed that they resulted from deformation of the specimen under water pressure, stress transfer, and hydraulic fracturing. After the initial leak developed, either the eroded material sealed the filter face and the leak stopped (successful filter) or it was carried through the filter without sealing (unsuccessful).

In these tests, a concentrated leak always developed at the higher pressures. In the tests with successful filters, erosion of the base specimen was slight, and only a small amount of colored water passed through the filter before the leak was sealed. In unsuccessful tests, the leak eroded a relatively large hole—5–10 mm in diameter—through the base specimen in a few seconds. Tests of this kind could have been used to define the filter boundary for various different fine-grained soils. It was decided, however, that a more direct approach would obtain the same result, using a test with a preformed slot in the compacted base specimen.

"Slot" Tests with High Water Pressure

After experimentation with various details, we adopted the test set-up of Fig. 2 for the next set of tests. Its main advantages over the kind of test described above were: (1) The concentrated leak could be always located in the center of the base specimen, discharging in the central portion of the filter; (2) a somewhat longer base specimen could be used; and (3) the "slot"—the initial channel for the leak—would have the same dimensions in all tests.

Fabrication of the specimen was prepared with the cylinder placed vertically as follows:

1. The first 3 in. (75 mm) of filter were placed in 3 equal, lightly tamped layers. Care was taken to minimize segregation. The material was then saturated and compacted on a vibrating table using a 10 kg surcharge weight.
2. A ring of children's modeling clay was placed around the inside of the cylinder wall (Fig. 2).
3. The final 1-in. (25 mm) layer of filter was placed and again vibrated, causing the filter material to penetrate the clay ring. As a result any suspended solids which tend to travel down the larger void channels along the cylinder wall, would be diverted at least 0.5 in. (12 mm) toward the center of the filter and back again.
4. A long strip of sheet metal 0.5 in. wide and 0.06 in. thick was wrapped in a plastic membrane and placed in the center of the filter. The base material was then placed in 6 equal layers. A tamper was used to compact each layer to about 95% of maximum Standard Proctor Density at near optimum water content, compacting around sheet metal strip. Finally the strip and plastic wrapper were withdrawn to form the slot for the initial leakage channel (Fig. 2).

With a few exceptions, the tests were run with the cylinder placed

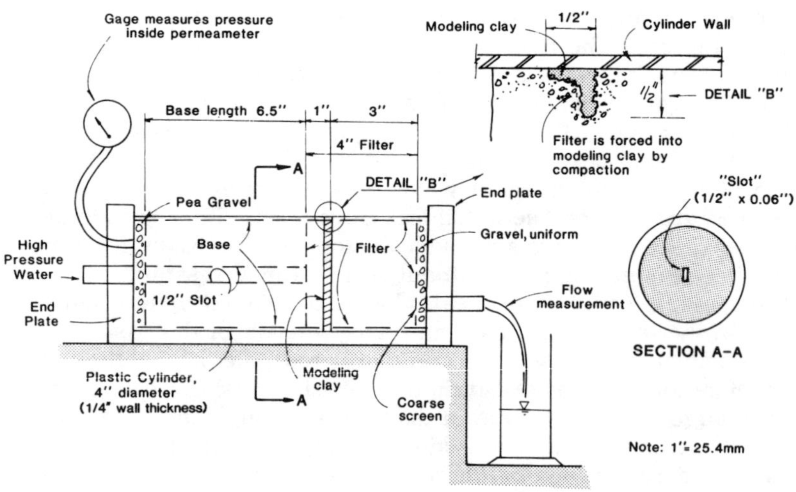

FIG. 2.—High Pressure "Slot" Test Apparatus Details (Schematic, No Scale)

horizontally (Fig. 2). To start the test, a water valve was opened abruptly, allowing the water to flow at about 4 kg/cm^2 pressure through the slot in the base specimen and discharge into the filter. A surge of dirty water soon emerged through the two pipes in the downstream end plate (Fig. 2). In most tests the initial quantity of dirty water ranged from 3–9 L in the first 30 sec (100–300 ml/s) giving a calculated velocity through the slot of about 5–15 m/s, a highly erosive velocity. In all tests, on all specimens, the initial surge of emerging water was highly colored with particles of base material.

For tests with successful filters, the flow rate rapidly decreased and the water became progressively clearer, finally sealing completely or stabilizing at a very small constant flow of clear water. It was generally possible to judge the test result within 2 or 3 min. When the specimen was examined after successful tests the slot generally was slightly enlarged and nearly full of soft mud. Sometimes only the downstream end was mud filled. In all successful tests, the amount of eroded base soil carried through the filter was very small (commonly 10–20 g), and the filter always appeared clean, with the seal made in the first few millimeters of the filter.

For unsuccessful filters, the surge of dirty water continued with no reduction in rate, and the test was commonly stopped after a few minutes. At the end of an unsuccessful test, there was commonly an open hole through the base specimen, with a diameter of 10–15 mm or larger. For the unsuccessful tests, hydrometer measurements of the particles carried through the filter showed generally the same particle size distribution as the original material. That is, the material was completely disaggregated by erosion and movement through the filter voids, and there was no significant quantity of aggregated clay particles remaining clinging together.

It was found that the results of the tests were reliably reproducible and that the boundary between successful and unsuccessful tests for any given base soil was narrow and well defined by the filter, D_{15}, size.

Before adopting this test procedure, we conducted a preliminary set of similar slot tests. The slot was larger and the water pressure lower (about 0.14 kg/cm^2 instead of 4.0 kg/cm^2). Initial velocity was commonly about 0.1 m/s. The results were not satisfactory for two main reasons: (1) For some clays the filter was not tested: the slot wall did not collapse, the concentrated leak did not have sufficient velocity to erode the compacted base sample, and the water emerged completely clear; and (2) more commonly, the walls of the slot slaked immediately and relatively large chunks of compacted clay fell into the water and sealed the filter. In the latter case, coarser filters were sealed than those for the same base soil later when using high pressure slot tests.

Because of these results, and the conclusion that a satisfactory critical filter should be able to withstand a test with high gradient over a short length of base, the low-pressure slot tests were abandoned.

"Slurry" Tests with High Water Pressure

After the high-pressure slot test was run on a considerable number of different clays and silts, it became apparent that the water was eroding the compacted base soil specimen to its basic particle sizes, i.e., the filter voids were not being plugged and sealed by small pieces of intact compacted clay equal to or larger than fine sand size (larger than 0.1 mm).

This observation led to consideration of another type of filter test in which clay-water slurries were used as the base specimen. After some experimentation with the details, the test set-up of Fig. 3 was adopted as the standard for subsequent testing. The same cylindrical plastic permeameters were used as for the slot tests, and the specimen was fabricated as follows:

1. The filter was placed and compacted as described for the slot tests above. The cylinder was left vertical for the test.
2. The slurry was mixed in a 2,000-ml beaker. Water was added gradually until the base slurry water content was about 2.5 times the liquid limit. The slurry was mixed with a laboratory mechanical mixer (milkshake mixer) progressively adding the water until its viscosity resembled that of automobile moter oil. Hydrometer tests on samples of these slurries, using no added chemical dispersant or further mixing, showed that the particle size distribution of the base soil in the slurry was very close to the basic particle size distribution as measured in a standard hydrometer test.
3. The slurry was poured on top of the filter. A deflector was used to prevent erosion of the filter surface. Next, the remaining space in the cylinder above the base slurry was filled with water. Filter paper was used as a temporary deflector to prevent disturbance of the slurry surface.
4. The upper end plate was attached, and the water valve was abruptly opened to a pressure of about 4 kg/cm^2.

The results of the tests were found to be reproducible. In successful

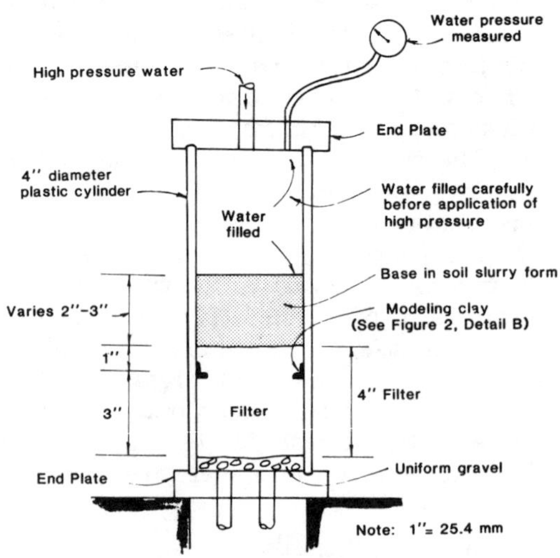

FIG. 3.—High Pressure "Slurry" Test Apparatus Details (Schematic, No Scale)

tests, the surface of the slurry abruptly settled a few millimeters and then stopped moving; a small amount of cloudy water emerged from the bottom (about 4–10 ml in 2 min), and the flow stopped and the test remained in equilibrium with the filter face sealed. For unsuccessful tests all the base slurry was forced through the filter in 2 or 3 sec and the upper filter surface was left clean.

Boundary between Successful and Unsuccessful Filters

In the first part of the main investigation, 15 different clay and silt base soils were tested with the slot test. Different filters were used with a total of 57 tests. Subsequently, 52 slurry tests were made using the same 15 base soils and, generally, the same base-filter combinations. For every base-filter combination for all 15 base soils, the two tests gave identical results.

Since the test results were identical, and the slurry test required considerably less laboratory effort to perform, slurry tests were used to complete the study. Twenty-one other fine-grained soils were tested in an additional 145 slurry tests with different base-filter combinations. Therefore, a total of 197 slurry tests were run on 36 base soils.

The 36 base soils ranged from nearly cohesionless silts to tough, highly plastic clays and included some highly dispersive sodium clays from dams which had failed by piping (7). The soils tested came from sites of Soil Conservation Service projects in various parts of the United States, with a wide range of geographic and geologic origins. Consequently, except for weathered residual soils of tropical areas, the samples represent the general range of fine-grained soils used in the impervious zones of embankment dams.

For each base soil, a series of tests were made with filters of different coarseness, finding for each the "filter boundary," D_{15B}. For filters finer

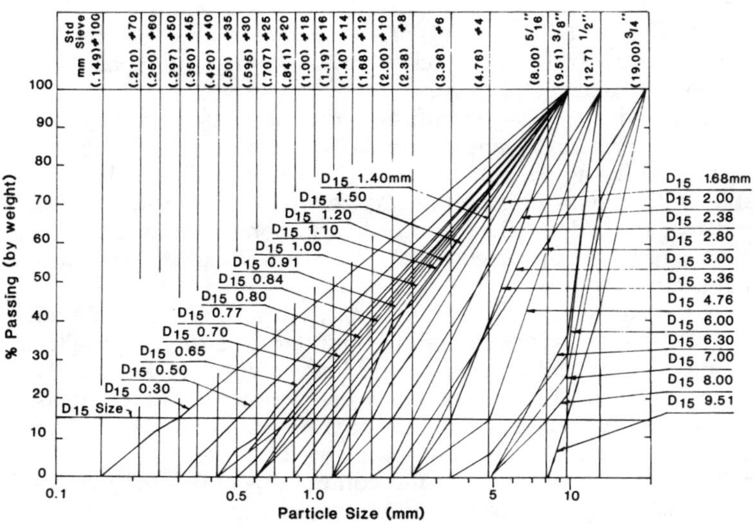

FIG. 4.—Gradations of Filters Used in Slot and Slurry Tests

than this boundary, the tests were successful; for coarser filters the tests were unsuccessful.

The filters used were subrounded to rounded, alluvial sands and sand-gravel mixtures. The filters were carefully fabricated by combining known weights of carefully sieved materials, using sieve sizes which ensured that the D_{15} size was reliably known. A total of 25 different filters were used with D_{15} ranging from 0.3–9.5 mm (Fig. 4).

Table 1 is a summary of the index properties of the clay and silt base soils tested and filter D_{15B} size at the experimentally determined failure-success boundary. In Table 1, the first 30 soils listed are generally clays (CL and CH) with a few clayey sands (SC); they are arranged in order of increasing, d_{85}, size. Soils 31–35 are silts of low cohesion (ML and CL-ML).

For some of the base soils, a large number of tests were made as the filter success-failure boundary was approached in small increments. With enough tests, the narrowness of the boundary became apparent, as illustrated by tests on a fine clay of high plasticity (CL-CH) and a fine silt of low plasticity (ML) (Table 2).

For most of the soils listed in Table 1, the filter boundary, D_{15B}, was defined with a precision of about 10%. In all tests, filters on the fine side of the experimentally determined boundary were successful, and those on the coarse side were not.

Analysis of Test Results

Fig. 5 presents, for all 36 base soils tested, the base soil, d_{85}, size and the experimentally determined filter boundary, D_{15B}. The following main relationships can be seen:

1. There is a general correlation between d_{85} and D_{15B}, but there is a wide scatter.
2. For all base soils the ratio D_{15B}/d_{85} is in excess of 9, as predicted from other research (8). For many clays, the D_{15B}/d_{85} ratio is about 25, and for some it exceeds 50.
3. For typical clays without significant content of sand-size particles with d_{85} in the approximate range from 0.03–0.08 mm, the filter boundary, D_{15B}, ranged from about 1.1–3.0 mm. In this group of 18 clays (No. 2–19, Table 1), there was no significant difference between the behavior of highly dispersive sodium clays and ordinary, nondispersive clays (Fig. 5). For this general group of clays, a filter with D_{15} of 0.5 mm has a substantial safety factor.
4. For the six silts tested (ML and CL-ML), with d_{85} of 0.04–0.08 mm, three soils (No. 31–33, Table 1) had filter boundaries, D_{15B}, ranging from 0.67–0.82 mm, somewhat lower than the lower boundary for the clays with similar d_{85} dizes. These three silts were typical weathered loessial soils of Iowa and Nebraska. For the other three silts (No. 34–36, Table 1), the D_{15B} was in the same general range as the clays.
5. For the clays with significant sand content (d_{85} of 0.1–0.5 mm, No. 20–30, Table 1), D_{15B} ranged from 2.4–6.0 mm, except No. 26, with D_{15B} of 10 mm. This range of D_{15B} clearly shows the influence of the sand-size particles in sealing the filters.

TABLE 1.—Summary of Base Soil Properties and Filter Test Results

Base soil number (1)	BASE SOIL INDEX PROPERTIES: Atterberg Limits LL (2)	PI (3)	Gradation: d_{85}, in millimeters (4)	d_{50}, in millimeters (5)	Percent < 0.002 millimeters (6)	Filter Test Results: D_{15B}, in millimeters (7)	d_{85}/D_{15B} (8)	D_{15B}/d_{85} (9)
1[a]	40	20	0.010	0.005	36	0.40	0.025	40
2	48	28	0.028	0.001	56	1.15	0.024	41
3	44	24	0.028	0.003	47	1.15	0.024	41
4[a]	42	19	0.031	0.004	35	1.6	0.019	52
5[a]	99	72	0.035	0.001	55	1.7	0.021	49
6[a]	32	10	0.038	0.008	22	1.25	0.030	33
7	22	4	0.041	0.021	9	2.3	0.018	56
8	34	12	0.048	0.020	21	1.1	0.044	23
9[a]	32	12	0.050	0.008	25	2.1	0.024	42
10[a]	37	20	0.050	0.009	30	1.2	0.042	24
11	38	17	0.052	0.015	31	1.1	0.047	21
12[a]	29	9	0.052	0.020	16	2.2	0.024	42
13	39	19	0.054	0.015	26	2.4	0.022	44
14	32	11	0.056	0.011	23	3.1	0.018	55
15	40	19	0.057	0.017	30	1.5	0.038	26
16	41	19	0.060	0.020	26	1.5	0.038	25
17	41	23	0.066	0.006	37	1.2	0.055	18
18[a]	40	18	0.074	0.023	24	1.4	0.053	19
19	33	12	0.080	0.030	18	2.3	0.035	29
20	28	10	0.11	0.034	24	3.1	0.035	28
21	33	16	0.12	0.015	34	5.2	0.023	43
22[a]	28	11	0.14	0.023	17	2.7	0.052	19
23	38	15	0.18	0.023	20	5.2	0.035	29
24	29	12	0.21	0.083	20	2.4	0.088	11
25	32	19	0.29	0.105	29	2.6	0.111	9
26	21	6	0.35	0.033	21	10.0	0.035	29
27	27	10	0.46	0.11	14	5.5	0.084	12
28	31	8	0.46	0.028	13	6.0	0.077	13
29	34	21	0.50	0.015	29	5.0	0.100	10
30	45	21	0.58	0.04	38	5.5	0.105	10
31	35	11	0.039	0.020	22	0.67	0.058	17
32	29	5	0.062	0.035	6	0.82	0.076	13
33	27	4	0.074	0.034	6	0.82	0.090	11
34	33	9	0.063	0.020	19	2.2	0.029	35
35	28	5	0.056	0.030	12	1.1	0.051	20
36	28	3	0.074	0.032	9	4.0	0.018	57

[a]Dispersive clays.

6. One very fine clay (No. 1, Table 1), was a highly dispersive sodium clay. Its d_{85} was about 0.01 mm, and D_{15B} was very low—0.4 mm. This soil was by far the finest soil tested. It was taken from Owl Creek Site 13 Dam in Oklahoma, a low homogeneous dam with no internal filter that failed by piping on the first reservoir filling 1957 (7).

TABLE 2.—Examples of Determination of Failure-Success Boundary, D_{15B}

Clay (Soil 2, Table 1)		Silt (Soil 34, Table 1)	
Filter D_{15}, in millimeters (1)	Test result (2)	Filter D_{15}, in millimeters (3)	Test result (4)
0.7	Success	0.7	Success
1.0	Success	0.77	Success
1.1	Success	0.80	Success
·	D_{15B} = 1.15 mm		D_{15B} = 0.82 mm
1.2	Failure	0.84	Failure
2.0	Failure	1.0	Failure

Fig. 6 presents the Atterberg limits and D_{15B}/d_{85} ratio for each of the fine-grained clays tested (soils with d_{85} less than 0.08 mm, No. 1–18, Table 1). The D_{15B}/d_{85} ratio ranged from 19 to 56. There was clearly no correlation (Fig. 6). Some property of the clay other than the d_{85} size influences the filter needed, but it is not the plasticity as measured by the Atterberg limits.

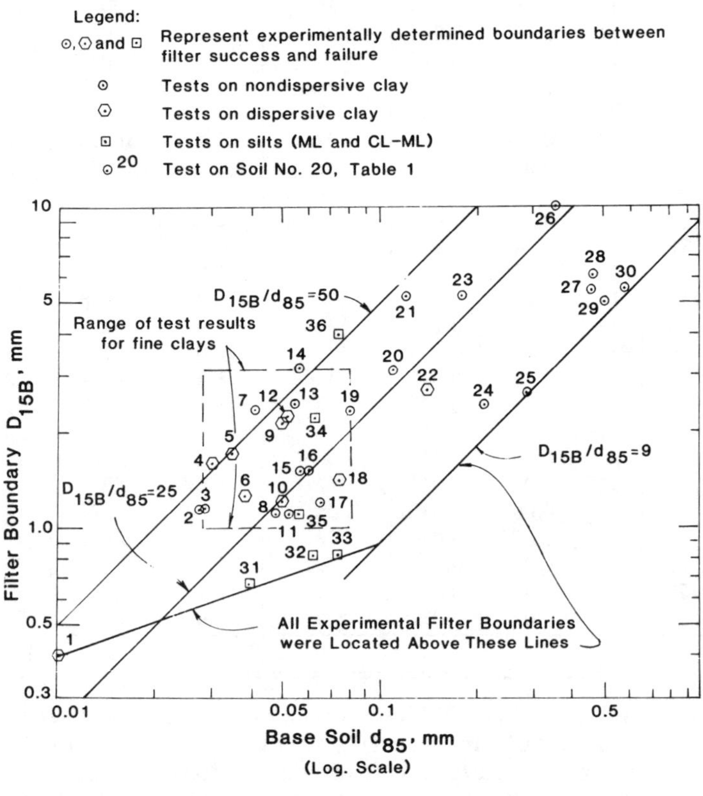

FIG. 5.—Summary of Test Results

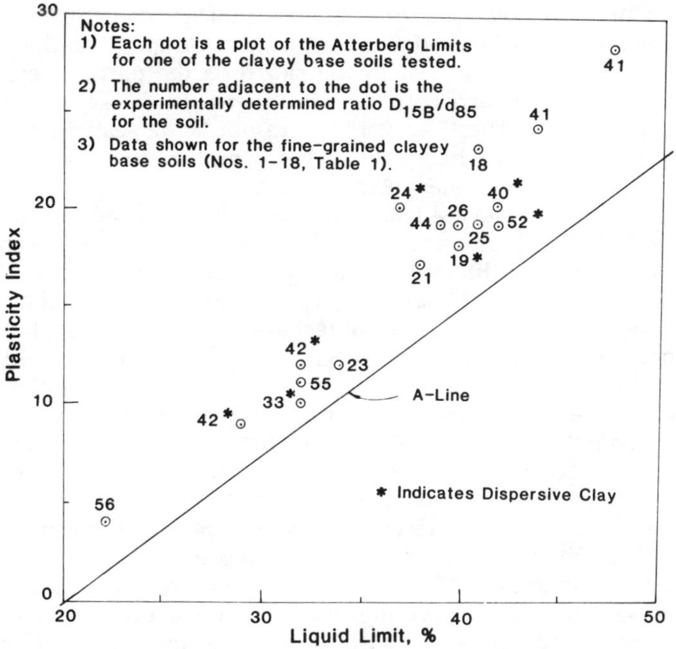

FIG. 6.—Plot Showing No Correlation between Base Soil Atterburg Limits and Experimentally Determined D_{15B}/d_{85} Ratio

Some tests (both slot and slurry tests) were made with distilled water, and some were made with the water from the local water supply system, which had about 8 meq/L of dissolved salts, primarily calcium and magnesium cations. Enough tests were made with both types of water on the same base soil to conclude that there was not a great difference in the results, but the D_{15B} determined using distilled water was consistently about 0.1 mm smaller than the value determined with the tap water.

For the filter boundaries determined experimentally, the ratio of average particle size of filter and base, D_{50}/d_{50}, ranged from approximately 100–2,500, with the most values in the range of 200–500. For most of the base soils it was not possible to determine the d_{15} size, since it was smaller than 0.001 mm—the finest particle that can be measured practically in the hydrometer test. These results support the conclusion that, for silts and clays, filter criteria employing the ratios D_{50}/d_{50} and D_{15}/d_{15} are not meaningful and should be abandoned.

A main aspect of the experimental results, which could not have been predicted, was that the slot test and slurry test gave identical results. In the slurry test the base soil is broken down by mechanical mixing in water to a state in which the particle size distribution is about the same as would be measured in a conventional hydrometer test without using a chemical dispersant. The result is that the largest particles in the slurry, when it is poured onto the filter, are the silt and fine sand particles contained in the base soil, i.e., there is no significant content of larger particles which are "clusters" of smaller particles clinging together.

The identical results show that in the slot tests the particles eroded from the slot walls and carried to the filter were also broken down to the basic particle sizes. The filter voids were not being sealed by small chunks of compacted clay which resisted being broken down by the erosive action of the flowing water in the test.

The slot test is a very severe test in the sense that the velocity of the initial concentrated leak imposed on the base specimen (average about 10 m/s) is much higher than will occur in a leak through a dam. The slurry test is also severe; the final pressure and gradient imposed on the stabilized base specimen in a successful test are very high (gradient about 4 kg/cm^2 across about 4 cm of base specimen thickness, or about 4,000/4 = 1,000). The final gradient in the successful slow test, where the seal is a thin skin of base soil particles on the filter face, is even higher.

Thus, the main elements of the two laboratory tests (high erosive velocity, complete breakdown of base soil to its basic particle sizes and high applied gradient) are conservative when compared to conditions that may exist in a dam which develops a concentrated leak through the core. More likely, the lower velocity of an actual concentrated leak would carry chunks of clay to the filter face, sealing it more readily. It is mainly for these reasons that we believe the results of the slot and slurry tests provide a conservative measure of the ability of a downstream filter to seal a concentrated leak through the core in a prototype dam.

A recent hypothesis (10) suggests that a more severe condition may develop from a slow leak in a clay core. It proposes that only the finest clay particles may be carried to the filter. Silt particles would be left behind as "debris" in the enlarged leakage channel. This hypothesis led to the suggestion that it is conservative, and probably desirable, to use a "perfect" filter which would catch the finest suspended particles. These are "clay flocs," often about 10 μ (0.010 mm) in diameter. We believe, however, that this approach is unduly conservative (6). The main reason is that nearly all fine clays contain a considerable amount of silt-size particles with diameter in the range from 50–80 μ. In any leak through a dam core which erodes particles, even a very slow one, these silt-size particles adjacent to the filter inevitably fall into the water and are available to seal the filter face. Thus, it is not necessary to provide a filter that will catch clay flocs of 10 μ. Silt particles with diameter of 50–80 μ sealed the filters in our tests.

TESTS ON SILT WITH SAND ADDED

As a peripheral experiment, soil No. 31 (Table 1), a fine-grained base silt, was mixed with various proportions of sand to create 3 progressively coarser sandy silts. Their particle size distributions are shown in Fig. 7 (soils 31A, 31B, and 31C). For each of these, a series of filter tests was made to determine the filter boundary, D_{15B} (Table 3). These results confirm that for a sandy silt, the filter boundary, D_{15B}, increases linearly with the d_{85} size.

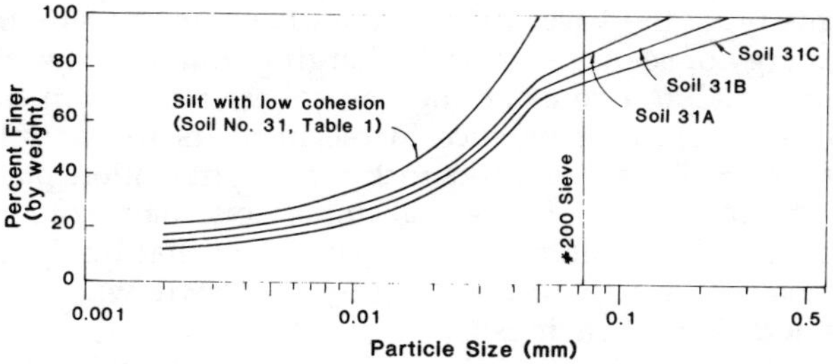

FIG. 7.—Silty Soil No. 31 (Table 1) Mixed with Varying Quantities of Sand to Create Sandy Silts with Increasing d_{85} Size

TABLE 3.—Filter Tests to Determine Filter Boundary D_{15B}

Soil (Fig. 7) (1)	d_{85}, in millimeters (2)	D_{15B}, in millimeters (3)	D_{15B}/d_{85} (4)
31	0.039	0.67	17
31A	0.074	1.1	15
31B	0.105	1.7	16
31C	0.150	2.2	15

Tests for "Noncritical" Filters

The slot and slurry testing was directed at "critical" downstream filters that may be required to control and seal a concentrated leak in the impervious core. For "noncritical" filters, such as those located upstream of the impervious core, a concentrated leak from the reservoir cannot occur. When water does seep toward the upstream filter, e.g., after reservoir drawdown, the maximum hydraulic gradient within the impervious embankment near the filter face is generally less than 1.0.

We conducted some tests to demonstrate that fine filters are not needed in the noncritical locations. One set of tests was made using a silt of low plasticity. This soil, No. 31 in Table 1, was a clayey silt that was one of the most erodible fine-grained soils we tested. Conventional downward-flowing filter tests were made with a compacted base specimen about 8 cm thick. No slot was formed in the base specimen. The filter was a compacted uniform gravel graded downward from 1-in. (25-mm) maximum diameter, and head was about 1.5 m for a gradient of about 150/8 = 19. In these tests the tailwater level was kept above the bottom of the base specimen so that the filter-base interface was always saturated. The relation between filter and base soil particle sizes in these tests was far outside current accepted filter criteria, about as follows: $D_{15}/d_{85} = 150$; $D_{50}/d_{50} = 410$; and $D_{15}/d_{15} = 1{,}360$.

Two separate tests were made with the same base and filter. One was continued for several months. The tests were stable. On dismantling the specimens at the end of the test, we found that the compacted silt had remained in a stiff state at the filter interface with no movement of silt particles into the filter. During the test a small flow of clear water emerged continuously (about 5 ml/hr).

In tests of this type, the base soil specimen is so impervious that the quantity and energy of seepage water discharging from the base soil into the filter are too small to overcome the low cohesive forces in the soil.

There has been a growing practice in recent years to use relatively coarse gravels or small-size quarried rock for the transition placed directly on the upstream face of a central impervious dam core of clay, with no reference to filter criteria. We are confident that this practice is safe and reasonable, even with rapid reservoir drawdown, frequently allowing considerable cost savings.

Segregation Problems in Coarse Filters

The compacted sand and gravel filters used in our tests (Fig. 4) ranged from fairly uniform to moderately graded (C_u generally 1.5–8). Their permeability generally is related to the D_{15} size in the approximate range of $k = 0.2\ (D_{15})^2$ to $0.6\ (D_{15})^2$, in which k is in centimeters per second and D_{15} is in millimeters (8). Their average permeability was about $0.35\ (D_{15})^2$.

The permeability of coarse, well-graded sandy gravels is frequently considerably less than the value calculated from this relationship. Fig. 8 shows the average gradation of a coarse sandy gravel filter (Curve A) used in a recent major dam. (Note that in Fig. 8: (1) Curve A is the average gradation of a processed sandy gravel used as the "critical" filter in a recent major dam; and (2) Curve B is the gradation of the finer half of Curve A.) Permeability tests on representative compacted specimens gave a considerable spread of results, with a median value of the order of $k = 2 \times 10^{-2}$ cm/s. For the filter of Fig. 8, $k = 0.35\ (D_{15})^2$ gives $k = 17 \times 10^{-2}$ cm/s, i.e., the permeability of the well-graded sandy gravel filter (Fig. 8) is only about 2/17 or 12% of the permeability of a uniform sand with the same D_{15} size (0.7 mm).

In a sandy gravel filter, the gravel particles have negligible influence on the filter action. They float in a mass of sand sized particles, and it is the pore sizes in the sand that govern the sizes of particles which will be caught and seal the filter. In the finer half of the sandy gravel filter (curve B, Fig. 8), the D_{15} size is about 0.35 mm (about half of the D_{15} size of the entire material). Using this value in the equation $k = 0.35$

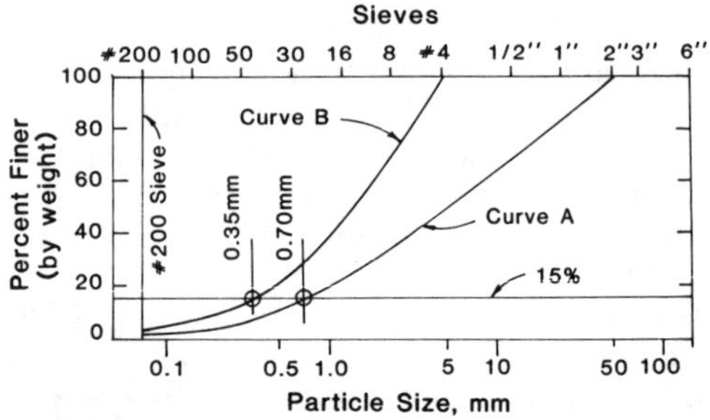

FIG. 8.—Comparison between Total Gradation of Gravelly Sand Filter and Sand Comprising Finer Half (by Weight)

$(D_{15})^2$, the computed k (4×10^{-2} cm/s) is of the same general magnitude as the measured k for the entire material ($2 \times 10 - 2$ cm/s), taking into account the fact that the sand portion (curve B, Fig. 8) occupies only a little more than half the total volume. Thus, the coarse, well-graded sandy gravel of Fig. 8 (with D_{15} of 0.7 mm and C_u about 20) should be able to catch particles about the same size as those caught by a uniform sand filter with D_{15} of 0.35 mm and C_u about 8.

Based on the reasoning and test results illustrated here, we believe that, for well-graded, coarse filters with maximum size particle of 1.5 in. or larger, it is appropriate to evaluate filter criteria as follows:

1. It is conservative to use the D_{15} size of the entire material when applying the filter criteria.
2. However, if permeability tests on compacted specimens show the average permeability is less than about $k = 0.05\ (D_{15})^2$, it is reasonable to use the D_{15} size of the finer portion of the gradation curve when applying the filter criteria. This can be the finer 50% or the sand portion finer than the No. 4 sieve.

Well graded sandy gravels are excellent filters for silts and clays up to a certain maximum coarseness beyond which segregation during construction becomes important. For excessively coarse materials, segregation during construction results in streaks or pockets of gravel particles, without sand in the voids, accumulating at the filter face. If such a segregated filter were used in a dam and an erosive concentrated leak developed, the voids in the segregated gravel pockets would have to fill with eroded fines from the impervious zone before the filter could function.

This segregation problem is difficult to evaluate except by experience. The amount of segregation depends both on the nature of the filter material and on the methods of handling and placing it. In our experience, sand-gravel filters with gravel content and maximum size shown in Fig. 8 (curve A) can be placed without significant segregation if considerable care is taken, but a filter of this gradation is near the limit of desirable coarseness for critical filters protecting fine-grained silts and clays. We have observed segregation problems in sandy gravel filters similar to Fig. 8, with only slightly larger maximum particle size and somewhat larger D_{50} size. These problems occurred despite careful handling and placement. For conservative practice, and to avoid excessive segregation, no more than 60% of a well-graded sand-gravel critical filter for a clay core should be material coarser than the No. 4 sieve. Also, the maximum particle size should be no larger than about 2-in. (50 mm). Many successful dams with fine-grained clay cores have had somewhat coarser down-stream filters, but we believe these criteria are reasonable and not too conservative.

Application to Practice (Critical Filters)

The current technical literature on dams says relatively little about filters for fine-grained clays. Only a few specialized publications offer filter criteria related to clay properties, and the difference between "critical" and "noncritical" filters is not considered.

In current design practice, however, many specialists agree that sand and gravelly sands containing fine sand sizes, and having D_{15} of about 0.5 mm or less, are suitable filters for even the finest clays. Such filters have been used for many major dams. Laboratory filter tests made for the design of individual dams, using various test procedures, have generally given confidence that such filters were conservative. For clays with some sand content ($d_{85} > 0.1$ mm), filters with $D_{15} = 0.5$ mm satisfy the most widely used filter criterion, $D_{15}/d_{85} \le 5$. For finer clays, such as soils 1–19 (Table 1), the criterion is not satisfied by a filter with D_{15} of 0.5 mm. It has been commonly assumed that a finer filter is not needed for these very fine clays because they are sufficiently cohesive and erosion resistant.

Laboratory tests of the types described in this paper can be used to study filters for any given dam design. However, we tested a large number of different clays and silts and believe that the range of results is valid for determining effective and economical, critical filters.

1. *Sandy Silts and Clays.* For silts and clays with significant sand content (d_{85} of 0.1–0.5 mm), the existing main filter criterion, $D_{15}/d_{85} \le 5$ is conservative and reasonable. Plasticity of the base soil does not affect the needed filter.

2. *Fine-Grained Clays.* For fine clays (d_{85} of 0.03–0.10 mm), sand or gravelly sand filters with average D_{15} not exceeding about 0.5 mm are reasonable and conservative. Plasticity or dispersibility of the clay do not affect the gradation needed. Fig. 9 shows our recommended ranges of gradation for coarsest acceptable filters. The bands of sand and gravelly sands of Fig. 9 are considered about equal in filter properties. Use of

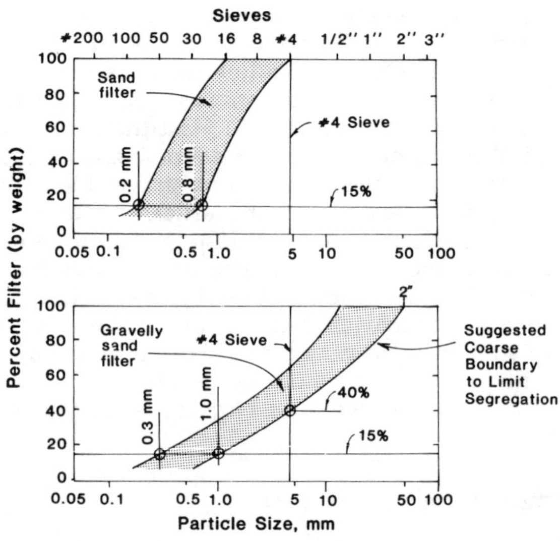

FIG. 9.—Coarsest (Largest D_{15}) Filter Bands Considered Desirable for "Critical" Filters for Very Fine Clays (CL and CH)

filter materials with a larger content of fine sand sizes than shown in Fig. 9 is more conservative.

3. *Fine-Grained Silts of Low Cohesion*. For fine silts without significant sand content (d_{85} of 0.03–0.10 mm) and low plasticity (plotting below the A-line and with liquid limit less than 30) sand or gravelly sand filters with average D_{15} not exceeding 0.3 mm are conservative.

4. *Exceptionally Fine Soils*. Clays and silts with d_{85} less than about 0.02 mm are not very common in nature, and only one was tested in this program (soil 1, Table 1). For soils in this category, laboratory filter tests are desirable, but a filter with average D_{15} of 0.2 mm or smaller probably is conservative for the finest silt or clay.

The particle size distribution curve of the filter does not need to be generally similar to the base soil gradation curve. Filter criteria employing ratios of D_{50}/d_{50} or D_{15}/d_{15} should not be used.

We did not test any tropical residual soils or highly plastic silts (MH), but would expect them to have about the same filter requirements as the clays tested with similar particle size distributions.

It should be noted that the base soils tested in this program were generally fine-grained silts and clays. The coarsest base material was a clayey sand with d_{85} of about 0.6 mm. No coarse internally unstable impervious soils were tested. For certain internally unstable, gravelly impervious soils the fines may enter the filter voids even if the coarser particles cannot (5,6). These soils include obviously gap-graded soils, deficient in sand size particles, and soils broadly graded from clay sizes to coarse gravels. For these base soils which were outside the scope of this study, different filter criteria need to be applied.

Summary and Conclusions

A "critical" filter, such as a downstream filter in a central core dam, should be capable of controlling and sealing a concentrated leak through the core. Critical filters should also be stable in conventional laboratory tests in which a relatively high gradient, such as 1,000, is applied over a short length of core, such as a few centimeters.

Two different types of laboratory tests (slot and slurry tests) were developed using relatively high water pressure. Both produced identical and reproducible results. The two types of tests give identical results. The results realistically and conservatively measure the ability of a downstream filter to seal a concentrated leak through the core of a prototype dam.

The boundary range of filter size between successful and unsuccessful tests for any given clay or silt was found to be narrow and well related to the filter D_{15} size. This failure boundary, D_{15B}, was determined for clays and silts of widely different geographic and geologic origin, including a number of highly dispersive and erodible sodium clays.

The needed critical filters for clays are not significantly influenced by cohesion or plasticity as measured by Atterberg limits. Needed filters are about the same for ordinary nondispersive clays and highly dispersive clays having similar particle size distribution.

Criteria based on our results are not greatly different than criteria in common use.

For noncritical filters, such as a filter upstream of a clay core, it is not necessary to apply quantitative filter criteria. Coarse gravel or small quarried rock can be used, even for a very fine clay core.

Appendix I.—References

1. Kjaernsli, B., and Torblaa, I., "Horizontal Cracks Through the Core at Hyttejuvet Dam," *Publication No. 80,* Norwegian Geotechnical Institute, Oslo, 1968.
2. Kulhawy, F. H., and Gurtowsky, T. M., "Load Transfer and Hydraulic Fracturing in Zoned Dams," *Journal of the Geotechnical Division*, ASCE, Sept., 1976, pp. 963–974.
3. Sherard, J. L., "Embankment Dam Cracking," *Embankment Dam Engineering,* John Wiley and Sons, New York, N.Y., 1973, pp. 272–353.
4. Sherard, J. L., Discussion of "Load Transfer and Hydraulic Fracturing in Zoned Dams," by F. H. Kulhawy and T. M. Gurtowsky, *Journal of the Geotechnical Division*, ASCE, July, 1977, pp. 831–833.
5. Sherard, J. L., "Sinkholes in Dams of Coarse, Broadly Graded Soils," 13th ICOLD Congress, India, Vol. II, 1979, pp. 25–35.
6. Sherard, J. L., Discussion of "Design of Filters for Clay Cores for Dams," by P. R. Vaughan, and H. F., Soares, *Journal of the Geotechnical Division*, ASCE, Sept., 1983, pp. 1195–1196.
7. Sherard, J. L., Decker, R. S., and Ryker, N. L., "Hydraulic Fracturing in Low Dams of Dispersive Clay," *Proceedings*, Specialty Conference on Performance of Earth and Earth-Supported Structures, ASCE, Vol. 1, Part I, 1972, pp. 563–590.
8. Sherard, J. L., Dunnigan, L. P., and Talbot, J. R., "Basic Properties of Sand and Gravel Filters," *Journal of Geotechnical Engineering*, Vol. 110, No. 6, June, 1984, pp. 684–700.
9. U.S. Army Corps of Engineers, "Drainage and Erosion Control—Surface Drainage Facilities for Airfields," *Engineering Manual*, Military Construction, Washington, D.C., 1955.
10. Vaughan, P. R., and Soares, H. F., "Design of Filters for Clay Cores of Dams," *Journal of the Geotechnical Engineering Division*, ASCE, Jan., 1982.
11. Wood, D. M., Kjaernsli, B., and Hoeg, K., "Thoughts Concerning the Unusual Behavior of Hyttejuvet Dam," 12th ICOLD Congress, Vol. II, Mexico, 1976, pp. 391–414.

Appendix II.—Notation

The following symbols are used in this paper:

C_u = coefficient of uniformity of filter = D_{60}/D_{10};

d_{85} = particle size in base soil for which 85% by weight of soil particles are smaller (similarly for d_{50} and d_{15});

D_{15} = particle size in filter for which 15% by weight of particles are smaller (similarly for D_{90}, D_{60}, D_{50} and D_{10});

D_{15B} = D_{15} size of filter found by testing to be boundary between successful and unsuccessful tests for any given base soil; and

k = coefficient of permeability, in centimeters per second.

Introduction to Paper 15

Trends and Debatable Aspects in Embankment Dam Engineering

by J.L. Sherard
Water Power and Dam Construction, Vol 36, No. 12,
December 1984, pp 26-32.

Prepared by Edward B. Perry[8], M. ASCE

In September 1984, Jim Sherard sent me a copy of his notes from a lecture presented at the 3rd Colombian Geotechnical Seminar, Bogota, Colombia, on August 7, 1984. An abbreviated version of these notes was published as this paper. Also, I was fortunate to attend the University of Missouri-Rolla 11th Annual Short Course on Embankment Dams Soils Aspects held in San Diego, California, in February 1985, where Jim, as the principal lecturer, covered most of the topics in this paper.

At the time Jim wrote this paper he had concluded his joint research with the Soil Conservation Service Soil Mechanics Laboratory in Lincoln, Nebraska, on dispersive clays, filter design, cracking in dams, and broadly graded core materials. Jim believed the downstream filter was the primary line of defense to control concentrated leaks through the core of an earth dam and other design measures to reduce differential settlement cracks were of secondary importance. Therefore, Jim noted a trend away from providing multiple lines of defense (advocated by Arthur Casagrande) against possible concentrated leaks in cracks through the core of an earth dam. While this has not been universally accepted (see Peck, R.B., 1990, "Interface Between Core and Downstream Filter," H. Bolton Seed Memorial Papers, BiTech Publishers Ltd., Vancouver), Jim would certainly have clarified the subject if he had lived to give the ASCE Karl Terzaghi lecture in August 1988.

[8]Research Civil Engineer, U.S. Army Engineer Waterways Experiment Station, Vicksburg, MS.

Construction sequence for placing flexible grout

- Drill hole, maintaining proper alignment.
- Clean grout hole of loose material.
- Prepare surfaces of hole according to the recommendation of manufacturer.
- Attach 2 ft × 2 ft (60 cm by 60 cm) pieces of burlap to a 45 ft (15 m) length of 2 in (5 cm) perforated p.v.c. pipe.
- Saturate the entire assembly with flexible grout and place down the hole.
- Pump in additional flexible grout down the 2 in (5 cm) perforated p.v.c. pipe to assure adequate filling of the hole.
- Fill top of hole with non-shrink grout.

Trends and debatable aspects in embankment dam engineering

By J. L. Sherard
Consulting Engineer*

Construction practice for embankment dams has undergone a number of changes over the last 10 to 20 years. Some new processes have evolved, others have disappeared. The changes discussed here by the author are based on current practice, although he cautions that not all of these are accepted or recognized by every engineer.

THE INCENTIVE for changes in practice comes primarily as an effort to reduce cost, but also in some cases from the desire to increase the basic margin of safety.

The most easily accepted trends are those for which the engineer can be convinced he will achieve both objectives. The most energetically debated trends are those for which the engineer has trouble evaluating the influence of the change in conservative attitudes.

One result of the gradually changing practice is that the specialized engineering literature dealing with these changes lags behind significantly. The busy specialist practitioner writes comprehensive articles or books only infrequently and it is often 10 to 15 years after a change in practice has been fairly well established, and used in projects in various parts of the world, before there is a reasonably satisfactory description in the literature.

Understanding of dam behaviour is the fundamental base from which changes in practice must be evaluated. In this article a number of different aspects of embankment dam engineering will be described where the author believes justified trends in practice to be under way. For these the relationship will be shown to lessons learned from dam performance, including in particular malperformance and failures.

The arguments will be limited to the problems of the embankment dam itself, assuming that the foundation is competent in all ways, such as hard, impervious rock. Also, in general, the article will focus on practice with moderately high to high zoned dams with internal earth cores and shells, and homogeneous dams, with chimney drains, generally used for reservoirs for hydroelectric projects. However, most remarks also apply generally.

Downstream filters

Until recently most of the research and testing which supports current filter design criteria for impervious embankment sections has been made in the laboratory where water was caused to seep slowly through the pores of the impervious (base) specimen and discharge into a filter. Because of this, there was a question as to whether a filter meeting these criteria would also be effective in controlling erosion in the more severe condition where a concentrated leak develops through the dam core from the reservoir and discharges into the filter.

There is much evidence that small concentrated leaks occasionally develop through the impervious sections of embankment dams designed and constructed according to good modern practice. The evidence is in the following main categories:

- Concentrated leaks have merged soon after the first reservoir filling on the downstream slope of small, well constructed homogeneous dams without internal drains. Such leaks are frequently at points of maximum differential settlement. They sometimes lead to breaching[1,2].
- In some recently constructed central core dams, piezometers at the downstream face of the core have registered nearly full reservoir pressure. Special borings have been made in the impervious cores of these dams to investigate the peculiar conditions. When these borings reached certain elevations in the impervious core, water entered the hole and rose rapidly to the pressure head of nearby piezometers. This rise shows conclusively that a concentrated leak was present[3,4].
- In several instances concentrated leaks have suddenly emerged at the downstream toe of modern central-core dams; these leaks have usually occurred during or soon after the first reservoir filling[5]. In some of these instances extensive investigations have concluded that hydraulic fracturing was the only likely cause of the leak[1,6,7,8].
- Before filling the reservoir, open cracks have been observed that extended below the full reservoir elevation.
- Calculations by the finite element method confirm that, for well constructed dams, the stress in the core is frequently low enough to allow hydraulic fracturing, even without unusual differential settlement[9,10].
- Small concentrated leaks can occur along the inevitable small open cracks at rock foundation contacts. At Yard's Creek upper reservoir dam, New Jersey, these leaks were observed in a test pit sunk through the core to the rock foundation[1].

There seems to be already sufficient evidence from dam behaviour, supported by theory, to require the designer to assume that small concentrated leaks can develop through the impervious section of most embankment dams, even those without exceptional differential settlement. Also, in areas of strong seismic activity, slumping caused by earthquake shaking can lead to differential settlement cracks.

Analysis of the overall experience with dam performance shows that there are almost no cases of damage or failure of embankment dams attributed to inadequate

*3483 Kurtz Street, San Diego, California 92110, USA.

filters, when the filters were designed according to accepted conservative practice. Almost all of the damages and failures as a result of piping have occurred in two special groups of dams:

- homogeneous dams without chimney drains (filter-drains) in which a concentrated leak develops through the dam and exits on the downstream slope without passing through a filter; and,
- a number of dams with impervious sections consisting of coarse, broadly graded soil, which had excessively coarse filters.

Hence, dam performance experience supports the conclusion that filters designed to meet current criteria will safely control concentrated leaks.

Also, as a part of design studies for many specific dams, filter tests have been made using compacted impervious embankment specimens with an artificial rack (a slot or circular hole) with water caused to flow through the crack and discharge into the filter. These generally have shown that filters meeting accepted conservative criteria would prevent significant erosion and safely control the concentrated leak in a stable equilibrium condition, or seal it. Recent concentrated laboratory research, by the staff of the Soil Conservation Service of the US Department of Agriculture, has confirmed that accepted criteria for filters under normal seepage conditions are generally also satisfactory to control and seal concentrated leaks, even with relatively high water pressures, velocities and gradients[11].

Hence, both our overall good experience of filters in dams and laboratory filter tests with concentrated leaks seem to show that downstream filters can be relied on to control safely concentrated leaks from any cause through impervious sections of embankment dams.

My confidence in the reliability of filters has been strengthened considerably following participation in the recent extensive Soil Conservation Service research over the past three years[11,12]. As a part of this work I assisted in the laboratory making many dozens of tests, observing directly the behaviour of the tests in which rapid flowing concentrated leaks were caused to flow through the compacted specimens of impervious soil and discharge into filters of varying coarseness. A main observation from this activity was that a conservative filter was remarkably effective in preventing erosion and sealing off the concentrated leak. With a conservative filter I saw that it is essentially impossible for any significant quantity of the impervious embankment core material to be carried into the filter. It is one thing to read about successful laboratory experiments, but it is much more convincing to see the action of the filter, repeated many times under very severe conditions, stopping the entry of soil particles and sealing and controlling the leak.

For all these reasons I support the trends in design practice which are based on the conclusion that "conservative downstream filters in embankment dams can be relied upon to control safely concentrated leaks".

Because of the important role of controlling possible concentrated leaks through the dam core, downstream filters have been characterized as "critical filters", as compared with filters in other locations within the dam. Research has confirmed that sands and gravelly sands with average D_{15} size of 0.5 mm or smaller are conservative filters for most fine-grained clays in nature, with d_{85} size of 0.03 mm or larger[11].

Cracking differential settlement

Starting about 1950, the profession began to overcome concerned about cracking observed in earth dams, even well constructed dams without unusually large settlements[1]. Previously it had been generally considered that embankment dams were too soft and deformable to sustain open cracks. In the following years various design measures were used to: (1) reduce differential settlement and the likelihood of a concentrated leak; and (2) control leaks in differential settlement cracks which develop in spite of preventative measures.

In the first category seven techniques are generally used:

- excavation of rock abutments to achieve flatter slopes and reduce abrupt changes in slope;
- use of upstream sloping impervious core instead of vertical for central core dams;
- compaction of the bottom half of the impervious earth core at relatively lower water content and to as high a density as practicable;
- compaction at higher water content for the upper part of the impervious core to make it flexible and capable of following imposed strains without cracking;
- use of a plastic clay in preference to silty sand for an impervious core, it is also more flexible;
- use of a strip of impervious core at especially high water content at the abutment contacts, sometimes also using a more plastic clay for this strip; and,
- arching embankment dams in plan, to increase the longitudinal stress when water pressure pushes the dam crest downstream.

There are two methods used in the second category:

- use of plastic clay for the impervious core believed to be most resistant to erosion of concentrated leaks; and,
- use of a conservative downstream filter.

In this second category the main defence against concentrated leaks in differential settlement cracks was the design effort to reduce the differential settlement and prevent the leak from occurring. The downstream filter was generally considered to be an essential secondary line of defence.

In the three decades since cracking first became of general concern, we have built higher embankment dams, at more rapid construction rates and in more narrow, steep-walled rock canyons, imposing increasingly severe differential settlement conditions. In spite of the various design measures to reduce differential settlement, cracks continue to develop fairly commonly on the dam crest, both transversely and longitudinally, soon after completion of construction. They are usually vertical and remain open only to depths of 3 to 6 m, below which they are closed by the internal embankment pressure.

Since about 1970 it has been speculated that in addition to the open cracks seen to appear on the dam crest, small leaks may develop through impervious earth dam sections by the process of hydraulic fracturing, also caused by internal stress transfer from differential settlement. Without going into this subject in detail, it is probable that small leaks do develop through embankment dam cores from this action fairly commonly, and must be anticipated.

In any event, as discussed earlier, experience with dam behaviour has shown practically no problem of erosion of concentrated leaks through dam earth cores where there were reasonable downstream filters. As a result of this experience, there has been less concern about differential settlement. The trend among many specialists, which I believe is justified, and in the correct direction toward both conservatism and economy, is to reverse the previous emphasis and consider that the downstream filter is the primary line of defence, and the other design measures to reduce the differential settlement are of secondary importance.

The influential technical literature on embankment dams of the last 20 years strongly emphasises that it is desirable to provide several lines of defence in important central core dams against possible concentrated leaks in cracks through the earth core. Multiple lines of defence were considered prudent because of the absolute need to prevent erosion of concentrated leaks and because no single technique was considered to be completely reliable. The current trend seems to move away from providing multiple lines of defence. As a result of this trend I note the following results:

- Less emphasis on the need for large excavations on rock abutments under the core foundation to regularize the shape or reduce the average slope. Much steeper rock abutments are accepted than were considered reasonable previously.
- The construction of more high embankment dams in steep narrow-walled canyons, where it would generally have been considered previously that only a concrete dam was suitable. An example is the 240 m-high Chivor (Esmeralda) dam, Colombia, built in a canyon with a crest length of about 320 m, completed in 1975. The dam has performed without problems since then[13]. This is currently being followed by others.
- Less emphasis on trying to make the lower part of the core more incompressible and the upper part more plastic (flexible). The trend even for high dams is to compact the entire core at the same water content in the same way.
- Fewer dams being built with a curve in plan for arching. Straight axes are considered satisfactory for even the highest embankment dams in narrow canyons.
- Less importance being accorded to the geometry and width of the impervious core in central core dams from the standpoint of differential settlement. These are generally chosen from other considerations.
- Increasing the safety margin of the main downstream filter, where high conservatism concerning differential settlement problems is considered desirable (because of steep abutments, great heights of dam, and so on). This is obtained by making it finer (smaller than D_{15} and more uniform particle size distribution), and thicker, and by increasing the width at rock foundation contacts. The cost of a substantial increase in the basic conservatism of the main downstream filter is usually not an important item compared with the other factors involved.

Impervious earth core material

The current trends to place more reliance on filters and to be less concerned about differential settlement has influenced opinions about appropriate criteria for impervious earth core materials.

In the past many engineers dealing with dams considered that clay was the best material for the impervious core. In some geographic areas this opinion was so entrenched that the terms "core" and "clay core" were practically synonymous. There are many examples in the last 20 years where clay has been hauled over great distances (30-40 km or more), at considerable cost, although large uniform deposits of excellent cohesion less silty sand were available immediately at the dam site. Clay was considered superior to cohesionless impervious materials because it would be more deformable and better able to follow imposed strains from differential settlement without cracking, and because it would have higher resistance to erosion if a concentrated leak developed from any source. Highly plastic ("tough") clay was particularly prized.

In view of the current trend, this preference for clay over cohesionless impervious soils is not considered valid. The two materials act differently in the presence of a concentrated leak in a crack. Each has theoretical advantages and disadvantages, and it is not clear that either is superior. The tough, highly plastic clay will definitely resist erosion of the walls of the concentrated leak much more strongly than a cohesionless silty soil. But, the relatively high unconfined compressive strength of the clay core provides more likelihood that the embankment material can arch around the leakage channel and keep it open. Cohesionless silty sands have the theoretical advantage over clay that they cannot sustain an open crack in an embankment, since they have no (or very little) unconfined compressive strength and there is a much greater tendency for any crack (which may be at first held open by water pressure) to collapse.

From the standpoint of resistance of clay to erosion of concentrated leaks, in the current trend it is believed that this is not important. All clays will erode under severe conditions and it is not considered reasonable or conservative to rely on the relatively feeble erosion resistance of a clayey soil to provide defence against piping of a concentrated leak. The principal job of the core is to be impervious. Prevention of piping is the job of the filter.

In the past, sometimes a good core material has been made by blending together two materials with radically different properties. For example, at some dams cohesionless fine silts (or fine clays) were mixed with sandy gravel to obtain a core material with the imperviousness of the silt and the high shear strength and low compressibility of the sandy gravel. This has been done only rarely, but for some major dams. The current trend would be away from this practice and to use the silt for the core and the sandy gravel in other zones of the dam.

In summary under this point, the current trend is to use the least costly of the available impervious soils for a central core, although it may be completely cohesionless and relatively erodible when exposed to concentrated water flows, and may be relatively compressible when compacted in the dam.

As historical background for this point, it is desirable to record that the older strong preference for plastic clay impervious cores was not at all universal. In some parts of the world there are no plastic clays available for dam construction. In those areas many major dams have been built over the last 30 years with impervious cores consisting of cohesionless silty sands, sandy silts, gravelly and sandy silts, and so on. In some areas these materials have been largely decomposed granites or similar rocks (usually cohesionless silty sand). In other areas great numbers of dams have been built with impervious cores of glacial moraines, often in the form of cohesionless gravelly silty sand. In other areas, impervious dam sections from weathered sandstone, often in the form of very fine, uniform silty sand, are common. A list of dozens of major dams built with cores of completely cohesionless silty soils in the last 30 years could easily be compiled.

Because of the above experience it may seem unnecessary to emphasize this point so strongly in this paper. I do it because I have observed so many heated debates on this subject only in the last few years. It is necessary to record that some engineers of long experience have very strong opposite opinions that it is worthwhile to spend a lot of extra money to obtain a clay core.

Seepage cutoff collars on concrete structures

Before about twenty years ago it was standard practice to provide cutoff collars on concrete conduits passing through the base of embankment dams, over the length of embedment in the impervious core. These were simply walls protruding out into the surrounding dam embankment. Similar collars were commonly provided on spillway walls and other concrete structures where embankment dams connected to them. The purpose was to force the particles of water which were seeping from upstream to downstream along the earth-concrete interface to have a longer path and not a straight line.

Over the years many engineers concluded that these cutoff collars had the strong disadvantage that they made it impossible to compact the impervious earth core directly against the concrete structure in a reliable fashion with heavy rollers. A large amount of compaction by hand labour was needed, which was clearly not desirable. Because of increasing concern over this problem, and at the same time increasing confidence in the efficacy and reliability of filters, over a period of years the number and size of these collars were gradually reduced, and special attention was paid to the downstream filters. This trend continued until the collars were eliminated altogether.

The common current practice, which I agree is appropriate, is generally as follows:

- No seepage collars provided.
- Sloping of the exterior concrete surfaces at the concrete-earth core interface slightly off vertical (such as about 1H:8V or 1H:10V) so that the earth can be compacted directly against the concrete with wheels of a rubber-tyred vehicle running parallel with the concrete face.
- Paying special attention to the downstream filter.
- The filter is often wrapped completely around the downstream portion of conduits; that is, underneath as well as on both sides and the top, so that all potential leakage water travelling along the concrete-earth core interface must exit in a controlled fashion in the filter.

I have not seen cutoff collars used on major projects for five years or so. However, the handbooks still show them as standard practice and they are still being used occasionally.

Downstream filters for impervious cores of coarse broadly graded soils

One of the more common soils used for impervious dam sections is relatively coarse, broadly graded material, with particle sizes ranging from clay sizes (0.002 mm) to coarse gravels (150 mm or larger), and including all intermediate sizes. Soils of this type are commonly of glacial origin but also are often from alluvial-colluvial and weathered rock origins. These materials frequently have a particle size distribution curve which is nearly a straight line on the commonly used semi-log particle size distribution plot.

For these materials there has been a question in the profession as to how to apply the main filter criterion ($D_{15}/d_{85} \leq 5$). Since the d_{85} size of the material is commonly of the order of 20 to 30 mm, a direct application of the above criterion would allow very coarse uniform gravel without sand sizes as a downstream filter, which was clearly not reasonable.

Since 1955 there have been about 15 dams built with cores of this type of material which have had bad experiences with concentrated leaks and extensive piping of the core[5]. This is the group of dams which is the exception to the general experience that filters have been successful.

In most of these damaged dams, relatively coarse downstream filters were employed, such as crushed or grizzled rock fragments (otherwise unprocessed), graded from 3 in to 6 in (7.5 cm to 15 cm) down to fines, or coarse river alluvium with a large content of coarse gravels. Such coarse filters commonly segregate badly during construction so that the interface between the earth core and downstream filter commonly consists of extensive pockets of segregated coarse gravels with no sand sized particles filling the voids.

These coarse, broadly graded impervious soils can be considered to be internally unstable, in the sense that if a sample is separated in two parts by screening at about the average particle size, the coarser portion is too coarse to act as a filter for the finer portion, according to our accepted filter criteria. The finer half is commonly a fine-grained silt or clay and the coarse half is a coarse gravel with only a small sand content. Recent research at the Soil Conservation Service Main Soil Mechanics Laboratory, Lincoln, Nebraska (unpublished) has shown that these soils act noticeably differently from ordinary materials in filter tests. In successful filter tests on ordinary impervious soils, the coarser particles plug the voids of the filter at the interface, preventing any significant further erosion. In similar tests with these broadly graded coarse soils, the coarser particles in the impervious soil which are much larger than the filter voids, accumulate on the face of the filter and interfere with the filter action.

Finally, it is now fairly well established from the evaluation of the records of the damaged dams and from laboratory experiments that the downstream filter for these materials must be fine enough to act as a filter for the portion of the total material which has particles smaller than about the fine sand size. A conservative and reasonable approach is to require a filter which will safely control erosion for the silt fraction of the material alone. A sand or gravelly sand filter with average D_{15} size not exceeding about 0.7 mm is conservative.

In summary for this section, both experience with dam performance and laboratory filter research show that dams with impervious cores of these coarse broadly graded soils need finer downstream filters than previously believed necessary. There is no way to apply the currently accepted filter design criteria for these materials. Sand filters with $d_{15} = 0.7$ mm or smaller are conservative.

Upstream filters for clay cores in central core rockfill dams

One of the most common dam types built in the last 30 years has been the central core rockfill dam, often with a clay core. In such a dam there are commonly two or three zones of progressively coarser filters (sometimes called transitions) between the core and the downstream shell. Twenty years ago the common design was to make the filters upstream of the core the same as those downstream; that is, the same number of zones and gradation of materials and the same widths of zones.

In the last twenty years there has been a gradual trend toward making the upstream filters in these dams more economical. This was motivated by the realization that the downstream filters were the critical defence against piping of the core from leaks from the reservoir, and consequently their conservatism is vital to the safety of the dam. The upstream filters, however, serve a much less important role.

The main function of the upstream filter is simply to prevent the clay core material from migrating in the upstream direction into the voids in the upstream rockfill shell and, theoretically, to act as a filter for the core following reservoir drawdown. In fact, consolidated under the pressure inside the dam the clay core generally remains relatively stiff, so that there is no tendency for it simply to migrate by falling under gravity action in the upstream direction. Also, for an imprevious clay core, the amount of water which seeps out of the core in the upstream direction following drawdown is very low. The maximum gradient for the seepage in the upstream direction following drawdown is less than 1, and the velocity and energy of the seepage entering the upstream fiter is too small to erode clay particles and carry them upstream. Such an upstream filter is considered a non-critical filter because it is never called upon to act to control a concentrated leak.

As a part of the recent filter research at the Soil Conservation Service Soil Mechanics Laboratory, tests were made with relatively erodible clays of low plasticity (using a gradient of about 20 and very coarse filters). These tests showed clearly that it is not necessary to apply the current fiter criteria to such non-critical filters as the upstream filter on a central core dam[11]. Hence, the laboratory research results support the trend is practice to make the upstream filter more economical.

The first steps in this trend resulted in making the upstream filters with less width than the downstream filters. The trend is continuing and the current designs for a considerable number of high dams retaining very large and important reservoirs have only one coarse transition between a very fine clay core and the upstream rockfill shell. This transition is commonly a zone of grizzled or crushed small rock, such as from 150 mm maximum size down. The relationship between the particle sizes of such a small rock transition zone and a fine clay core is completely outside, and bears no resemblance to our current general criteria for acceptable filters.

The above conclusion is valid for clayey cores because their unconfined compressive strength and low permeability eliminate any significant tendency for the material forming the upstream face of the core from migrating in the upstream direction. I believe the conclusion applies to any core material which 30 per cent or more of fines (passing the no. 200 sieve) of clay or cohesive silt with plasticity index of 8 or higher. For cores of cohesionless silty sands and clayey silts of low plasticity, with plasticity index of less than 8, the material of the upstream face of the core can be reasonably imagined to migrate upstream into transitions with large voids, under the combined action of gravity and seepage following reservoir drawdown. Especially for dams retaining reservoirs which are frequently lowered (such as pumped-storage reservoirs), and for dams with thin, near vertical central earth cores, with steep upstream faces, it is conservative to provide an upstream filter meeting the well-established main filter criterion, $D_{15}/d_{85} \leq 5$, for cores of low cohesion, and to limit D_{15} to maximum of about 10 mm for cores of coarse silty, sandy gravel.

Upstream sand zones in central core dams as crack fillers

A fairly common recent design detail in central core dams is a sand zone (filter) at the top of the dam on the upstream face of the impervious core, such as those with a width of 2 to 4. The principal purpose is to provide a reservoir of sand which would be washed into a crack which might develop through the upper part of the impervious core from any cause, such as differential settlement or earthquake shaking.

To serve the intended purpose, the maximum particle size of the sand needs to be smaller than the width of the crack which might develop. Hence, sand with a maximum size not larger than about 1 mm has commonly been considered desirable, so that it could reliably be washed into cracks with widths of several millimetres without bridging over and blocking the entrance. In some cases designers have provided coarser sand, such as one with maximum size of 0.25 in or even 0.5 in (6 mm or 13 mm), considering that it might be useful if a major crack opened.

The potential value of this upstream sand "crack filter" is debated at the present time, with strong opinions on both sides. Many important central core dams in recent years have been built without these zones, even though suitable sand was available at reasonable cost.

The several main arguments raised against this upstream sand zone are:

- The filter downstream of the core is the main line of defence for controlling and sealing concentrated leaks, and it must be relied upon. Hence, if there is any question about the adequacy of the downstream filter, it should be made more conservative and the question eliminated, rather than providing another line of defence of questionable efficacy upstream.
- It is difficult to be confident that the upstream sand would fill the leakage channel. The velocity of flow in a concentrated leak, even in a fairly wide open crack, is not likely to be very high because it is limited by the permeability of the downstream filter, the upstream face of which should be progressively clogged by eroded debris from the core.
- There is an argument as to whether it is desirable for the upstream sand to enter a crack, because a limited amount of sand may enter and tend to prop the crack open and tend to prevent it from sealing itself off by softening and collapse or swelling of the crack walls.
- In addition to the unit cost of the upstream sand zone itself, the sand zone upstream usually requires another band of coarser transition to hold the sand in place, which complicates the construction and adds more cost.

I believe that there is more of a debate than a trend for this design feature. I never propose to add such a zone to a design, but would not object if someone strongly proposed to use it.

Dispersive clay: present status

Largely because of the pioneer research efforts of Australian workers and the efforts of the US Soil Conservation Service, it is now widely recognized that clays in nature can be roughly grouped into two main categories, having fundamentally different properties with respect to erodibility:

- ordinary clays, which are relatively erosion resistant; and,
- dispersive clays, which are highly erodible by a process in which clay particles on the surface of a clay mass immersed in water are repelled from their fellows and go off into suspension in practically still water with negligible erosive energy.

This great difference is caused by differences in the dissolved salts in the clay pore water. Ordinary erosion-resistant clays have a preponderance of divalent calcium and magnesium salts in solution. The dispersive clays have primarily sodium salts in solution in the pore water.

This is a relatively new discovery for the practising dam engineer. The first early research results relating to behaviour of dispersive clays in dams were published in the mid-1960s. These led to research in the USA, published in the early 1970s and finally to an international symposium sponsored by the ASTM in 1976. The proceedings of the ASTM symposium are still the best single literature source and a good starting point for study of the subject[14].

Although relatively recently introduced into the dam building business, the concept has been accepted rapidly around the world and engineers working with dams in almost every country are aware of dispersive clay and making appropriate laboratory tests of their clay embankment materials.

The following is a brief summary of the present status and the trend in practice regarding the problem of dispersive clay related to embankment dam engineering:

• Dispersive, sodium clay is found much less commonly in nature than ordinary clays. It is found much more commonly in some areas and geological settings than others, but at the present time the engineer must anticipate he could find it anywhere.
• Dispersive clay can only be reliably recognized from the results of laboratory tests. One of the apparant paradoxes associated with this subject is that some of the most erodible sodium dispersive clays have the physical appearance and consistency of clays of high plasticity, which are tough and hard when remoulded at the plastic limit, and would formerly have been considered as being among the least erodible core materials for embankment dams.
• The best two laboratory tests are the so-called "pinhole" test, which is a direct measure of the tendency of the clay to disperse in water and the test measuring the relative quantities of dissolved salts in the pore water. The two different types of tests generally give the same result, but not always. Therefore, it is desirable to make both.
• Recent concentrated laboratory research has shown that sand or gravelly sand filters with average D_{15} = 0.5 mm or smaller will safely control and seal concentrated leaks through compacted specimens of the great majority of dispersive clays (with d_{85} larger than about 0.03 mm)[11]. Sand filters with average D_{15}=0.2 mm or smaller are conservative for the very finest dispersive clays.
• Several major dams have been built in the last few years with impervious sections which were identified as being of dispersive clay. This is generally considered reasonable practice for the future.
• Dispersive clay is changed completely to its non-dispersive state and made relatively resistant to erosion by blending with lime. In the laboratory 1 per cent lime by weight of clay is usually sufficient. In construction a minimum of 2 per cent has been generally used.
• For dams with dispersive clay cores on rock foundations, special consideration must be given to preventing the clay penetrating small rock cracks. This is the main difference in design from that with cores of ordinary clay. A thin layer of lime-treated clay (or other non-dispersive impervious soil) placed directly on the rock foundation is considered a satisfactory detail.
• The existence of dispersive clay is particularly important for small, low-cost homogeneous dams for which no vertical chimney drain-filter might ordinarily be used. For such a dam of dispersive clay there is a much greater likelihood of breaching than when constructed of ordinary clay.

Divergent and convergent foundations

Another somewhat debatable item has to do with the slope of the foundation surface (especially rock foundations) under central core dams. Historically it has been considered desirable that the foundation surface under the core (as seen on cross-sections drawn perpendicular to the dam axis) be horizontal or sloping upstream (convergent). The idea behind this was that if the core tended to move slightly downstream under the push of the reservoir, it would move on a horizontal surface or would have to climb uphill and wedge, creating a better seal with the foundation.

Occasionally a situation arises in which the only suitable location for the dam is where the foundation rock surface under a substantial portion of the core slopes away downstream.

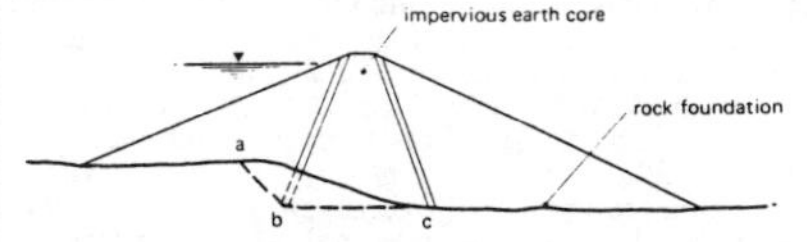

Foundation treatment for an embankment dam on a sloping foundation.

When this condition arises there is sometimes a debate as to what to do. A few specialist engineers at the present time would argue that it is desirable to excavate the rock under the core (see a, b, and c in the Figure) to provide a horizontal surface when seen on a transverse section.

Assuming that the rock under the core is of suitable quality for the foundation, the current design trend is to accept the natural shape of the rock surface and build the dam on top of it. There is no significant tendency for the lower part of the impervious earth core to move or slide with respect to the rock foundation surface. Hence, excavating the rock to a horizontal surface would not increase the dam safety margin. It would only remove a certain quantity of good hard rock and replace it with embankment or much inferior average properties.

Concrete face rockfill dams

One of the strongest trends in the industry is the recent rapidly increasing use of the rockfill dam with upstream facing of reinforced concrete.

This kind of dam has had a long and turbulent history, briefly as follows:

1920-40. Embankments of dumped rockfill reached a maximum dam height of about 100 m. concrete slab in separate square panels used with water stops on all sides. Dams commonly experienced considerable long-time settlement, openings in water stops and large leakages.

1940-65. More or less the same practice continued. Quite a few dams in the range of height from 50 to 100 m, and one of 150 m (New Exchequer, California), were built. Bad experience with settlement continued, broken water stops and high leakages. Many engineers have concluded that this is not a satisfactory dam type because of the leakage problem, and it is not considered at most sites. At the end of this period, the practice of placing rockfill in thin layers commenced.

1965-80. (Period of rapid development.) Almost all the dams were built with rockfill placed in thin layers and compacted with smooth steel drum vibrating rollers, and post-construction settlements occurred less than previously. The upstream slab details changed, eliminating horizontal joints, running reinforcing steel continuously from bottom to top of dam, and gradually reducing the slab thickness. There was less trouble with leakages. Gradually more engineers became convinced that this was a good type of dam. Many dams of this type built in Australia. The 160 m-high Areia dam in Brazil was completed, having the biggest area of concrete face and largest reservoir; its excellent performance attracted much attention from the profession internationally.

1980-present. This type of dam is now generally accepted and being seriously considered as an option at most sites with rock foundations. It generally competes with the rockfill dam with a central earth core, as well as the concrete arch dam. More dams of this type than ever before are under construction at the present and many are in various stages of design, all over the world.

The bedding layer under the concrete slab is commonly crushed rock (minus 3 or 6 in (7.5 or

15 cm) heavily compacted in thin layers. There is a trend to requiring that this material have an increasing content of sand sizes and a few per cent of silt or clay fines, to make it semi-pervious, and sufficiently well-graded to be reliably internally stable. This semi-pervious layer (k about 10^{-4} to 10^{-5} cm/s) assures that the total leakage through dam cannot be large, even if large cracks developed in the concrete slab, such as from a strong earthquake. It also makes the dam completely safe even if there is no concrete slab on the upstream face, because the amount of water which can seep through the upstream semi-pervious gravelly zone is far less than the capacity of the rockfill dam to carry away the seepage safely. I believe that more attention will be paid to design details of this layer below the concrete slab in the future. With relatively small changes, this layer can be made relatively more impervious and reliably self-healing against leaks in the slab while still retaining high shear strength needed for the steep upstream slope.

Some of the principal advantages of the concrete face rockfill dam, compared with the central earth core rockfill dam, are:

- smaller embankment volume, simpler construction without internal core and filter zones;
- more rapid construction;
- shorter hydraulic structures; diversion and outlet tunnels, spillways;
- foundation grouting independent of embankment construction;
- fewer problems with rainy weather;
- fewer problems with steep canyon walls, overhangs, etc.;
- easier to use rock directly from required excavation without stockpiling; and,
- generally lower construction costs.

The concrete faced rockfill dam is considered to have high inherent resistance to strong earthquake shaking. Since the water pressure acts on the upstream face, the entire rockfill embankment mass acts to provide stability. Since there is no water inside the embankment there cannot be any significant deformation of the compacted rockfill caused by the earthquake shaking. The concrete facing could be cracked in a strong earthquake causing leakage, but it is believed essentially impossible that this type of dam on a good rock foundation could be so severely damaged as to threaten failure.

It is to be expected that this type of dam will be used with increasing frequency in the future. Until now its maximum height has been 160 m, but higher structures have been considered at a number of sites in recent years. There is no good reason that it cannot be built to the same height as rockfill dams with central earth cores (now up to about 250 m). Because of the intrinsic conservatism offered by the fact that the entire embankment is downstream of the plane of application of the reservoir pressure, it is possible that in the near future it will be considered appropriate for the very highest dams.

The current widespread acceptance of the concrete faced rockfill dam and the remarkable increase in the frequency of its use all over the world in the last few years is a result largely of its fundamental technical and cost advantages. It is also a result of the influence of J. B. Cooke, Consulting Engineer, San Rafael, California, who has been a strong and persuasive advocate of this type of dam for more than 40 years, and who has been associated with the design and construction of a large majority of these dams in this period.

The best overall piece of literature describing the development and present status of these dams is the 1982 Terzaghi Lecture by J. B. Cooke, scheduled for publication in the ASCE Geotechnical Engineering Division Journal in the next few months. □

References

1. Sherard, J. L. "Embankment Dam Cracking", Embankment Dam Engineering, John Wiley and Sons, New York, USA; 1973.
2. Sherard, J. L., Decker, R. S. and Ryker, N. L. "Hydraulic Fracturing in Low Dams of Dispersive Cay", *Proceedings*, Specialty Conference on Performance of Earth and Earth-Supported Structures, ASCE, Vol. I, Part I: June, 1972.
3. Dascal, O. "Peculiar Behavior of the Manicouagan 3 Dam's Core", *Proceedings* of the ASCE Conference on Case Histories in Geotechnical Engineering, St. Louis, Missouri, USA; May, 1984.
4. Flores Calcaño, C. E. and DeFries, K. "Incidents in Earth and Rockfill Dams in Venezuela", 25th Anniversary Volume, Venezuelan Society of SMFE, Caracas, Venezuela; November, 1983.
5. Sherard, J. L. "Sinkholes in Dams of Coarse, Broadly Graded Soils", XIII ICOLD Congress, India, Vol. II; 1979.
6. Kjaernsli, B., and Torblas, I. "Horizontal Cracks through the Core of Hyttejuvet Dam", Publication No. 80, Norwegian Geotechnical Institute, Oslo, Norway; 1969.
7. Vaughan, P. R. and Soares, H. F. "Design of Filters for Clay Cores of Dams". *Journal*, Geotechnical Engineering Division, ASCE January; 1982.
8. Wood, D. M., Kjaernsli, B. and Hoeg, K. (1976). "Thoughts Concerning the Unusual Behavior of Hyttejuvet Dam", XII ICOLD Congress, Vol. II, Mexico; 1976.
9. Kulhawy, F. H. and Gurtowsky, T. M. "Load Transfer and Hydraulic Fracturing in Zoned Dams", *Journal*, Geotechnical Division, ASCE; September 1976.
10. Sherard, J. L. Discussion of "Load Transfer and Hydraulic Fracturing in Zoned Dams", by F. H. Kulhawy and T. M. Gurtowsky. *Journal*, Geotechnical Engineering Division, ASCE; July, 1977.
11. Sherard, J. L., Dunnigan, L. P. and Talbot, J. R. "Filters for Silts and Clays", *Journal*, Geotechnical Engineering Division, ASCE; June, 1984.
12. Sherard, J. L., Dunnigan, . P. and Talbot, J. R. "Basic Properties of Sand and Gravel Filters", *Journal*, Geotechnical Engineering Division, ASCE; June, 1984.
13. Hacelas, J. E. and Ramirez, C. A. "Interaction Phenomena Observed in the Core and Downstream Shell of Esmeralda Dam, Chivor Hydroelectric Project", Specialty Conference Session 8, Ninth International CSMFE, Tokyo, Japan; June, 1977.
14. "Dispersive Clays, Related Piping, and Erosion in Geotechnical Problems", ASTM Special Technical Publication 623; 1977.

Introduction to Paper 16

Filters and Leakage Control in Embankment Dams

by J.L. Sherard and L.P. Dunnigan
Proceedings, Symposium on Seepage and Leakage from Dams and and Impoundments, Volpe, R.L., and Kelly, W.E., eds., American Society of Civil Engineers, 1985, pp 1-30.

Prepared by Lorn P. Dunnigan[9], M. ASCE

In 1977, Dr. Sherard contacted me and expressed an interest in working together again on a research project of mutual interest. He said he was at a point in his career where he wanted to give up some of his practice and devote more time to research. By coincidence we were both working on projects involving coarse grained soils. I was making some studies involving filter requirements for broadly graded glacial tills from New York. Jim was working on his "Sinkholes in Dams of Coarse, Broadly Graded Soils" paper. We agreed that a study involving filters would be an interesting and worthwhile project.

In the next few years, before the start of our laboratory research effort, Jim spent a good deal of time researching literature and thinking about the subject. By the time we started our laboratory testing program in 1981 he had prepared a guide entitled "Study of Filters for Dams and Ideas on Potentially Useful Tests to Clarify Questionable Points." This guide was a summary of previous research, design practices, personal experience and observations. It guided our initial laboratory testing which finally led to the test and design recommendations in this paper. Jim's extensive literature review, consulting experience, communication with colleagues and observations made during our laboratory study provided the background information and the stimulus for this paper.

[9]Former Head, Soil Mechanics Laboratory, Soil Conservation Service, U.S. Department of Agriculture, Lincoln, NE.

Filters and Leakage Control in Embankment Dams

James L. Sherard, F. ASCE*
Lorn P. Dunnigan, M. ASCE**

Abstract

In the practice of earth dam engineering there has been some doubt that downstream filters could be relied upon completely to control and seal concentrated leaks through impervious dam sections (cores). This doubt led to the "multiple lines of defense" concept, particularly for high, central core dams, in which several independent lines of defense are provided to prevent concentrated leaks through differential settlement cracks and to control the leaks, if they should develop in spite of efforts to prevent them. This situation is currently changing. Recent laboratory research shows conclusively that filters with properties similar to those used in current common practice will reliably control and seal concentrated leaks. This conclusion is supported by experience with dam behavior. This increased confidence in filters is stimulating changes in practice, especially with regard to design and construction of the impervious earth core, frequently allowing lower cost. Recommended design criteria are given for filters needed to control concentrated leaks, based on results of a recently completed, 4-year laboratory research program carried out by the Soil Conservation Service.

Introduction

This paper presents a description of changing practice and opinions regarding design measures in embankment dams for leakage control. For such important structures as dams, changes in practice are only accepted gradually and remain debatable for a number of years. The main opinions presented regarding changes in practice are currently controversial but they have been generally accepted and applied by the engineering teams responsible for a considerable number of major dams in recent years. Hence, while the points presented may seem moderately surprising to some readers they have already been tempered by a number of years of exposure and debate in the profession, and are widely accepted by many specialists. The paper considers primarily the problem of leakage through the dam itself, assuming an impervious and erosion resistant foundation.

*Consulting Engineer, San Diego, California
**Head, Soil Mechanics Laboratory, National Technical Center, Soil Conservation Service, U.S.D.A., Lincoln, Nebraska.

Several causes of concentrated leaks through embankment dams can be imagined, such as through inadequately compacted layers. However, for practical purposes the only likely cause of a concentrated leak through the impervious earth section of a reasonably well constructed dam is a crack. In low dams in arid climates, cracks from drying and shrinkage have been troublesome, but cracks from differential settlement are the primary cause of concentrated leaks which must be considered by dam designers. Differential settlement causes open cracks on the surface and also creates deeper within the embankment conditions of low stress which allow the water to open and enter cracks which were not in existence before the water in the reservoir rose above the level of the crack, an action generally called "hydraulic fracturing".

In a companion paper (11) prepared for this conference, current available experience concerning hydraulic fracturing is summarized. This experience leads to the conclusion that hydraulic fracturing is much more common than was previously considered likely. Concentrated leaks occur in most embankment dams of all types and sizes without being observed. This is somewhat surprising; however, the evidence available for this conclusion is now so strong that it must be accepted. There can only be some remaining questions about details.

Leaks caused by hydraulic fracturing usually do not cause erosion, either because the velocity is too low or because they discharge into effective filters. Occasionally these concentrated leaks have caused erosion damage in central core dams in which the filters provided were too coarse, and have led to breaching failures of homogeneous dams that had no filters.

Both experience with dam behavior and recent laboratory research show conclusively that concentrated leaks are reliably controlled and sealed by adequate filters, even under the most extreme conditions. In the past, designers were not wholly confident about the efficacy of filters designed according to current criteria and they were reluctant to rely on them completely as the sole line of defense against possible concentrated leaks.

This paper scrutinizes the changes in design practice for embankment dam leakage control, which are currently being generated by the gradual acceptance of the two main conclusions described above; i.e., (1) concentrated leaks develop through most embankment dams, even without large differential settlement, in spite of design efforts to avoid them, and (2) filters with appropriate properties can be relied upon with complete confidence to control and seal concentrated leaks.

The changes consist of less emphasis on the necessity for keeping the dam core watertight by avoiding cracks and more emphasis on the importance and details of the downstream filter. In the past practice the designer held as an axiom that:

> The impervious core is the most important element in the dam. As long as the impervious core remains intact, with no cracks or other concentrated leaks, the dam will be safe. Therefore, the primary and most important objective of the design is to provide measures

> which will minimize the likelihood of a concentrated leak to the greatest extent possible.

Based on the current available experience the designer is now inclined to see the situation differently:

> We have been deluded in the past thinking that the impervious sections of our dams remain intact. Evidence now shows that concentrated leaks commonly develop in well designed and constructed dams. It is now clear that the most important element in the dam is the filter (or transition zone) downstream of the core. By providing a conservative downstream filter, we can quit worrying about possible concentrated leaks through the core.

Brief Review of Development of Current Practice

History of Concern about Cracking

Starting about 1950 the profession began to be actively concerned that open cracks were seen to be developing in embankment dams, even in well constructed dams without unusually large settlement (9). Previous to this time it had been generally assumed that impervious sections of embankment dams compacted near Standard Proctor Optimum water content were sufficiently flexible and deformable that open cracks would not occur.

Particularly in the period 1960-75 a lot of attention was paid to potential leaks in embankment dam cracks, both by researchers and practicing engineers. In a 1963 reference book on embankment dams, it was found that only six articles in the literature contained most of the information on cracking published at that time (12). From 1965 to 1975 many dozens of articles were published on various aspects of dam cracking, generally expressing concern that the problem was not adequately understood.

In 1967 the ASCE Committee on Embankment Dams published a list of the ten outstanding problems concerning design and construction of embankment dams for which research was considered urgently needed (1). The problem of cracking was rated second highest in priority on this list. Similar concern developed internationally about the same time. At the 1970 ICOLD meeting in Montreal one of the General Reports (7) and a considerable number of other papers and discussions were devoted to cracking.

"Multiple Lines of Defense"

This concern led to the adoption of various design measures for the two basic purposes of (1) reducing the likelihood of differential settlement cracking, and (2) sealing and controlling concentrated leaks if they developed in spite of the design measures provided to prevent them. The situation existing in the early 1970's as seen by many specialist engineers was described by A. Casagrande as follows (3);

"... I have been given credit for having invented cracks

> in dams ...; i.e., for imagining something that does not happen in reality. But as the height of dams increased, the frequency of cracking increased also, and cracking is no longer a figment of my imagination. ... To ensure safety, the designer will provide several lines of defense against cracking. The most important defense lies in the selection of materials which are self-healing, particularly in a wide zone downstream of the core, and if at all possible, also in the core. In addition, a designer may employ such defensive measures as upstream arching of the dam, widening of the core ... etc."

Since about 1970-75 it has been fairly widely accepted that for important dams "multiple lines of defense" should be provided against leaks in cracks. This idea was applicable especially for high dams in steep-walled canyons or other situations where large differential settlement was inevitable. The various design measures making up these lines of defense fall into two main categories:

(1) Methods to Reduce Differential Settlement and the Likelihood of a Concentrated Leak

(a) Excavation of rock abutments to achieve flatter slopes and reduce abrupt changes in slope.

(b) For central core dams, using upstream sloping impervious core instead of vertical core.

(c) Compacting the bottom half of the impervious earth core at relatively lower water content and to as high density as practicable, to reduce its compressibility.

(d) For the upper part of the impervious core, compacting at higher water content, to make it flexible and capable of following imposed strains without cracking.

(e) Using plastic clay instead of silty sand for impervious core, also because of flexibility.

(f) Using strip of impervious core at especially high water content at the abutment contacts, sometimes using also a more plastic clay for this strip. (This is used to 1) obtain a good seal with the soft, relatively erosion resistant soil against the rock, and 2) allow the embankment to settle with respect to the rock abutment.)

(g) Arching embankment dams in plan, to increase the longitudinal stress when water pressure pushes the dam crest downstream.

(2) Measures to Control Erosion by Concentrated Leaks in Cracks That Develop in Spite of Preventive Measures

(a) Use of conservative downstream filters (transitions)

to prevent leaks from carrying eroded soil out of the core.

(b) Use of plastic clay for impervious core, believed to be more resistant to erosion of concentrated leaks.

(c) Use of sand in zones upstream of the core, the sand would be washed into and fill concentrated leakage channels.

Multiple lines of defense were considered necessary because of lack of confidence that any one of the lines of defense was completely reliable. This conclusion that multiple lines of defense were required has been widely accepted and applied, but has always been less than completely satisfying for dam designers. Its use leaves hanging the unspoken question, "If we are not wholly confident of any single line of defense, how can we be completely confident of any combination?"

Doubt About Filter Reliability

The adoption of the multiple line of defense practice was stimulated at least partially by doubt that downstream filters could be relied upon with complete confidence to prevent erosion of concentrated leaks through cracks in the core. This doubt reflected two main groups of questions: (1) questions about the validity of the early laboratory research on which current filter design criteria are based; and (2) questions as to whether experience with dam performance supported the conclusion that filters can be relied upon.

In the early research that led finally to our current accepted filter criteria the laboratory tests generally were made by the "conventional filter test procedure" in which water is caused to seep gradually through the pores of the base (protected) soil specimen and discharge into the filter. Also, with a few exceptions the pioneer research programs (2, 5, 17) were carried out on filters needed for cohesionless, relatively pervious sands. This older research generally did not include many tests on impervious base soils because the results of conventional filter tests on impervious soils were inconclusive. When a clayey soil is tested in a conventional filter test using a reasonable hydraulic gradient (2 to 20), the quantity of water which seeps through the clayey soil base specimen and discharges into the filter is so small (drops of water per hour) that there is no tendency for erosion. The significance of such tests was considered problematical at best.

As part of the design effort on specific projects, many individual engineers have made laboratory tests in which concentrated leaks were caused to flow through the impervious soil and discharge into the filter. The results of these tests are not generally published and there is no useful summary available in the literature.

As discussed later, the performance of dams with reasonable downstream filters has been generally very good. This was widely considered favorable experience supporting the conclusion that filters are successful; however, it was also realized that this generally good experience could be due primarily to the fact that no concentrated leaks developed in the dams; i.e., perhaps the filters in our dams are not tested by concentrated leaks?

SCS Filter Research

The Soil Conservation Service is just completing a 4-year comprehensive research program devoted primarily to understanding the action of downstream filters for controlling concentrated leaks in dam cores. The work was carried out by SCS's National Soil Mechanics Laboratory, Lincoln, Nebraska.

In this paper, we are discussing only the role of the filter in sealing leaks and preventing erosion of the protected soil; i.e., we are not discussing the needed permeability of the single band filter which also functions as a drain.

In this research program a large effort was devoted to experimentation with different types of laboratory tests having concentrated leaks. Several different tests were employed using the same soil types allowing an evaluation of the influence of the test procedure.

SCS researchers concluded that downstream filters with properties not greatly different from those used in current common practice would reliably control and seal concentrated leaks, even when the severity of leakage was very much more severe (higher gradients and velocities of flow) than could possibly exist in a dam. It was also concluded that one of the several different types of tests employed was superior for evaluating critical downstream filters.

Results of the first 2 years of the SCS research effort, devoted primarily to fine silts and clays, were published in 1984 (13). During the second 2 years the effort was broadened to include tests on filters for coarse-grained impervious soils, but also substantial advances were made in the understanding of filters for fine clays and silts.

Technical papers summarizing the results of the entire 4-year long program are in preparation. Because these summary papers will probably not be published for at least another 2 years, we are presenting here the main results in brief form without supporting data. We expect no fundamental change in the finally published results.

The paper describing the first 2 years of research on filters for silts and clays presented two different types of laboratory tests -- slot tests and slurry tests (13). In both of these tests there was a small amount of erosion and penetration of fines into the successful filter before the filter sealed. Later in the research program another test was adopted in which it was found possible to define a filter boundary size (separating successful and unsuccessful tests) at which "no visible erosion" of the walls of the leakage channel through the base specimen took place during the test. Tests with filters slightly coarser than the boundary had visible erosion. This test defined a somewhat finer filter boundary than was defined for the same soils by the previous slot and slurry tests. For example, for a common fine clay in which the slot and slurry tests defined a filter boundary of 1.5mm (13), the new test gave a filter boundary of 0.5mm. This test, which we have named the No Erosion Filter Test, was also found to work very well for coarse-grained impervious soils, whereas the slot and slurry tests were not satisfactory for impervious soils with d_{85} exceeding about 0.1 mm.

No Erosion Filter Test Procedure

The details of the laboratory test set-up are shown in Fig. 1. The test procedure and criteria for judging the results are as follows:

(1) After preparation of the sample as shown in Fig. 1, with cylinder in the vertical position for downward flow, fill the voids in the gravel above the base specimen with water, letting the air escape through the open hole in the top plate where pressure gage attaches.

(2) Attach pressure gage and open the valve to water source, applying full pressure (about 4 kg/cm^2) to top of specimen rapidly. For most soils water from the municipal supply was used. For dispersive clays, distilled water pushed by compressed air was used.

(3) Observe behavior for 5 to 10 minutes; this includes measuring of quantity of water coming through the filter and observing of turbidity (color of water).

(4) Shut off water flow, dismantle apparatus, and observe erosion of base specimen.

For each impervious (base) soil a series of these tests are made using filters (sands and gravelly sands) with gradually increasing coarseness, ranging from filters fine enough to prevent all visible erosion of the hole in the base specimen through coarse filters that permit considerable erosion. As previously reported (13, 14), it was a primary and definite conclusion of the SCS research program that the D_{15} size of the filter is a very good quantitative measure of the pore size of the compacted filter that allows particles to pass through and of its ability to act as a filter for base soils of different coarseness. Hence, normal (reasonably well graded) sand and gravel filters are well defined by the single parameter (D_{15}), in millimeters. The D_{50} size is not a good measure of the pore diameters of filter ability of a sand or gravelly sand and filter criteria using D_{50} can be abandoned.

These tests always allow definition of a "boundary" filter (D_{15b}) for any given base soil. For all tests on a given base soil with filters finer than D_{15b} there is no visible increase in the diameter of the initial hole through the base specimen. There is a very slight erosion of base soil that seals the face of the filter quickly but the quantity is too small to be observed. For tests with filters progressively coarser than the boundary (D_{15b}) the initial preformed leakage channel (hole) in the base specimen, Fig.. 1, is eroded progressively larger; however, for tests on fine-grained soils using filters only slightly coarser than the boundary, with D_{15} in the range

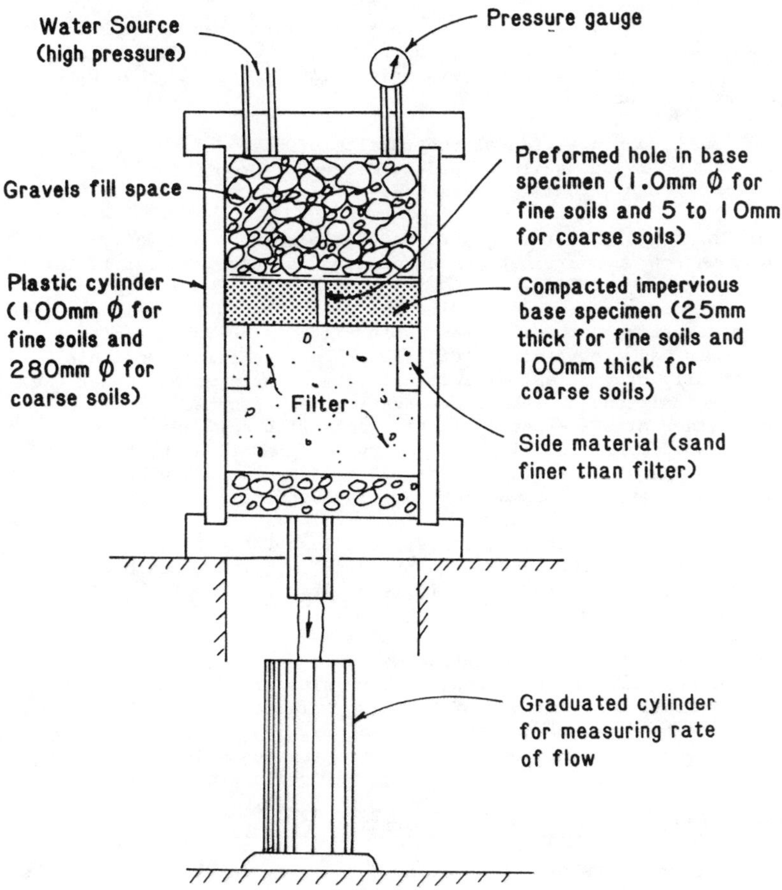

Fig. 1: No Erosion Filter Test Details
(No Scale -- Schematic)

between 2 D_{15b} and 4 D_{15b}, the filter is generally sealed rapidly and completely during the test without a large amount of erosion (a few grams of base material seal the filter face).

For tests on sandy or gravelly sands with less than about 20% fines (passing the No. 200 sieve), the boundary D_{15b} was determined to be abrupt. In tests with filters only slightly coarser than D_{15b} a large amount of erosion occurs.

After much consideration of the results of tests of this kind performed on impervious soils of widely different properties, the following main conclusions were reached:

(1) The tests give reliably reproducible results and define a relatively narrow boundary between successful and unsuccessful filters;

(2) For most fine-grained soils the filter boundary defined (D_{15b}) gives a conservative but still reasonable filter size for direct use in design. It has a built-in safety factor because filters of this size essentially prevent all visible erosion of the specimen under the most severe conditions of laboratory testing, and because filters somewhat coarser also are sealed in the tests with only small erosion of the base specimen. The filter boundary (D_{15b}) defined by this test for typical silts and clays is roughly 20 to 40% of the boundary (D_{15B}) defined by the previously used slot and slurry tests (13).

(3) For coarse impervious soils, with less than about 20% of fines passing the No. 200 sieve, the boundary (D_{15b}) defined by the laboratory tests is so abrupt that it is desirable to apply a safety factor for use in design.

(4) The test results are influenced only by the grain size distribution of the base soils tested. The results are not influenced by the inherent erosion resistance of the base soil, as related to plasticity of the fines or potential for dispersive erosion, because the erosion conditions imposed in the laboratory tests (velocity, pressure and gradient) are so severe that all soils erode.

Recommended Design Criteria for Critical Filters

On any given job where it is desired to carry out laboratory tests to confirm the suitability of a proposed downstream critical filter, the test procedures and the evaluation criteria described above are considered suitable and appropriate. Testing using these procedures was carried out on a sufficient number of impervious soils, covering a wide range of coarseness and plasticity, to establish general criteria for evaluating critical filters for all types of soils used for the impervious sections of embankment dams.

The criteria depend only on the grain size of the impervious soil and are independent of the plasticity and of the dispersion potential of the soil fines. For gravelly impervious soils the gravel particles should be ignored in the application of the filter design criteria presented below (except as discussed below for gravelly sands of

Group 3). For soils with gravels a new particle size distribution curve should be calculated for the portion of the soil that is finer than the No. 4 sieve and this reconstructed gradation curve should be used to apply the filter criteria. The criteria recommend below already include a conservative safety factor.

(1) Impervious Soil Group 1 (Fine Silts and Clays): For fine silts and clays that have more than 85% by weight of particles finer than the No. 200 sieve, the allowable filter for design should have $D_{15} \leq 9\ d_{85}$ (where d_{85} is the size of the silt or clay for which 85% is finer).

(2) Impervious Soil Group 2 (Sandy Silts and Clays and Silty and Clayey Sands): For sandy (and gravelly) impervious soils with 40 to 85% by weight (of the portion finer than the No. 4 sieve) finer than the No. 200 sieve, the allowable filter for design should have $D_{15} \leq$ 0.7mm. For these impervious soils, the influence of the fines dominates in the filter test and the results are not influenced by the existence of the sand particles.

The recommended criteria for this group of soils is different from that presented in our earlier paper (13). Research carried out after completion of the earlier manuscript showed that the difference between results of the No Erosion Filter Test and the previously used slot and slurry tests was increasingly great for sandy impervious soils with increasing d_{85} size. Using the slot and slurry tests the criterion $D_{15}/d_{85} \leq 5$ appeared to be conservative, for soils in range of d_{85} from 0.1 to 0.5 mm (13). Using the No Erosion Filter Test, it has now become apparent that for soils in this range of gradation it is desirable to use a filter with D_{15} not exceeding about 0.7 mm as the sample erosion for these soils becomes large before sealing with coarser filters.

The recommended D_{15} size criteria for Groups 1 and 2 (D_{15} always $\leq$ 0.7 mm) apply to filters composed wholly of sand or gravelly sand in which the sand fraction predominates and there is not a sufficient content of coarse gravel to cause excessive segregation during construction. As discussed in more detail elsewhere (13), it is considered reasonable and appropriate to require that the coarsest allowable gravelly sand filters for fine-grained soils should have more than 40% sand sizes (smaller than the No. 4 sieve) and maximum gravel size not exceeding about 50mm.

(3) Impervious Soil Group 3 (Sands and Sandy Gravels with Small Content of Fines): For silty and clayey sands and gravels with 15% or less by weight (of the portion finer than the No. 4 sieve) finer than the No. 200 sieve the allowable filter for design should have $D_{15} \leq 4\ d_{85}$, where d_{85} can be the 85% size of the entire material including gravels. For soils in this category we have tested filters only up to about D_{15b} = 30mm, but we believe that the rule is applicable for coarser base-filter combinations.

(4) Impervious Soil Group 4: For coarse impervious soils intermediate between Groups 2 and 3 above, with 15 to 40% passing the No.

200 sieve, the allowable filter for design is intermediate, inversely related linearly with the fine content and can be computed by straight line interpolation. As an example, for an impervious sandy soil with 30% of silty or clayey fines and d_{85} = 2mm, the allowable filter for design is in between the value of D_{15} = 0.7mm (for soils of Group 2) and D_{15} = 4 (2) = 8mm (for soils of Group 3), and is calculated:

$$D_{15} = \frac{40 - 30}{40 - 25} (8 - 0.7) + 0.7 = 3.6 \text{ mm} \qquad \text{(See also Fig. 2)}$$

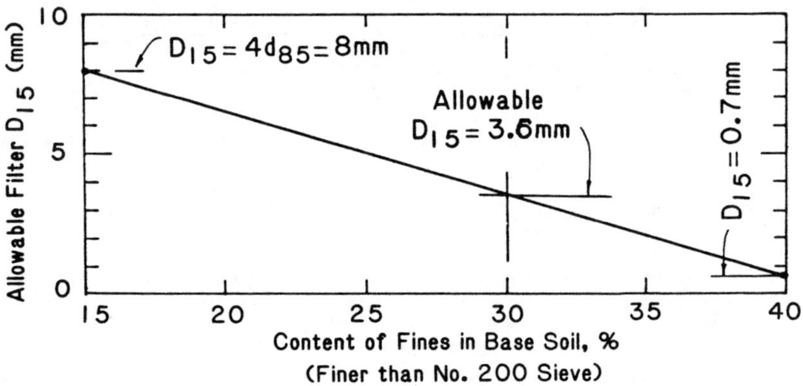

Example of Determination of Allowable Downstream Filter for Impervious Earth of Clayey Sand or Silty Sand with 30% Fines and d_{85} = 2.0 mm.

Fig. 2: Determination of Allowable Downstream Filter for Impervious Soil Group 4 (having between 15% and 40% finer than No. 200 Sieve).

Soil Groups 1 through 4 include all soils used for the impervious sections of embankment dams. The above criteria can be applied for all soils in Groups 1 and 2 regardless of the shape of the particle size distribution curve. For soils of Groups 3 and 4 the criteria apply to reasonably well-graded soils: for soils in Groups 3 and 4 which are highly gap-graded it desirable to provide a filter for the finer portion of the gap-graded soil, or make No Erosion Filter Tests in the laboratory to select the appropriate filter.

Comparison with Current Criteria

Soil Group 1: For fine-grained silts and clays the criterion $D_{15} \leq 9\ d_{85}$ gives about the same filter in common use; that is, sand (or gravelly sand) filters with D_{15} of 0.2 to 0.7mm. Although there is no general rule widely quoted in the literature for filters for these fine soils, such sand filters are commonly used in practice.

Soil Group 2: For the sandy soils in Impervious Soil Group 2 above, the results of the laboratory tests were dominated by the fines, with filter boundary defined always of the order of 0.7mm, or a little larger. From the standpoint of behavior at the filter face, when tested as described above, these soils acted in the same manner as fine silts and clays which do not contain sand sizes (Soil Group 1). It was concluded from the observations of many tests on soils of this type that the fine particles dominate the behavior because the sand sized particles are simply floating in a matrix of the fines and it is the fine matrix which must be prevented from moving at the filter face in order to control safely and seal the concentrated leak.

For soils of Group 2 there is some question in the profession at the present time as to what filter criteria to use and how to apply the criteria. As an illustration, Fig. 3 shows the particle size distribution of a sandy clay, a soil representative of materials used for the impervious sections in many embankment dams all over the world. Soils of gradation similar to Fig. 3 are of all geologic origins from alluvium to weathered in-situ residual soils. In current practice there are two main criteria that would be widely considered for evaluating a filter for a sandy clay of this gradation:

Criterion 1 $D_{15}/d_{85} \leq 4$ (or 5)

Criterion 2 $D_{50}/d_{50} \leq 40$ (or 25)

Criterion 1 is the oldest criterion employed and is universally accepted in the profession. Over the years it was observed that Criterion 1 permitted filters that appeared visually to be coarser than considered reasonable for protecting some impervious soil types. This observation and some laboratory tests led to the proposal for Criterion 2 (16), which is currently accepted and applied widely but not universally.

As seen in Fig. 3, if only Criterion 1 were applied for this common sandy clay, it would allow a coarse gravel filter to be used, with $D_{15} \leq 4$mm. However, if Criterion 2 is applied it would require that $D_{50} \leq 0.24$mm; i.e., it would require a filter with more than 50% fine sand. It is seen, therefore, that when applied to evaluating a needed filter for this common sandy clay, Criteria 1 and 2 provide radically different answers. In common practice the filter allowed by Criterion 1 would be considered too coarse, but the $D_{50} \leq 0.24$mm required by Criterion 2 would be considered unnecessarily fine and generally impractical.

In practice in this situation an intermediate filter is commonly used, such as a sand or gravelly sand with average D_{15} size not exceeding a maximum of the order of 0.5 or 1.0mm. Hence, the criterion based on the results of the recent SCS research, specifying a sand or gravelly sand with $D_{15} \leq 0.7$mm, is generally consistent with the current practice, and is intermediate between existing filter Criteria 1 and 2.

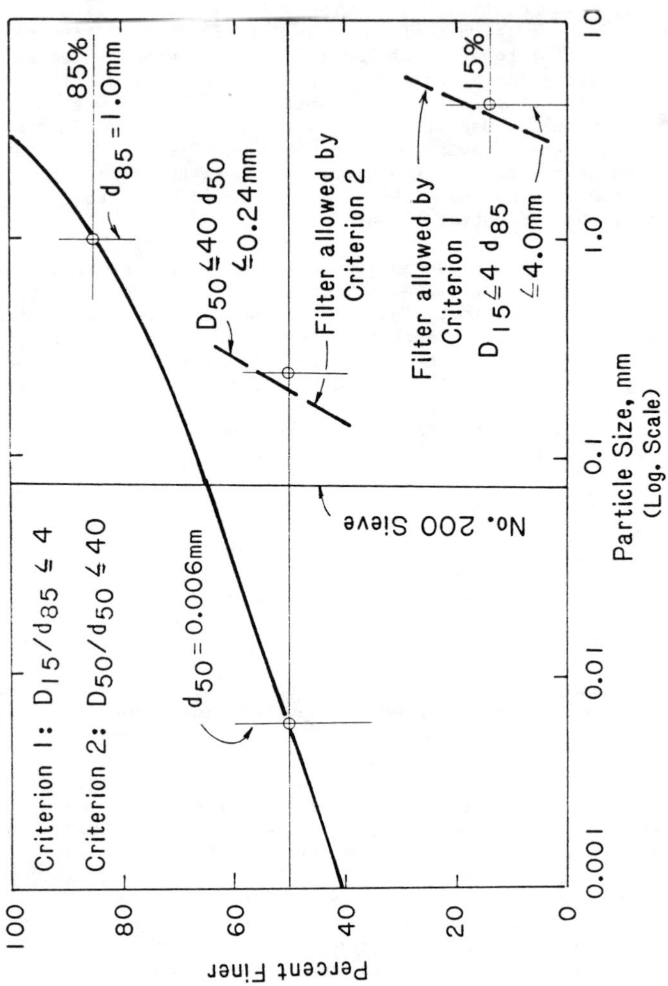

Fig. 3: Comparison of Filters for Sandy Clay as Required by Criteria 1 and 2.

In the practice of the last few years there has been a serious question among specialists as to what filter is needed for broadly graded impervious soils, such as those typical of glacial moraines, which contain all particle sizes from cobbles down to clay sizes, with a grain size curve that commonly plots as a near straight line on the conventional semi-logarithmic chart of particle size distribution. For these soils there has been no general agreement on the appropriate way to apply current filter criteria and there has been considerable trouble with erosion of cores of these coarse soils in which coarse filters were used (10). In the recent SCS research a considerable amount of effort was spent testing these typical glacial moraines and similarly graded, coarse, impervious soils. Results of the research show conclusively that the particles coarser than fine sand in these soils cannot be relied upon to help seal the filter and that it is necessary to provide a critical downstream filter that will retain the fines. The typical coarse glacial moraines fall in Impervious Soil Group 2 and require a sand or gravelly sand filter with $D_{15} \leq 0.7mm$.

Soil Group 3: For sandy soils with a small fine content, the criterion $D_{15} \leq 4\ d_{85}$ is the same as the current practice. For these soils the laboratory tests show the failure boundary generally at about D_{15b} = 7.5 to 9, so that the design criteria ($D_{15} \leq 4\ d_{85}$) has a safety factor of the order of two.

All these filter tests and criteria are for "critical" filters, which are located downstream of the dam core and are needed to control and seal concentrated leaks. For upstream filters and other filters placed adjacent to impervious embankment sections where it is not possible for a concentrated leak to develop, coarser materials can be safely used, as discussed later.

Experience with Filters in Dams Is Generally Good

In the last 30 years or so thousands of dams have been designed and built with criteria and procedures not greatly different from the current practice. A small fraction have had behavior problems of various kinds. Some of these problems never become known widely in the profession. Even for those cases which are thoroughly studied, with results presented in the literature, it is frequently difficult for the investigators to be completely confident that the cause of trouble was reliably understood.

In spite of these difficulties it is possible to extract from the mass of experience with dam behavior some conclusions on which there should be wide agreement in the profession. One of these is that there are practically no records of leakage or internal erosion (piping) through earth cores that is attributed to filters considered reasonable according to current practice. Almost all records of damage or failure from leakage and piping have been for dams in two specific groups: (1) homogeneous dams without chimney drains (vertical filter-drains), in which a concentrated leak develops through the dam and exits on the downstream slope without passing through a filter; and (2) dams with impervious sections consisting of coarse, broadly graded soil for which excessively coarse filters were used (10).

As discussed above, there has been a major problem in evaluating the "lessons learned" from experience with dam behavior. Problems with erosion (piping) practically never have developed in dams provided with reasonable downstream filters. However, it was not known whether this good experience could generally be attributed to the action of the filter in sealing concentrated leaks, or whether no leaks developed and the filters were not tested by concentrated leaks.

In the last few years much evidence has been accumulating to support the conclusion that concentrated leaks commonly develop without being observed, due to differential settlement. The evidence and experience is now so great that there is no longer any reasonable doubt: concentrated leaks by hydraulic fracturing from differential settlement occur in most well designed and constructed dams, even those without unusual settlement. These leaks are not observed and usually their existence is not suspected because they do not erode and they become sealed before much water passes through, either because the velocity is not high enough or because the leak discharges into an effective filter. These unobserved leaks develop in cracks during or soon after the first reservoir filling and then later the cracks close by swelling or softening and squeezing shut. The evidence that concentrated leaks by hydraulic fracturing are common, too extensive to present adequately here, is described in another paper prepared for this Symposium (11).

In summary, concentrated leaks develop commonly. In dams of moderate height or greater these leaks are sufficiently large to cause erosion but they do not erode because the downstream filters (or transitions) commonly provided in current practice are effective. The writers have no doubt that the overall lesson learned from the mass of experience available with dam behavior is that concentrated leaks commonly develop in well designed and constructed dams and that "filters work satisfactorily" to control the concentrated leaks.

Impervious Earth Core Properties

In the past many engineers considered that clay was the superior material for the impervious core. In some geographic areas this opinion was so strongly entrenched that the terms "core" and "clay core" were practically synonymous. There are many examples in the last 20 years where clay has been hauled great distances (30 to 40 km or more), at considerable cost although large uniform deposits of excellent cohesionless silty sand were available immediately at the dam site. Clay was considered superior to cohesionless impervious materials because it would be more deformable and better able to follow imposed strains from differential settlement without cracking, and because it would have higher resistance to erosion if a concentrated leak developed from any source. Highly plastic ("tough") clay was particularly prized.

In the current trend, supported by increasing confidence in downstream filters, this preference for "clay" over cohesionless impervious soils is not considered valid. The two materials act differently in the presence of a concentrated leak in a crack. Each has theoretical advantages and disadvantages, and it is not clear that either is superior. The tough, highly plastic clay will definitely resist erosion of the walls of a leak in an open crack better than a cohesionless silty soil. But the higher unconfined compressive strength of the clay core provides more likelihood that the embankment material can arch around the leakage channel and keep it open. In impervious embankments of cohesionless silty sand the walls of initially open, water-filled cracks quickly become saturated, soften and tend to squeeze shut.

The greater resistance of clay to erosion of concentrated leaks is believed not important. All clays will erode under severe conditions and it is not considered reasonable or conservative to rely on the relatively feeble erosion resistance of a clayey soil to provide defense against piping of a concentrated leak. The principal job of the core is to be impervious. Prevention of piping is the job of the filter.

In the past, sometimes a superior core material was manufactured by blending together two materials of radically different properties. For example, at some dams cohesionless fine silts (or fine clays) were mixed with sandy gravel to manufacture a core material with the imperviousness of the silt or clay and the high shear strength, low compressibility and "self-healing" properties of the sandy gravel. This has been done only rarely, but for some major dams. The current trend would be away from this practice and towards the use of the fine silt or clay for the core and the sandy gravel in other zones of the dam, even for high dams.

In summary, the current trend is to use the least costly of the available impervious soils for a central core, although it may be completely cohesionless and relatively erodible when exposed to concentrated water flows, and may be relatively compressible when compacted in the dam.

As historical background for this point it is desirable to record that the previous strong preference for impervious cores of plastic clay was not at all universal. In some parts of the world there are no plastic clays available for dam construction. In these areas many major dams have been built over the last 30 years with impervious cores consisting of cohesionless silty sands, sandy silts, gravelly, sandy silts, etc. In some areas these materials (usually cohesionless silty sand) have been largely decomposed granites or similar rocks which are frequently considered among the best impervious earth materials available for embankment dams. In other areas great numbers of dams have been built with impervious cores of glacial moraines, often in the form of cohesionless gravelly silty sand. Impervious dam sections from weathered sandstone, often in the form of very fine, uniform silty sand, are common. A list of dozens of major dams built with cores of completely cohesionless silty soils in the last 30 years could easily be compiled.

Because of the above experience it may seem unnecessary to emphasize this point so strongly. Presentation of this subject is needed in the literature at the present time because there have been so many debates on this subject in connection with major projects only in the last few years. It is necessary to record that some specialized engineers of long experience have the very strong opposite opinion that it is worthwhile to spend a lot of extra money to obtain a clay core.

An exception to the general practice described above of using the least costly impervious material available arises for the strip of impervious core material directly in contact with hard rock foundations. There is always concern that erodible impervious earth core material can be carried into cracks in the foundation rock, where there is no filter to prevent the erosion. Cracks in the rock surface are sealed by consolidation grouting and surface treatment (slush grout, shotcrete, etc), but there is always a possibility that some cracks are not adequately sealed and, for high dams at least, there is also a possibility that sealed cracks may open again in the treated rock surface by the foundation deformations caused by the weight of the embankment. There have been many cases where erosion of core material into rock cracks has caused serious problems.

For these reasons it is not desirable to place very erodible fine-grained soils (fine, cohesionless silts or dispersive clay) directly on rock foundations. For dams on rock foundations where fine, highly erodible soils are used for the main impervious core, it is desirable to place a thin layer of a different impervious material directly against the foundation rock surface, the material having properties so that it will not be easily eroded into small open rock cracks.

Earth Core Width, Slope and Compaction

The main decisions relating to the earth core have generally been influenced by the designer's concern over the risk of differential settlement cracks. For rockfill dams with central cores of compressible earth, the core is often made to slope upstream, instead of vertically. The upstream sloping core is generally considered to be less likely than the vertical core to develop low internal stresses from arching of the top between the upstream and downstream shells.

The minimum acceptable core width of a central core dam at a given site is often a subject for debate. The greater the potential differential settlement, generally the less likely that a very thin core would be used.

Particularly for high dams in narrow valleys, some engineers insist on making the lower half of the dam less compressible by heavy compaction at relatively low water content. Sometimes also the upper part of the core is compacted at a relatively high water content in order to make it flexible and able to follow the strains resulting from the differential settlement without cracking.

The current design trend, based on increased confidence in the reliability of downstream filters is changing practice on these points. The geometry and width of the core are generally chosen on the basis of other considerations. Less emphasis is placed on reducing differential settlement and cracking by varying the construction water content: the trend, even for high dams, is to compact the entire core at the same water content with the same compaction effort.

The width of the earth core in a dam with rockfill or gravel shells is generally chosen arbitrarily on the basis of precedent, commonly in the range between 30 and 60% of the head. There are some well known examples of completely successful dams with thinner cores; for example, at the 80 m. high Nantahala Dam, completed in 1939, in Tennessee, with sloping earth core, rockfill shells and an excellent downstream filter of sand, the width of the core was about 11% of the head (12). In the 45 years since the construction of Nantahala Dam there have not been many dams built with cores thinner than about 25% of the head.

In the last few years thinner cores have been used at a few dams with good filters. The Svartevann Dam in Norway, completed in 1976, is a rockfill dam about 130 m. high, with upstream sloping core of cohesionless glacial moraine (6). The core width was made about 18% of the dam height, using a good downstream filter of gravelly sand with average D_{15} about 0.1 mm. Because the Svartevann Dam is the highest dam in Norway it was provided with more instruments than commonly used to allow detailed understanding of its behavior, which is shown to be satisfactory in every respect.

The main function of the core is to make the dam relatively watertight. Most compacted impervious soils have a coefficient of permeability of 10^{-6}cm/sec., or less. Hence, theoretically a core width of 2 or 3 meters would be adequate to reduce the total quantity of seepage to a small value, even for relatively high dams.

In the laboratory it is shown conclusively that a compacted silt or clay specimen with thickness of only a centimeter or two, when it has an appropriate filter under it, will permanently and securely act as an impervious barrier to hold back water with pressures only limited by the capacity of the laboratory equipment. The writers have made many such tests with compacted specimens of various types of impervious soils about 2 centimeters thick with appropriate sand and gravelly sand filters, using pressures of about 4 kg/cm^2, giving an hydraulic gradient of about 2000. Nothing happens in these tests, even when holes are punched through the specimen causing a concentrated leak as shown in Fig. 1.

These experiences show clearly that as long as the water seeping through the earth core material discharges into an adequate filter, there is no practical limit on the tolerable hydraulic gradient that can be safely imposed on the impervious earth core. Therefore, it is reasonable in the future to contemplate making earth cores in dams much thinner than the minimum widths used in current practice.

Junctures Between Embankment Dams and Concrete Structures

Before about 1960-65 it was standard practice to provide cutoff collars on concrete conduits passing through the base of embankment dams, over the length of embedment in the impervious core. These were simply walls protruding out into the surrounding dam embankment. Similar collars were commonly provided on spillway walls and other concrete structures where embankment dams connected to them. The purpose was to force the particles of water that were seeping from upstream to downstream along the earth-concrete interface to have a longer path and not a straight line.

Over the years many engineers concluded that these cutoff collars had the strong disadvantage that they made it impossible to compact the impervious earth core directly against the concrete structure in a reliable fashion with heavy rollers. A large amount of compaction by hand labor was needed, which was clearly inferior. Because of increasing concern over this problem, and at the same time increasing confidence in the efficacy and reliability of filters, over a period of years the number and size of these collars were gradually reduced, and special attention was paid to the downstream filters. This trend continued until the collars have been eliminated altogether.

A common current practice is generally as follows:

(1) No seepage collars provided.

(2) The exterior concrete surfaces at the concrete-earth core interface are made smooth and are sloped slightly off vertical (such as about 1H:8V or 1H:10V) so that the earth can be compacted directly against the concrete by the wheels of a rubber-tired vehicle running parallel with the concrete face, Fig. 4. The earth core is placed against the normal smooth, concrete surface; i.e., the concrete surface is not sandblasted, chipped, painted or treated in any way.

(3) Special attention is paid to the downstream filter. The filter is often wrapped completely around the downstream portion of conduits; i.e., underneath as well as on both sides and the top, so that all potential leakage water travelling along the concrete-earth core interface must exit in a controlled fashion in the filter.

The above practice has been used successfully at many projects. One interesting example is at the large Guri Hydroelectric Project, recently completed in Venezuela, where a 125-meter-high central core rockfill dam is connected to a concrete gravity dam, Fig. 5. The smooth concrete face of the gravity dam over the area of contact with the earth core was made at 7V to 1H, and the earth was compacted against the concrete as shown in Fig. 4. The angle between the longitudinal axis of the rockfill dam and the face of the concrete dam was made 80° (instead of 90°), with the idea that downstream deflection of the dam under the push from the reservoir water would wedge the core against the concrete. On similar connections on other jobs the earth core has been brought in at 90°, which the writers believe is also reasonable.

At the concrete face the downstream critical filter at the Guri connection was made several meters wider than the normal width in the rockfill dam, a common desirable detail. For embankment dams in which the critical downstream filter contains any significant quantity of gravel-sized particles, there is a danger that the gravels could segregate at the concrete-earth core interface forming pockets of gravel without sand in the voids. In this situation, in addition to increasing the width of the filter at the concrete surface, it is desirable to consider using a thin layer of filter material consisting only of sand at the interface.

Fig. 4: Compaction of Impervious Clay Core Against Smooth Sloping Concrete Surface with Wheels of Heavy, Rubber-Tired Vehicle.

Fig. 5: Connection Between Rockfill Dam and Concrete Dam (Guri Hydroelectric Project) Showing Smooth Concrete Surface Under Impervious Earth Core.

High Dams in Narrow Canyons

Starting about 1950 when increasing concern was developing about cracking, there was debate in the profession about the maximum permissable steepness of rock abutments. There was no general agreement on this point, but some engineers in those years held the opinion that rock abutments steeper than about 1:1 were probably undesirable. The steeper the abutments, the narrower the canyon and the higher the dam, the more concern was felt about the suitability of an embankment dam, and the more likely a concrete dam would be chosen.

This concern over high steep rock abutments has been gradually relaxing as good experience has accumulated and as confidence has developed in the reliability of filters. In the last 30 years, an increasing number of embankment dams have been built successfully with increasing height in canyons with increasingly steep rock valley walls. At the present time very high embankment dams are being built in narrow, steep-walled canyons where only concrete dams would have been considered acceptable previously.

Two notable examples of the current practice are the Chicoasen Dam in Mexico (8) and the Chivor (Esmeralda) Dam in Colombia (15). The Chicoasen Dam, a central core rockfill dam, 260 m. high, completed in 1980, has nearly vertical rock abutments over most of the dam height. The 240 meter high Chivor Dam, a sloping earth core rockfill dam completed in 1975, was built in a narrow, rockwalled canyon with width of only about 320 m. at top (crest length). The behavior of the Chivor Dam as recorded by an extensive system of instrumentation has been completely normal in all ways (4). At present in Colombia at a site not far from the Chivor Dam, the Guavio Dam, designed to have a maximum height of 250 m., and crest length of about 360 m., is being built in another steep-walled canyon.

This trend will probably continue until it is generally agreed that the highest embankment dams can be safely built with earth cores in narrow canyons with near-vertical abutments. It is only necessary that the rock canyon walls are not overhanging. This conclusion is based firmly on confidence in the reliability of filters to control and seal concentrated leaks.

Arching Dams in Plan

Concern over cracking in the 1950's caused some engineers to design embankment dams with the longitudinal axis curved. The purpose was to cause the length of the dam to decrease if any downstream deflection occurred, thereby tending to close cracks.

In the current trend, curved axes are not considered necessary. Straight axes are considered satisfactory for even the highest embankment dams in narrow canyons. The Chivor, Chicoasen and Guavio Dams described above, all have straight axes.

Scarification of Compacted Layers and Slickensides

In current practice there are strong differences of opinion about the need for, and amount of, scarification between compacted layers to achieve bond, especially for impervious sections compacted with smooth rollers. Also, however, modern tamping rollers with short feet of fairly large area (c ompared to the older type sheepsfoot rollers with long feet and small area) commonly leave the final compacted layer surface relatively smooth and hard. The practice for bonding between layers on various jobs in recent years has ranged from no scarification (usually when the impervious soil being compacted is fairly wet and the two layers are shown to fuse together naturally) to the use of several passes of a disc harrow which cuts nearly through the entire thickness of the previously compacted layer.

Another common problem is the development of "slickensides" or "laminations" in the compacted soil, caused by shearing movements under the rubber tires of trucks or other heavily loaded equipment. This occurs most frequently in fine clayey soils with water content near or above Standard Proctor Optimum, but it can occur in heavily travelled areas in embankments of all types of soils (clayey sands, etc.). Troubles have been caused on quite a few jobs where these slickensides develop and it is considered necessary to stop the work and study the problem. The typical conclusion from these studies is that the slickensides have no significant influence on thepermeability. The shear strength of the mass may be decreased moderately but this is usually not an important factor in a rockfill or gravel fill dam on a sound rock foundation.

For the central core dam the worst conceivable result of not scarifying to obtain bond between compacted layers, or of slickensides within compacted layers, is a condition in which it can be imagined that a leak by hydraulic fracturing could develop more easily than otherwise. In the current trend in practice with increased confidence in the reliability of downstream filters there is much less concern about inadequate bonding between layers and slickensides, and less effort and money is spent to eliminate them.

The writers believe that in future practice routine scarification of the top of compacted layers can be abandoned, even using smooth steel-drum rollers. If the soil is reasonably moist the two layers will bond together adequately without scarification. If the top of the previous layer has dried out, satisfactory bonding can be obtained by sprinkling the surface before the next layer is placed on top.

Upstream Sand Zones as "Crack Fillers"

A fairly common design detail in central core dams is a sand zone (filter) 2 to 4 m. wide at the top of the dam on the upstream face of the impervious core. The purpose is to provide a reservoir of sand that would be washed into a crack that might develop through the upper part of the impervious core from any cause, such as differential settlement or earthquake. This detail was developed as one of the multiple defense lines, and has been used fairly commonly since about 1965.

In order to serve the intended purpose, the maximum particle size of the sand needs to be smaller than the width of the crack that might develop. Hence, sand with a maximum size not larger than about 1.0 mm has commonly been considered desirable, so that it could reliably be washed into cracks with widths of several millimeters without bridging over and blocking the entrance. In some cases designers have provided coarser sand, such as with maximum size of 6 mm., or even 12 mm., considering that it might be useful if a major crack opened.

The potential value of this upstream sand "crack filler" is debated at the present time, with strong opinions on both sides. In recent years many important central core dams have been built without these zones, even though suitable sand was available at reasonable cost.

The several main arguments raised against this upstream sand zone are:

(1) The filter downstream of the core is the main line of defense for controlling and sealing concentrated leaks, and it must be relied upon. Hence, if there is any question about the adequacy of the downstream filter, it should be made more conservative and the question eliminated, rather than providing another line of defense of questionable efficacy upstream.

(2) It is difficult to be confident that the upstream sand would fill the leakage channel. The velocity of flow in a concentrated leak, even in a fairly wide open crack, is not likely to be very high because it is limited by the permeability of the downstream filter, the upstream face of which should be progressively clogged by eroded debris from the core.

(3) There is an argument as to whether it is desirable for the upstream sand to enter a crack, because a limited amount of sand may enter and tend to prop the crack open and tend to prevent it from squeezing shut by softening or swelling of the crack walls.

(4) In addition to the unit cost of the upstream sand zone itself, the sand zone upstream usually requires another band of coarser transition to hold the sand in place, which complicates the construction and adds more cost.

The writers believe that this upstream sand zone will be used less frequently in the future and probably will be finally abandoned as a design practice.

Upstream Transitions for Clay Cores

One of the most common dam types built in the last 30 years has been the central core rockfill dam, often with a clay core. In such a dam there are commonly two or three zones of progressively coarser filters (sometimes called transitions) between the core and the downstream shell. Twenty years ago the common design was to make the filters upstream of the core the same as those downstream; i.e., the same number of zones and gradation of materials and the same widths of zones.

In the last 20 years there has been a gradual trend toward making the upstream filters in these dams more economical. This was motivated by increasing realization that the upstream filters serve a much less important role than the downstream filters.

The main function of the upstream filter is simply to prevent the clay core material from migrating in the upstream direction into the voids in the upstream rockfill shell and, theoretically, to act as a filter for the core following reservoir drawdown. In fact, consolidated under the pressure inside the dam the clay core is in a relatively stiff state, so that there is no tendency for it to simply migrate, by falling under gravity action in the upstream direction. Also, for an impervious clay core, the amount of water that seeps out of the core in the upstream direction following reservoir drawdown is very low. The maximum gradient for the seepage in the upstream direction following drawdown is less than 1.0, and the velocity and energy of the seepage entering the upstream filter is too small to erode clay particles and carry them upstream. Such an upstream filter is considered a noncritical filter because it is never called upon to act to control a concentrated leak.

As a part of the recent SCS filter research tests were made with relatively erodible clays of low plasticity (using a gradient of about 20 and very coarse filters). These tests showed clearly that it is not necessary to apply the current filter criteria to such noncritical filters as the upstream filter on a central core dam (13). Hence, the laboratory research results support the trend in practice to make the upstream filter more economical.

The first steps in this trend resulted in making the upstream filters narrower than the downstream filters. The trend is continuing and the current designs for a considerable number of high dams retaining very large and important reservoirs have only one coarse transition between a very fine clay core and the upstream rock fill shell. This transition is commonly a zone of grizzlied or crushed rock, such as from 150 millimeters maximum size down. The relationship between the particle sizes of such a small rock transition zone and a fine clay core is completely outside our current criteria for acceptable filters.

As an example, Fig. 6 shows the cross-section of the 100-m high rockfill dam used in the main river section for the large Tucurui Hydroelectric Project, recently completed in Brazil. The earth core is an extremely fine clay of medium to high plasticity. As seen, the design has four bands of filters and transitions downstream and only one band of grizzlied rock upstream in the lower part of the dam. Fig. 7 shows the typical appearance during construction of the contact between the clay core and the upstream zone of grizzlied rock. Inevitably there is segregation and the upstream zone of small rock in no way can be considered a filter for the fine clay core: it is simply acting to retain the upstream face of the stiff clay core. The clay is prevented from migrating upstream into the voids of the grizzlied rock zone because it has sufficient stiffness to arch across the voids. The clay core is stiff already when it is compacted (unconfined compressive strength of the order of 3 kg/cm^2) and then it becomes stiffer when consolidated under the weight of the overlying embankment.

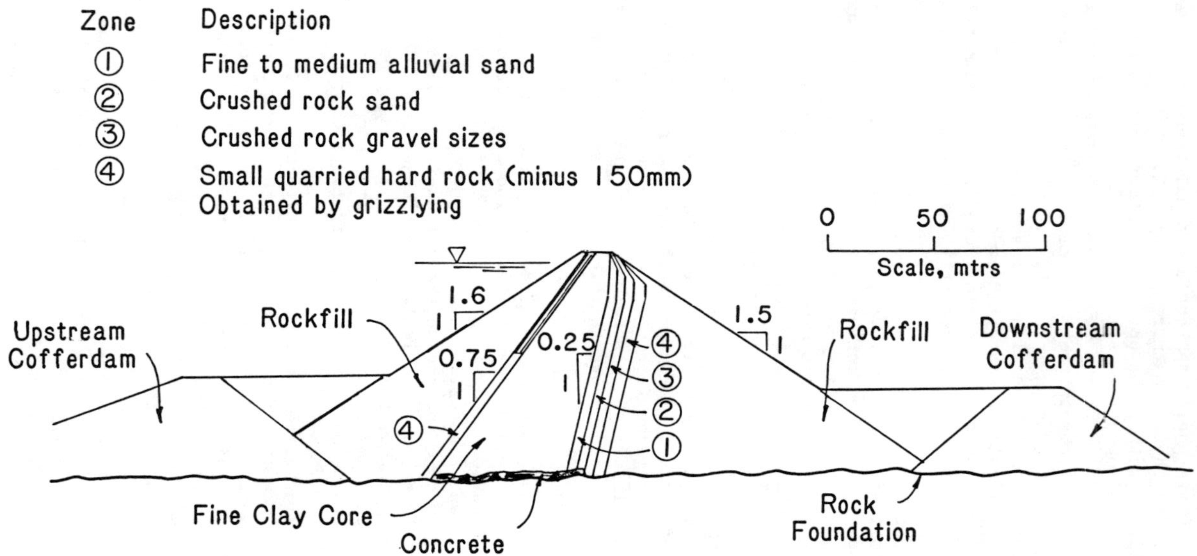

Fig. 6: Cross-Section of Tucurui Rockfill Dam in Main River Section Showing Difference Between Upstream and Downstream Filters

Fig. 7: Typical appearance contact between fine clay core and upstream transition of small (grizzlied) rockfill during construction at Tucurui Dam (see Fig. 6). Notebook in foreground is 18 by 25 cm.

The above conclusions are valid for clayey cores because their unconfined compressive strength and low permeability eliminate any significant tendency for the material forming the upstream face of the core from migrating in the upstream direction. We believe the conclusion applies to any core material that has 30% or more of fines passing the No. 200 sieve) of clay or cohesive silt with plasticity index of 8 or higher.

For cores of cohesionless silty sands and sandy silts, the material of the upstream face of the core can be reasonably imagined to migrate upstream into transitions with large voids, under the combined action of gravity and seepage following reservoir drawdown. Especially for dams retaining reservoirs that are frequently lowered (such as pumped storage reservoirs in the extreme), and for dams with thin, near-vertical central earth cores, with steep upstream faces, it is desirable to use a reasonably conservative upstream filter.

Summary and Conclusions

1. Concentrated leaks by hydraulic fracturing occur in most embankment dams without being observed, even in dams without large settlements. Usually the leaks do not cause erosion either because the velocity is too low or they discharge into effective filters.

2. Both recent extensive laboratory research and evaluation of experience with dam behavior support the conclusion that adequate filters will reliably seal and control concentrated leaks through earth cores

of embankment dams.

3. A main conclusion from the recently completed, 4-year long SCS filter research program is that the No Erosion Filter Test, described herein, is superior to the slot and slurry tests used earlier in the research for silts and clays (13). It was found to be the best test for the general filter research and also the best available test for routine laboratory evaluation of filters for specific projects. It is applicable for tests on coarse impervious soils as well as fine clays and silts.

4. Filters needed for control of concentrated leaks through impervious dam cores are not greatly different than those used in current common practice. Design criteria for various impervious soils are shown to be primarily dependent on the content fines (passing the No. 200 sieve) in the following main categories:

Soil Group	% Fines* (< No. 200 Sieve)	Recommended Design Criteria (already incl. adequate safety factor)
1	85 to 100%	$D_{15}/d_{85} \leq 9$
2	40 to 85%	$D_{15} \leq 0.7$ mm
3	0 to 15%	$D_{15}/d_{85} \leq 4$
4	15 to 40%	Intermediate between Groups 2 and 3

*Where % Fines is the fine content of the sand fraction (< #4 Sieve)

5. Increased confidence in filters is causing a change in practice. Filters are being relied upon to seal concentrated leaks and less money and effort is being spent on the various design measures formerly provided to reduce differential settlement.

6. Impervious soils with cohesionless silty fines are equally satisfactory for impervious core materials as clayey soils. As a general rule the least costly available impervious soil should be used for the dam core.

7. Central vertical cores are equally satisfactory as upstream sloping cores, even for high dams in steep canyons.

8. High embankment dams can be safely built in very narrow canyons with steep, near vertical, rock abutments.

9. Dams with straight longitudinal axes are as satisfactory as dams that are curved upstream in plan.

10. It is reasonable conservative practice to compact the entire impervious section of a high dam at the same water content for the full height. It is not necessary to compact the lower part at higher density and lower water content or the upper part at higher water content.

11. Seepage collars on conduits and at connections with concrete dams are no longer used.

12. Upstream sand zones provided for the purpose of being carried into and sealing cracks are not favored.

13. The increasing confidence in filters allows consideration of thinner earth cores in the future practice.

14. At sites where relatively high differential settlements are inevitable, the most effective method of ensuring a high margin of safety for the dam is to increase the conservatism of the critical downstream filter.

15. For typical coarse glacial moraines, graded from cobbles to fines, and other similarly graded impervious soils, there has been an unresolved question in the profession about the needed filter. The recent SCS research concentrated on this problem, concluding that a sand or gravelly sand with $D_{15} \leq 0.7$ mm., is needed for a conservative downstream filter.

16. Adequate bonding between compacted layers can be obtained without the need for routine scarification of the top of the previously compacted layer.

17. For the transition between an impervious core of fine clay and an upstream rockfill shell, a single band of relatively coarse material is adequate, such as grizzlied hard quarried rock.

. . . .

Appendix I.--References

1. ASCE (1967), "Problems in Design and Construction of Earth and Rock-fill Dams," Journal of the Soil Mechanics and Foundation Engineering Division, ASCE, May, 1967, p. 129.

2. Bertram, G. E., "Experimental Investigation of Protective Filters," Soil Mechanics Series No. 7, Graduate School of Engineering, Harvard University, 1940.

3. Casagrande, A., "Transcript of Informal Discussion on Embankment Dam Design," 7th International Conference on Soil Mechanics and Foundation Engineering, Mexico, Vol. 3, 1969, p. 301.

4. Hacelas, J. E., and Ramirez, C. A., "Interaction Phenomena Observed in the Core and Downstream Shell of Esmeralda Dam, Chivor Hydro-electric Project," Specialty Conference Session 8, 9th International CSMFE, Tokyo, June 1977.

5. Hurley, H. W., and Newton, C. T., "An Investigation to Determine the Practical Application of Natural Bank Gravel as a Protective Filter for an Earth Embankment," Thesis, Dept. of Civil Engineering, Massachusetts Institute of Technology, 1940.

6. Kjaernsli, B., et al, "Design, Construction Control and Performance of the Svartevann Earth-Rockfill Dam," 14th ICOLD, Rio de Janeiro, Vol. IV, 1982, pp. 319-337.

7. Lowe III, J., "Recent Development in the Design and Construction of Earth and Rockfill Dams," Proc. 19th ICOLD, General Report, Question 36, Montreal, Vol. V, 1970, pp. 1-28.

8. Moreno, E., and Alberro, J., "The Behavior of the Chicoasen Dam: Construction and First Filling," 14th ICOLD, Brazil, Vol. I, 1982, pp. 155-182.

9. Sherard, J. L., "Embankment Dam Cracking," Embankment Dam Engineering, John Wiley and Sons, New York, 1973, pp. 272-353.

10. Sherard, J. L., "Sinkholes in Dams of Coarse Broadly Graded Soils," 13th ICOLD Congress, India, Vol. II, 1979, pp. 25-35.

11. Sherard, J. L., "Hydraulic Fracturing in Embankment Dams," Paper prepared for Symposium on Leakage Control in Embankment Dams, ASCE Denver, Colorado, May 1985.

12. Sherard, J. L., Woodward, R. J., Gizienski, S. F. and Clevenger, W. A., Earth and Earth-Rock Dams, John Wiley and Sons, New York, 1963.

13. Sherard, J. L., Dunnigan, L. P. and Talbot, J. R., "Filters for Clays and Silts," Journal of the Geotechnical Engineering Division, ASCE, June 1984, pp. 701-718.

14. Sherard, J. L., Dunnigan, L. P. and Talbot, J. R., "Basic Properties of Sand and Gravel Filters," Journal of the Geotechnical Engineering Division, ASCE, June 1984, pp. 684-700.

15. Sierra, J. M. and Buendia, J., "Esmeralda Dam, Chivor Project," (Spanish) 5th Panamerican Conference on Soil Mechanics and Foundation Engineering, Buenos Aires, Vol. II, 1975, pp. 277-293.

16. U. S. Army Engineers, "Filter Experiments and Design Criteria," Technical Memorandum No. 3-360, Waterways Experiment Station, Vicksburg, Mississippi, 1953.

. . . .

Appendix II--Notation

The following symbols were used in this paper:

d_{85} = particle size in base soil for which 85% by weight of particles are smaller (similarly for d_{50}).

D_{15} = particle size in filter for which 15% by weight of particles are smaller (similarly for D_{50})

D_{15b} = D_{15} size of filter found to be boundary between successful and unsuccessful tests for any given base soil, using no-erosion filter tests.

D_{15B} = Similar filter boundary for a given base soil as determined by the slot and slurry tests used earlier in the research program (13).

Introduction to Paper 17

Hydraulic Fracturing in Embankment Dams

by J.L. Sherard
Proceedings, Symposium on Seepage and Leakage from Dams and Impoundments, Volpe, R.L.,. and Kelly, W.E., eds., American Society of Civil Engineers, 1985, pp 115-141 (also published in Journal of the Geotechnical Engineering Division, Vol 112, No. GT10, October 1986, pp 905-927).

Prepared by Edward B. Perry[10], M. ASCE

One measure of the greatness of a person is the size of the problems they tackle. Sooner or later in their careers, most of the great minds in geotechnical engineering have been drawn to earth dams. Jim Sherard spent his entire career advancing the state-of-the-art of earth dams.

Jim received his graduate education at Harvard University. Upon completing his course work, his major advisor, Professor Karl Terzaghi, suggested he do his Doctor of Science dissertation on the "Influence of Soil Properties and Construction Methods on the Performance of Homogeneous Earth Dams." Jim went to work with the Earth Dams Section of the Bureau of Reclamation and conducted an investigation of 60 homogeneous earth dams in 12 western states. His dissertation covered earth dam cracking, earth dam piping, and embankment slope slides. These three topics would remain of paramount importance to Jim for the remainder of his career.

In May 1985, a symposium on Seepage and Leakage from Dams and Impoundments was held in conjunction with the ASCE national convention in Denver, Colorado. This symposium provided an opportunity for Jim to present his third and final paper on hydraulic fracturing (and cracking) in embankment dams. Jim was awarded his second Norman Medal by ASCE in 1986 for the following paper.

[10]Research Civil Engineer, U.S. Army Engineer Waterways Experiment Station, Vicksburg, MS.

Hydraulic Fracturing in Embankment Dams

James L. Sherard*

There is now sufficient evidence to conclude that concentrated leaks occur commonly through the impervious sections of embankment dams by hydraulic fracturing without being observed, even in dams which are not subjected to unusually large differential settlement. Usually these concentrated leaks do not cause erosion, either because the velocity is too low or because the leak discharges into an effective filter. Subsequently the leakage channel is squeezed shut by swelling or softening of the embankment material forming the walls of the crack. In the typical case no measurable leakage emerges downstream and there is no other indication that a concentrated leak developed and was subsequently sealed. This action probably occurs to some degree in most embankment dams.

Introduction

Grouting of dam foundations of both rock and soil using pressures high enough to cause the grout to flow out of the grout hole in concentrated streams by hydraulic fracturing has been understood and applied practically since before 1960. Since about 1970, it has been known that hydraulic fracturing commonly has occurred inadvertently when borings are drilled in completed embankment dam cores and the pressure in the drilling fluid at the bottom of the hole exceeds the adjacent embankment earth pressure (15)(16).

There is a substantial difference, however, between hydraulic fracturing at the bottom of boreholes by pressurized drilling fluid and hydraulic fracturing by reservoir water acting on the upstream face of the dam core causing concentrated leaks of water to enter the core. In 1970 there was no mention of this possibility in the literature and little or no discussion or recognition of the phenomenon among specialists. In a general study of dam cracking in 1972, the writer found only three pieces of literature describing experiences in which it was concluded that concentrated leaks had probably developed through embankment dams by hydraulic fracturing and had caused erosion damage or failure (9, 20, 24, 27).

Since about 1975 there has been a large change, with wide acceptance that hydraulic fracturing causes leaks through dams under certain conditions. In the last few years there has been a considerable volume of literature published on various aspects of the subject, among which the following are representative (5, 7, 8, 10, 14, 17, 25, 26).

At the present time, however, it is still not completely accepted

*Consulting Engineer, San Diego, California.

in the profession that concentrated leaks can develop through dams by hydraulic fracturing. Also, among specialists who accept the idea in principle, there is no widespread agreement on the conditions needed for, or the probable frequency of, concentrated leaks by hydraulic fracturing.

Evaluation of hydraulic fracturing in dams has an inherent, important difficulty. Hydraulic fractures, by definition, are only open when the water pressure acts and they are inside the dam, so it is not possible to verify the action by direct visual observation. It is necessary to infer that concentrated leaks from hydraulic fracturing develop through embankment dams by showing that there is no other reasonable cause for the leaks which are seen to develop, and by showing that the hydraulic fracturing is readily predicted by theory.

In any given case where a dam is breached or where an erosive leak develops short of failure, there is nearly always a difference of opinion among investigators regarding the origin of the initial leak causing the trouble. It is usually impossible to determine for sure what caused the initial leak because the erosion destroys any evidence which might have existed. Another common problem is that different interested parties have different vested interests in the outcome of the investigations. It is natural that the designer desires to show that the design was not at fault, that the engineer who controlled the construction wants to show that the construction supervision was done carefully, and that the owner wants to establish that someone is clearly responsible, etc.

Because of these problems, with rare exceptions, investigations of dams which have developed erosive concentrated leaks have not generally reached a definite conclusion, agreed upon by all parties, with regard to the cause of the initial leak. Frequently the majority of investigators of any given case conclude that differential settlement cracking is the only reasonable explanation for the initial leak, but there is enough smoke thrown up by the heat of the arguments presented by the minority to prevent a clear or strong conclusion.

In spite of these problems with the analysis of individual dams, as the years have gone by a fund of experience is developing which shows definite patterns. There have been quite a few records of troubles of different kinds caused by concentrated leaks through dams designed and constructed by current accepted practice. With increasing frequency the investigators have concluded that the leak developed by hydraulic fracturing.

The writer believes that there is now essentially incontrovertible evidence available to the profession supporting the conclusion that small concentrated leaks by hydraulic fracturing develop commonly in well designed and constructed dams without being recognized, even in dams not subjected to large differential settlement. This paper presents this evidence, in several main categories: (1) concentrated leaks which appear soon after the first reservoir filling through well constructed homogeneous dams with no internal filter or drain, sometimes leading to breaching failure; (2) erosive leaks through central core dams with inadequate filters, usually occurring soon after the first relatively rapid reservoir filling; (3) records of well constructed and designed central core dams in which nearly full reservoir pressure was measured in piezometers at the downstream face of the core, and in which exploratory borings in the dam made to study this condition showed conclusively that near horizontal, water-filled cracks existed in the core; (4) FEM calculations showing that only relatively small differential settlements will easily create stress conditions in most dams under ordinary conditions which will allow hydraulic fracturing to occur; and (5) the discovery of "wet seams" inside impervious dam sections, with water content higher than can be accounted for reasonably by any other mechanism than the entry of water into an open crack.

Before proceeding to the presentation of the experience and evidence showing hydraulic fracturing in dams is common, it is desirable to review briefly the basic mechanics by which hydraulic fracturing can create a leak in a dam. It has long been observed that differential settlement open cracks appear fairly commonly on the surface of embankment dams. These open cracks do not usually extend into the embankment to depths of more than a few meters, because the internal compressive stresses in the embankment, which increase with depth, squeeze them shut.

In addition to open cracks the differential settlement creates strains in the compacted embankment, changing the internal stress distribution (sometimes called "stress transfer"). In this action the internal pressure is increased in some locations and decreased in others. In localized, but fairly large, zones within the dam the minor principal stress is reduced to near zero, or to even tensile stress if the dam has sufficient cohesion to withstand tension.

When the reservoir rises above one of these surfaces with low stress in the impervious section of the dam, water can enter along the surface from the upstream face in a concentrated thin layer. When the reservoir water pressure becomes slightly greater than the embankment stress, there is no resistance to the entry of water into the embankment in a concentrated leak. As the reservoir continues to rise, increasing the pressure in the water-filled crack, the stress conditions in the embankment are changed, the impervious embankment material is deformed and the crack is jacked open wider.

There has been some speculation that the reservoir water could not penetrate the upstream face of the dam core and open a crack by hydraulic fracturing unless the reservoir water pressure were considerably higher than the earth pressure existing in the embankment on the plane of the crack. This speculation was based on the idea that the upstream face of the impervious embankment section could be considered a smooth, uniformly impervious wall, in which there is no opportunity for water penetration to start.

Experience with dam behavior has shown that this idea is not correct. Experience shows that concentrated leaks enter cracks in the impervious core, as the reservoir is rising, when the reservoir level is only a meter or two above the elevation of the entrance to the crack. This shows that no significant "excess pressure" is needed to initiate the fracturing action. The behavior of the Wister Dam, discussed later, where near-horizontal concentrated leaks passing with a length of more than 200 meters through a clay dam appeared when the reservoir was being filled for the first time and the water level was only about 4 meters above the level of the leak, shows clearly that the water must have entered the dam by hydraulic fracturing when the pressure was only 1 or 2 meters of water head.

This lack of resistance to the initial penetration of water into the upstream face of the impervious dam section is not difficult to understand when the physical conditions are considered. All impervious sections of embankment dams contain a multitude of cracks formed during construction by drying, deformation under wheels of equipment and incompletely bonded layers. These cracks are squeezed shut by the weight of the overlying embankment but they remain nevertheless as closed cracks in the fabric of the compacted soil which the reservoir water can probably enter when the pressure is greater than the earth pressure, starting the hydraulic fracturing process. After the initial penetration of the water for a few centimeters, into these small closed cracks or fissures at the upstream face of the impervious section, the water pressure starts to act on the surface of least resistance (which has total stress acting on it less than the water pressure) and the advancing water jacks open a crack (fissure) which was not in existence before the reservoir was raised. Probably the initial penetration of water for a few centimeters into the upstream face of the impervious section is facilitated by the small existing closed cracks, but the main hydraulic fracture extending through the dam core is generally in a new crack jacked open progressively by the penetrating water.

Experiences with Homogeneous Dams

Downstream Slope Wet Spots

At a number of small well-constructed homogeneous dams of clayey soil (without internal drains) over the years the writer has seen "wet areas" appear on the downstream slope, usually in the upper portion, soon after the first rapid reservoir. These dams usually retained small reservoirs which filled completely for the first time in a few hours (or a day or two) following a rainstorm. In these "wet areas" there was generally only enough water to cause the soil surface to be completely saturated in spots, with small surface indentations accumulating some free water, rarely with a trickle of water running down the slope. Within a short time (days or weeks) these wet areas on the downstream slope often dried up completely, with no subsequent reappearance. Often these wet spots occurred near the abutments where the dam height was changing relatively rapidly and where tensile strains would be expected in the upper part of the dam embankment.

After puzzling over these temporary wet spots the writer concluded that they must have been caused by small concentrated leaks from hydraulic fracturing. In these experiences the channel (crack width) must have been very small with the leakage velocity far too low to erode soil from the crack walls. In such a case, softening or swelling of the embankment material in the crack walls probably acted to close the leakage channels in a short time, readily explaining the observed disappearance of the wet spots.

The above experiences taken by themselves are not highly persuasive. However, they are completely consistent with the other evidence for hydraulic fracturing in homogeneous dams as discussed below.

Visible Open Cracks

Fig. 1 shows the typical pattern of open cracks seen fairly frequently in the crest of low homogeneous dams, caused by differential settlement between the portion of the dam at the abutment and in the main dam. The width of the crack at the crest depends primarily on the magnitude of the settlement and on the properties (stiffness) of the embankment, and varies quite a bit. In the writer's experience an open crack with a surface width of 5 to 10mm can easily occur as the result of the post-construction differential foundation settlement is 20 to 30 cm. Frequently the cracks are increased in width by drying and embankment shrinkage.

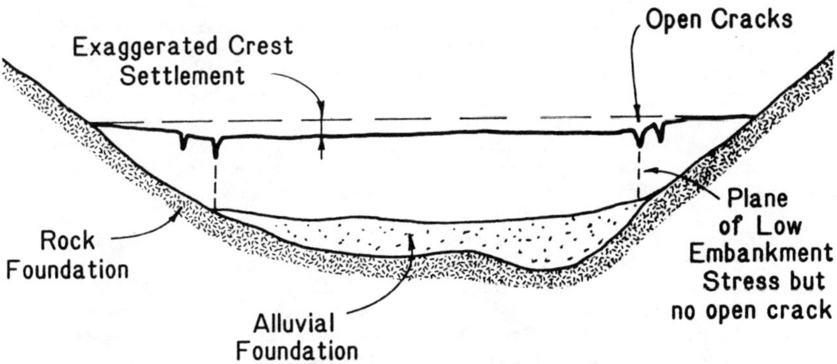

Fig. 1: Longitudinal Section of Typical Small Dam Showing Common Open Cracks on the Crest and Planes of Likely Hydraulic Fracturing.

These cracks are commonly sealed by grouting or trenching and refilling the trench with compacted material before the reservoir is filled. The internal embankment stress acting on the transverse plane extending down below the bottom of the open crack (dotted lines, Fig. 1) is clearly not very high, and probably is tension rather than compression in many cases. A leak by hydraulic fracturing on this plane obviously could be expected.

There are many examples where low homogeneous dams, without internal drains or filters, failed rapidly by breaching (washing out of a section of the dam) immediately following the first rapid reservoir filling. There are dozens of such experiences and frequently the break occurred over the abutment, where longitudinal stretching of the dam and hydraulic fracturing is most probable (Fig. 1). For many of them there was every reason to believe that the dams were well constructed and there was no likelihood of a concentrated leak from any other source. Over the years the writer has had the opportunity personally to investigate a number of homogeneous dam breaching failures in each which it was considered highly probable the initial leak occurred in a crack which was never seen.

Stockton Creek Dam

One representative, fairly well studied, example was the breaching failure of the Stockton Creek Dam in California immediately following the first rapid (few hours) reservoir filling in 1950. This was a homogeneous dam about 25 m. maximum height, founded on hard, incompressible bedrock, Fig. 2. It was well-constructed under constant, competent engineering supervision, with the construction also inspected by the State of California Division of Dam Safety, the strongest state control group in the U.S.A. No open cracks were seen during or after construction. The failure occurred when the reservoir filled rapidly for the first time at night with no eyewitness so that it was not known how the initial leak started. The failure breach found to exist the next morning was about 12 m. wide at the top and was located at the right end at a point where there was a small but near vertical step in the sound rock abutment, where stress transfer and hydraulic fracturing would have been most probable, Fig. 2b. After thorough study of this experience the writer concluded that there was no other reasonable explanation for the breaching failure than erosion of a concentrated leak which developed in a crack through the embankment from differential settlement, although no crack was ever seen.

Since the dam foundation was relatively incompressible rock, if hydraulic fracturing occurred the internal stress transfer must have been caused by the differential settlement of the dam itself, which was very low. But the material was rocky and the compacted embankment was relatively hard and stiff, not able to adjust easily to movements without relatively high resistance and internal stress transfer. A more complete description with photographs is given in Ref. (16).

Oklahoma and Mississippi Failures

Another group of 14 similar breaching failures in Oklahoma and Mississippi which occurred between 1957-70 studied by the writer are described in Refs. (19) and (20). These were all low homogeneous clay dams, 7 to 20 m. high, without vertical chimney drains, which were well-constructed with competent engineering supervision and moisture-density control. Essentially all of these failures occurred immediately after the rapid first reservoir filling, and the initial leak appeared at the downstream side within a few hours (or a day or two) after the reservoir filled. In almost all cases the breach occurred at a point along the dam length where the maximum differential settlement was expected. As discussed in detail in Ref. (20), these experiences left little doubt that the initial leak which eroded to failure developed in a hydraulic fracturing crack.

The main difference between this group of dams which failed by breaching and hundreds of other similar dams built by the same engineering organization which did not fail was the fact that the failed dams

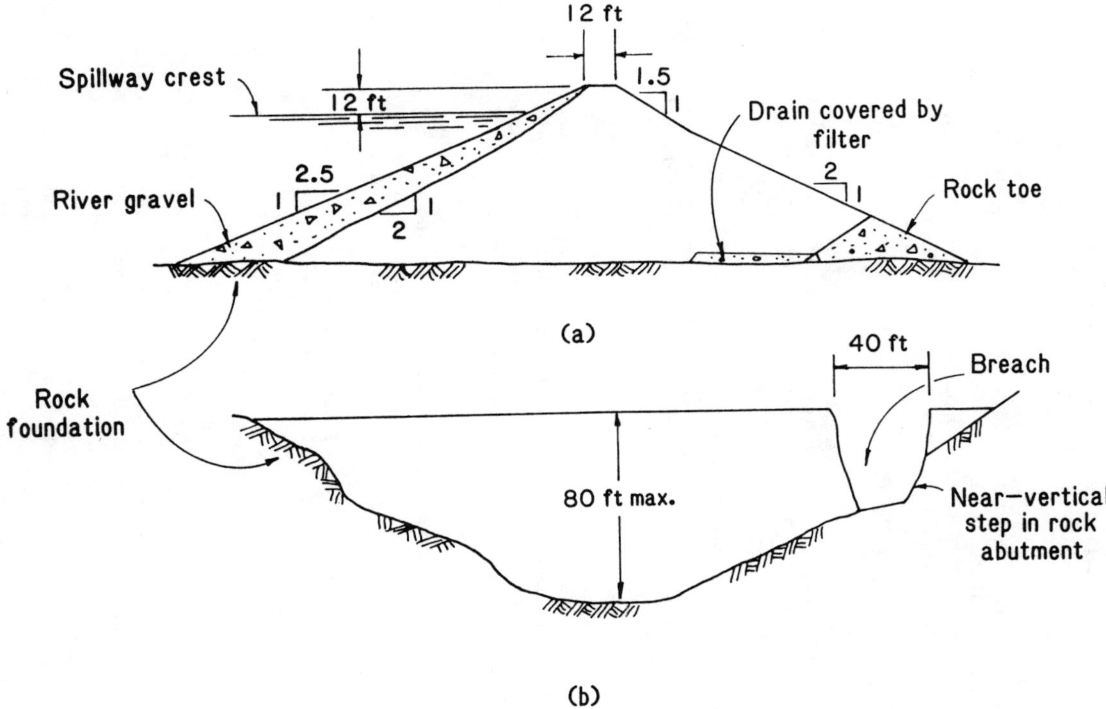

Fig. 2: Stockton Creek Dam. Cross-section (a) and Longitudinal Section (b) Showing Breaching Failure.

were constructed in areas of highly erodible dispersive clay. For dams of dispersive clay even a very small initial concentrated leak with very low velocity may erode the material faster than the leakage channel is closed off by swelling of the crack walls. For homogeneous dams constructed of nondispersive impervious soils evidently the initial leak needs to be considerably larger and have greater velocity in order to start erosion than for dams of dispersive clay. Therefore, everything else being equal the likelihood of breaching failure by hydraulic fracturing is considerably higher for homogeneous dams of dispersive clay. This conclusion is well-supported by experience. There are many fewer cases of breaching of this kind in homogeneous dams of ordinary, non-dispersive soils (such as the Stockton Creek Dam failure) than in dams of dispersive clay.

There are many other similar examples. In all these cases the failure nearly always occurs within a very short time after the reservoir is filled rapidly, usually in one rainstorm. In these cases, the appearance of the initial leak on the downstream slope within hours after the reservoir rises above the elevation of the leak leaves no doubt that the water was traveling through the dam as a concentrated leak. The typical dam embankment is far too impervious to allow water to travel completely through the dam in such a short time other than in a concentrated leak.

Wister Dam

The near failure by piping of the Wister Dam in Oklahoma in 1949 is among the more important case histories in this category, and the significance of the experience has not been widely recognized (3, 16, 23). The Wister Dam experience was very similar to the 14 failures in Oklahoma and Mississippi described above, but occurred in a much more important dam of the U. S. Army Corps of Engineers. The main details were generally as follows: homogeneous clay dam (no vertical drain), 30 m. high, 1800 m. long, with reservoir volume of about 500 million m^3, well-constructed by experienced engineers during 1946-48, Fig. 3. In January 1949 a large storm caused the water to rise rapidly for the first time to a maximum depth of about 18 meters on January 28, 1949. Then, with the large bottom outlets open, the reservoir water level started to drop slowly. On January 30, seepage carrying eroded embankment material was first noticed on the downstream slope with estimated quantity of about 150 l/sec. This increased to more than 500 l/sec., in the following several days, always very dirty. By February 3, with bottom outlets discharging at full capacity the reservoir level had dropped about 4 m. (maximum water depth about 14 m.) exposing several erosion tunnels with diameter of about 0.6m entering the upstream slope, all at about the same elevation, over a length of about 90 meters. Subsequently, the reservoir water level was lowered below the tunnels and the dam was saved. Studies with dyes showed that the average leakage water travelled a distance of about 225 m. through the dam on a gradient of about 1:50, in about 13 minutes. The leaks travelled diagonally through the dam on a path nearly parallel with the original river bed, through a length of the embankment which had been constructed last as a closure section (Fig. 3).

The results of thorough studies made of the cause of the Wister

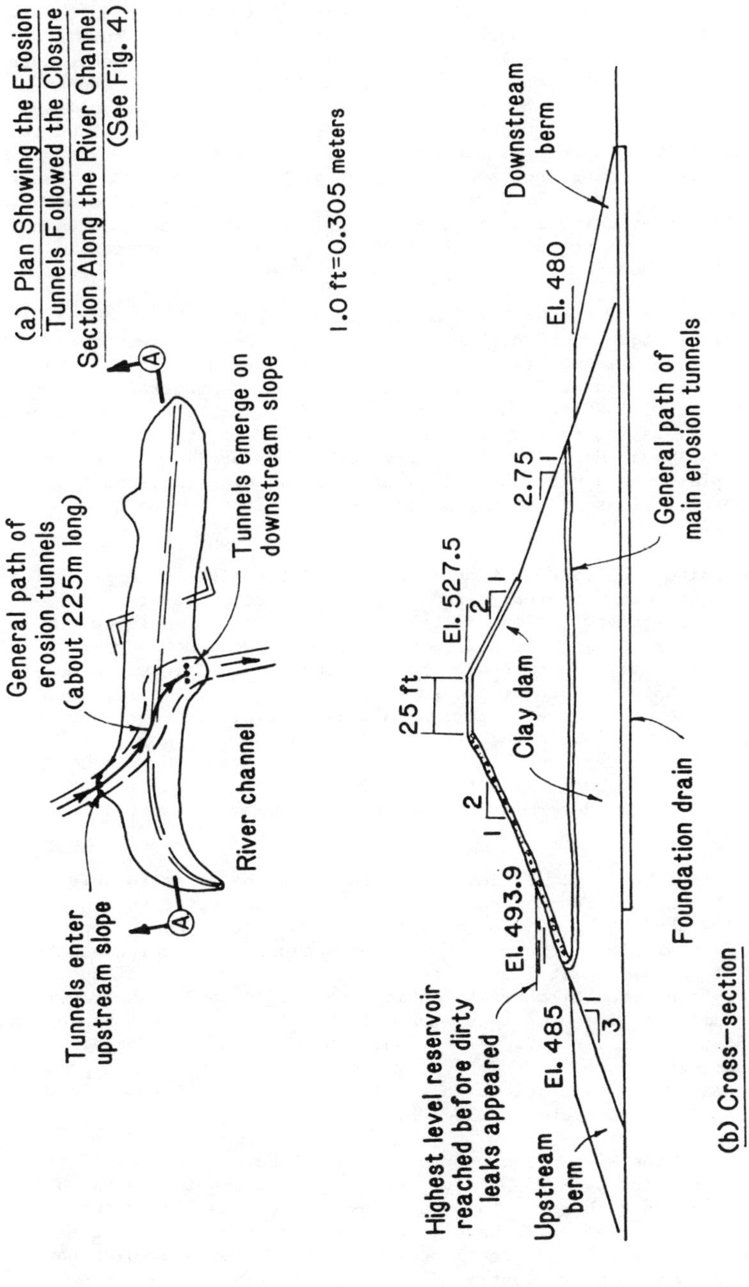

Fig. 3: Wister Dam Cross-section and Plan Showing Location of Erosion Tunnels from Hydraulic Fracturing.

Dam leakage by experienced specialized engineers were inconclusive. Finally, it was generally agreed that the initial leak must have been travelling in a differential settlement crack, even though the measured foundation settlements were not large (about 25 centimeters) and no cracks were seen. It was considered beyond the realm of possibility that the near horizontal concentrated leak entering the dam with only 4 m. head could have been caused by a poorly compacted layer or any other conceivable cause but a crack. Soon after the event, Casagrande wrote (4), "I am seriously disturbed by this case, because it demonstrates that clay which is compacted at optimum water content does not necessarily possess the ability to follow even small differential settlements without cracking."

At the time of the Wister Dam experience studied about 30 years ago, the investigators were thinking only of cracks which were open before the reservoir was filled. There was no consideration of the phenomenon of hydraulic fracturing. Later it was demonstrated conclusively that the Wister Dam was constructed of highly erodible dispersive clay, making the experience easier to understand (16)(20). During the several days when the dirty leaks were eroding the embankment material, the dam crest settled about 5 cm. over the length of the closure sections, evidently as the result of the removal of eroded clay.

The writer believes that near failure of the Wister Dam by piping is one of the more important single experiences to guide present thinking about the evolution of leakage control design measures. This was not a dam retaining a small farm pond, built with methods which could be reasonably questioned as being inadequate. This was a major dam designed and built under the supervision of the U. S. Army Corps of Engineers, one of the most experienced organizations in this type of work. All the facts were documented. The dam was well built by conventional methods from silty clay near optimum water content. The differential settlement was not high. The path of the concentrated leak was definitely determined. The entire drama was observed by experienced engineers.

The writer strongly believes that there can be no reasonable doubt that a concentrated leak developed by hydraulic fracturing completely through the dam for a distance of more than 200 m. on a near horizontal plane. This must have been caused by arching of the portion of the dam built last in the closure section between the two previously completed embankment sections on both sides of the river. The maximum total measured settlement of the alluvial foundation on the left side of the river (Fig. 4) was only about 25 cm. The details of Wister Dam experience are practically identical to the details of the erosion and breaching problems in the 15 small dams in Oklahoma and Mississippi described above (19, 20).

Probably the Wister Dam problem would not have developed if the embankment had not been dispersive clay. This conclusion is based on the knowledge that there are a great many homogeneous dams constructed of nondispersive clay which have been subjected to much more severe conditions of differential settlement than the Wister Dam which have not developed erosive concentrated leaks. Probably if the Wister Dam

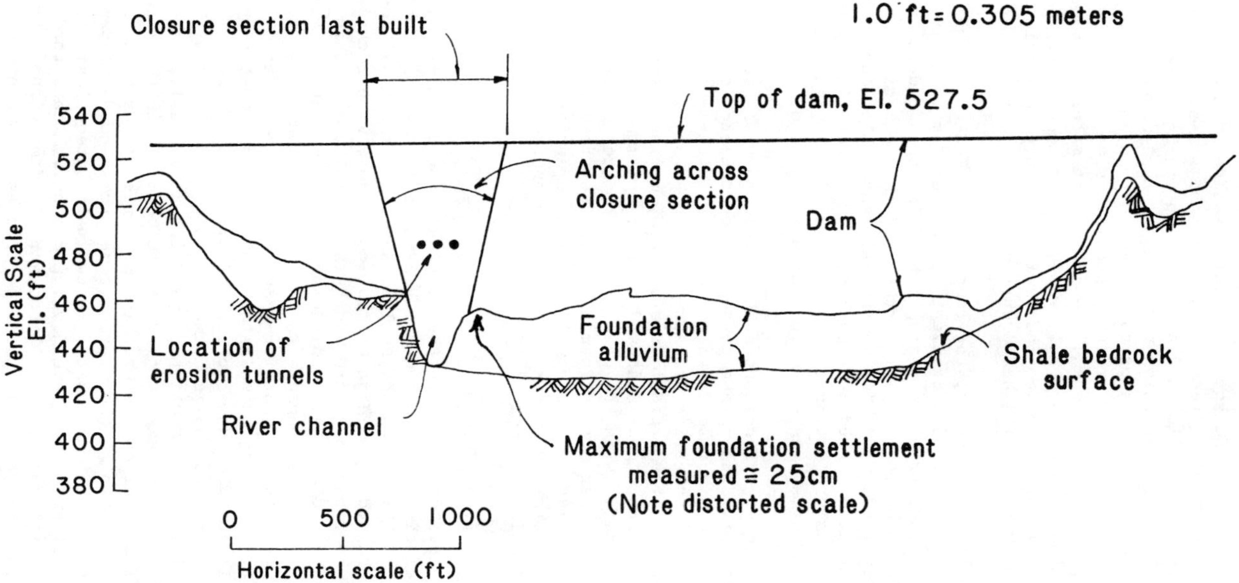

Fig. 4: Wister Dam. Longitudinal Section A-A (Fig. 3) Showing Location of Erosion Tunnels with Respect to the Closure Section.

had been constructed of nondispersive clay the same initial leak would have developed by hydraulic fracturing, since the stress condition in the embankment needed to allow hydraulic fracturing has nothing to do with the erodibility or dispersion potential of the embankment material. The initial leakage channel was probably very narrow and the initial velocity insufficient to erode ordinary nondispersive clay. Therefore, if the initial leak had not eroded, the embankment material forming the walls of the leakage channel probably would have squeezed shut by swelling, and there would have been no indication to alert engineers at the site that hydraulic fracturing had occurred.

In summary in this section, the concentrated leak and erosion tunnels 225 m. long under 50:1 gradient, in Wister Dam is a vitally significant experience from the standpoint of improved understanding of embankment dam behavior. The importance of this experience has not been widely recognized. It shows beyond all reasonable doubt that hydraulic fracturing occurred on a path over 200 meters long when the reservoir water pressure acting on the upstream face of the dam which initiated the fracturing was not more than 4 m. of head.

The Wister Dam experience shows the development of hydraulic fracturing on a near horizontal plane caused by the tendency of the top of the dam to create a horizontal arch, Fig. 4, in the portion of the dam built last in the river closure section. The Wister Dam embankment was constructed of a typical fine-grained, medium plastic clay (CL) at average water content near Standard Proctor Optimum. The embankment was not hard and stiff as in the case of the Stockton Creek Dam described above, but rather more deformable than the average impervious dam zone. The experience shows how little settlement is needed even in a relatively deformable embankment, to cause enough stress transfer to allow hydraulic fracturing.

Experiences with Central Core Dams

The experiences described in the previous section show hydraulic fracturing in relatively low homogeneous dams where the width of the impervious section through which the leak develops is 5 to 7 times the head of water retained. For high central core dams, with rockfill or gravel shells, the impervious core width is generally less than the head of water retained and the conditions for stress transfer are much more severe. This is true because the total settlements are greater, and because internal stress transfer takes place in both directions, i.e., by transverse arching between the upstream and downstream shells and longitudinally between dam sections of different height. Therefore, if hydraulic fracturing is common in homogeneous dams with their relatively wide impervious sections, even without unusually large settlements, it should be expected even more commonly in the relatively thinner cores of central core dams. The writer believes that there is ample evidence available at the present time from records of dam performance to support this conclusion, in two main categories: (1) dams with inadequate filters which have developed erosive leaks through the core on the first reservoir filling, and (2) dams without erosive leaks in which exploratory borings in the core have encountered water-filled cracks.

Starting as early as 30 years ago a few engineers made measurements on dams leading to serious concern that "silo action" in zoned dams could reduce the total pressure acting on horizontal planes through the core to dangerously low values, or even create open cracks on horizontal planes. One interesting early example was at the U. S. Army Engineers' John Martin Dam where total pressure cells installed during construction in 1942 measured pressures on horizontal planes in the core which were less than 30 percent of the weight of column of embankment above (13).

Similar low total pressures were measured in two thin core dams in Sweden about the same time (12). In one of these Swedish dams it was estimated that the post-construction crest settlement should have been about 60 cm., and the actual measured settlement was only about 5 cm. It was the opinion of the engineers that the thin central core had probably arched between the less compressible upstream and downstream shells and that an unseen horizontal crack or series of cracks with a combined width of the order of 30 cm. could exist through the core. This hypothesis created enough concern to cause the engineers to drive a row of steel sheet piling through the core before the reservoir was filled for the first time (11).

Piping in Dams with Coarse Broadly Graded Cores

There have been more than 15 important central core dams, designed and constructed according to good current practice, in which concentrated erosive leaks have emerged suddenly at the downstream toe, usually during or soon after the first reservoir filling, discussed in more detail in Ref. (18). These are dams in many parts of the world designed and built by different experienced engineers. Most of these have been dams in which the core material was a coarse, broadly graded mixture of gravel, sand and clayey or silty fines, with many being glacial moraines. For core materials of this type, there was no general agreement on how to apply our current accepted filter design criteria, with the result that most of these dams with erosive leaks had been provided with coarse downstream filters, commonly unprocessed crushed rock or coarse alluvial gravels. These coarse filters inevitably segregate badly during construction at the core-filter interface so that they offer practically no filter action for the fine-grained portion of the material. As discussed in a companion paper (21), recent research shows that in order for a filter to be effective in sealing concentrated leaks through such a core material it must contain sufficient fine to medium sand to give a D_{15} size of 0.7 mm or smaller. Therefore, on the basis of this new understanding of filters, the coarse, segregated filters provided at the typical dam which developed erosive concentrated leaks could not be expected to seal the leaks.

In most of these cases the results of the investigations of the cause of the initial leak were inconclusive. In some of these dams extensive investigations led to the conclusion by different groups of experienced specialists that hydraulic fracturing was the only likely cause (9, 24, 25, 27). Since publication of the initial summary of this experience (18) the writer has had the opportunity to be part of the investigating team studying several other almost identical experiences, concluding that it is overwhelmingly probable that the initial leak causing the troubles in this group of dams was usually caused by hydraulic fracturing. The evidence for this conclusion consists primarily of the following elements: (1) the erosive leaks have usually occurred in conditions where low internal embankment stresses would be especially likely, such as in dams with unusually narrow, vertical cores and in locations along the length of the dam where the height is changing rapidly; (2) the initial leak appeared downstream almost immediately following the first reservoir filling, indicating that there had to be a concentrated leak through the core (there wasn't time for general seepage through the voids of the core material to have caused the leak by progressive backward piping); (3) calculations of the stress conditions (FEM) show that the low stress conditions needed for hydraulic fracturing are generally present; (4) there is no other tenable hypothesis to explain the development of the initial leaks.

In the group of damaged dams described in this section above, the initial leak was able to erode the core because the downstream filter was too coarse to prevent erosion of the broadly graded impervious core material. In the world today there are many dozens of completely successful dams with the same type of core material and designs very similar to those of the group of damaged dams, except they had generally finer downstream filters. Undoubtedly initial leaks by hydraulic fracturing developed in many of the successful dams without being known because the filter successfully prevented erosion. The development of initial leaks by hydraulic fracturing is not limited to dams with inadequate filters.

Discovery of Water-Filled Cracks in Dam Cores

There have been a number of experiences in which piezometers in the upper half of dams located near the downstream edge of central cores have registered full, or nearly full, reservoir pressure.

Manacouagan 3 Dam: One of the most instructive of these is the experience at the Manicouagan 3 Dam (5), in Quebec, with main details as follows:

1. A central core dam with gravel shells, about 105 m. high and 400 m. long, and relatively thick vertical core of cohesionless, sandy silt, completed in 1975.

2. On the first relatively rapid reservoir filling, piezometers at the downstream face of the core registered immediately nearly full reservoir pressure, and continued to show the same pressure as the reservoir was kept nearly full for several years. Many piezometers on four instrumented sections all showed the same pattern of measurement so that it was apparent that the situation was not isolated, but a general condition.

3. To study this situation three special borings, about 75 m. deep, were drilled through the core near the downstream edge of the crest using casing without drilling fluid (mud or water), removing cuttings with compressed air. The casing was driven progressively in increments of about 2 to 4 m. At each stage in the casing descent a hole in the core was drilled 1.5 to 3.0 meters below the casing bottom and a waiting period of 1 to 2 hours was allowed during which the rising water level and elapsed time were recorded. Generally no

measurable inflow of water entered because the embankment was sufficiently impervious to limit the seepage inflow in two hours to a negligible quantity, but at certain levels water entered rapidly. For example, in the first boring drilled no water entered until the hole had reached a depth of about 21 m., at which time water rose in the casing about 11 m. in 60 minutes. Below this depth for some distance there was again no inflow, and then again at depth about 39 m. water rapidly entered again. The other borings had similar action.

4. There has been no measured leakage through the dam and the piezometer readings are stable. The downstream filter is conservative, with average D_{15} about 0.2mm. The condition is considered completely safe.

The investigators at Manicouagan 3 concluded that "The presence of a set of cracks (filled with loose muddy material) or decompressed layers distributed erratically, constitutes the main cause of the high pore pressure transmission from upstream to the downstream part of the main dam's core ... can be attributed to arching of the dam's embankment and differential settlement of its various zones ...". (5)

The writer believes that the Manicouagan 3 Dam experience must be recognized as providing basic proof that concentrated leaks from hydraulic fracturing develop through central core dams designed and constructed according to good current practice. The experience also shows that hydraulic fracturing can develop in embankments of cohesionless sandy silt.

Manicouagan 3 Dam was built and the problem later studied and reported by experienced engineers, with the assistance of an eminent Board of Consultants. The design would be generally considered very conservative. The central core was relatively wide for this kind of dam, about 75% of the water head. The investigations left no doubt that there were thin cracks through the core: there had to be at least thin open cracks to provide the needed water supply to allow the water to rise in the casing 11 m. in 60 minutes.

Because the core material was a cohesionless sandy silt it is probable that the roof (or hanging wall) of the crack collapsed and the sandy silt fell down into the space being jacked open by the increasing water pressure generated by the rising reservoir. This action would create a thin near horizontal layer which is filled or nearly filled with very loose saturated sandy silt. The effective stress in this layer must be essentially zero and the water pressure, about equal to reservoir head, is the total embankment stress acting on the plane of the layer. From the borings made in the core it was not possible to get undisturbed samples of the embankment in the vicinity of the water-bearing cracks, but the investigators definitely developed the idea that there were "soft and permeable layers...composed of loose muddy material..."(5).

As the reservoir level rose above the elevation of the piezometers the downstream piezometers started to record nearly full reservoir pressure with practically no time lag. This shows that the crack developed rapidly. It was impossible for the measured pressure at the downstream edge of the thick impervious core to have responded so rapidly by seepage of water through the pores of the compacted core material.

El Guapo Dam: There has been another very similar experience at the El Guapo Dam, a central clay core, gravel-fill dam, about 65 m. high, completed in 1978 in Venezuela, designed by experienced engineers and well-constructed by good modern practice (6). At the El Guapo Dam some of the piezometers near the downstream face of the clay core at about midheight of the dam registered pressure heads only slightly less than the reservoir level. Special borings were put down in the suspect area, using drilling techniques similar to those at Manicouagan 3 Dam (without drilling water or mud). In one of these borings, peetrating the clay core near the downstream face, water rose in the casing fairly rapidly, stabilizing at about the pressure head measured by the nearby piezometers.

This experience also can only be reasonably explained by a concentrated leak. This dam was of conservative design with excellent clayey core material and sandy gravel shells. The total settlements of the foundation (about 10 meters of dense gravel overlying bedrock) was small and there was no tendency for unusual differential settlement.

"Wet Seams" at Yard's Creek Dam

The experience of the Yard's Creek Dam, New Jersey, constructed 1963-65, was representative of the erosion damages in the group of dams with cores of coarse broadly graded soils described in the last section. This is a long, low rockfill dam with central vertical core of broadly graded glacial moraine, with maximum height about 25 meters, length about 2500 m., and with the earth core founded on unweathered, impervious, hard quartzite bedrock, Fig. 5. The dam forms the upper reservoir for a hydroelectric pumped storage project. The dam was designed and constructed under the supervision of experienced engineers.

During the first rapid reservoir filling very dirty leakage emerged abruptly at the downstream toe at one location over a length of about 100 m., reaching a maximum rate of about 100 l/sec. The experience is described in detail in Ref. (16), pp. 317-23. The problem was thoroughly investigated with generally inconclusive results as was typical for many of these investigations, because it is impossible to determine for sure what caused the initial leak, and because different memebrs of the investigating teams had different vested interests in the outcome of the investigations.

During the Yard's Creek investigation the writer had the rare opportunity to enter test pits in the core of the dam, put down to the bedrock foundation, in locations where there was no leakage emerging at the toe. In these, some peculiar "wet seams" were observed.

As described in Ref. (16), p. 320, the glacial moraine core generally appeared uniformly well-compacted, homogeneous and highly impervious. In the typical compacted core exposed in the walls of the

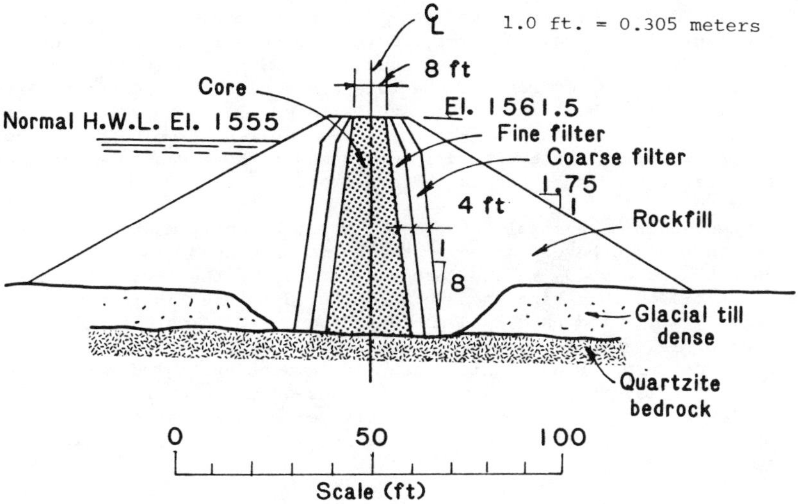

Fig. 5: Yard's Creek Upper Reservoir Dam Cross Section

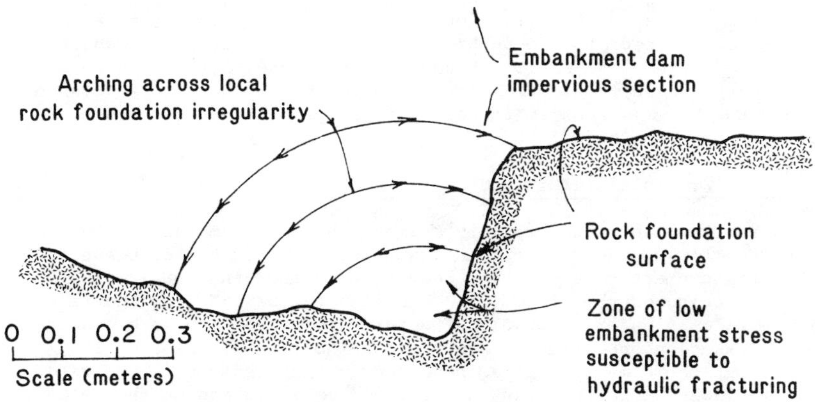

Fig 6: Zones of Low Embankment Stress Caused by Localized Arching Adjacent to Small Irregularities in Foundation Rock Allow Hydraulic Fracturing.

test pits, it was difficult to push in a sharp pointed steel rod more than a few millimeters. However, there were occasional thin layers, one to two meters apart which were much softer, obviously at a considerably higher water content than the compaction water content, into which it was possible to push the steel rod easily. These layers were similar in appearance to the rest of the embankment except for the soft, wet consistency. They appeared impervious and there was no evidence that concentrated leaks had caused any erosion.

These thin horizontal "wet seams" were obviously much softer and wetter than could be explained by the normal action of water seeping through the pore voids in the core and displacing the air. A specimen of soil of this kind when well compacted in the laboratory is relatively stiff and hard, and contains about 3 or 4 percent air. If a specimen is saturated in the laboratory at near constant volume, as by backpressuring in a triaxial test, it remains stiff and hard. The softening of layers in the dam must be caused by the entrance of a significant quantity of new water associated with a near complete elimination of the effective stress acting in the soil, as would occur adjacent to a waterfilled, open crack.

The origin of these "wet seams" was not understood at the time but the writer now believes, supported by the other evidence subsequently accumulated as discussed herein, that they were caused by hydraulic fracturing, by actions approximately as described below.

Stage 1: Arching of the Core and Stress Transfer Before Reservoir Filling.

During and after construction the more compressible central core settles with respect to the upstream and downstream filters, causing arching and stress transfer. At the time the reservoir begins to fill, the total vertical stress acting on some near horizontal planes through the core is reduced to nearly zero, probably to the order of 1 to 2 m. of water head.

Stage 2: Initial Thin Hydraulic Fracture.

When the water rises rapidly against the upstream face of the core during the first reservoir filling, at a certain horizontal plane the water pressure acting on the upstream face exceeds the total earth pressure acting on the plane. At this time the water from the reservoir is able to enter (hydraulic fracture) the core in a thin concentrated leakage channel (crack) on the near horizontal plane of low stress. At the time of initial penetration the pressure in the water is only slightly greater than the previous total stress on the plane, so that there is no significant change in the overall stress conditions in the core. The width of the initial crack is so small that the velocity is too low to cause erosion. The fact that the water pressure in the concentrated leak is higher than the pore water pressure in the immediately adjacent embankment creates seepage from the crack into the embankment, acting to hold the crack open temporarily.

Stage 3: Increasing Pressure Jacks Crack Open Wider.

As the reservoir level continues to rise, the pressure in the concentrated leak also rises, increasing the total stress on the horizontal plane, by 10-ton/m^2 for each 10 m. of rise in the reservoir level. This increase in pressure on the near horizontal plane of the crack causes a change in the stress conditions in the core and an increase in the width of the crack. Rough computations using reasonable assumptions for the deformation modulus of the core, and the typical dimensions of a central core dam of moderate height, show that if only a single crack developed, it could be jacked open to a width of the order of 5 to 10 cm. by a pressure increase of 10 to 20 m. of water head. If several parallel fractures developed, this computed width would be the sum of the individual crack widths. At Yard's Creek Dam, where the depth of water at the level of the several "wet seams" was only of the 5 to 7 m., the calculated aggregate width of cracks computed in this way would be about 1.0 cm. During this action the water pressure in the crack is always equal to the total embankment stress and the effective stress in the embankment material in the crack walls is zero.

When the near horizontal crack is jacked open to any significant width the roof must inevitably soften and collapse, tending to fill the space created with chunks of softened embankment material, so that the resulting condition would be a horizontal zone partially filled with loosened embankment material surrounded by free water. The width of such a zone depends on the dimensions of the dam and also on the distance between hydraulic fractures, when more than one develops.

Subsequently two alternate actions can occur. First, the eroded material carried to the face of the downstream filter may seal it, causing the water pressure in the crack for the full width of the core to be nearly equal to the reservoir head, so that the entire head loss occurs through a few centimeters of choked filter face. This is the action which must have occurred at Manicouagan 3 and El Guapo Dams. In the second alternative, the average velocity of flow in the collapsed zone which forms along the initial leak on the horizontal plane may be relatively low (a few centimeters/sec.) with no significant erosion or tendency for sealing the downstream filter. In this case the head loss will occur in hydraulic friction, and the water pressure in the crack will vary approximately linearly across the core from upstream to downstream. Apparently this second condition is the more common situation which develops in dams, since the high pressures measured in downstream piezometers (as at Manicouagan 3 and El Guapo Dams) are not common.

Stage 4: Action When the Reservoir Lowers Again.

After the reservoir is lowered again below the crack level, the water pressure in the concentrated leak reduces to zero (gage pressure), causing the crack to squeeze shut again. Some of the free water in the former concentrated leak is forced out, but most of it is trapped inside the core. This trapped water and the chunks of loosened embankment material which had fallen into the previously open crack are squeezed together, finally creating the soft "wet seams" seen in the Yard's Creek Dam test pits.

Only some action of this kind can result in such a soft, wet layer being found deep inside a well-constructed dam. There has to be a significant amount of water entering from the reservoir and there has to be a cavity created (open crack) in the embankment to accommodate this free water.

"Wet Seams" at Other Dams

If the mechanics of hydraulic fracturing described above are generally valid, "wet seams" similar to those seen in the Yard's Creek Dam core should develop fairly commonly in many dams. It is relatively rarely that engineers have the opportunity to examine the inside of the impervious core of a dam in test pits after the reservoir has been in operation. It is unlikely that occasional exploratory borings put down in dam cores would reliably identify wet seams. Therefore, if "wet seams" exist fairly commonly in certain locations in a significant fraction of our dams, it would not be known.

If it had not been for the investigation of the leakage problem at Yard's Creek Dam there would have been no opportunity to see the "wet seams". And in a career of specialized engineering practice devoted to embankment dams, the writer has not had many other similar opportunities.

The writer believes that it is probable that "wet seams" originating from the general action described above are not uncommon. It is impossible to prove this, but there is evidence to support the hypothesis.

The first type of evidence is from the experiences such as those of Manicouagan 3 and El Guapo Dams, described earlier. In these dams it was demonstrated conclusively that there were open water-filled cracks in the core. These cracks are in contact with the reservoir and carry nearly full reservoir pressure to the downstream side of the core. It is inconceivable that these water-bearing cracks remain completely open. The embankment material forming the roof of the crack, with zero effective stress, must collapse and fall down into the temporarily opened crack. The observation made by the investigators at the Manicouagan 3 Dam, describing layers of "loose muddy material" (5) are consistent with this hypothesis.

A second piece of evidence is the extensive series of "wet seams" discovered in the remaining intact portion of the embankment of the Teton Dam after the failure (7, 14). These are on more or less horizontal planes and are readily explained as being caused by hydraulic fracturing of the general type which occurred at Wister Dam, i.e., arching of the upper part of the dam in the longitudinal direction between the rock abutments, creating zones in the lower part where there are low total stresses on near horizontal planes.

When the Teton "wet seams" were exposed in the walls of exploratory excavations made into the dam embankment, small streams of free water dribbled out of the wet seam in the impervious compacted embankment. The time that this occurred was about 12 months after the dam failure, during all of which time the "wet seam" had not only not consolidated under the weight of the overlying embankment, it still had a lense of free water in it which would flow out when the wet seam was exposed in the exploratory excavation. This surprising behavior must be explained by the fact that a substantial volume of free water in a concentrated leak was trapped inside the dam at the time of the breaching failure.

Evidence from FEM Computations

The finite element method when applied to embankments is still in the developmental stage, in the sense that there are still doubts about how reliably embankment stresses can be predicted by the calculations. But the calculation probably does give a rough approximation of the general stress distribution and the average values. Calculations have been made to study the influence of moderate settlements on the stress distribution and the average values. These confirm generally that only relatively small differential settlements are sufficient to create large zones within dam embankments where the minimum principal stress is below the values which permit hydraulic fracturing.

As an example, in 1972 (20) some FEM calculations were made for the writer by Drs. J. M. Duncan and G. Lefebvre at the University of California (Berkeley) of a 2-dimensional mathematical model simulating one end of a low homogeneous dam of compacted clay. The end of the dam was assumed to be on an incompressible rock foundation and the post-construction foundation settlement about 150 feet away was assumed at about 30 to 50 cm. The calculations showed that even this moderate settlement created a large embankment zone near the abutment where the stresses on vertical transverse planes were in tension, and where hydraulic fracturing would be predicted, confirming the cracking commonly seen in dams in this area, Fig. 1.

In another interesting study Kulhawy and Gurtowski (10) showed that the computed stresses in the central core were generally conducive to hydraulic fracturing, even assuming: a 2-dimensional condition, an incompressible foundation and uniform properties of each embankment zone from top to bottom. Clearly similar calculations including such actual conditions as steep rock abutments, compressible foundations, and impervious cores with different compressibility at different elevations (constructed at different times with different borrow at different average water content) would obviously show much larger zones with lower internal stresses developing commonly.

In summary, embankment stress computations give clear support for the conclusion that hydraulic fracturing should be expected to be common even without unusually large settlements. The calculations show that impervious zones in embankment dams are commonly sufficiently stiff and rigid such that only small strains imposed from the outside will cause large redistributions of stress within the dam, frequently with compression changed to tensile stress.

Hydraulic Fracture Leaks Along Rock Foundation Surfaces

There is much evidence for the conclusion that small concentrated leaks commonly develop along the interface between embankment dam impervious cores and rock foundations and concrete sturctures (conduits, etc.). There are many experiences in which erosion tunnels have developed through small dams just at the foundation contact or directly adjacent to conduits passing through the base of the dam.

Probably in many cases small concentrated leaks can pass along the contact because the total pressure acting between the core and the rock foundation (or concrete conduit) is less than the reservoir water pressure (hydraulic fracturing). In one case the writer had the opportunity to see water travelling at a rock foundation contact. The test pits put down through the core at the Yard's Creek Dam in lengths of the dam with no trouble described above, were carried down to the hard rock foundation, which was cleaned off for inspection. During these studies, with the test pits still open, the reservoir level was occasionally raised 2 or 3 meters above the bottom of thepits. At this time water entered the pits in small concentrated streams, completely clear, at the rock surface with a total volume of a few l/min.

From these observations it must be concluded that the pressure existing between the dam core and the rock surface at the Yard's Creek Dam along the leakage path must have been less than 2 or 3 meters of water head. This may have been either general low pressure over the entire area caused by arching of the core between the less compressible filters, Fig. 5, or localized areas of low contact pressure caused by smaller scale arching between local irregularities in the rock surface as shown schematically in Fig. 6. Although impossible to prove or be completely confident, the writer believes it is nearly sure that small localized leaks by hydraulic fracturing of the type shown in Fig. 6 occur commonly in many dams.

At quite a few dams with leakage troubles, steps or irregularities in the rock foundation have been suspected as the most probable cause of the initial leak. One example is the Stockton Creek Dam problem discussed above, where the breach occurred at the exact location of a near vertical cliff in the hard rock abutment, about 6 m. high.

Another example is the problem of the Yard's Creek Dam, described above: erosive leakage occurred only in one section of length of the long, low dam, and the surface of the rock foundation under the core in this length was very irregular with 1 to 2 meters high near vertical steps (Ref. 16 shows photographs). Over most of the rest of the length of the dam, where no erosive leakage occurred, the rock surface under the core was more regular or smooth.

Influence of Drying and Shrinkage

Hydraulic fracturing is normally made possible by differential settlement and resulting development of zones of low internal total embankment pressures by stress transfer. Low internal embankment pressures can also be caused by drying and shrinkage, both during and after construction, especially for low dams of clay in areas of arid climate. There are many cases of erosive leaks through low dams attributed to drying cracks. In some cases these cracks may be open cracks, but more commonly they are opened by the reservoir water pressure.

A spectacular example was given by the performance of the long, low La Escondida Dam, built with homogeneous section of dispersive clay with length of 2400 meters. About 50 separate individual erosion tunnels developed at the same time in a few hours after the reservoir filled in 1972 (1). No open drying cracks were seen, but the development of so many initial leaks simultaneously must heav been caused by low internal stresses from drying and shrinkage: no other explanation is possible.

Drying and shrinkage was considered to be a contributary factor to the hydraulic fracturing of the 14 dams of dispersive clay in Oklahoma and Mississippi, described above herein. This is discussed in more detail in Ref. 20, pp. 661-665.

Influence of High Construction Pore Pressures

In dams which develop pore pressures during construction, from consolidation under the weight of overlying embankment, which are higher than the subsequent reservoir pressure, hydraulic fracturing cannot occur. In such dams the minimum principal stress always exceeds the construction pore pressures which are too high to allow hydraulic fracturing.

Relationship to Earthquake Resistance

This paper is devoted essentially entirely to the conclusion that both experience with dam behavior and theory require the dam design to assume that concentrated leaks by stress transfer and hydraulic fracturing develop through impervious sections of most dams, even without exceptional settlement. The initial leak generally occurs during or soon after the initial reservoir filling, when the reservoir water first encounters the embankment zone of low stress.

During earthquake the embankment may be again distorted by slumping. This again causes stress transfer and a second opportunity for hydraulic fracturing, which can be more extreme (larger zones of lower stress) than the conditions existing during first reservoir filling.

Summary and Conclusions

1. Experience with dam performance and evidence from theory is presented to support the conclusion that concentrated leaks by hydraulic fracturing occur frequently without being recognized.

2. This action occurs inside the dam only after the reservoir rises and cannot be demonstrated by direct visual observation. The evidence available is nevertheless sufficiently strong that the main conclusion must be accepted: there can only be some remaining questions about the details.

3. Hydraulic fracturing is made possible by differential settlement and internal stress transfer. This occurs even in dams without unusually large settlement. The hydraulic fracturing may develop on near vertical transverse planes caused by longitudinal stretching (Fig. 1) on near horizontal planes caused by horizontal arching (Fig. 4), or a more complex combination of actions.

4. These conclusions are based primarily on analyses of dams

which have developed concentrated leaks, many of which leaks cannot be otherwise explained reasonably. The conclusions are also unequivocally supported by FEM computations which show that only relatively small settlements of the foundation, or the lower part of the dam itself, are needed to create zones of low total stress in the dam capable of allowing hydraulic fracturing.

5. The concentrated leaks are not usually recognized because they do not erode the impervious embankment material, and because the water pressure in the leak (crack) across the core is generally about the same as would be measured with piezometers if the concentrated leak did not exist.

6. The action occurs in dams of all sizes, ranging from low homogeneous dams to high central core dams.

7. The width of the hydraulic fracturing crack depends largely on the pressure of the water entering the crack (and the reservoir depth above the crack). For low dams the cracks are probably generally very narrow, such as a small fraction of a millimeter. For high dams calculations show that the cracks can be jacked open several centimeters.

8. In the general case the concentrated leaks do not erode the core either because the velocity is too low or because they discharge into a filter which effectively prevents erosion.

9. If no erosion occurs the initially open crack is filled by swelling or softening of the adjacent embankment material, creating soft layers ("wet seams"). After the crack is thus filled, the water pressure distribution through the "wet seam" is generally linear from reservoir pressure to zero (gage) pressure at the downstream end. In a few cases, as demonstrated by the Manicouagan 3 and El Guapo Dam experiences, the hydraulic control develops at the downstream end of the crack by sealing the filter, in which case the water in the crack for its full length through the core has nearly full reservoir pressure.

10. Occasionally in low homogeneous dams without internal filters the initial leak is able to start eroding and the dam fails by progressive erosion. This happens much more easily and commonly in dams built of highly erodible dispersive clay. Only rarely do failures of this kind occur in low homogeneous dams of ordinary non-dispersive impervious soil; but it is impossible to predict under what circumstances they will fail. The Stockton Creek Dam failure occurred in a well-built homogeneous dam which had very low settlement.

11. The leak through the Wister Dam could not have developed by any other mechanism but hydraulic fracturing. This experience is completely and reliably documented. The Wister Dam experience alone is sufficient to support the conclusion that concentrated leaks develop by hydraulic fracturing even in dams with low to moderate settlement.

12. Concentrated leaks along the contact between rock foundations and impervious cores may be caused by hydraulic fracturing. The low stress allowing fracturing may be general on the surfaces or it may be

locally developed adjacent to irregularities in the rock foundation surface.

13. Hydraulic fracturing can occur in dams with impervious sections of cohesionless silt as well as sections of clayey soil, as demonstrated by the Manicouagan 3 Dam behavior.

Appendix 1 -- References.

1. Benassini, A. and Casales, V., "Discussion of Piping Failure of La Escondida Dam, Mexico," Specialty Conference on Performance of Earth and Earth-Supported Structures, ASCE, June 11-14, 1972, Vol. 3, pp. 135-141.

2. Bertram, G. E., "Experimental Investigation of Protective Filters," Soil Mechanics Series No. 7, Graduate School of Engineering, Harvard University, 1940.

3. Bertram, G. E., "Experience with Seepage Control Measures in Earth and Rockfill Dams," 9th ICOLD, Istanbul, Vol. III, 1967, pp. 91-109.

4. Casagrande, A., "Notes on the Design of Earth Dams," Journal of the Boston Society of Civil Engineers, Vol. 37, Oct. 1950.

5. Dascal, O., "Peculiar Behavior of the Manicouagan 3 Dam's Core", International Conference on Case Histories in Geotechnical Engineering, May, 1984, University of Missouri-Rolla, Vol. II, pp. 561-569.

6. Flores, C. E. and DeFries, C. K., "Incidents with Earth and Rockfill Dams in Venezuela", (Spanish), Proceedings of the 25th Anniversary, Venezuelan Society of Soil Mechanics and Foundation Engineering, Caracas, 1983, pp. 243-251.

7. Independent Panel to Review Cause of Teton Dam Failure, "Report on Failure of Teton Dam," U. S. Government Printing Office, Washington, D. C., 10401, 1976.

8. Jaworski, G. W., Duncan, J. M. and Seed, H. B., "Laboratory Study of Hydraulic Fracturing", Journal of the Geotechnical Engineering Division, ASCE, June 1981, pp. 713-731. (Discussion in Nov. 1982 Journal.)

9. Kjaernsli, B. and Torblaa, I., "Horizontal Cracks Through the Core at Hyttejuvet Dam," Publication No. 80, Norwegian Geotechnical Institute, Oslo, Norway, 1968.

10. Kulhawy, F. H. and Gurtowsky, T. M., "Load Transfer and Hydraulic Fracturing in Zoned Dams," Journal of the Geotechnical Division, ASCE, Sept. 1976, pp. 963-974.

11. Lofquist, B., "Earth Pressure in a Thin Impervious Core," 4th ICOLD New Delhi, Vol. 1, pp. 99-109, 1951.

12. Lofquist, B., Discussion of Cracking in Dams, 5th ICOLD, Paris, 1955, Vol. III, pp. 21-22.

13. Penman, A. D. M., "Materials and Construction Methods for Embankment Dams and Cofferdams" General Report, Question 52, 14th ICOLD Rio de Janeiro, 1982, Vol. IV, pp. 1105-1228.

14. Seed, H. B. and Duncan, J. M., "The Teton Dam Failure -- a Retrospective Review," 10th ICSMFE, Stockholm, Vol. 4, 1981, pp. 214-238.

15. Sherard, J. L., "Loss of Water in Boreholes in Impervious Embankment Sections," Proceedings, 10th ICOLD Congress, Montreal, Vol. VI, 1970, pp. 377-381.

16. Sherard, J. L., "Embankment Dam Cracking," Embankment Dam Engineering, John Wiley and Sons, New York, 1973, pp. 272-353.

17. Sherard, J. L., Discussion of "Load Transfer and Hydraulic Fracturing in Zoned Dams," by F. H. Kulhawy and T. M. Gurtowsky, Journal of the Geotechnical Engineering Division, ASCE, July 1977, pp. 831-833.

18. Sherard, J. L., "Sinkholes in Dams of Coarse, Broadly Graded Soils," 13th ICOLD Congress, India, Vol. II, 1979, pp. 25-35.

19. Sherard, J. L., Decker, R. S. and Ryker, N. L., "Piping in Earth Dams of Dispersive Clay," Proceedings of the Specialty Conference on Performance of Earth and Earth-Supported Structures, ASCE, June 1972, pp. 589-626.

20. Sherard, J. L., Decker, R. S. and Ryker, N. L., "Hydraulic Fracturing in Low Dams of Dispersive Clay," Proceedings of the Specialty Conference on Performance of Earth and Earth-Supported Structures, ASCE, June 1972, Vol. 1, Part I, pp. 563-590.

21. Sherard, J. L. and Dunnigan, L. P., "Filters and Leakage Control in Embankment Dams, ASCE, Denver, Colorado, May 2, 1985.

22. U. S. Army Engineers, "Filter Experiments and Design Criteria," Technical Memorandum No. 3-360, Waterways Experiment Station, Vicksburg, Miss. 39180, 1953.

23. U. S. Army Engineers, "Review of Soils Design, Construction and Performance, Wister Dam, Oklahoma," Technical Report 3-508, June 1959, Waterways Experiment Station, Vicksburg, Miss.

24. Vaughan, P. R. et al, "Cracking and Erosion of the Rolled Clay Core at the Balderhead Dam," Proceedings of the 10th ICOLD Congress Montreal, Vol. 3, 1970, pp. 73-93.

25. Vaughan, P. R. and Soares, H. F., "Design of Filters for Clay Cores of Dams," Journal of the Geotechnical Engineering Division, ASCE, January 1982.

26. Widjaja, H., Duncan, J. M., and Seed, H. B., "Scale and Time Effects in Hydraulic Fracturing," Research Report to U. S. Army Engineers, Waterways Experiment Station, P. O. Box 631, Vicksburg, Miss. 39180, 1984.

27. Wood, D. M., Kjaernsli, B., and Hoeg, K., "Thoughts Concerning the Unusual Behavior of Hyttejuvet Dam," 12th ICOLD Congress, Mexico, Vol. II, 1976, pp. 391-414.

The Upstream Zone in Concrete-Face Rockfill Dams

James L. Sherard, F.ASCE*

Abstract

The upstream zone of small rock, the supporting zone for the slab (Zone 2), at many recent concrete-face rockfill dams has contained less than 20% of particles smaller than the No. 4 sieve. Such materials cannot be placed practically without substantial segregation of the coarser particles, causing the zone to have relatively high permeability. On a few recently designed CFRDs the specifications have required a higher content of fines in this upstream zone, the main purpose being to provide an upstream zone which is free of significant segregation and has reliably low permeability. Data and arguments are presented to support the conclusion that the upstream Zone 2 should have an average of at least 40% finer than the No. 4 sieve. For some dams this does not increase the cost. For others the requirement of 40% finer than the No. 4 sieve causes a small cost increase in the upstream Zone 2, which is justified. There are significant advantages in a grading with about 40% minus No. 4 sieve sizes.

Introduction

Figure 1 shows the general cross-section of the modern concrete-face rockfill dam (CFRD), typical of many built in the last few years, using zoning designations as suggested by Cooke (2), which parallels zoning designations for the earth core rockfill.

Ideas concerning the purpose and the design of Zone 2 have evolved considerably since 1965-70. Since 1970-75 most dams of this type have had a Zone 2 consisting of grizzlied or crushed rock, graded down from a maximum size of 6-inches (150 mm) or 3-inches (75 mm). This direct method of processing did not produce a high quantity of fines smaller than the No. 4 sieve.

On some recent jobs this Zone 2 has been specified to have a considerably larger quantity of particles finer than No. 4 sieve for the purpose of eliminating segregation and assuring that the compacted material is not highly pervious. This paper presents data and theoretical considerations which support the conclusion that Zone 2 should have an average content of about 40% finer than the No. 4 sieve. A few related details are discussed.

Brief History of Zone 2 Design Evolution

Almost all CFRDs before 1950 were constructed as dumped and sluiced

*Consulting Engineer, San Diego, California

Zoning Designation:

Zone (2) Small rock, usually processed, placed in 0.4 or 0.5m thick layers.

Zone (3A) Quarry run small rock placed in layers with same thickness as Zone 2.

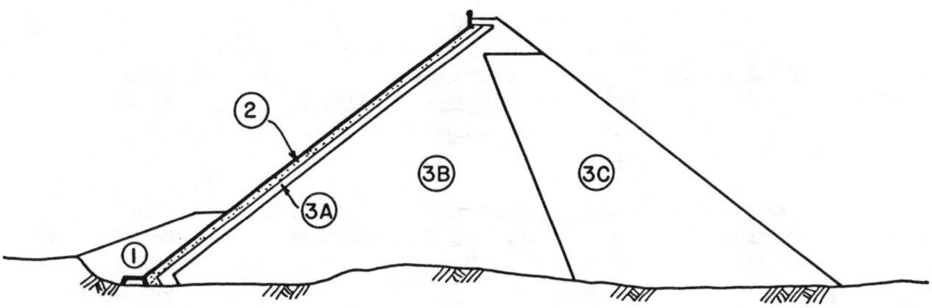

Fig. 1: Typical CFRD Section Showing Zone 2 Located Directly Under Concrete Slab

rockfill with a thin zone of crane-placed large rocks on the upstream face, just under the concrete slab (1). This practice changed in the period 1955-60 when rockfill dams started to be placed and compacted in layers and the zone of crane-placed, large rock was abandoned in favor of an upstream zone of small rock, placed and compacted in thin layers. The purpose of this zone was to provide good support for the concrete slab, by using relatively small rocks so that the water load acting on the slab would be uniformly distributed over the face of the dam; that is, to eliminate the possibility that a single large rock under the slab would attract a high concentrated load.

Also, for the dams constructed during the 1960's, generally the Zone 2 material was screened to remove small rocks and soil, such as to remove all particles smaller than 2-inches (50 mm) or 1-inch (25 mm). The main design reason for removing these small rocks and fines was to eliminate concern that leaks through the joints in the concrete slab might erode the soil fines and carry them out of the Zone 2 layer into the voids of the main rockfill embankment, creating an undesirable localized loss of support under the concrete slab, causing cracking.

Some of the larger dams built with Zone 2 consisting of pervious, screened small rock during this period were (1)(11):

Dam	Year Completed	Height	Zone 2
New Exchequer (California)	1966	150 m	2 to 15 inches
Cabin Creek (Colorado)	1967	76 m	2 to 10 inches

Piedras Dam (Spain)	1967	41 m	3/4 to 12 inches
Pindari Dam (Australia)	1969	45 m	1 to 12 inches

(1 inch = 2.54 cm)

The responsible engineers for each of the above projects, which were being studied about the same general time, were aware of the designs of the others, largely through the activities of J. B. Cooke, Consulting Engineer, who was advisor on most of the projects of this kind during the last 30 years.

An abrupt change was made in Zone 2 at the Cethana Dam, 110 m high completed in Australia in 1971 (12). At Cethana Dam it was decided to employ the same principle of building Zone 2, but to use the rock without processing to remove the soil and small rocks (smaller than 1 or 2 inches). The specifications required that the material be sound hard rock having maximum size of 9-inches (22.5 cm), with range of particle size distribution as shown in Table 1. This specification allows a wide range of particle size distribution which will essentially permit the use of any quarried hard rock with maximum size smaller than 9-inches (22.5 cm), except materials with significant soil fines (smaller than No. 200 sieve).

Table 1
Zone 2 Gradation Specifications

	Allowable Percent Fines		
Size	Cethana	Alto Anchicaya	Foz do Areia
12 in.		100	
9 in.	100		
6 in.	80-100		100
3 in.	51-100	50-100	55-100
3/4 in.	15-63	25-60	18-65
No. 4 Sieve	0-40	0-30	5-22
No. 30 Sieve	0-17	0-5	0-8
No. 200 Sieve	0-5		0-1

At Cethana Dam the engineers did not consider that there was any real probability of the fines being washed out by a leak through a slab joint with consequent slab settlement. In addition to saving the cost of grizzlying the material to remove rocks smaller than 2-inches (50 mm), the Cethana designers believed that it was technically advantageous to leave the rock fines in the Zone 2 material for two main reasons: (1) it was desirable that the material be less pervious, so that less water would get through the dam in the event of a leak through the dam in the event of a leak through a slab joint or crack and in the circumstance that a construction flood may back up water behind the dam before the concrete slab is in place; and (2) with the small rock and fines included the exterior surface could be more easily formed to a smooth plane, with less concrete placed outside the design thickness, and the sloping surface would be more stable (less tendency to ravel) under the feet workers.

Cethana Dam was provided with more than usual instrumentation which confirmed that the behavior was excellent in all respects. The Cethana Dam construction and performance experience (3(12) received much attention worldwide and influenced subsequent designs. On most concrete-face rockfill dams after Cethana Dam the same general type of Zone 2 material was specified, without removal of small rock and earth fines. The next high dam of this type was the Alto Anchicaya Dam built in Colombia during 1970-74, with maximum height of 140 meters. As seen in Table 1 the specified range of particle sizes for Zone 2 was very similar to that at Cethana Dam. At Alto Anchicaya Dam large leakage developed at perimeter joints, reaching a maximum of about 1800 liters/sec., on the first reservoir filling (8). This experience showed that the Zone 2 layer at Alto Anchicaya Dam was not very effective as a "secondary water barrier" in controlling leaks through the concrete slab.

For the 160-meter high Foz do Areia Dam, completed in Brazil in 1982, the Zone 2 layer of crushed rock was also very similar to that at Cethana Dam. As seen in Table 1, the Foz do Areia specification required a maximum size of 6-inches and required at least 5 percent sand-sized particles (finer than No. 4 sieve): the previous Cethana and Alto Anchicaya specifications allowed 9" and 12" maximum sized particles and did not require any minimum quantity of sand-sized particles. The average Zone 2 material used at Foz do Areia was very similar in gradation to that used at Alto Anchicaya; i.e., about 12 percent finer than the No. 4 sieve and 30 to 40 percent finer than 3/4 inches, Fig. 2. Multiple water stops were introduced at Areia and leakage was not high.

At another 130-meter high dam, the Zone 2 specifications were about the same as at Cethana Dam, except that the maximum size was limited to 6-inches. When the reservoir filled for the first time in 1984, considerable leakage emerged downstream, reaching a maximum of about 1600 liters/sec., coming through cracks in the concrete slab. By dumping silty sand on the upstream face the leakage was reduced to about 100 liters/sec. This experience confirms the earlier conclusion from the Alto Anchicaya experience; i.e., small rock which meets the gradation specifications used first at the Cethana Dam, and followed subsequently for many other dams, has sufficiently high permeability such that it cannot be relied upon to control leaks through the concrete slab to small quantities.

The Zone 2 gradation specification used at Cethana, Alto Anchicaya and Foz do Areia Dams defines a range of particle size distribution which experience shows can be obtained during construction without screening the small rock into various sizes and recombining them. For most hard, unweathered quarried rocks, the material obtained by crushing (at 4 to 8 inches) without further processing, by grizzlying to remove the smaller sizes from the quarry run rock, or by simply special blasting and selective excavation to get small rock, will have particle size distributions falling within the specified range. The

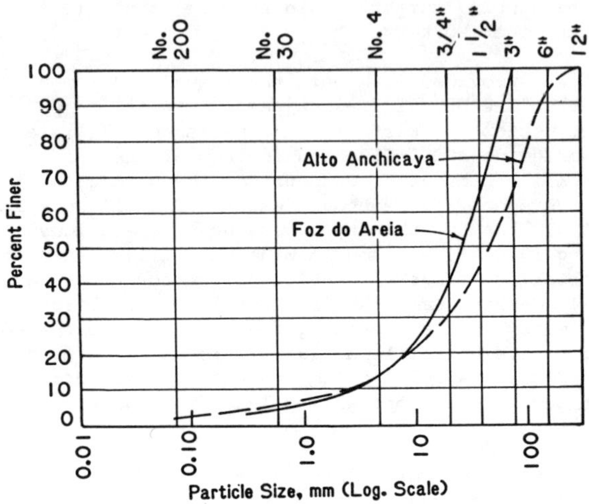

Fig. 2: Average Materials Placed in Zone 2 (7)(8)

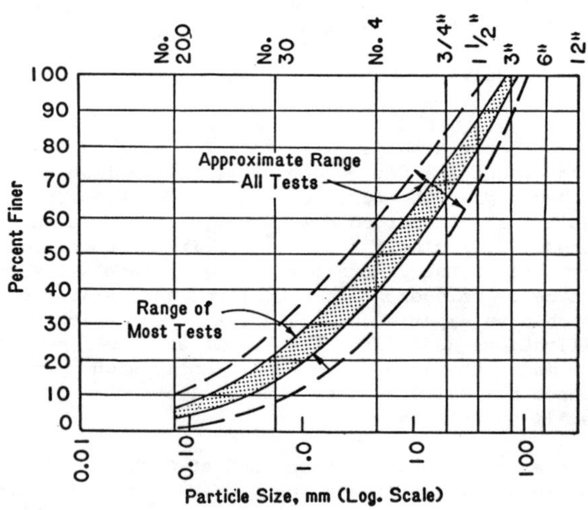

Fig. 3: Main Downstream Filter Used at Itaparica Central Earth Core Rockfill Dam

specification limits were chosen with this knowledge, as an effort to keep the unit cost of Zone 2 relatively low.

Compacted zones of small rock with sizes falling within these specifications are relatively pervious even if the material is placed carefully in a homogeneous mass because there is not a sufficient quantity of fines smaller than the No. 4 sieve to fill the voids between the larger particles. But more important it is very difficult to handle and place material of this type in such a way that it is a homogeneous mixture. Inevitably, during the handling and spreading processes the coarser particles segregate into pockets and streaks composed of rocks of 1-inch diameter and larger which have little or no finer particles in the voids. This segregation is commonly so important for the thin strip of embankment Zone 2 that the permeability is not much lower than it would be if the material did not contain any fine particles. For large parts of the segregated Zone 2 embankment the effective coefficient of permeability is probably of the order of 10 cm/sec., or higher. This segregated nature of the small rock meeting the common Zone 2 specifications, Table 1, is the main reason that leaks of more than 1000 liters/sec., coming through torn water stops or cracks in the concrete slab can pass through the Zone 2. If the material had a sufficient quantity of fines to fill the voids between the larger rocks, and the material was not segregated, the permeability of the compacted mass would be determined by the properties of the fine fraction and would have a coefficient of permeability (k) of the order of 0.001 cm/sec., or less. Thus, the coefficient of permeability of a small rock Zone 2 which is not segregated and has sufficient fines to fill the voids between the larger rocks is of the order of 0.001/10, or 1/10,000, of the permeability of Zone 2 material used in the recent large dams, Table 1 and Fig. 2.

Segregation Control in Sand-Gravel Filters

Filter thinking and experience for earth dams and earth core rockfill dams also applies to Zone 2 of the CFRD. In dams with central earth cores the main filter downstream of the core is often made of sandy gravel or gravelly sand with maximum size of 2 or 3 inches, sometimes larger. Gravel sizes may be crushed rock. A well-graded sand-gravel mixture is an excellent filter to protect even very fine-grained impervious core material. The main problem in specifying such a coarse sandy gravel filter is the possibility that during handling and placement on the construction surface the coarse gravels will segregate into pockets in which there is little or no sand in the voids, which pockets frequently accumulate at the interface between the core and filter. When the impervious core consists of a fine-grained soil it is necessary that at the core-filter interface the sand-gravel filters should be a homogeneous mixture, with the gravel particles always surrounded by sand.

The amount of undesirable segregation of sandy gravel filters which occurs on a given job is related both to the gradation of the material and the care taken during construction to eliminate segregation. If the material has barely sufficient sand to fill the voids, then a great deal of construction care and effort is required to avoid important segregation. The construction care and effort consists of avoiding handling methods which cause segregation and using hand labor on the construction surface to remove segregated pockets. At the other extreme, for materials with an excess of sand, the gravel particles are floating in a matrix of sand and only little effort is required during construction to avoid significant gravel segregation.

The relationship between the amount of sand in a sand-gravel mixture and the construction effort and care need to avoid important segregation cannot be determined by analysis or theory. The relationship can be evaluated by direct observation of the magnitude of the segregation problem which develops with different sand-gravel mixtures being handled by different construction methods on different jobs. On visits to dam construction sites the writer has endeavored to evaluate the degree of visible segregation in sandy gravels used for filters and other embankment zones and compare with the material properties and construction methods. These efforts were supplemented by simple laboratory experiments in which samples (about 20 kg) of reasonably well-graded sand-gravel mixtures with different proportions of sand and gravel were manipulated and the relative tendency for segregation observed. These experiments were made as part of a comprehensive research program on filters carried out recently by the U.S. Soil Conservation Service (9)(10). The results of these field observations and laboratory experiments support the following general conclusions for reasonably well-graded sand-gravel mixtures containing all sizes from a maximum of about 3-inches (or 6-inches) down to fine sand sizes (a few percent passing the No. 200 Sieve):

1. For mixtures with 20% sand (finer than No. 4 sieve) or less, it is practically impossible to avoid important segregation, even using great care during construction.

2. For mixtures with 25-35% sand it is possible to achieve reasonable freedom from important segregation but continuous large effort is required during construction.

3. For mixtures with 40% sand, or moderately more, important segregation can be avoided with reasonable construction efforts. For processed sand-gravel filters, where freedom from segregation is important, an average of about 40% sand is desirable (9). The permeability of compacted zones of these materials is governed by and is roughly the same (a little lower) as the permeability of the fraction smaller than the No. 4 sieve.

4. From the standpoint of the segregation problem the content of particles smaller than sand sizes is the most important factor. As long as a sand-gravel mixture has an average of about 40% of particles finer than the No. 4 sieve, the maximum size and distribution of the coarser 60% of gravels is of less importance. A maximum size of 2 or 3 inches is commonly specified for critical downstream filter, but material with 40% by weight (of the total material) of particles finer than the No. 4 sieve, can still be placed easily without important segregation even if it contains a few pieces up to 6 or 8 inches; i.e., the problem of avoiding segregation is about the same when the maximum rock size is 2-inches as when it is 6-inches. This is true because

when the 60% of coarser particles is graded up to 6-inch maximum size, instead of 2 or 3 inches, the 60% by weight of particles coarser than the No. 4 sieve (gravel or rocks) is comprised of a much fewer number of individual coarse particles and the 40% of sand particles has sufficient excess volume to surround and embed the coarser particles.

Itaparica and Xingo Dams, Brazil

The above experiences with sand-gravel filters for fine-grained earth cores, showing that a minimum average of about 40% of sand-sized particles was desirable, led the writer to the general conclusion that a similarly graded material for the Zone 2 under the concrete slab in the CFRD has technical advantages. This opinion was strengthened by observations during construction of the Itaparica Dam earth core rockfill, Brazil, starting about 1980. At the Itaparica HE Project, where construction started slowly, there were several million m^3 of required excavation for the powerhouse and spillway in weathered granite, consisting partly of hard rock and partly of decomposed granite (sandy soil). The material, excavated by blasting, was placed in large stockpiles before a decision was made on the final design details for the rockfill dam.

Finally on the left bank the dam was built as a rockfill dam with central earth core of decomposed granite (silty and clayey sand). The main downstream filter was made from the stockpiled material, removing rocks larger than 4-inches (10 cm) with a vibrating grizzly. As seen in Fig. 3, the typical filter material had 40 to 50% finer than the No. 4 sieve and about 5% finer than the No. 200 sieve. During construction of the Itaparica rockfill dam the writer paid careful attention to the problems of handling this material and of the properties of the compacted embankment zones obtained using it, with the following main conclusions:

1. There was significant tendency for segregation during the grizzly operation in which the material was dropped in free fall from the end of a conveyor belt. Pockets of rocks from 1-inch to 4-inches in size without fines accumulated in the stockpile but it was relatively simple to mix the material again into a homogeneous blend by working the material in the stockpile with a small bulldozer, and with the front-end loader while loading.

2. The resulting embankment zone seen in many test pits excavated for the main purpose of visual examination for potential segregation, was uniformly free of any sign of pockets of segregated coarse particles. The coefficient of permeability, as determined by many tests made on the embankment during construction, ranged approximately between 1×10^{-5} cm/sec., and 4×10^{-4} cm/sec., with a median of the order of 5×10^{-5} cm/sec. It was apparent from these observations that this material (Fig. 3) was satisfactory for construction of the impervious zone of an embankment dam.

This experience led the writer to conclude that materials similar to the Itaparica filter (Fig. 3) had significantly better properties for the Zone 2 placed directly under the concrete slab of the CFRD

(Fig. 1) than the coarser material used in the earlier dams, Table 1. The shear strengths of both types of material are about the same, being dominated by the gravel-sized particles. The main advantage of the material with gradation such as Fig. 3 is that it can be placed without significant segregation and, consequently, it has reliably low permeability, which is highly desirable as discussed later to provide a back-up impervious zone behind the concrete slab.

A second advantage of the material with gradation such as Fig. 3 is that the upstream slope of the dam embankment provides better support for workers because it is more cohesive than the earlier materials with lesser fine contents. A disadvantage is that the material is more erodible and the need for erosion protection of the upstream face during construction is greater.

Starting about 1983, final designs were prepared for the Xingo HE Project, also on the Sao Francisco River, Brazil, downstream from the Itaparica Project for the same owner. The final dam design is a large CFRD, about 130 m. high. The granite rock formation at the site is similar to that at the Itaparica Dam. Because of the Itaparica filter construction experience described above, it was easily agreed by the designers and Consulting Board to specify that the Zone 2 under the slab should be similar to the Itaparica Dam filter, Fig. 3. This material could be made in a manner similar to that at Itaparica (excavating unweathered and decomposed granite together and grizzlying) or by blending together crushed hard granite with decomposed granite (silty sand). The Xingo Dam specification was a substantial departure from the Zone 2 specification used at the Foz do Areia Dam, the only other CFRD in Brazil, completed a few years earlier, Table 1. The Xingo Project Construction Documents are completed but construction has been delayed.

Other Recent Dams Having Zone 2 with High Sand Content

Similar thinking is influencing current practice in other parts of the world. At the 125 m. high Cirata Dam, with construction currently under way in Indonesia, the Zone 2 is to be manufactured by processing crushed hard rock (andesite). The specified gradation, Fig. 4, with 30 to 55% finer than the No. 4 sieve, was recommended by the Consulting Board with the intention to "provide a grading which will not segregate on placing"(6), showing independent development of the 40% sand size criterion. For the Cirata Dam the specifications also included, "The intent of this gradation is to obtain a permeability equal to or less than 1 x 10^{-4} cm/sec."

The Macagua Project, located downstream from the Guri Dam on the Caroni River in Venezuela, will be built in the next few years. This will have a long, low CFRD built of hard unweathered granitic gneiss. The designers have recently made experiments with various gradations of crushed rock for the Zone 2 layer under the concrete slab, using the Guri Dam concrete aggregate plant for crushing, screening and reblending. Finally, after permeability testing, a Zone 2 specification was chosen which would give a coefficient of permeability of 1 x 10^{-3} cm/sec., or less. The specification calls for 40 to 65% finer than the No. 4 sieve and 5 to 15% finer than the No. 200 sieve,

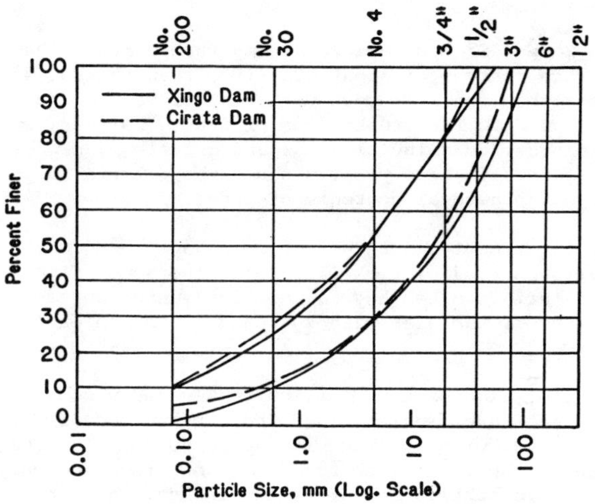

Fig. 4: Zone 2 Gradation Specifications for Xingo and Cirata Dams

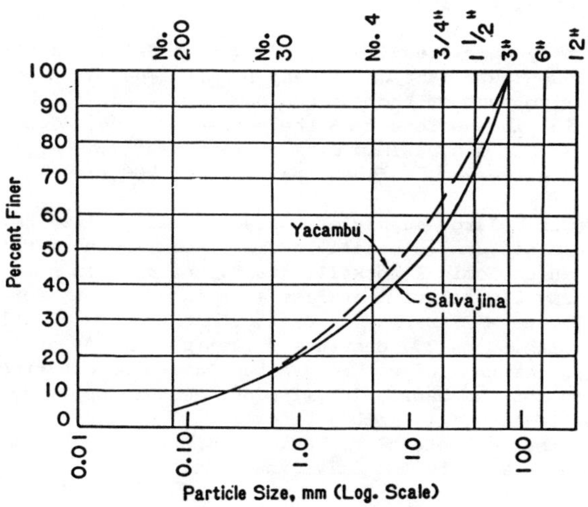

Fig. 5: Average Material Placed in Zone 2 at Yacambu and Salvajina Dams (4)(5)

another example of the trend toward a Zone 2 material with more than 40% sand sizes.

When the Zone 2 is manufactured from crushed rock, the cost to obtain an average content of about 40% finer than the No. 4 sieve is somewhat higher than the unit cost with smaller sand content, since fines must be added to the crusher run product. However, where the Zone 2 is obtained from alluvial sands and gravels by processing or not, there is generally less problem since the alluvial deposits often contain fairly high natural contents of fines.

A few of the recent concrete-face dams have been constructed of alluvial gravels instead of rockfill, including some of the major dams. These materials often have a relatively high content of sand so that a zone under the slab with 40% sand sizes can easily be obtained economically.

At the 160 m. high Yacambu Dam, currently under construction in Venezuela, the entire embankment is to be constructed of coarse river gravels and cobbles, with maximum size up to 60 cm (5). The zone under the slab is the typical coarse alluvium, processed only by screening to remove cobbles larger than 3 inches. As seen in Fig. 5, the average material has about 38% finer than the No. 4 sieve and about 4% finer than the No. 200 sieve, and the particle size distribution falls nearly in the center of the specified range for the Xingo Dam, Fig. 4. The Yacambu Zone 2 material was easy to handle and place without significant segregation and had uniformly low permeability, probably less than 1 x 10^{-4} cm/sec.

The recently completed (1985) 150 m. high Salvajina Dam in Colombia is a concrete-face gravel and rockfill dam with embankment zoning similar to that of the Yacambu Dam. As seen in Fig. 5, the average material in the Zone 2 of the processed river alluvium under the concrete slab has a maximum size of 3-inches (7.5 cm), 35% finer than the No. 4 sieve and 5% finer than the No. 200 sieve (4).

If there is a large supply of coarse gravelly alluvium available, the most economical dam is usually a gravel dam, even if there is a location for an economical rockfill quarry nearby. In some cases the quantity of gravelly alluvium available is insufficient for the entire dam, in which case a CFRD may be used with the zone directly under the concrete slab, Zone 2, constructed from processed, or unprocessed, river gravels. At the 100-m. high Laiban Dam to be constructed in the next few years for the Manila (Philippines) water supply system, the CFRD is to be of rockfill, with a thick Zone 2 of processed river gravels under the concrete slab, having properties very similar to this zone used in Yacambu and Salvajina Dams.

Influence of Zone 2 Permeability on Leaks through Concrete Slab

When the embankment zone directly under the concrete slab is very pervious, the quantity of a leak through a ruptured water stop or crack in the slab is limited by the crack dimensions and hydraulic friction acting on the leak passing through the concrete slab; i.e., the "hydraulic control" determining leak quantity is in the short path which the water travels through the crack in the concrete slab. When the embankment Zone 2 directly under the slab is semipervious, the resistance to flow (head loss) caused by the seepage through Zone 2 is considerably greater than the pressure head lost in travelling through the crack in the concrete slab. In this latter condition the quantity of leakage which can get through a crack in the concrete slab is limited by the permeability of the Zone 2 under the slab.

Assuming that the underlying embankment Zone 2 is sufficiently impervious so that the entire head is lost in seeping through the voids of the compacted Zone 2 (no head loss flowing through the crack in the concrete slab), the quantity of leakage can be estimated roughly by simple flow nets. Fig. 6 shows schematically a leaking crack in the concrete slab running up the slope, such as a torn waterstop. As seen in Section A-A, Fig. 6, the water seeps through the Zone 2 under the concrete slab and discharges into the underlying rockfill.

Fig. 7 shows the mechanics of obtaining the flow net, starting with a large scale (Fig. 7a) for the seepage pattern at the bottom of the crack and then using progressively smaller scales (Fig. 7b). For this work it is assumed that there is no "privileged path" for seepage along the underside of the concrete slab; i.e., the permeability along path from "a" to "b" in Fig. 7a is the same as along any other seepage path in the compacted Zone 2. This is reasonable since upstream face is thoroughly compacted, the concrete is poured directly on it, and the reservoir water pressure is pushing the concrete slab against the Zone 2.

Fig. 8 shows the final flow net for an assumed width of crack (through the slab) of 1.0 mm. and 100 meters of head. Similar flow nets were also made for assumed crack widths of 10 mm. and 10 cm. (not shown). The overall results of these calculations can be written as follows:

Crack Width	Leakage
1.0 mm	$Q = 0.33$ kH
10 mm	$Q = 0.44$ kH
100 mm	$Q = 0.64$ kH

where Q = the leakage rate per meter of crack length, k = coefficient of permeability and H is the head loss.

These results show that the width of the open crack in the concrete slab has a relatively small influence on the total leakage, when it is governed by the permeability of the Zone 2 layer. The leakage quantity through an open crack with width of 10 cm (4-inches) is only about twice (0.64/0.33 = 1.9) the leakage through a crack with width of 1.0 mm. This is true as seen in Fig. 7b because most of the head is lost by the water seeping into the upstream face of Zone 2, a distance equivalent to about 15 crack widths. Beyond 15 crack widths the rate of head loss is much less important. Therefore, even for an assumed crack with a width of 4-inches (10 cm.), most of the head is lost in seeping through the upstream 60-inches (1.5 m.) of Zone 2.

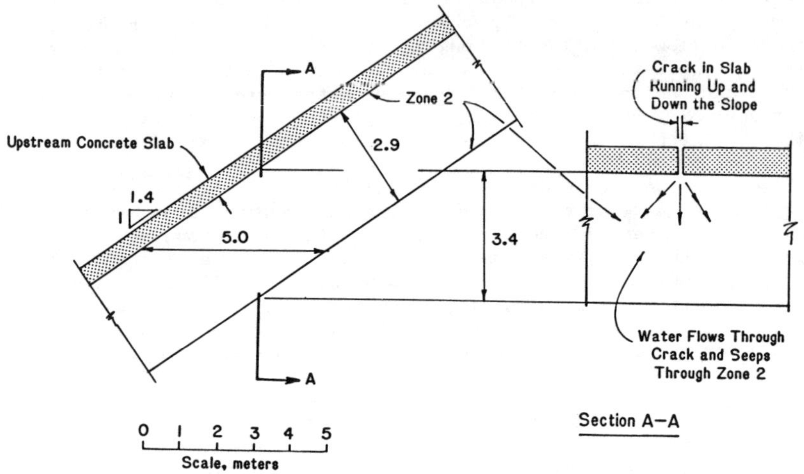

Fig. 6: Leakage through Crack in Concrete Slab Seeps Down through Zone 2

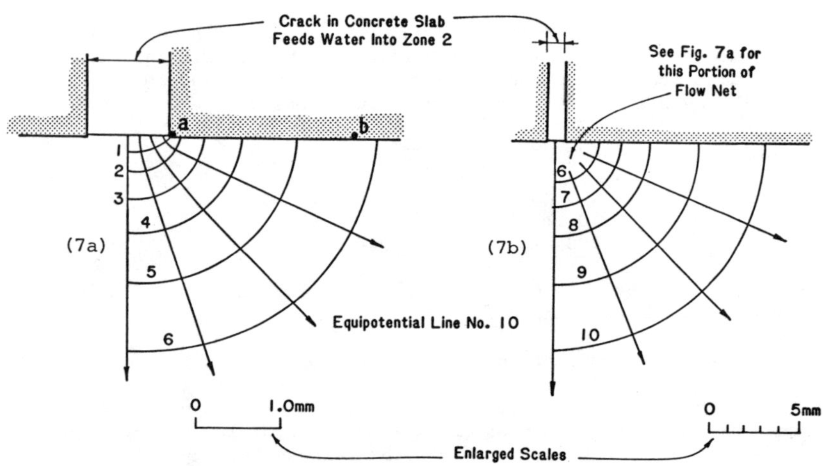

Fig. 7: Progressive Development of Flow Net of Fig. 8 Using Sketches of Different Scales

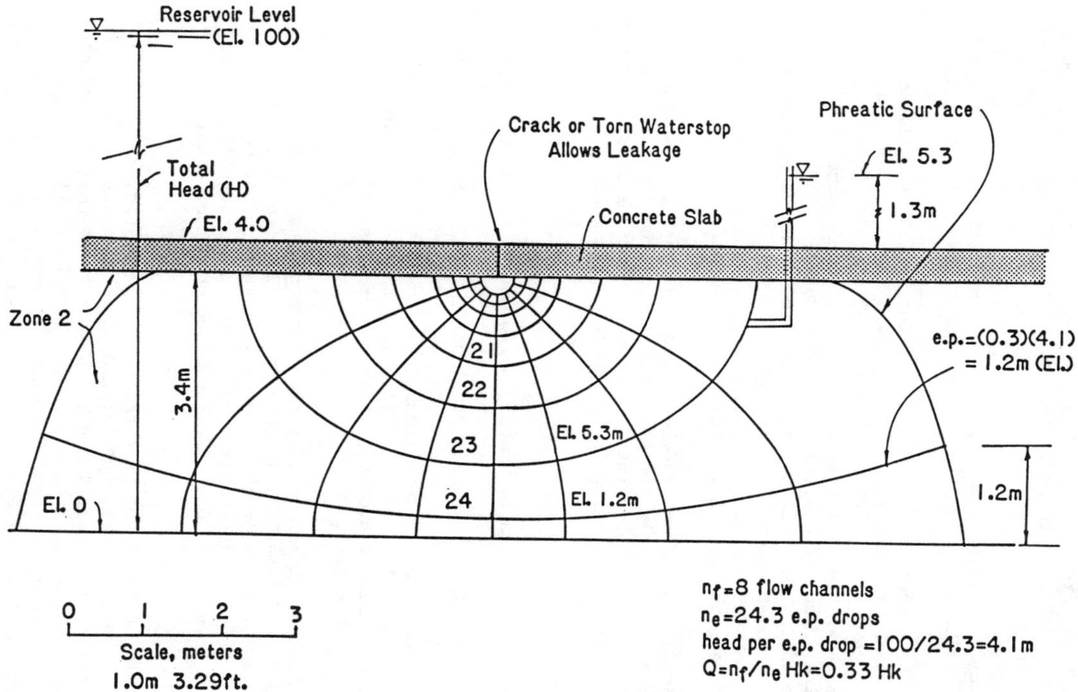

Fig. 8: Seepage Flow Net for Zone 2, Drawn for a 1.0mm Crack in Concrete Slab

The shape of the flow net and location of the phreatic surface is influenced to a small degree by the magnitude of the head loss (H), which is different from most flow nets for civil engineering purposes. For H less than 330 ft. (100 m.), such as 82 ft. (25 m.), the seepage lines do not travel so far horizontally under the concrete slab away from the crack as shown in Fig. 8. For purposes of calculating the seepage quantity, however, the influence of the head on the flow net shape is negligible; the equations above can be used for all heads.

Fig. 9 shows computed leakage quantities for 330 ft. (100 m.) of head, crack widths varying from 0.04 to 4 in. (1 to 100 mm.), and coefficients of permeability for Zone 2 of 1×10^{-2} and 1×10^{-3} cm/sec. This summary shows that for $k=1x10^{-3}$ cm/sec., the leakage will be about 0.5 liters/sec. of crack length. For Zone 2 with $k = 10^{-4}$ cm/sec., the calculated leakage would be 1/10 of the quantity for $k = 10^{-3}$ cm/sec., or about 0.05 liters/sec./meter of crack. These results show the desirability of providing a Zone 2 layer with coefficient of permeability of 10^{-3} cm/sec., or lower.

The calculations presented above relating seepage quantities and crack widths are valid only for the condition of the assumption that the concrete slab remains in intimate contact under pressure with the upstream face of Zone 2, which is the normal situation.

Zone 2 Internal Stability

Because the Zone 2 layer is only a few meters thick, the average gradient of leakage travelling through it is relatively high. The question arises as to whether a typical sandy Zone 2 material, such as that proposed for the Cirata and Xingo Dams (Fig. 4) is "internally stable" under such seepage conditions, that is, whether the compacted Zone 2 will permanently resist the high gradient seepage without change in permeability caused by migration (erosion) of the finer particles out of the mass.

Because of the coarse well-graded nature of the material and because the volume of fines (finer than the No. 200 sieve) is smaller than the volume of voids in the sand-gravel skeleton, it has been generally assumed that materials of this kind are internally stable, but there has been little direct evidence which supports this con-conclusion.

The writer has no doubt that these materials (Fig. 4) are completely stable, with "stability" defined as meaning that when the compacted material is placed in such a way that the coarser particles in the compacted mass (3/4" to 3") cannot move then all the finer particles in the mass will be held in place. Seepage through the material will cause no particle movement, regardless of the magnitude of the pressure head and seepage gradient. Each smaller particle is wedged between larger particles and all are held securely in place by intergranular friction.

As part of a recently completed comprehensive research program by the U. S. Soil Conservation Service on filters for dams (10) extensive

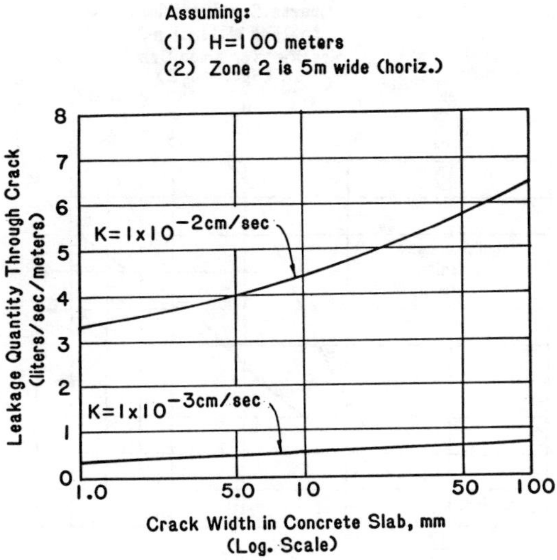

Fig. 9: Computed Seepage Quantity Coming through Cracks in Slab as Limited by the Permeability of Zone 2 (from Flow Nets Such as in Fig. 8)

tests were made of the internal stability of broadly graded silty sands. An example was a typical decomposed granite with gradation as shown by Specimen 1 in Fig. 10. Many filter tests were made with this material, all showing conclusively that as long as the filter prevented the coarser particles from moving, there was absolutely no internal erosion or movement of the finer particles. Tests were made with gradients up to 40-meters of head on samples 4-inches (10 cm.) thick for a gradient of 40/0.1 = 400.

As a part of this program a test was also made on a coarser sample (Specimen 2, Fig. 10) which was made by mixing sands and gravels together to approximate the average material of a Zone 2 material of the type specified for Xingo Dam (Fig. 4). This test was made in a 10-inch (25 cm.) diameter cylinder with a specimen length of 7.5 in. (19 cm.) and a filter consisting of coarse gravels (passing a 3-in. (7.5 cm.) sieve and retained on a 2-in. (5 cm.) sieve). The test was completely stable. Under a pressure of about 130 ft. (40 m.) of water head, with gradient of 40/0.19 = 210, the seepage through the specimen was constant at about 40 ml/sec., giving a calculated $k = 4 \times 10^{-4}$ cm/sec. For the first 30 to 60 seconds the seepage emerging had a slight color, after which the seepage became and remained completely clear. This was considered an extreme test because the short length of sample (only about 2.5 times the size of the largest size gravel in the specimen) and the high gradient.

These tests confirm the conclusion that the sandy gravel Zone 2 in a CFRD with particle size distribution as being discussed herein

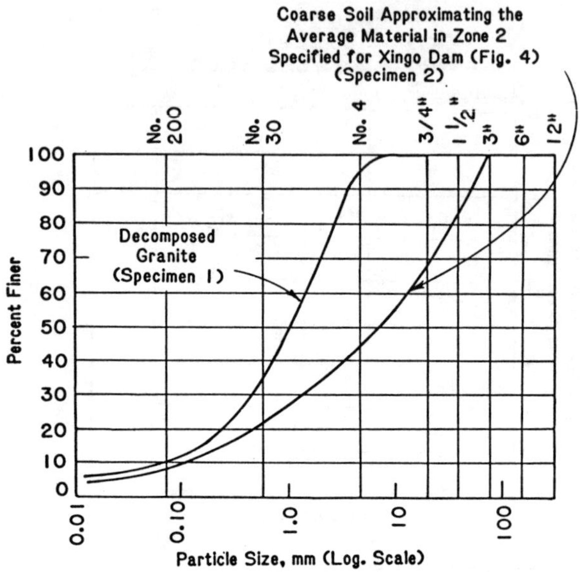

Fig. 10: Soils Demonstrated by Severe Laboratory Testing to Be Internally Stable, Regardless of Seepage Gradient Imposed

(Fig. 4) is internally stable. As long as it is underlain by a layer of smaller rockfill, such as Zone 3A rockfill (Fig. 1) placed in 0.4 or 0.5 m. layers, there is no possibility that there will be erosion of the Zone 2 layer. The writer has confidence that Zone 2 material of the type of Fig. 4 can be placed safely directly against the main Zone 3B rockfill embankment (placed in layers about 1.0 meters thick); however, it is more conservative and desirable for construction to use a thin layer of smaller rockfill just adjacent to Zone 2 (Zone 3A, Fig. 1).

Possible Uplift of Concrete Slab

With a Zone 2 of reliably low permeability under the concrete slab it is necessary to examine the question as to whether high pore water pressures existing in Zone 2 following rapid reservoir lowering could threaten to lift the slab.

While it is somewhat difficult to analyze reliably, the writer agrees completely with the general conclusion that there is no possibility of slab uplift. A main reason is that the Zone 2 material is relatively impervious, drains downward and tends to hold pore water in its voids by capillarity. Therefore, even if the Zone 2 in a given dam were nearly completely saturated at full reservoir, because of many leaks through the concrete slab, when the reservoir lowered the quantity of water from the Zone 2 voids which would have the tendency to flow in the upstream direction toward the slab would be very low, probably a few drops per square meter of area of the upstream dam face. Further, current design and construction practice should not result in significant face cracking or leakage into Zone 2.

Theory shows that there is no significant tendency for seepage from Zone 2 to flow in the upstream direction. Seepage theory shows horizontal equipotential lines at atmospheric pressure and vertically downward seepage. The natural tendency of the water in the pores of the thin inclined Zone 2 above the lowered reservoir is to seep vertically downward into the underlying free draining rockfill under gravity gradient. Even if the Zone 2 permeability is assumed to be higher in the horizontal direction than the vertical direction, the seepage pattern remains generally the same, with vertically downward seepage.

In summary, classic seepage theory shows no tendency for flow of pore water in Zone 2 toward the concrete slab following reservoir lowering. Also, Zone 2 is relatively thin and the amount of water which will seep out of each cubic meter after reservoir lowering is very small (because the pore water is held by capillarity) even if there were a small component of seepage in the upstream direction. For these reasons the writer is in complete agreement with the general conclusion that it is impossible for the slab to be lifted.

Sealing Leaks by Dumping Soil

The design change for Zone 2 material at Cethana Dam, discussed earlier, leaving in the rock fines, had the advantage over the immediately prior practice that it was easier to seal leaks through cracks or torn water stops by dumping earth (silty sand) into the reservoir on top of the slab. In the earlier practice, in which all the rock fines, smaller than about 2 in. (5 cm.) were removed, the Zone 2 voids were so large that it was not easy to make a seal. Sand particles dumped into the reservoir on top of the slab which were able to pass through the opening in the concrete slab were of a size that they could be carried on through the voids in the plus 2-in. (5 cm.) Zone 2.

Dams built with Zone 2 having 40% sand sizes, or more, will not only have much smaller leaks through cracks and defective water stops, the leaks that develop will inevitably be reduced naturally by plugging with small particles in suspension in the water. A Zone 2 compacted embankment with gradation such as that of Fig. 4 will have $k = 1 \times 10^{-3}$ cm/sec., or smaller. Such a sandy embankment has voids which will be plugged by particles in suspension with diameter of the order of 10 to 50 microns (9)(10). Hence, any even slightly turbid water will carry in suspension soil particles which are large enough to seal voids in the Zone 2 material.

General Use of Zone 2 with High Sand Content

The writer believes that it will be generally accepted that the

Zone 2 with about 40% sand-sized particles (Fig. 4) is technically superior to the Zone 2 made using the broader specification allowing less sand, such as used at Cethana, Alto Anchicaya and Foz do Areia Dams (Table 1). The only question remaining is, "If the sandier Zone 2 will cost more, should it be specified; i.e., is the extra cost justified in face of the knowledge that Cethana and Foz do Areia Dams, and many others, have performed very well with no significant leaks through cracks or defective water stops?"

This question only applies to the circumstance where the specification for the 40% sand content will significantly increase the cost. This is generally the case where the Zone 2 material must be made from unweathered, hard, crushed rock. In those situations where the Zone 2 material with about 40% sand sizes will not be more costly (where there is sandy alluvium available), it should be used automatically.

Where Zone 2 is to be made of crushed rock generally less cost is required to produce a material which will meet the specified gradation of Table 1 but will not meet the preferable specification of Fig. 4. This is true because frequently "crusher run" material will meet the specifications of Table 1 but the material will generally have less than 20% finer than the No. 4 sieve. Hence, in order to make a material with 40% finer than the No. 4 sieve it is necessary to spend more effort at crushing to obtain a larger fraction of sand sizes, and on screening and blending.

The extra cost is mainly for energy for the processing plant, but a somewhat larger plant investment may also be needed. The extra processing cost should be generally in the range from US$1.00 to US$3.00 per cubic meter of Zone 2 material. Since the Zone 2 is commonly made of the order of only 15 ft. (5 m.) wide, horizontal dimension, or less than 10 ft. (3 m.) normal to the slope, this extra cost is not great. Compared to the average cost of the reinforced concrete slab, commonly of the order of US$100/m^2 of slab area, the extra cost of processing the Zone 2 will be of the order of US$4 to US$8/m^2 of slab area, or roughly the order of 6% of the cost of the concrete slab.

The writer believes it is desirable to specify a Zone 2 material of the general nature of that of Fig. 4, even when the cost is higher than for the less sandy material. The main reasons are:

1. Avoid Accidental Leaks. It eliminates the possibility that large leaks (several m^3/sec.) will develop for whatever reason. Although these large leaks have not threatened failure and can usually be controlled finally by dumping sandy soil into the reservoir, they nevertheless involve cost. Also, because of the thousands of lineal meters of water stops used in a large CFRD, there is always the possibility of a defective installation and large leaks are possible on the first reservoir filling of each new dam.

2. Reduce Future Slab Costs. With the sandy Zone 2, Fig. 4, behind the slab acting as a reliable second water barrier, future designs can reasonably consider significant cost savings in the

slab. Consideration can be given to decreasing the slab thickness and reinforcing and to using simpler, less costly, water stops in joints.

The writer believes that it is likely that the not distant future evolution of the CFRD could arrive at a constant slab thickness of the order of 8 or 10 inches, even for high dams, with simpler and more economical joint seals. There has been successful experience with durability of concrete (and gunite) on CFRDs for more than 50 years in the California Sierra. With air entrainment, pozzalan and modern concrete technology, thinner slabs and higher gradients would be reliably acceptable. This experience, combined with the uniform support, semi-pervious nature and filter sealing actions of the improved Zone 2 layer, justifies the use of thinner face slabs.

Zone 2 Details

Gradation Specification

On the basis of the current knowledge, the writer believes that the following is an appropriate and desirable Zone 2 specification:

Size	Percent Finer
3 inches	90-100
1-1/2 inches	70-95
3/4 inches	55-80
No. 4 Sieve	35-55
No. 30 Sieve	8-30
No. 200 Sieve	2-12

The critical factor is the requirement for 35 to 55% finer than the No. 4 sieve, assuring that the average material meeting the requirement will have at least 40% finer than the No. 4 sieve. That is, if all samples must have more than 35% sand sizes, then because of the natural range in grading the average will have at least 40%.

The requirement of minimums of 8 and 2% finer than the No. 30 and 200 sieves, respectively, assures that the compacted material will be relatively impervious, probably with $k = 1 \times 10^{-4}$ cm/sec., or lower. For Zone 2 material manufactured by crushing hard rock, the fines can be obtained by secondary crushing to sand sizes without washing, or by blending in a small quantity of silty sand or sandy silt from a soil borrow area.

The requirement that 55 to 80% shall be finer than 3/4 in., assures that the material will have at least 25% of particles coarser than 3/4", which is desirable to make it compatible with the underlying small rockfill, Zone 3A, Fig. 1.

Zone 2 Width and Foundation Contact

Zone 2 has generally been made with a constant width (horizontal dimensions) of about 13-16 ft. (4-5 m.) or made somewhat wider at the bottom. There is no current evidence to indicate that it is necessary or desirable to increase these common widths in general, though ob-

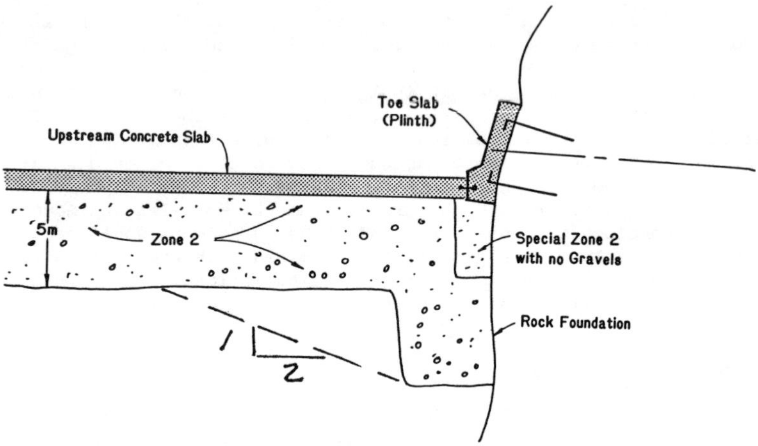

Fig. 11: Horizontal Section for Zone 2 at Contact with Foundation and Abutments

viously the wider the Zone 2 is, the less likely that an accidental pocket of segregated coarse gravels will extend all the way through it. The writer believes that with Zone 2 material having at least an average of 40% sand sizes, the 13-16.5 ft (4 to 5 m.) horizontal width should be generally adequate, but a few meters more of width for high dams seems reasonable, such as used at Foz do Areia.

At the contact between Zone 2 embankment and the rock foundation there is an increased tendency for undesirable segregation of the coarse particles directly at the rock surface. For this reason it would be conservative design to provide special details at the rock contact, such as using a thin strip of Zone 2 with smaller maximum size (finer than 1/2 inch), increasing the width of Zone 2 locally, or both, Fig. 11. The writer believes that the cost of increasing the conservatism at the foundation contact, such as in Fig. 11, is better justified than that of a general increase in the Zone 2 width.

Toe Wall Height

The perimeter line should be generally about 3 ft (1 m.) above the rock surface; however, at some sites it is necessary for localized lengths along the contact between upstream concrete slab and the foundation to build a relatively high toe wall. This may be for a short length across a sharp irregularity in the fresh rock surface, such as across a notched road cut excavation in a steep rock abutment, or fault zone.

Such a wall must resist full reservoir pressure, without depending on support on its downstream face from the rockfill. This is true because in order to develop any supporting passive pressure from the rockfill, the toe wall would have to move a significant amount, which is not allowed for a concrete structure.

There have been cases, for dams with asphaltic concrete upstream facing as well as with concrete slabs, where high toe walls have been designed and built generally as shown in Fig. 12, with near vertical face upstream and a downstream slope of about 1.0(V) to 0.8(H). At first glance a design of this kind appears stable, since it has the approximate dimensions of a concrete gravity dam. However, the relationship of the concrete weight and water pressures are completely different for such a toe wall than for a gravity dam.

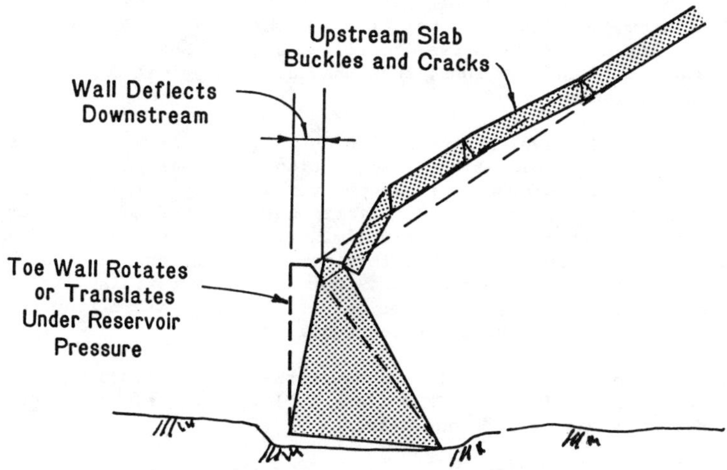

Fig. 12: Schematic Detail of Unsatisfactory High Toe Wall

At dams with high toe walls of the type shown in Fig. 12, the toe wall is pushed toward the dam, either by rotating or by sliding on the foundation. This movement is of sufficient magnitude to cause buckling and cracking of the concrete slab as shown schematically in Fig. 12.

At localized points where it is desirable to design a high concrete toe wall, it is necessary to build this as a massive concrete block in which the water pressure acting on the toe block is passed straight through to the rock foundation, without calling on the rockfall dam for any stabilizing force.

Summary and Conclusions

1. In recent practice two main different types of material have been used or specified for the Zone 2 directly under the concrete slab:

(1) crushed or grizzlied small quarried rock generally with less than 20% of particles smaller than the No. 4 sieve (Cethana, Alto Anchicaya and Foz do Areia Dams); and (2) crushed rock or alluvial sandy gravel with an average of about 40% of particles finer than the No. 4 sieve (Cirata, Xingo, Yacambu and Salvajina Dams).

2. Experience with dam performance shows that Zone 2 material with 20% or less of particles finer than the No. 4 sieve does not have sufficiently low permeability to limit the quantity of leaks through slab cracks or torn or defective water stops to small quantities. Leaks of more than one m^3/sec. have occurred.

3. The coefficient of permeability of a Zone 2 material which is placed without significant segregation and which has sufficient fine sizes (smaller than No. 4 sieve) to fill the voids between the larger rocks is about 1/1000, or less, than that of the typical Zone 2 used at many dams, Table 1.

4. For sand-gravel mixtures with an average of 40% or more of sand-sized particles significant segregation of gravel particles into pockets and streaks without sand in the voids can be reliably eliminated with moderate and reasonable efforts during construction to avoid segregation. For similar sand-gravel mixtures with an average of about 30% sand-sized particles, large efforts are required during construction to avoid important segregation. For similar materials with 20% sand-sized particles or less, it is essentially impossible to place the material without significant segregation.

5. A good general specification for the Zone 2 placed directly under the concrete slab in a CFRD is as follows (from either processed quarried rock or alluvial sandy gravel):

Size	Percent Finer
3 inches	90-100
1-1/2 inches	70-95
3/4 inches	55-80
No. 4 sieve	35-55
No. 30 sieve	8-30
No. 200 sieve	2-12

Such a compacted zone will have a coefficient of permeability lower than 1×10^{-3} cm/sec. The two main advantages are: (1) elimination of the possibility of large leaks; and (2) potential future cost savings in the concrete slab.

6. For Zone 2 materials comprised of alluvial sandy gravel, it is usually not more costly to specify 40% minus #4 mesh. For Zone 2 obtained from processed hard rock the unit cost to obtain this is somewhat greater, since more fines must be produced or added. The extra cost is considered easily justified by the advantages. For the contractor, extra construction cost is offset by lower placement cost and less excess concrete.

References

1. Cooke, J. B., "Progress in Rockfill Dams," Journal of the Geotechnical Engineering Division, ASCE (18th Terzaghi Lecture), October 1984, pp. 1381-1414.

2. Cooke, J. B., "CFRD-Zone Under Face - Designation and Grading", Memo No. 74, J. B. Cooke, 1050 Northgate Drive, San Rafael, California, April 5, 1985, version.

3. Fitzpatrick, M. D., et al, "Ten Years of Surveillance of Cethana Dam", 14th ICOLD, Rio de Janeiro, I, 1982, pp. 847-866.

4. Ingetec, "Salvajina Works, Dam Statistics until August 1984", Engineering Report to CVG, Cali, Colombia, 1984.

5. Ortega, J. M. and Carrero, L., "Design of Yacambu Dam," 14th ICOLD, Rio de Janeiro, Vol. III, 1982, pp. 1263-86.

6. Pinkerton, I., Personal correspondence, 1985.

7. Pinto, N. L. S., Materon, B. and Lagos, P. M., "Design and Performance of Foz do Areia Concrete Membrane as Related to Basalt Properties", 14th ICOLD, Brazil, 1982, Vol. IV, pp. 873-906.

8. Regalado, G., et al, "Alto Anchicaya Concrete Face Rockfill Dam -- Behavior of the Concrete Face," 14th ICOLD, Brazil, Vol. IV 1982, pp. 517-535.

9. Sherard, J. L., Dunnigan, L. P., and Talbot, J. R., "Filters for Clays and Silts," Journal, Geotechnical Engineering Division, ASCE June 1984, pp. 701-718.

10. Sherard, J. L. and Dunnigan, L. P., "Filters and Leakage Control in Embankment Dams," Proceedings, ASCE Symposium on Seepage and Leakage from Dams and Reservoirs, Denver, May 1985, pp. 1-30.

11. Steele, I. C. and Cooke, J. B., "Concrete-Face Rockfill Dams," Chapter 19 in Handbook of Applied Hydraulics, McGraw-Hill Book Company, New Yor, 3rd Edition, 1969.

12. Wilkins, J. K., et al, "The Design of the Cethana Concrete-Face Rockfill Dam," 11th ICOLD, Madrid, Vol. III, pp. 25-44.

Introduction to Paper 19

Concrete Face Rockfill Dam: I. Assessment

by J.L. Sherard and J.B. Cooke
Journal of the Geotechnical Engineering Division,
American Society of Civil Engineers, Vol. 113, No. GT 10,
October 1987, pp 1096-1112

Prepared by J. Barry Cooke

Twenty years before preparing this paper Jim's interest in the impervious face rockfill dam is evidenced by his paper, **Rockfill Dams: Steel Faced Dam**, presented at the ASCE Symposium on Rockfill dams held at Portland, Oregon in June 1958, published in ASCE Transactions, Vol. 125, Part II, 1962; an exclusive volume of 710 pages. Most of the world's high rockfill dams of all types were presented in that Symposium. In Jim's paper he points out the advantage of the upstream impervious membrane type of dam in inherent stability with water pressure upstream from all the rockfill and no internal pore pressure.

This paper, I. Assessment, is the first of two papers, co-authored by Jim Sherard and I. We were co-chairmen of the symposium committee for the symposium "Concrete Face Rockfill Dams - Design, Construction, and Performance" held at the ASCE Convention in Detroit, Michigan, October 1985. The Proceedings contain 37 papers by 97 authors, on existing dams, dams in design and on specific features. Discussions were invited and arranged to update and to supplement the Symposium Volume. The many discussions and the two papers by the Symposium editors were published in a special ASCE-GT Journal devoted exclusively to the CFRD. The papers were closely tied to the Symposium by specific references to the Proceedings Papers. The Proceedings and the special GT Journal can be considered a two-volume comprehensive and practical reference work on the CFRD.

As ASCE-GT Journal Vol. 113 No. 10 October 1987 was being assembled, we lost Jim. The GT Executive Committee and the ASCE Committee on Publications elected to make the Journal a Special Memorial Issue. The biography in the Special Memorial Issue, "In Memoriam, James L. Sherard, Fellow, ASCE, 1925-1987", begins:

> "Jim Sherard made many contributions to dam engineering; and this issue of the Journal of Geotechnical Engineering, in which he co-authored two papers and reviewed the discussions, contains the last of them. It is therefore thought appropriate to dedicate this issue to his memory, particularly to his energetic and constant technical efforts and to his warm and enthusiastic presence."

Concrete-Face Rockfill Dam: I. Assessment

By James L. Sherard[1] and J. Barry Cooke,[2] Fellows, ASCE

Abstract: The concrete-face rockfill dam (CFRD) is currently being used with increasing frequency throughout the world. It has substantial advantages over the earth-core rockfill dam, the common alternate dam type considered. The CFRD is considered to have high fundamental safety, especially against strong earthquake shaking, and to be appropriate for use for very high dams. Placing the rockfill embankment in stratified layers, with the larger rocks in the bottom of each layer, is desirable practice. Crest settlements are relatively low, and decrease in rate rapidly after the first few years. Static stability analyses are not used for CFRD design. It is believed safe and reasonable to build spillways over CFRDs.

Introduction

The papers in the symposium proceedings, *Concrete Face Rockfill Dams—Design, Construction, and Performance* (ASCE, 1985), are of particular value in presenting, in one volume, records of many recent dams, and design thinking of many engineers. It is apparent that there has been much progress in recent years. This paper presents the views of the editors of those symposium proceedings (see Appendix. References) on some fundamental aspects of the CFRD and probable future development, with references to the proceedings papers. A companion paper published in this issue presents our views on the main details of design and construction. For brevity, "CFRD" is used to denote the concrete-face rockfill dam, and "ECRD" denotes the earth-core rockfill dam.

Increased Use of CFRD

The many CFRDs constructed since about 1970, and the large number currently in the design stage, as illustrated by the valuable contributions to the 1985 ASCE Symposium on the CFRD (1–37), show the increasing adoption of this dam type by engineers and owners throughout the world. This widespread current use of the CFRD results from increased recognition that: (1) The modern CFRD is a very good dam type from all technical standpoints; and (2) the CFRD is often the lowest-cost dam type. The previous problems of high settlement and leakage of the older, dumped (uncompacted) rockfill dam with concrete face have been overcome.

The CFRD is being used for reservoirs for all purposes and sizes, from small irrigation projects to large reservoirs on major rivers, including pumped storage. Because of the ability to seal all important leaks in the concrete slab by placing silty sand in the reservoir (14,26), it is not

[1]Consulting Engr., P.O. Box 9138, La Jolla, CA 92038.

[2]Consulting Engr., 1050 Northgate Dr., Suite 400, San Rafael, CA 94903.

Note. Discussion open until March 1, 1988. Separate discussions should be submitted for the individual papers in this symposium. To extend the closing date one month, a written request must be filed with the ASCE Manager of Journals. The manuscript for this paper was submitted for review and possible publication on March 19, 1987. This paper is part of the ***Journal of Geotechnical Engineering,*** Vol. 113, No. 10, October, 1987. ©ASCE, ISSN 0733-9410/87/0010-1096/$01.00. Paper No. 21852.

necessary to lower the reservoir if leaks develop in the concrete slab, or to use divers at excessive depths for repairs. Recognition of this aspect has allowed the selection of the CFRD for very large reservoirs, for which low-level release facilities are neither feasible nor necessary. The Foz do Areia Project has a reservoir volume of about 6,000 million m^3. The Xingó Dam (32) on the Saõ Francisco River, a major river in northern Brazil (about 3,000 m^3/s average annual flow) will retain a reservoir for a 5,000 MW hydro plant.

Comparison of CFRD and ECRD

There is no significant difference in the site conditions needed for the CFRD and the ECRD. Both dam types require a reasonably good rock foundation. At sites where CFRDs have been chosen and built in recent years, the designers generally also studied ECRD alternatives. The CFRD is a natural choice where earth core material is not available. However, in many cases where the CFRD was chosen the design studies indicated that it was the least-cost alternative feasible dam type (5,15,18,19,20, 21,23,31,32,35,36,37), even when suitable material was available for an ECRD.

Comparing only the cost of the dam itself, the CFRD usually has significantly less cost than the ECRD. The volume of rockfill embankment in the two types is roughly the same, but the unit cost of the rockfill is less in the CFRD. This is mainly because of the greater freedom allowed in the order of construction of various parts of the rockfill embankment, with easier use of internal haul roads, (11,12,13). The total cost of the earth core and filters in the ECRD will almost always be greater than the cost of the concrete face in the CFRD. The cost of the CFRD foundation treatment (rock surface treatment and blanket and curtain grouting) is generally significantly less, both because the work can be done independently of the embankment construction, and because the treated rock surface area is smaller. For sites with especially rainy weather, the CFRD has an even greater cost advantage (5,13,15,24,31,35,36).

In addition to the cost advantage of the dam itself, the smaller base width resulting from the steeper slopes of the CFRD (1.3:1) frequently allows shorter tunnels, penstocks, and spillways. For projects with big spillways, large hydroelectric projects, and important river-handling problems (11, 35,38) during construction (with several large tunnels), the total cost advantage given by the CFRD over the ECRD may be much greater than the cost difference for the dam alone.

The CFRD has a substantial advantage from the standpoint of schedule. The CFRD can be built more rapidly because of independent work on the grouting, less trouble with rainy weather, smaller embankment volume, and the greater opportunity of placing rockfill embankment sections before river diversion. At both Khao Laem Dam (19) and Kotmale Dam (21) it was estimated that the adoption of CFRDs instead of ECRDs allowed projected completion one year earlier. For all these same reasons, the risk of unanticipated construction delays is considerably less for the CFRD.

The rapid acceptance of the CFRD in Brazil illustrates the currently increasing recognition of these advantages over the ECRD. In the decade before the first modern CFRD was completed in Brazil in 1980, the Foz do

Areia Dam (11,12), more than a dozen major ECRDs were built, primarily for hydroelectric projects. Consequently, by 1980, Brazilian engineers had great experience and familiarity with the ECRD. Since 1980, many Brazilian engineers and electric companies have been strongly impressed with the technical and cost advantages of the Areia Dam (11,12), causing a remarkable change in practice. At the present time there are nearly a dozen CFRDs in various stages of design or early construction in Brazil (30,32,34,35,36), with some planned to heights of more than 200 m. A similar rapidly increased use of the CFRD is currently developing in China, as indicated by the discussion of Ref. 32 in this Journal, by Cao Keming.

Leakage through CFRD Has Different Basic Significance than Leaks in Dams with Earth Cores

Leakage emerging at the downstream toe of a CFRD is traveling either through openings in the face slab or through the rock foundation. In both cases, there is no soil in the dam which can erode. Therefore, leakage through a CFRD has a basically different significance than leaks through other embankment dams.

When water emerges at the toe of the ECRD, or an earth dam, it is always possible that the leak could eventually increase in size by erosion of the earth core. Even if the leak is completely clear it is generally necessary and good practice to monitor it for the life of the dam (or to seal it off in some manner). Hence, for the ECRD and earth dam, leakage has two aspects: (1) It is a potential threat to the dam safety which must be watched, and (2) an economic loss of water. For the CFRD the significance is quite different. There is no threat to the dam safety, only an economic problem of water loss. For most reservoirs on rivers of any significant size, a leakage of a few tens of liters per second has negligible economic value. Consideration of action to reduce leaks through CFRDs is usually desirable only when the value of the water loss justifies the expenditure.

This conclusion strictly applies to the typical situation where the CFRD foundation is a hard, nonerodible rock. At sites with rock foundations containing faults with soillike gouge, or zones of badly weathered rock, some limited foundation erosion is always possible.

CFRD Has Advantages for Both Low and High Dams

The CFRD has generally been used for dams of moderate height (40 m) or higher. It is also, however, suitable and often has less cost than the ECRD for long low dams for two main reasons:

1. At a site for a long, low rockfill dam, either CFRD or ECRD, the cost of foundation treatment relative to the total dam cost is greater than for a high dam. This gives the CFRD the advantage for the low dam since the rock foundation area needing expensive treatment (under the toe slab) is generally significantly smaller than that for the ECRD (under the earth core).
2. The unit costs of all materials in a long, low ECRD are greater than in a higher dam, both because the filters constitute a larger percentage of

the dam volume, and because the zones are smaller and the construction more restricted.

An example is the adoption of the CFRD alternative for the Macagua Dam with a length of 2,800 m and average height of about 20 m (37). This is the main dam for a major hydroelectric project (2,400 MW) currently under construction on the Caroni River in Venezuela, downstream from the Guri Project. The dam crosses a wide river running on hard rock with highly irregular surface (white-water rapids). Rockfill is coming mainly from the required excavations for powerhouse and tailrace canal. The CFRD was chosen as less costly primarily because of the lesser cost of foundation surface treatment and the relatively high volume and cost of filters processed from crushed rock for the ECRD alternative.

For high ECRDs in narrow canyons with steep abutments there has been concern that arching (stress transfer) in the earth core might result in concentrated leaks through the earth core in differential settlement cracks, or by hydraulic fracturing. From this standpoint the CFRD has the advantage that there is no earth core to erode. For these reasons the CFRD has been used for high dams in some very narrow canyons where an ECRD would probably have not been considered suitable (5,7).

For a high ECRD with steep rock abutments, the blanket and curtain grouting and foundation treatment under the core on the abutments are frequently done in stages of height above the rising embankment construction, working from scaffolding. This work often gets on the critical path and causes schedule delays and technical problems. It is preferred to do all the foundation treatment for ECRDs in narrow steep-walled canyons before the embankment construction is started, but this usually requires an additional year of construction time. The CFRD has the advantage for this important aspect of the work because the area of foundation treatment (under the toe slab) is small and the work is independent of the embankment construction.

Stratified Rockfill Is Desirable

When compacted rockfill was first being used widely for dams, about 1965–1970, several different practices were used to dump and spread the rock. Also, different ideas emerged with regard to the maximum rock size which should be allowed, relative to the layer thickness. In the last 10 yrs, experience has led to general agreement on these points.

The preferred method for rockfill placement, for both the ECRD and CFRD, is to dump the trucks on the surface of the layer being placed, and then to push the rock over the advancing face. This action creates significant segregation (38), with the larger rocks in the bottom of the layer and the smaller rocks and most of the fines in the upper part of the layer.

The main advantage comes from the relatively smooth upper surface created by the action of pushing the dumped rock a short distance on top of each layer being placed, causing surface depressions and voids between larger rocks to become progressively filled with small rocks and quarry fines. The smooth layer substantially reduces cost by reducing tire wear, allowing rapid truck travel and providing a better surface on which to operate the vibrating roller.

Earlier attempts were made to avoid this segregation by dumping the truck loads in spaced piles on the previously compacted layer, and then spreading the piles to form the desired layer thickness. This procedure results in an irregular upper surface, making travel of the equipment difficult. The vibrating roller pounds on high points of larger rocks, resulting in harmful shaking and excessive maintenance. These rollers are not strong enough to operate continuously on an irregular surface, where the drum is usually pounding on a few high points of hard rock. The irregular surface of rock points, which are frequently sharp, also requires trucks to move slowly, with rapid wear and damage to tires. This method of rockfill placement is now considered obsolete by most specialists, but it is still proposed with sufficient frequency to justify discussion here.

There are no technical disadvantages to the preferred method of placement in segregated layers. In addition to lower cost, there are several other advantages. The stratification assures that any flow through the rockfill embankment will travel much more easily in the horizontal direction than vertically. This has the technical advantage that the downstream slope stability is increased for the condition when the reservoir is partially filled during construction (or the dam is overtopped) before the concrete face is in place. Also, for rock containing a relatively high content of fines, the average permeability of the rockfill is considerably greater when placed in stratified layers than when placed in a more homogeneous fashion.

The current general practice is to specify the maximum-size individual rock as that which is embedded so that a reasonably smooth layer surface is available for compaction; i.e., the maximum rock size can be the layer thickness. This does not have any detrimental influence on the compaction obtained or final compressibility of the rockfill embankment (38).

Compacted Rockfill Is Stable against Water Flow-Through

Using the preferred construction method, the top of each layer contains smaller average sized rock and more quarry fines than the lower part. When this practice was first being used, some engineers had concern that water flowing through the rockfill voids, such as from floods during construction or from a leak in the concrete face during reservoir operation, might cause erosion of some fines separating larger rocks, resulting in a "loosening" of the rockfill internal structure and large additional settlement.

There have now been many experiences in which large quantities of water have passed through compacted rockfill dams, both from construction floods (38) and leaks through the concrete face (5,7,14), with no observations of significant settlement caused by flow-through.

This experience confirms that compacted rockfill has a structural skeleton, with rock to rock contact carrying the weight of the overlying embankment. During construction, the rocks are jammed into solid contact by the heavy vibrating roller. Subsequently as the dam is constructed the imposed weight is carried through the skeleton. The areas of contact between the rocks comprising the skeleton are small and the contact compressive stresses developed are high, approaching (and exceeding) the compressive strength of the rock. These high compressive stresses at the rock contacts create high frictional resistance, precluding any tendency for one rock in the skeleton to move with respect to its neighbors. These frictional forces holding the rocks in place are much greater than the hydraulic drag forces caused by water flowing through the voids.

In addition to the rocks forming the main structural skeleton, there are other unstressed rocks in the rockfill—i.e., small rocks and soil-sized fragments which simply sit inside larger voids. These idle fragments can be moved within the voids by flowing water without causing settlement. Small rock fragments and soil particles are carried completely out of the dam, as is sometimes evidenced by the dirty water emerging downstream, though this water usually clears rapidly.

Static Stability Analyses Are Not Made

For most of the CFRDs constructed in recent years, the designers have not made stability analyses. The side slopes were selected based on precedent, without calculations, at about 1.3H : 1V (about 37°), roughly the angle of repose of loose dumped rock. This 37° slope compares with about 45°, or more, for compacted rockfill strength (28: Figure 6).

In conventional stability analyses (often called "limit equilibrium" analyses), commonly used for natural slopes and earth dams, it is postulated that a "trial sliding mass" could move by sliding on curved or plane shearing surfaces. The driving and resisting forces are calculated and compared to define a safety factor. For the CFRD which has no water in the voids and is founded on a sound rock foundation, all experience and theory confirm that such a shear slide cannot occur. Therefore, the conventional stability analysis, when applied to the typical CFRD, calculates the safety factor against a type of movement known to be physically impossible. This fact has caused general recognition that the conventional slope stability analysis is not applicable for the CFRD. As an example of this practice, the Hydro-Electric Commission of Tasmania, which has built nine CFRDs since 1966 and has five more in design, does not carry out stability analyses (24: p. 417).

Occasionally CFRDs are built on rock foundations with unfavorably located joints or other planes of weakness. For such a case, a shear slide is conceivable, with most of the sliding surface passing through the foundation rock. Conventional stability analyses are useful in these cases, as illustrated by the studies made for the Kotmale Dam (21).

Postconstruction Settlement

Compaction of rockfill embankments has been shown to be very effective in reducing settlements. Fig. 1 shows representative crest settlement measurements for five compacted CFRDs and for three older dams of the dumped-rock type.

The rate and magnitude of crest settlement is influenced by both the rockfill properties and the shape of the valley. In narrow canyons, the compacted rockfill arches between the abutments, reducing the crest settlement in the early years. Both Cethana and Alto Anchicaya Dams are built in relatively narrow canyons and the measured crest settlements apparently continue for many years at a decreasing rate, probably due to slow progressive release of the arch action by creep. Areia Dam (11) is

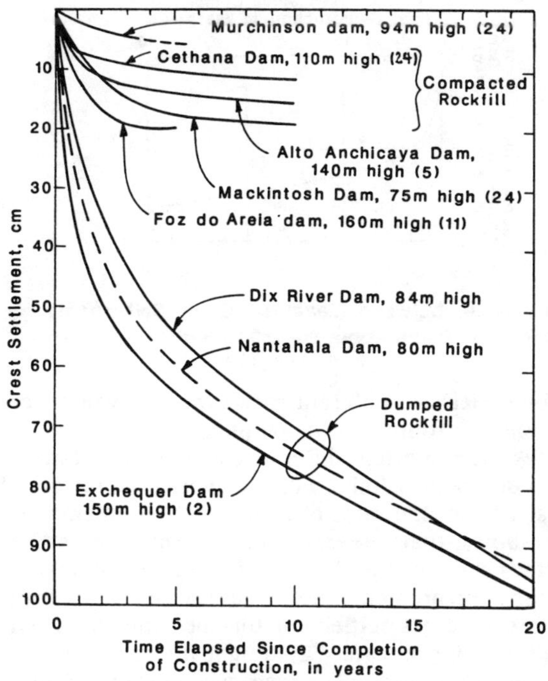

FIG. 1. Comparison of Crest Settlements for Compacted and Dumped Rockfill Dams

TABLE 1. Rates of Measured Settlement in CFRDs

	Approximate Rate of Crest Settlement for 100-m-High CFRD (mm/yr)		
Type (1)	After 5 yrs (2)	After 10 yrs (3)	After 30 yrs (4)
Compacted rockfill	3.5	1.5	0.6
Dumped rockfill	45	30	10

constructed in a broad valley where no significant arch action between abutments would be expected, and the measured crest settlement practically stopped after three years (Fig. 1). Also, the Areia Dam rockfill had larger average rock size and less rock fines than the average quarried hard rock used for CFRDs (11). Measurements during construction show that the Murchison and Mackintosh rockfill embankments are probably the most incompressible and compressible, respectively, of the modern CFRDs for which good records are available (24). Therefore, the range of crest settlement–time records shown in Fig. 1 should be near the total

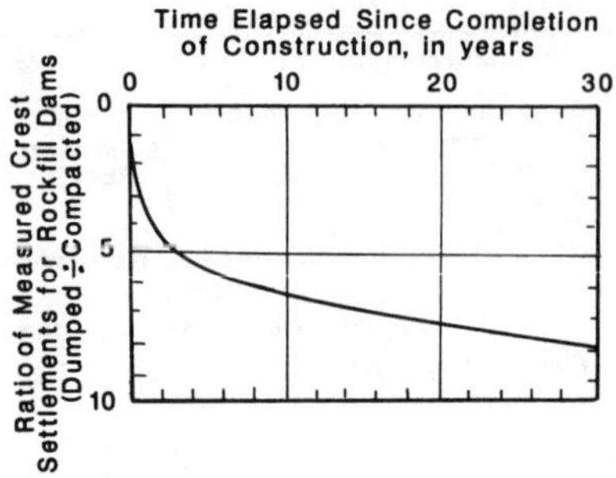

FIG. 2. Approximate Ratio between Measured Crest Settlements for Dumped versus Compacted Rockfill Dams About 80–180 m High

range expected for CFRDs of different rock types, in valleys of different width, with dam heights from 75 m to 160 m.

Fig. 1 also shows the important differences in the long-time settlement for compacted and dumped rockfills. For the average of the curves for the two types of construction, the rates of measured settlement are shown in Table 1. Fig. 2 shows that the ratio of measured crest settlements (dumped/compacted) is about 5 at the end of 3 yrs, and almost 8 in 30 yrs.

From the existing measurements there is every reason to conclude that the future time rate of crest settlement for the compacted CFRD in a narrow valley will be approximately a straight line on a semilog plot, as exemplified by the Alto Anchicaya Dam measurements (5: Figure 12). This assumption has been used to predict that the crest settlement for the 110-m-high Cethana Dam would reach about 16.0 cm at the end of 100 yrs (24: p. 430).

Use of CFRD for Very High Dams

The fact that the entire embankment weight is downstream of the plane of application of the water pressure naturally gives the CFRD a high degree of stability and inherent conservatism. The CFRD has the fundamental advantage over the ECRD that there is no possibility of erosion (piping) of the impervious earth core. The CFRD has the advantage over the concrete arch dam that it relies on gravity for stability, and does not need high-strength rock abutments. The large rockfill mass of the CFRD supports potentially unstable steep rock abutments instead of stressing them. All these factors support the conclusion that the CFRD should be considered in the future as a suitable design for the very highest dams.

This conclusion is already being rapidly accepted. At present, about a half dozen CFRDs with heights in the range of 180–220 m are in final design in various parts of the world, some of which are described in the discussions published in this issue. The writers believe that this trend will continue in the future, with justified confidence.

In the recent past, during the period of rapid acceptance of the modern CFRD, it has been generally considered reasonable and prudent to limit height increases for new, higher dams to a relatively small increment over the past precedent. This led to the progress from the Cethana Dam (24) (110 m high), to Alto Anchicaya (5) (140 m high), and finally, to Foz do

Areia Dam (11,12) (160 m high), all of which were completed between 1971 and 1980.

As the result of all the experience accumulated in the last 15 yrs, it is now quite well known how the compacted CFRD behaves. From this fund of experience it is possible to predict with good reliability how dams of much greater height will behave. As already discussed, the entire range of crest settlements for many dams of different rock types and valley shapes has not been great. The range of most 5-yr crest settlements for dams 75–160 m high has been about from 10–20 cm (Fig. 1), with little continuing settlement after 5 yrs. For the same group of dams, the maximum deflection of the face slab after 5 yrs has been approximately in the range of 20–75 cm, or roughly 2–4 times the crest settlement.

A rough and conservative prediction of settlements for future higher dams can be made on the assumption that the settlements are proportional to the square of the height. Using this assumption, if the 160-m-high Foz do Areia Dam had been 300 m high in a similar wide valley, using the same rockfill and construction methods, the predicted settlements for the higher dam are $(300/160)^2 = 3.5$ times greater than those actually measured; i.e., a 5-yr crest settlement = (3.5) (20) = 70 cm; and a 5-yr face slab deflection = (3.5) (75) = 260 cm. Similar crest settlements and smaller face deflection are predicted from similar calculations assuming 300-m-high Cethana and Alto Anchicaya Dams. These predictions are considered conservative (high) because the compressibility modulus has been considered constant. The modulus can be increased by placing and compacting the rock in thinner layers with a greater number of roller passes.

Based on these extrapolations from existing dam settlements, a 300-m-high CFRD of most types of rockfill constructed with current common practice can confidently be expected to have 5-yr measured movements in the range of 0.5–1.0 m of crest settlement, and a maximum face slab deflection of 1.3–3.0 m. In the writers' opinion, minor changes from current good CFRD design details will assure that such crest and slab movements can be accommodated easily in a 300-m-high dam without risk of significant leakage.

Conservative perimeter joint details would be used for the very high CFRD to allow larger movements than in lower dams. Probably more important, for very high future dams, more conservative details will be used for the semipervious, processed material placed directly under the face slab (26). Such semipervious upstream embankment zones of processed material (from crushed rock or alluvial gravels), about 5 m wide, are already being included in the design of recent large CFRDs, including the Cirata Dam in Indonesia (23) and the Xingó Dam in Brazil (32). In the future, this zone will probably be a standard design feature for important high dams. For very high CFRDs (200 m and higher), this zone can be made more conservative (wider, and more carefully processed and constructed to assume homogeneity, freedom from segregation, and reliably low permeability). At relatively small added cost, this zone can be counted upon with complete assurance to eliminate any possibility of important leakage through the dam, for any conceivable reasonable size of openings which may develop through the concrete face slab or ruptured waterstops.

Under this subject it is interesting and pertinent to record that an abrupt jump in maximum dam height was made for ECRDs starting in the early

1970s. This led to the Esmeralda Dam (235 m high), completed in 1975 in Colombia, followed by the Chicoasen Dam (260 m high), completed in 1980 in Mexico, and the Rogun Dam (330 m high), currently under construction in Russia. At the time that this evolution in ECRD maximum height started—about 1970—there were only three ECRDs in the world in the range of height of 150–160 m with a few years of performance experience behind them (Cougar, Goeschenenalp, and Gepatsch Dams). Based on the relative advantages of the CFRD over the ECRD, as discussed earlier, the writers believe it is equally reasonable to build CFRDs to these greater heights.

SPILLWAYS OVER CFRDs

In the last 10 yrs, many engineers have seriously considered placing large spillways on top of CFRDs. These have ranged from service spillways designed for large discharges each year, to auxiliary spillways designed to operate only at rare intervals. These designs frequently result in important potential savings compared to other spillway locations. Only a very few such studies have been carried forward to final design and construction. This design alternative has often been abandoned, in spite of the fact that the designer is confident that it is reasonable and safe, because there is not sufficient precedent for such a spillway.

As discussed, it is now well established that the total settlements of the modern CFRDs are relatively small. Most of the movement takes place in the first few years, and the rate of crest settlement after 5 yrs is only a few millimeters per year. Such settlements can easily be tolerated without damage in ungated spillways.

In order to proceed with the reasonable practice of placing spillways routinely over the top of CFRDs, it is necessary to relax the idea that direct precedents are required. The typical compacted CFRD embankment provides a spillway foundation which is as basically secure and conservative as many of the steep and weathered rock valley walls, where spillways are commonly constructed.

Except for low dams, it would be desirable to consider compacting the rockfill mass comprising the downstream slope of the dam and the spillway foundation in the same layer thickness as used for the upstream half of the dam which supports the concrete face slab. Also, it would probably be generally desirable to place a layer of smaller rock directly under the spillway and to compact and protect the sloping embankment surface as done for the upstream slope under the concrete face slab. These actions will assure that the spillway chute will be subjected to approximately the same type and magnitude of strains as those which are imposed in the upstream face slab before reservoir filling: i.e., generally a relatively low compressive strain in the longitudinal direction, and a small, gradually distributed, dish-shaped settlement. The subsequent increment of spillway settlement caused by the weight of the water in the chute will be negligible, since the stresses added to the massive rockfill slope by the small water weight will be insignificant.

The expected movements are sufficiently small so that the spillway on top of the CFRD can be designed using about the same details as would be used for a more conventional foundation. The chute slab should be placed continuously from bottom to top as for the upstream slab. Reinforcing in the center of the slab should pass through all construction joints in both directions and extend into the side-wall footings, to tie the entire bottom slab together. Expansion joints would be provided in spillway walls to absorb the small continuing settlements expected.

Since the bottom slab is completely tied together with continuous steel, and the underlying embankment is very pervious, there would be no significant water pressures developing under the spillway chute. Therefore, steel anchor bars extending into the underlying rockfill should not be needed to resist uplift. Anchors may be considered desirable as a conservative effort to minimize possible vibration created by the water flow, or to resist earthquake shaking, but the need for anchors is considered debatable.

For high dams, air grooves should be provided in the chute slab. These would also serve as contraction joints to absorb any significant compressive strain which may develop in the longitudinal direction as the result of the continuing embankment settlement.

Reviewers of this paper in manuscript generally agreed with the basic idea but cautioned that it should be made clear that these assessments do not apply to large gated spillways on very large rivers which must discharge continuously several months per year. The opinions in this section refer to the typical moderate-sized spillway, generally ungated, such as those carrying a maximum peak discharge of the order of about 25–30 m^3/s per meter of chute width, and generally on rivers with flood flows of a few weeks or less. It is for this relatively common situation that the spillway over the CFRD commonly offers large potential cost savings, and for which it is agreed it can be designed conservatively. Also, of course, the design for energy dissipation downstream must receive appropriate consideration, as for a spillway on the abutment discharging near the dam.

Earthquake Considerations

Since the entire CFRD embankment is dry, earthquake shaking cannot cause pore pressure in the rockfill voids. The CFRD foundation is rock, which does not magnify the incoming acceleration forces. The embankment is heavily compacted in thin layers to a dense state with vibrating rollers. Earthquakes can only cause small deformations during the short period of strong shaking. After the earthquake is over the CFRD is as stable as before.

In very strong earthquakes the concrete face may be cracked, increasing the leakage. The potential cracking and leakage cannot threaten the overall safety of the dam, because the amount of leakage which can get through the cracks and the zone of small rock under the face slab can easily be passed safely through the main rockfill embankment.

For these reasons the CFRD is considered to have the highest fundamental conservatism against earthquakes, and the same basic design has generally been used in regions of high seismicity as in nonseismic areas.

Until the present, there have been few serious efforts during CFRD designs to include earthquake analyses using modern theory. Since the much simpler analyses of static stability used for decades in dam design are generally considered not practical or useful for the CFRD, as discussed, there is understandable reluctance to place confidence in the results of the newly developed dynamic analyses. The main reason is that the new dynamic theories necessarily include assumptions whose validity and applicability are difficult to evaluate. Also, no large fund of experience with dams shaken by earthquake is available to compare with results of dynamic calculations, and the few available experiences seem to indicate that the deformations being calculated using current dynamic theories are considerably higher than those measured (see discussion by Sherard of Ref. 28 published in this issue).

The few dynamic analyses made as part of the design activity of CFRDs built in recent years have been primarily justified as potentially useful research efforts. The results have not had significant influence on the main design decisions. Recently, two interesting and valuable independent general efforts have been made at seismic evaluation of the CFRD (27,28) using state-of-the-art procedures. The main results of the two efforts are not greatly different, and support the conclusion that the modern CFRD with 1.3H:1V side slopes will withstand very strong earthquake shaking without serious damage.

One paper (27) summarizes calculations of movement of individual rigid "trial sliding wedges," using the breakthrough concept of Newmark (that permanent deformation occurs in a series of pulses when the acceleration forces acting momentarily during each pulse exceed the shearing strength), with current application procedures. The other (28) introduces a new, powerful finite difference program (DSAGE) capable of computing permanent deformations at each point in the dam and including the influence of the reservoir water pressure. Detailed evaluation of the assumptions and mechanics of these analyses is beyond the scope of this paper, but it is valuable to summarize some of the main findings of practical interest:

1. Both methods of calculation give roughly the same calculated deformations for strong earthquakes.
2. For shaking expected at sites near the epicenter of earthquakes with Richter Magnitude of about 7.5, the calculated crest settlements for CFRDs with heights of the order of 100–150 m are relatively small and easily tolerable, of the order of 0.5–1.0 m.
3. For shaking of lesser intensity, the new DSAGE computations give lower calculated deformations than those computed from rigid block analyses. The DSAGE computed deformations are apparently more consistent with the few available records of measured dam settlements caused by earthquakes.
4. Both procedures show that most of the deformation occurs in the upper half (or third) of the dam.
5. The deformation pattern from DSAGE computations show a gradual deformation (slump) of the crest caused by permanent shear strains distributed uniformly throughout the top half of the dam. For a very strong earthquake the computed maximum settlement of the crest is less than 0.5 m (for a 100-m-high dam).

An excellent summary of currently available experience with measured settlement of rockfill dams caused by earthquakes is given in Ref. 28. This confirms two main points: (1) Very few dams have been strongly shaken; and (2) the measured settlements to date are small. Seven records considered reliable are presented (28: Table 2). For six of these records, the crest settlements during the earthquakes are less than 0.1% of the dam height. For the seventh—Cogoti Dam, in Chile, judged to have been shaken considerably more strongly than any of the others—the measured crest settlement was about 0.4 m or about 0.45% of the 84-m dam height during a 1943 earthquake with Magnitude variously reported in the range of 7.9–8.3, and epicenter located about 89 km away. Hence, the Cogoti Dam record is the only dam until the present time that has had fairly large earthquake-induced settlement, and is consequently a large part of the total experience fund currently available for very strong shaking (1).

The Cogoti Dam is a CFRD constructed in 1935–1938 by the older procedure of dumping the rock in high lifts without compaction. As discussed, it is known from many measurements that settlements (without earthquake) of the older CFRDs of dumped rock are commonly many times greater than for the modern, compacted CFRD. Investigators of the Cogoti Dam concluded that the 0.4-m settlement which occurred abruptly during the 1943 earthquake was probably due mainly to compression caused by shaking the rockfill into a denser condition which would have occurred anyway later, since the rate of crest settlement after the earthquake was abruptly reduced (1: Figure 6). Hence, it is not known how much, if any, of the measured 0.4-m crest settlement was caused by pulses of shear strain of the Newmark mode as postulated in the dynamic theory.

As shown in Figure 1 of Ref. 28, the Cogoti Dam crest settlement caused by the 1943 earthquake nevertheless approximated very well the value calculated by modern dynamic theory, assuming similar earthquake shaking (Earthquake Severity Intensity of 11–12). The computed results were based on the assumption that the dam had the high shear strength of compacted rockfill (about 45° to 50°; see Figure 6 of Ref. 28). Since the Cogoti Dam was constructed as dumped rockfill, with rock in its loosest possible state, the actual shear strength available for resisting pulses of acceleration forces was much lower (perhaps about the angle of repose of dumped, loose rock, or about 37°). There can be little doubt that if the Cogoti Dam had been constructed as a modern compacted CFRD, the 1943 earthquake-induced crest settlement would have been a small fraction of the 0.4 m measured. This record is an important part of the evidence presently available to show that dynamic analyses, using current procedures and assumptions, seem to overestimate significantly the settlements caused in compacted rockfill by strong earthquake shaking.

In Ref. 28, the authors show that the measured earthquake-induced settlements for all of the seven dams subjected to strongest shaking approximate fairly well the calculated values using current dynamic theory (8: Figure 1). But none of these seven records are compacted CFRDs. In addition to the Cogoti Dam, one other (Minase Dam, Japan) is a CFRD constructed many years ago by loose dumping in high lifts. The other five dams cited are ECRDs, which behave quite differently from CFRDs from the standpoint of postconstruction crest settlement.

There is much evidence from measurements to show that the postconstruction crest settlement of the typical high compacted CFRD is less than for the ECRD of similar height and other conditions, because of the gradual postconstruction relaxation of temporary arching between the earth core and shells of the ECRD. The postconstruction settlement of the entire upper part of the typical high ECRD is larger than for a CFRD of similar dimensions because of the initial arching of the more compressible earth core during construction between the less compressible filters upstream and downstream, which delays compression of the earth core. This arching creates high shear stresses in the earth core, approaching the earth shear strength, which high stresses are partially released later by gradual creep. Hence, the normal postconstruction crest settlement of the ECRD under static conditions has a significant component due to this arch relaxation, causing the postconstruction settlement to be larger and continue longer than for the CFRD, where this action does not exist. Earthquake-induced crest settlements of the ECRD undoubtedly also include a significant component of this arch collapse due to pulses of earthquake stress momentarily overcoming the earth core shear stresses supporting the arch.

Until recently, the technology for dynamic analyses was limited to calculations of movements of individual rigid trial sliding blocks (27), obviously a very crude approximation of actual conditions. Only with the recently developed DSAGE program has it been possible to compute the internal distribution of earthquake-induced deformations (28: Figure 9). Until this time, there have been few attempts at general studies to determine the influence of the various main underlying assumptions to give confidence that small and reasonable differences in the basic assumptions do not produce greatly different calculated results. Hence, the applicability of the dynamic analysis in design is presently somewhat problematical. Substantial advances will be made in the future with the rapidly developing computing technology to allow confidence that the improved dynamic analyses can be relied upon as firm design tools, but these improved analyses may give significantly lower compacted deformation than those presently computed. At this point, the writers believe that the following general conclusions are appropriate for CFRD design in seismic areas:

1. For the great majority of sites which may be very strongly shaken, such as near the epicenter of a Magnitude 7.5 quake, or at sites with calculated Earthquake Severity Index (28) in the general range of 10–15, the same CFRD design can be used as in nonseismic areas. For these sites, all present experience with dam behavior and the overall results of current dynamic calculations give confidence that the worst earthquake-induced crest settlement will be substantially less than 1% of the dam height. A sudden crest settlement of 1% of the dam height will not threaten the safety of a modern CFRD.

2. For the smaller number of sites where the world's strongest ground shaking could occur, within a short distance of major faults capable of generating earthquakes of Magnitude 8 or greater, there is very little evidence from any source to guide judgment about the maximum probable earthquake-induced crest settlement of a modern CFRD. At these sites, it is desirable for the designer to make every effort to obtain as much help as is available from the most advanced dynamic analyses. It is essential in such an effort to make parametric studies of the several main assumptions, including the method of feeding the earthquake energy from the foundation

to the dam and the shear strength and damping characteristics of the compacted rockfill. Probably it will be concluded in the future that the maximum possible earthquake-induced crest settlements of a compacted CFRD cannot exceed 1% or 2% of the dam height under the most severe conceivable conditions of earthquake shaking. The provision of conservative extra freeboard is probably the appropriate and economical seismic design provision. Since the worst damage to the dam during strong earthquake shaking is a general slumping of the upper part of the embankment, as shown quantitatively by the results of the new DSAGE calculations (28: Figure 9), there can be no threat to the basic dam safety unless the crest settlement exceeds the freeboard. In the extreme case, some provision to prevent individual rocks near the top of the dam on the downstream sloping surface from being dislodged may be justified.

CONCLUSIONS

The following conclusions are reached:

1. The CFRD is currently being chosen as the preferred dam type with increasing frequency for reservoirs of all sizes and purposes throughout the world.
2. The CFRD has advantages over the ECRD of cost and schedule, even when economical sources of soil are available for an impervious earth core. The CRFD has these advantages for dams of all heights, including long, low dams.
3. The shorter base width of the CFRD often results in savings in the auxiliary structures at the dam, compared with the ECRD.
4. River handling costs and risks are less for the CFRD than for the ECRD at many sites.
5. Leakage emerging downstream of the CFRD has a basically different significance than leaks through dams with earth cores, because there is no possibility of earth core erosion and no potential threat to the dam safety.
6. The preferred method of rockfill placement is to dump on the layer being placed and to push over the advancing face. The stratified rockfill obtained, with larger rocks in the bottom of each layer and smaller rocks in the upper part of each layer, is desirable. The maximum allowable rock size should be equal to the layer thickness.
7. Compacted rockfill has a highly stressed skeleton. Intergranular friction holds the rocks comprising the skeleton in place so that the rockfill is stable and does not settle significantly under the action of water flow-through.
8. Conventional (limit equilibrium) stability analyses are not used for the CFRD.
9. The crest settlement for the compacted CFRD is a small fraction of that previously measured on the dumped (uncompacted) rockfill CFRDs of the older practice. For a modern compacted CFRD of 100 m height, the expected crest settlement will generally be about 10–15 cm in 5 yrs, and 15–25 cm in 100 yrs.
10. The CFRD is an appropriate type in the future for the very highest dams. For a 300-m-high CFRD constructed of most rock types, acceptable

performance can be predicted, based on reasonable extrapolation of measurements on existing dams.

11. It is safe and reasonable to build moderate-sized spillways over CFRDS.

12. There is strong evidence from the experience now available that current dynamic analyses substantially overestimate crest settlements of CFRDs. Parametric studies are needed to investigate the influence of the various main assumptions on deformations calculated in dynamic analyses.

13. The CFRD is inherently safe against strong earthquake shaking. For the great majority of sites in areas of strong seismic activity, the same CFRD design can be used as in nonseismic areas. For the few sites where the world's strongest shaking is possible (within a short distance of master faults capable of generating earthquakes of Magnitude 8 or larger), the provision of extra freeboard is appropriate to compensate for earthquake-induced permanent crest settlement, guided by dynamic analyses.

Acknowledgment

The writers are indebted to the 96 authors and coauthors who prepared the 37 papers for the October 1985 Symposium on the Concrete Face Rockfill Dam, and to the 40 discussers of the Symposium Papers.

Appendix. References

The following references (1–37) appeared in *Concrete face rockfill dams—Design, construction, and performance*, edited by J. Barry Cooke and James L. Sherard, Proceedings of a symposium held at the ASCE Convention, at Detroit, Mich., on October 21, 1985, ASCE, New York, N.Y., 1985:

1. Arrau, Luis, Ibarra, Ismael, and Noguera, Guillermo, "Performance of Cogoti Dam under seismic loading," pp. 1–14.
2. Leps, Thomas M., Cashatt, C. A., and Janopaul, R. N., "New Exchequer Dam, California," pp. 15–26.
3. Millet, J. C., Louis, F., and Robert, F., "Design, construction & performance of Fades Dam," pp. 27–39.
4. Good, R. J., Bain, D. L. W., and Parsons, A. M., "Weak rock in two rockfill dams," pp. 40–72.
5. Materon, Bayardo, "Alto Anchicaya Dam—Ten years performance," pp. 73–87.
6. Salembier, M., "Membrane sealing of Rouchain Dam," pp. 88–97.
7. Amaya, Fabio, and Marulanda, Alberto, "Golillas Dam—Design, construction and performance," pp. 98–120.
8. Dascal, O., "The Outardes 2 concrete-faced rockfill dam," pp. 121–139.
9. Casinader, R., and Watt, R. E., "Concrete face rockfill dams of the Winneke Project," pp. 140–162.
10. Beene, Ralph R. W., and Pritchett, Edward C., "The R. D. Bailey Dam—A concrete-faced, earth-rockfill," pp. 163–172.
11. Pinto, Nelson L. de S., Marques Filho, Pedro L., and Maurer, Edilberto, "Foz do Areia Dam—Design, construction, and behaviour," pp. 173–191.
12. Materon, Bayardo, "Construction of Foz do Areia Dam," pp. 192–207.
13. Mackenzie, P. R., and McDonald, L. A., "Mangrove Creek Dam: Use of soft rock for rockfill," pp. 208–230.
14. Bodtman, William L., and Wyatt, John D., "Design and performance of Shiroro rockfill dam," pp. 231–251.

15. Perdomo, D. A., Vasilescu, M. S., and Ferro, E. A., "Fortuna concrete face rockfill dam," pp. 252–265.
16. Sierra, Jesus M., Ramírez, Carlos A., and Hacelas, Jorge E., "Design features of Salvajina Dam," pp. 266–285.
17. Hacelas, Jorge E., Ramírez, Carlos A., and Regalado, Guillermo, "Construction and performance of Salvajina Dam," pp. 286–315.
18. Rogers, R. L., "Boondooma Dam," pp. 316–335.
19. Watakeekul, Somkuan, Roberts, Gordon J., and Coles, Andrew J., "Khao Laem—A concrete face rockfill dam on karst," pp. 336–361.
20. Morris, Michael, "Design and construction of Terror Lake Dam," pp. 362–378.
21. Gosschalk, Edward M., and Kulasinghe, A. N. S., "Kotmale Dam and observations on CFRD," pp. 379–395.
22. Phillips, P. R., "Batang Ai—Transition zone," pp. 396–409.
23. Pinkerton, Ivor L., Siswowidjono, Soetomo, and Matsui, Yutaka, "Design of Cirata concrete face rockfill dam," pp. 642–656.
24. Fitzpatrick, Mike D., Cole, Bruce A., Kinstler, Frank L., and Knoop, Bram P., "Design of concrete-faced rockfill dams," pp. 410–434.
25. Varty, Alan, Boyle, Rick J., Pritchard, Edward D., and Gill, Roger E., "Construction of concrete face rockfill dams," pp. 435–458.
26. Sherard, James L., "The upstream zone in concrete-face rockfill dams," pp. 618–641.
27. Seed, H. Bolton, Seed, Raymond B., Lai, S. S., and Khamenehpour, B., "Seismic design of concrete faced rockfill dams," pp. 459–478.
28. Bureau, Gilles, Volpe, Richard L., Roth, Wolfgang H., and Udaka, Takekazu, "Seismic analysis of concrete face rockfill dams," pp. 479–508.
29. Tsai, Chan-Feng, Yen, Bing C., Tucker, Keith D., and Askari, Shahen, "Design and stability evaluation of Balsam Meadow Dam," pp. 509–527.
30. Imaizumi, Hideo, and Sardinha, Álvaro E., "A study of deformations in concrete faced rockfill dams," pp. 528–540.
31. Hollingsworth, Harold, Conner, Timothy R., and Victor E. Anderson, "Design of Deer Creek Dam," pp. 541–558.
32. Vasconcelos, Aurélio A., and Eigenheer, Luiz Paulo, "The Xingó rockfill dam," pp. 559–565.
33. Pineros, Juan E., "Rockfill compression tests and other aspects of the design of Miel I Dam," pp. 566–586.
34. Pinto, Nelson L. de S., Marques Filho, Pedro L., and Maurer, Edilberto, "Segredo Dam—Basic design aspects," pp. 587–593.
35. Corrêa, Regis Danton, and Thielen, Fernando, "Machadinho Dam in Brazil," pp. 594–600.
36. Antunes Sobrinho, José, and da Cunha, Joaquim C., "Itá Dam—Selection of type," pp. 601–607.
37. Prusza, Z., De Fries, K., and Luque, F., "The design of Macagua concrete face rockfill dam," pp. 608–617.

Other References

38. Cooke, J. Barry. (1984). "Progress in rockfill dams," 18th Terzaghi Lecture. *J. Geotech. Engrg.*, 110(10), 1381–1414.
39. Discussion of "Progress in rockfill dams." (1986). Discussions by Casinader, R.; Chadwick, W. L.; Fetzer, C. A.; Fitzpatrick, M. D.; Fucik, E. M.; Hacelas, J. E., and Ramírez, C. A.; Houlsby, A. C.; Marulanda, A., and Ospina, C. S.; Materon, B.; Merritt, A. H.; Murti, N. G. K.; Pinkerton, I. L.; De Porcellinis, P.; Ripley, C. F.; Sherard, J. L.; Strassburger, A. G.; Swiger, W. F.; and Taylor, H. Closure by Cooke, J. Barry. *J. Geotech. Engrg.*, 112(2), 217–253.
40. Hacelas, J. E., and Ramírez, C. A. (1986). "Salvajina: A concrete face dam on a difficult foundation." *Water Power and Dam Constr.*, 38, Jun.
41. *Symposium on rockfill dams.* (1960). *Trans.*, ASCE, 125(II).

Introduction to Paper 20

Concrete Face Rockfill Dam: II. Design

by J.B. Cooke and J.L. Sherard
Journal of the Geotechnical Engineering Division,
American Society of Civil Engineers, Vol. 113, No. GT 10,
October 1987, pp 1096-1112.

Prepared by J. Barry Cooke

This paper, II. Design, is the second of two papers co-authored by Jim and I, as the closing papers to the ASCE Symposium "Concrete Face Rockfill Dams - Design, Construction, and Performance", Detroit, Michigan, October 1985. The first of the two papers is I. Assessment, the Proceeding Paper 19 of this volume.

It was decided that two papers, one a general assessment of the type of dam and one on design, would best cover the subject. Preparation took place as discussions were arranged and reviewed, and included occasional weekend coordination meetings. Jim was principal author of the Assessment Paper and I of the Design Paper. The Journal, with the two papers and the many discussions, concluded five years of work on the ASCE Symposium on the Concrete Face Rockfill Dam.

Reference is made to the Introduction of "Concrete Face Rockfill Dam-Assessment", Paper 19.

Concrete-Face Rockfill Dam: II. Design

By J. Barry Cooke[1] and James L. Sherard,[2] Fellows, ASCE

Abstract: The writers discuss and give opinions on the main details of design and construction for the concrete-face rockfill dam, including toe slab foundation treatment, dimensions, reinforcing, and grouting; face slab thickness, concrete quality, reinforcing, joints, and construction; the parapet wall and camber; and rockfill embankment foundation treatment, zoning, rock grading, rock quality, and construction. Current practice and recent trends are assessed and commented upon. Empirically derived specifications are offered.

Introduction

In the symposium proceedings, *Concrete Face Rockfill Dams—Design, Construction, and Performance* (ASCE, 1985), 96 authors present 37 valuable papers (Refs. 1–37). These authors showed general agreement on the main design aspects, but illustrated that there was a range of views on many important details. In this paper, the writers—the editors of the symposium proceedings—present their opinions on most of the main details of design and construction. A companion paper published in this journal gives a general assessment of the concrete-face rockfill dam, and includes stability and seismic considerations, two subjects not included in this design paper. In this paper, the abbreviations CFRD and ECRD are used for the concrete-face rockfill dam and earth-core rockfill dam, respectively.

All of the design features that have been developing on CFRD dams, and that have been implemented on existing dams, have been safe. The criteria guiding the content of this paper are proven features of design and of requirements in specifications that provide for a safe and satisfactory dam at lowest cost.

Toe Slab

Foundation Treatment. The toe slab is usually on hard, nonerodible fresh rock which is groutable. For less favorable foundation rock, after a trench is made to an estimated acceptable foundation, many methods are available to treat local imperfections (17,24,39,40). The criterion is to eliminate the possibility of erosion or piping in the foundation. Careful excavation is used to minimize fracturing of the rock surface on which the toe slab is placed (9,24). Air or air-water cleanup, just prior to placing concrete, is required to obtain a bonded contact of the concrete to the foundation.

[1]Consulting Engr., 1050 Northgate Dr., Suite 400, San Rafael, California 94903.
[2]Consulting Engr., P.O. Box 9138, La Jolla, CA 92038.

Note. Discussion open until March 1, 1988. Separate discussions should be submitted for the individual papers in this symposium. To extend the closing date one month, a written request must be filed with the ASCE Manager of Journals. The manuscript for this paper was submitted for review and possible publication on March 19, 1987. This paper is part of the ***Journal of Geotechnical Engineering***, Vol. 113, No. 10, October, 1987. ©ASCE, ISSN 0733-9410/87/0010-1113/$01.00. Paper No. 21853.

While most CFRDs have erosion-resistant hard rock formations, many have been built on rocks with faults or badly weathered seams, or on fairly large areas of softer rock of questionable erosion resistance (16,19,40). For these, the length of the foundation seepage path has been increased to one-fourth, or more, of the water head by placing a layer of shotcrete (or concrete) on the final foundation surface downstream of the toe slab, with a filter placed for a distance downstream of the shotcrete, and sometimes a filter placed also on top of the shotcrete to anticipate possible cracking (9,13,16,24,39,40). Reinforced shotcrete has sometimes been used (13,24) in an effort to distribute any settlement cracking into a series of small cracks rather than a few large cracks.

At the 148-m-high Salvajina Dam, where the upper half of the dam foundation contained large areas of intrusion dikes severely weathered to soil-like properties, the foundation was covered with a two-layer filter, the finer filter being a concrete sand. This sand filter can be relied upon to prevent the foundation material from being eroded into voids in the dam embankment. This is considered good conservative practice at locations where there is any possibility that soil-like foundation material (weathered rock, faults, and soil-filled joints) could be caused to migrate by erosion into the embankment voids.

Dimensions. Toe slab widths for hard and groutable foundation rock are made of the order of 1/20–1/25 of the water depth (5,7,11,15,24). At a given dam, the width is changed in several steps and is not tapered, mainly for construction convenience. The minimum width has generally been 3 m (24), though 2 m can be used for dams less than 40 m high on very good rock (37). For a poor local rock condition, or for a site in poor rock, greater widths are used (16).

The design thickness has been frequently made about equal to the face slab thickness. Excavation overbreak and irregular topography usually provide significantly greater thickness, so that a minimum design thickness of 0.3–0.4 m is generally reasonable for most toe slabs (24). Since the toe slab uniformly is supported on sound, hard, thoroughly grouted rock, it is difficult to justify the value of a thickness much greater than 0.3 m. A design thickness of 0.4–0.5 m may be considered for the lower toe slab of high dams.

Stability. The toe slab must resist the high horizontal water thrust without support from the rockfill. For the toe slab of normal thickness, there is ample frictional resistance to resist the water thrust, unless the foundation has unfavorably oriented planes of low shearing resistance just below the toe slab. For high toe slabs, and usually in local areas, stability analyses are required (24,26). Uplift pressure under the slab is assumed to vary linearly from full reservoir pressure to zero over the width. Water loading on the face slab opens the perimeter joint so there is no interaction between toe slab and face slab. The rockfill may exert a resisting force on the downstream side of the toe slab, but this cannot be relied upon, and is neglected in stability computations.

In plinth alignment, it is attempted to minimize the height of the toe slab, not only from the standpoint of stability, but to limit the thickness of rockfill that must resist water loading in the starter slab area, and thus minimize perimeter joint offset.

Layout. Various excavation methods and toe slab orientations are used depending primarily on the abutment steepness, and the thickness and nature of soil and weathered rock overlying a satisfactory foundation. Rock excavation is made by blasting. Occasionally, only earth moving is needed to expose satisfactory foundation along parts, or all, of the length. Sometimes a continuous trench is made along the plinth, digging to refusal with earth-moving equipment, before a decision is made to locate the toe slab rock excavation reference line. In steep canyons with suitable hard rock near the surface, the slab is sometimes placed on or very near the existing surface in order to minimize difficult excavations (5,7).

The toe slab is laid out as a series of straight lines. The angle points are selected to suit the foundation conditions and topography, and have no required relation to the vertical joints of the face slab. The reference line for the excavation is at the bottom of the toe slab, and for forming the toe slab, in the plane of the bottom of the face slab (33: Figure 8).

In each straight segment of length, the toe slab is sometimes made horizontal in a direction at right angles to the reference line. This puts the upstream edge at the deepest line of excavation, and gives the most convenient surface for toe slab construction and grouting (11,24). However, the toe slab plane can be at any angle in plan to the reference line. For very steep abutments, the toe slab may be nearly normal to the dam axis (7).

Reinforcing, Joints, and Anchor Bars. The toe slab details used in common current practice have been proven satisfactory by much experience. Recent improvements have been in simplified construction and in design changes that lower the cost.

The main purpose of the reinforcing is the same as in the face slab; i.e., to function as temperature steel, and to spread out and minimize the widths of any cracks which may tend to develop from the small bending strains imposed. While two layers of longitudinal steel have sometimes been used in the past, it is now generally accepted that a single layer should be used (16,18,37,19). The steel is put 10–15 cm clear of the upper surface as temperature steel, where it is hooked by the anchors: 0.3% each way is adequate. A double row of longitudinal steel has a slight theoretical disadvantage because it makes the toe slab stiff and less able to adjust to any small differential settlements of the underlying rock, which has no tensile properties. Practically, a lower layer of reinforcement makes it more difficult to clean the rock surface just before pouring concrete, and to get the optimum concrete bond to the rock. For the construction service as a grout stop, the one upper layer has been adequate.

In the past, construction joints with waterstops located at predetermined distances were commonly used. They were needed, since the joints opened. The toe slab waterstop connection to the perimeter waterstop was inconvenient. At the Salvajina Dam this practice was changed, allowing the contractor to choose the locations for construction joints (with reinforcing passing through) based on topography and on construction convenience (16). Today longitudinal reinforcing is continued through the joints without waterstops. This is considered good practice, is more economical, and has been adopted for more recent dams (31,34,36,37). With the longitudinal steel passing through the construction joint, there is no need to use waterstops.

The purpose of the anchors is simply to pin the concrete to the rock. The anchors are not to resist any given uplift loads. Lengths, spacing, and bar diameters should be chosen on the basis of precedent and the characteristics of the rock foundation. The anchors and temperature reinforcing do improve the slab as a grout cap.

It should be emphasized that the anchors are not needed for resisting uplift by grout pressures acting on the slab bottom, a common misconception. When a concrete slab is poured on a cleaned hard rock foundation, the concrete is bonded intimately to the cleaned rock surface, and stable cement grout injected in the rock below cannot travel along the concrete-rock interface exerting general uplift. The only uplift exerted on the slab by the grout is the pressure acting over the area of any small open cracks in the rock; therefore, the uplift forces are negligible. This concluison is supported by many experiences where thin, unreinforced and unanchored, concrete slabs poured on rock have been used successfully for grout caps, through which grout at high pressures is injected with packers placed in or just below the slab. The only requirement is to obtain a good concrete-rock bond, by thoroughly cleaning the rock surface and wetting it with water or grout just before placing the concrete. The anchors may have some value in preventing uplift during grouting if there are open, near-parallel bedding planes or rock joints within a short distance below the bottom of the toe slab.

Anchors used in common practice have generally been 25–35-mm-diameter bars (#8–#11), spaced about 1.0–1.5 m each direction, with lengths usually of 3–5 m (7,8,16,20,23,31,35,37). The anchors are commonly simple dowels of reinforcing steel, grouted full length in the rock, and hooked (90°) on the one layer of reinforcing. They are usually installed by partially filling the drilled holes with thick cement-water grout and then inserting the bar, displacing the grout and ensuring that the steel bar is grouted for the full length. Anchor-pulling tests to determine uplift capacity are not necessary since there is no design load.

Grouting through Toe Slab. The toe slab is placed before the adjacent fine-rockfill Zone 2 is placed (5,12,13,25). Grouting is done with the toe slab as a grout cap, and at any time during construction. Both points are important for the shortest schedule and lowest cost.

In grouting, there is an interest in being able to see grout return on the adjacent rock surface. This has led engineers on a few jobs to specify that the grouting be done before the adjacent rockfill is placed, and even to clean to rock upstream from the toe slab. This is unnecessary and costly. Any advantage of grouting in advance of placing rockfill adjacent to a concrete toe slab can only relate to the cost of the grouting; i.e., there is less wasted grout from surface losses. Current grout practice of beginning with a thick mix and of early further grout thickening makes losses low in any event. The toe slab confines grout where it is needed. The procedure for obtaining a finally satisfactory grout job is the same whether the grout return to the surface is seen or not; i.e., grouting in an area is continued until closure is obtained in the final grout holes and the rock will no longer take any significant quantity of grout.

The independence of grouting from the dam schedule is important. The specifications should not require the grouting to be done in advance of the adjacent rockfill placement. Especially, it should always be required that the grouting be done through the toe slab, most effectively grouting the upper zone of rock under the toe slab (24).

Commonly the grouting consists of a central line of deeper curtain grouting holes, plus upstream and downstream lines of the blanket (consolidation) grouting holes of lesser depth. For the CFRD, the blanket grouting is of special importance because of the relatively short seepage path through the rock directly under the toe slab. The blanket grouting should be carried to depths sufficient to penetrate any surface zone of open fissures or higher permeability. Realization of the greater relative importance of the blanket grouting has led to a desirable trend to greater conservatism in the blanket grouting relative to the curtain grouting (24), i.e., less emphasis on depth and closure criteria for the primary curtain.

Concrete Face Slab

Concrete. For the concrete, durability and impermeability are more important than strength, a 28-day compressive strength of about 20 MPa (3,000 psi) being adequate (11,16). Maximum-size aggregate of 38 mm (1.5 in.), air entrainment, and use of pozzolan (8,11) are common features in current practice. However, a maximum size of 64 mm (2.5 in.) has sometimes been used, and is satisfactory with special care being taken at construction joints and waterstops.

There has been no record of alkali aggregate reaction in face slabs for CFRDs, and use of pozzolan in many dams may be a factor. The writers believe it should be considered good practice generally to use pozzolan, even with apparently nonreactive aggregates, for providing a more impervious and durable concrete, and for reducing the risk, albeit rare, of concrete deterioration from some as yet little-understood and possibly slowly occurring cement-aggregate reaction.

Thickness. The concrete face has demonstrated its durability under high hydraulic gradients (2) and extreme weather conditions (8,38), even for early dams (1920–1930) of dumped rockfill, and those constructed before the use of pozzolans and air entrainment. Compacted-rockfill dams, with continuous placement of vertical slabs, have experienced no cracking except for horizontal temperature hairline cracks which are of no consequence. The several cases of cracks near the perimeter joint involved departure from general practice. Current design practice provides a permanent and watertight face.

The thickness of slab on the early, dumped-rockfill dams was traditionally 0.3 m + $0.0067H$. The face was underlain by a layer of large, derrick-placed rocks. For the CFRD with compacted rockfill and a compacted upstream face, the thickness increment was decreased to $0.003H$ (5,13,16,19,20,23,33), and even to $0.002H$ or less (9,21,24). These slabs have given satisfactory performance, and there is a current general trend toward thinner slabs.

In Tasmania, where the Hydro-Electric Commission has built nine CFRDs since 1966, and has five more scheduled for completion by 1992, the designers favor the use of a thinner slab of constant thickness (24). For four dams, 40–75 m high, they have used a uniform design thickness of 0.25 m, with thickening at the perimeter joint. For the 94-m-high Murchison Dam, completed in 1982, the slab has a constant design thickness of 0.30

m. For the 122-m-high Lower Pieman Dam, completed in 1986, the design thickness is 0.30 m + 0.001*H*.

Another recent example is the long, low dam for the major 2,600-MW Macagua hydro plant on the Caroni River in Venezuela, where the face slab will have a constant design thickness of 0.25 m (37).

The trend toward thinner slabs is reasonable, and probably will continue. Experience shows that the older slabs have been remarkably satisfactory. With modern concrete technology, the quality and durability of concrete is uniformly higher and more reliable than in the past. There is a certain thickness, such as 0.25 or 0.30 m, which would be generally considered as a minimum needed for covering the steel, for good construction, for making minimal surface shrinkage cracking, etc.

Based on the presently available experience and current practice, it is reasonable to use slabs with constant thickness of 0.25–0.30 m for dams of low-to-moderate height (say 75–100 m), and to use an incremental thickness of about 0.002*H* for the very important and high dams.

Reinforcing. The use of 0.4% reinforcing, in each direction, for compacted rockfill, has been an economical and successful change from the traditional 0.5% used with dumped-rockfill dams (38). At Salvajina, 0.4% was used near the perimeter and in predicted tension areas, and 0.35% was used in compression areas (16). For several recent dams, 0.3% steel is planned in the large central area of known compression, with 0.4% being retained near the perimeter (34,35,36). The steel area should be calculated on the basis of the design concrete thickness.

A trend which appears to be desirable and economical is to carry horizontal reinforcing through the vertical joints, reverting to a practice used successfully on some older CFRDs (38). The trend is based on the fact that the major and lower area of the face is under horizontal compression. Several vertical joints near abutments are contraction joints to minimize perimeter joint opening. Where reinforcing passes through vertical joints, a bottom waterstop has sometimes also been used (19,20) as a carryover from the earlier practice. With the steel passing through, there is little or no more tendency for a crack to open at the construction joint than at other locations in the slab. Bonded joints are assumed in reinforced concrete design. Waterstops are not used in CFRD face-slab horizontal construction joints.

A cost advantage of running the steel through the vertical joints is the elimination of the waterstops and antispalling steel. There is less risk of local poor concrete due to waterstops. A somewhat different form is needed when the steel passes through the joint (19: figure 8) but there is no significant added cost of form or steel placement. When the lower copper waterstop is not used, a mortar pad is not needed.

For the great majority of CFRDs constructed of ordinary rockfill on a normal rock foundation and up to about 120 m high, the face slab under initial reservoir filling is in compression, except for small tensile strain from the perimeter joint at the base of the dam and the upper abutments. Also, small vertical tensile stresses develop in the area near the crest. For very high dams (11) and dams with more than usual expected foundation settlement, significant horizontal tension strains develop in the rockfill near the upper abutment (24), and several vertical joints have been left with no horizontal steel passing through, and have been provided with double waterstops (19).

Reinforcing steel is always placed in one mat in the center of the slab, or a little above the centerline. The purpose is to make a slab of given thickness as flexible as possible, allowing it to follow small differential settlements without developing high bending stresses, and to provide equal bending resistance in both directions. Even for slabs thicker than 0.6 m, the steel in the center functions satisfactorily to control shrinkage cracks, but may be several centimeters above. Temperature shrinkage cracks have been horizontal, and with surface widths of the order of 0.1–0.2 mm, which are self-healing. They can be sprinkled with cement just underwater as the reservoir fills.

On almost all the recent CFRDs, a double row of small bars (antispalling reinforcement) has been used at the perimeter joint (7,11,13,14,16). All experience shows that there is no possibility of high contact stresses and spalling at perimeter joints in dams of low to moderate height, since there is little compression in the slab before the reservoir is filled and the joint opens under the reservoir load (11,24). While there has been no trouble with joint spalling in CFRDs of compacted rockfill, it is probably desirable to continue with this edge reinforcement at the perimeter joint for high dams. However, it would be adequate to use alternate bars only (16,19). No anti-spalling steel is needed in the toe slab below the copper waterstop.

Usually ordinary reinforcing steel is used, but on a few recent jobs, including the Areia Dam, high-yield steel has been used, without changing the amount of steel (24). The decision to make the change from the traditional 0.5–0.4% was made before the selection of the lower-cost high-yield reinforcing bars. High-yield reinforcing is desirable if available at equal or lower cost.

Perimeter Joint. The perimeter joint always opens and offsets moderately when the reservoir is filled, and is a potential source of leakage if not well designed, inspected, and constructed. For dams of low to moderate height (less than about 75 m), the joint movement has commonly been only a few millimeters (24), and joints with current waterstop details have usually remained watertight. For some of the higher dams, the joint openings and displacements have been several centimeters. At the 160-m-high Areia Dam, the opening in one area was 2.5 cm and the offset 5 cm: No joint leakage occurred, but it is probable that the central bulb waterstop was ruptured.

Because of the leakage history at perimeter joints, the trend has been to install from one (3,5,6) to two and then three separate waterstops (8,12, 15,19,24,36). There are many views on the desirable details for the perimeter joint. It is probably not necessary to include the central bulb waterstop, even for high dams in steep canyons. Its presence makes it difficult to place good concrete reliably and continuously at the joint, and the upper and lower waterstops have given satisfactory service. The writers believe that the detail with two waterstops is adequate; that is, the surface mastic securely covered with a PVC or Hypalon sheet, and the copper waterstop at the bottom of the slab, underlain by the asphalt-impregnated-sand bedding or concrete mortar.

The upper mastic waterstop, covered with a membrane, was first devised for the design of the 160-m-high Yacambu Dam, in Venezuela, but

FIG. 1. Perimeter Top Seal, Fortuna Dam

first constructed at the Areia Dam (11). In principle, the upper mastic should reliably seal the joint and prevent leaks, even with very large joint displacements. For this design to function as planned, it is necessary that the covering membrane be securely bolted down to make a watertight seal against the concrete, so that reservoir water pressure is mobilized on top of the membrane and the pressure in the mastic is equal to the reservoir pressure (Fig. 1). This will force the mastic into a joint opening. To accommodate movement, the membrane should be convex upward (11: figure 8), i.e., not flat.

In order to assure that the upper membrane is sealed against the concrete, it is desirable to use segments of a stiff steel beam between the bolts, such as the steel angle used at Areia (11), and not a flat bar. However, a more frequently pinned flat bar is an alternative, if the concrete is irregular. The larger the opening foreseen in the perimeter joint, the more conservative should be the details of the mastic waterstop; i.e., the larger the opening expected, the larger the volume of mastic needed to fill the space created, and the tougher the upper plastic sheet desired. Hypalon rubber is longer-lasting and tougher than other available membranes, and should be considered the preferable material (19) for the higher dams. To avoid or minimize damage during construction, the upper seal should be placed as late as practicable in the schedule.

The mastic waterstop and the covering membrane are required to function only for the first several years of reservoir operation, during which time most of the face-slab movements take place. It will, of course, last longer. After this initial period, the mastic will be jammed into the joint opening, stopped either against the copper waterstop or against the lower asphalt-impregnated-sand layer (if the copper waterstop ruptures). For the rest of the dam life, the situation should be in equilibrium.

A wood plank about 12.5–20 mm thick, or a similar thickness of some other compressible joint filler, is usually placed at the interface between the toe slab and face slab (7,11,16,20,24,34,35,37). The purpose is to prevent edge concentrations of compressive stress in the joint during construction and before reservoir filling, during which time the face slab is pushed against the toe slab by embankment settlement. After reservoir filling, the joint opens.

Parapet Wall and Camber. The early dams had a parapet wall of 1.2 m. A higher 3–5-m wall has become an economical and desirable practice (7,11,14,19,23,24,31,33–36,38). The economy is in saving a slice of upstream rockfill when the saving exceeds the cost of the wall. The rockfill volume saved in this slice increases with the height of dam. The wide surface of rockfill at the base elevation of the wall provides desirable surface width for the slip form operation and parapet placement (24). The higher parapet automatically is best for wave runup, although wave splash does not adversely affect the rockfill. The stiff vertical parapet wall requires soft joints to allow for settlement and temperature variations.

The freeboard for the CFRD is calculated from the top of the parapet wall, rather than from the top of the dam, with the walls being extended into the abutments.

The small postconstruction and reservoir-filling settlements require only nominal camber (7–11,14,17,19). At the 160-m-high Areia Dam, the crest settled 20 cm in the first 2 yrs, and essentially zero in the next three years (11). Though camber may be provided in the embankment by variable oversteepening of the upper slopes of rockfill, it is more conveniently obtained by variable height of the parapet wall.

Construction. Beginning with the Piedras Dam in Spain in 1967, all face slabs have been placed in vertical strips by slip form continuously from bottom to top, using simple horizontal construction joints, with only a few exceptions (3,6,8). The detail for horizontal construction joints with the steel running through and no waterstop, has been used on almost all CFRDs with complete success (7,16,19,20,39). Construction joints are used as required by stage construction or by the contractor. Any honeycomb at the joint is chipped out, and the rough concrete surface below the steel is carefully cleaned before continuing concrete placement above.

The face slab has usually been placed in 12–18-m-wide strips, 15 m being most commonly used. The dimension should be left to the contractor (20). For long dams of uniform height, the slab should be equally satisfactorily placed in horizontal strips, at the contractor's option. This latter possibility is currently being foreseen in the specification for the long Macagua Dam (37), and the Jurua (40 m high) and Babaquara dams (80 m high) in the Amazon region of Brazil, each of 3 km length and of uniform height.

There has been some thinking that the rockfill dam should be completed to full height before starting the placement of the concrete face. The construction and performance experience at the Areia, Salvajina, and Khao Laem Dams (11,16,19) conclusively shows that the face slabs can be placed in any sequence convenient to the contractor to obtain maximum schedule and cost benefits.

Khao Laem (Fig. 2) is the prime example of the use of temporary rockfill slopes and staged face construction. There were two transverse temporary slopes and seven independent stages of concrete-face placement. It was

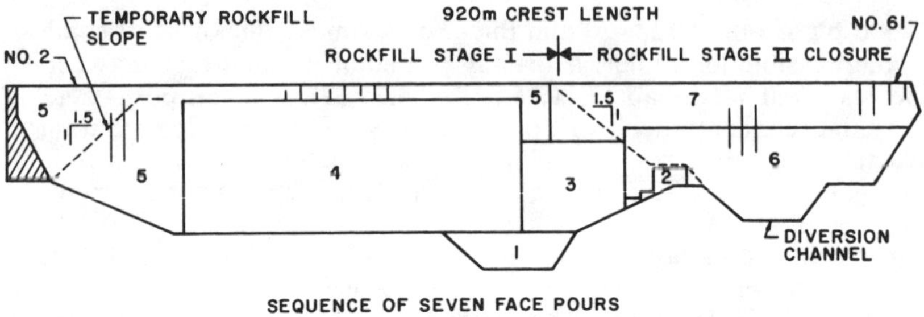

FIG. 2. Khao Laem: Longitudinal Rockfill Construction Slopes and Concrete Face Pours

important that the reinforcing pass through the vertical joints to allow rockfill movements parallel to the face to occur without opening the vertical joints in the previously placed stage.

The trapezoidal starter slabs, to provide a starting line for the slip form, can be placed at any time. They may be screeded or formed (12,17), screeding being most economical. When screeded, it is convenient to screed in one-half the standard slab width. The reinforcing passes through the vertical joint between the two halves and also continues vertically into the slab above to be slip-formed. The contractor should be free to select the height of the starter slab to accommodate his equipment and operation.

The rockfill upstream slope is built to the design line. By the time the face is compacted and the embankment is completed, the upstream slope has moved. The concrete slab with required minimum design thickness is placed on the then existing upstream slope. The small difference in the face slab position compared to the design line has no influence on the construction, performance, or appearance of the slab. Measurements show that with careful construction the excess concrete is generally in the range of 5–10 cm more than the design thickness (11,12,16,19,24). It is considered good contract practice to acknowledge this by making the pay quantity for concrete about 7.5 cm more than the design thickness of slab.

At the Areia Dam—the highest CFRD to date (160 m), and with the highest settlement—the concrete slab was placed on the lower 80 m of the dam height before the rest of the embankment was completed. The top of the first-stage face slab in the center of the valley moved downstream normal to the slope about 0.6 m while the rest of the embankment was being completed (11,38), causing no problem.

ROCKFILL EMBANKMENT

Foundation Excavation and Treatment Criteria. For the CFRD, all the rockfill is downstream from the water thrust. The base width is more than 2.6 times the height and essentially all of the water load is taken into the rock foundation upstream from the dam axis. Measurements on many dams have confirmed that the reservoir filling causes very little movement in the downstream shell (11,14). The overall sliding factor (weight divided by horizontal reservoir thrust) is about 7.5. The criterion for foundation excavation and treatment under the downstream half of the dam is therefore lower than for the ECRD.

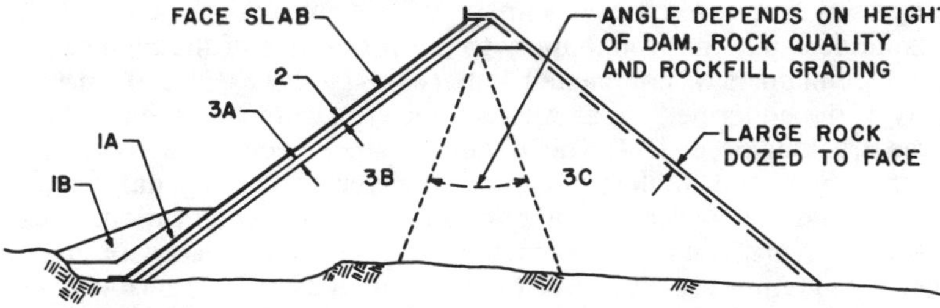

IA IMPERVIOUS SOIL

IB RANDOM

2 PROCESSED SMALL ROCK TRANSITION

3A SELECTED SMALL ROCK PLACED IN SAME LAYER THICKNESS AS ZONE 2

3B QUARRY RUN ROCKFILL, ABOUT 1 m LAYERS

3C QUARRY RUN ROCKFILL, ABOUT 1.5 TO 2.0 m LAYERS

FIG. 3. Zone Designations for CFRD of Sound Rockfill

The foundation excavation criterion employed at Khao Laem is an example of good current practice for a CFRD on a deeply weathered and highly compressible foundation (19). At the dam axis, the foundation settled 1.25 m during construction, and only 0.025 m during and since reservoir filling.

Typical current accepted good practice requires trimming overhangs and vertical faces higher than about 2 m for a horizontal distance downstream of the toe slab equal to about 30% of the dam height, or about 10 m minimum. In the rest of the foundation area, cliffs and overhangs are left in place. In this main foundation area, abutment voids or zones of lower density under overhangs are arched over by the highly frictional rockfill embankment, and have no influence on the face-slab movements or CFRD performance.

Over most of the foundation area, the excavation is made only with earth-moving equipmnent, removing soil-like surface deposits and exposing the points of in situ hard rock. Dozer removal is adequate; ripping seldom being required. Under an upstream portion of high-dam foundations, most of the soil and soft weathered rock between hard rock points is excavated, with a backhoe or similar equipment, but final cleanup by hand labor is not required. Under the downstream portion of the foundation, and most of the upstream portion for dams of low to moderate height, soil and surface material between the hard rock points is left in place. Alluvial gravel deposits, not sand, in the riverbed are commonly left in place except for a short distance downstream from the toe slab. Such gravel deposits usually have a high modulus of compressibility, far exceeding that of well-compacted rockfill. Settlement is not a problem, but alluvium that is judged to be possibly subject to liquefaction is removed.

Zoning Designations. It is useful to use standard zoning designations and to adopt those common for the ECRD materials (Fig. 3); i.e., Zone 1 for impervious, Zone 2 for the filter or transition zone directly under the concrete slab, and Zone 3 for the main rockfill (31). Fig. 3 does not show the zoning used in special cases of impermeable or weak rock (4,7,13,16), but the zone designations would apply to the properties of the material.

Zone 1. A blanket of compacted impervious soil was placed on the lower part of the concrete face at Alto Anchicaya Dam (5), since the dam height was breaking precedent. This detail has since been repeated on the Areia, Khao Laem, and Golillas Dams and on several other high dams. The purpose is to cover the perimeter joint and slab in the lower elevations with impervious soil, preferably silt, which would seal any cracks or joint openings. A minimum practical construction thickness of impervious soil can be used directly adjacent to the concrete slab and rock foundation, covered with less costly waste material for stability (Zone 1B, Fig. 3). As suggested by Nelson Pinto, a strip of silty fine sand could be located over the perimeter joint.

Dams without this upstream Zone 1 have been completely successful, indicating that this Zone 1 is not necessarily useful. It is useful only if a problem develops. Hence, there is a question in any given case whether the cost is justified, except perhaps for the highest dams. The writers believe that it is generally desirable to consider placing Zone 1 to a level several meters above the original riverbed. This limits the accumulation of pervious debris on the base of the face slab. With the development of a finer Zone 2 of approximate dirty filter grading, a high Zone 1 becomes less justified, since leakage can be sealed by use of dirty fine sand.

Zone 2. The early and primary purpose of the thin zone of finer rock directly under the slab was to provide uniform and firm support for the concrete slab (38). Crusher-run minus-15–7.5-cm rockfill has been used. Recently there has been a trend toward making Zone 2 grading have a sufficient quantity of sand-sized particles and fines to improve workability, reduce excess concrete and have reliably low permeability (23,26,32,37), and have an approximate filter grading. Zone 2 shear strength is always adequate. Such select grading of Zone 2 provides a semi-impervious barrier, preventing any large leakage, even if a leakage path develops through a crack in the concrete slab or a defective waterstop. The semipervious property is of value near the perimeter joint, and to an elevation where flood retention during diversion could rise, before the placement of the concrete face. The approximate filter grading makes the sealing of any-size opening, by applying dirty fine sand, possible. For the major central area of the face, there have been no crack or joint problems, and the grading may be crusher-run minus-7.5-cm material.

Horizontal shrinkage cracks, usually less than 0.1 mm and a maximum of 0.2 mm wide, seal themselves with time. However, they can be sealed underwater with dirty fine sand, sieved to a maximum size just smaller than the crack opening, or with cement. The fracture is very irregular, the tension failure being essentially in the mortar. At any angle of an aggregate particle not parallel to the crack, the local opening is smaller than the crack opening. The crack is a maze of short cracks of zero to full width, and it functions as a filter. Cracks of 3–7 mm were sealed within the concrete by a small amount of dirty fine sand, minus 2 mm, at Khao Laem. The incident is presented in the closure, published in this issue, to the Khao Laem paper (19).

For the select type of Zone 2, with 40% sand sizes and fines, the material is placed and compacted in layers of 0.4 or 0.5 m thick, with a smooth steel-drum vibratory roller. For this kind of material the compaction obtained is not strongly influenced by the water content. Generally it is delivered to the site in a "damp" condition, commonly with water content (of the minus No. 4 fraction) in range of 4–10%. In this condition, it compacts to a satisfactory high density without the need for quantitative water-content specifications. Since this semipervious type of Zone 2 is especially sensitive to an excess of water (13), the specification needs to indicate that the water content shall not be so high that the compaction equipment does not operate on firm material. A method specification of four coverages of a 10-t smooth drum vibratory roller is used, and density tests are taken for the record.

For face compaction, the roller is first pulled up the slope without vibration for several passes, and then given four upward passes with vibration (11,24,38). A new and promising development is the use of a backhoe-mounted plate vibrator (22). On several new dams, the plate vibrator is being specified for Zone 2 compaction adjacent to the toe slab, and optional for face compaction.

At sites with heavy rainfall, it is also necessary to protect the Zone 2 sloping surface from runoff. At the Alto Anchicaya and Salvajina Dams (5,17), erosion gullies adjacent to the toe slab reached 5–8 m depth in a few hours of heavy rainfall and required important repairs. Similar damage occurred at Shiroro and Cirata. Damaging erosion is caused primarily by concentrations of runoff from the abutments or from the horizontal construction surface of the main embankment. Hence, the most effective erosion protection is obtained by capturing and diverting these flows. Rain falling directly on the upstream face of Zone 2 is a minor problem requiring nominal sealing. At rainy sites, it is desirable to place the erosion protection as soon as possible after embankment placement. Surface protections comprised of about 50–76 mm of shotcrete or a sprayed sand-asphalt skin (11) have been equally satisfactory for general erosion protection (8,9,10,11,14,17,19,20,24), but cannot be relied upon alone to resist large concentrations of runoff from the abutments (17).

Zone 3. The rockfill embankment is in three zones of increasing layer thickness to give a desirable transition of compressibility and permeability from upstream to downstream. Lowest compressibility is desirable in the portion of the upstream shell which transmits water load to the foundation. The increasing permeability from Zone 2 progressively through Zones 3A, 3B, and 3C (Fig. 3) is desirable during construction, in the event of a flood before the concrete face is placed. After the concrete face is placed there is no credible face problem that could cause more leakage than the rockfill could handle without damage.

Zone 3A is a transition between Zone 2 and the main rockfill, placed and compacted at the same time and in the same layer thickness (0.4–0.5 m) as Zone 2. The main purpose is to limit the size of voids and ensure that Zone 2 material could not be washed into large voids in the main rockfall. Similar zones of small rockfill placed in thin layers are commonly used in ECRDs at the filter-rockfill interfaces for the same purpose. Usually Zone 3A rock is obtained by selected loads; however, on some jobs the Zone 3A is obtained by grizzlying.

Because most of the water load passes into the foundation through the upstream shell (11,17,38,40) it is desirable that the compressibility of Zone

3B be made as low as practical to minimize slab settlement. Much experience has shown that embankments placed in layers about 1 m thick and compacted with four passes of a 10-t (static) smooth steel-drum vibratory roller give satisfactory performance.

The downstream Zone 3C takes negligible water load, and its compressibility has little influence on the settlement of the face slab. Consequently, Zone 3C is commonly placed satisfactorily in thicker layers, usually about 1.5–2 m and also compacted with four roller passes (15,18,19,34–36). It is desirable to specify the thicker layers for Zone 3C because it will be highly permeable and there is a substantial cost saving due to less equipment wear. Also, the thicker Zone 3C layers provide a location into which maximum size rocks are placed, thickening the layer if necessary.

Rockfill Grading and Quality. For most hard-rock formations, the quarried rockfill has less than about 30% smaller than 2.5 cm. Because of this, it has been common to specify that rockfill should not contain more than 30% (or 40%) of sizes smaller than 2.5 cm, but quarried rock with up to 50% sizes smaller than 2.5 cm has produced excellent rockfill. At Salvajina 80% was accepted with thinner layers and more compaction. Also, in recent years many hard rocks with excellent properties for rockfill dams have been encountered which shatter into sizes smaller than 10 or 15 cm, with few larger pieces. These small-sized rocks are often similar to sandy gravel river alluvium, and make excellent compacted embankments with low compressibility and high shear strength (7,16,38).

The most important properties of the CFRD embankment are low compressibility and high shear strength. Usually rockfill is highly pervious, but less pervious rockfill can be used in CFRDs by providing special interior drainage zones (4,7,10,13,16). As a general rule any quarried hard rock with an average particle size distribution having 20% or less finer than the No. 4 sieve, and 10% or less finer than the No. 200 sieve, will have the needed rockfill high shear strength and low compressibility. These limits may be a better means of defining rockfill by sizes than the common specification requiring a maximum percentage of particles smaller than 2.5 cm.

When a rockfill contains a fines content exceeding these limits, commonly the final evaluation of suitability can be made on the basis of the trafficability of the rockfill surface when the material is thoroughly wetted. A stable construction surface under travel of heavy trucks demonstrates that the wheel loads are being carried by a rockfill skeleton. An unstable construction surface, with springing, rutting, and difficult truck travel, shows that the volume of soil-like fines is sufficient to make the rockfill relatively impervious. Where the surface is unstable, the fines dominate the behavior and the resulting embankment may not have the properties desired for a pervious rockfill zone.

The rockfill placement and zoning practices described above are equally applicable for essentially all hard quarried rock used for CFRDs, regardless of the hardness or strength of the rock particles. The unconfined compressive strength of intact specimens of rock used for rockfill dams generally ranges from 100–200 kg/cm^2 (very low) to more than 2,500 kg/cm^2 (highest), with the majority in the range of 500–1,500 kg/cm^2. As a general rule, any rock with compressive strength of 300 kg/cm^2 or more is an adequately hard rock.

For the CFRD (and ECRD) there is no technical need for using rocks with high compressive strength, i.e., compacted embankments from rocks with strengths of 300–400 kg/cm^2 are not more compressible in the completed dam than those of much harder rocks, and are just as satisfactory in every way. On the contrary, rockfill from rock of low to moderate compressive strength has several cost advantages, being less costly to drill and blast and being considerably less damaging to rubber-tired equipment. Rock with high to very high compressive strength (1,000 kg/cm^2 and higher) has higher quarry and handling cost, and often breaks with sharp, strong edges and points which damage rubber tires and increase equipment wear.

Many types of rock with high absorption, or low compressive strength when tested dry, will give considerably lower strengths when tested saturated; saturated strengths 20–40% of the dry strength are frequently found (38). When these rocks are compacted with a heavy steel-drum vibratory roller, considerable crushing of the larger rocks occurs. Nevertheless, these materials can be used in CFRD embankments with appropriate zoning and construction methods (4,13).

The acceptability of poor rock in the downstream shell, Zone 3C, is demonstrated by the well-documented construction and performance of Salvajina (17,40). This sandstone-siltstone had 80% passing 2.5 cm, whereas specifications called for a maximum of 30%. This was compensated for by using 1-m layers and six passes of the vibratory roller. Though construction settlement in the downstream shell was high, the movement of the downstream shell under water load was negligible (17,40).

One measure of the quality of the rockfill in the dam is its vertical modulus of compressibility as measured during construction by water-level settlement devices or crossarms. It is of particular interest and useful for the CFRD that the apparent modulus perpendicular to the slab under water loading is two to three times greater than the vertical modulus (11, 24,38).

Adding Water to Rockfill. The addition of water always improves the rockfill properties by reducing compressibility (18,38). However, for hard rock with low water absorption (less than 2% saturated surface dry water content), the improvement is small and probably does not justify the cost, especially for dams of moderate height and for Zone 3C.

The need for adding water is debatable. The writers believe that the following are reasonable general guidelines:

1. For most hard rocks and CFRDs of low to moderate height, the addition of water has negligible influence on the dam behavior.
2. For high dams and for rocks which have significantly lower unconfined compression strength when tested in saturated condition, water should probably be added routinely for the upstream shell (Zone 3B).
3. For rocks with questionably high contents of earth and sand-sized particles, water should nearly always be used. For dirty rock, the water softens the fines so that the larger rocks can be forced into contact with each other by the vibrating roller.

The main object of adding water is to wet the material, to soften the fines, and to reduce the unconfined compressive strength of the coarser rock particles. The purpose is principally to minimize the postconstruction settlement. There is no need or intention to use the added water to wash or sluice the fines into the larger rockfill voids. Therefore, it is not necessary that the water be delivered with high-pressure monitors. It is satisfactory that the water be added by any means which thoroughly wets the material, preferably before being compacted, though the benefit continues as the rockfill is subject to percolating water and humidity during construction.

The quantity of water has been commonly 10–20% of the rockfill embankment volume, occasionally 30%. For almost all rocks, 10% (100 L per cubic meter of embankment) is more than adequate. The practice of adding the water to the rock in trucks just before dumping on the construction surface is economical and assures thorough rock wetting (25).

Downstream Rockfill Embankment Face. Depending on the construction method employed, the downstream face of a given rockfill dam (both CFRDs and ECRDs) can have an aesthetically attractive appearance, or can be relatively unattractive. In recent dams, exemplified by the practice at the Esmeralda Dam (ECRD) (38), and the Areia Dam (see the photograph on the cover of the symposium proceedings volume, and Ref. 11, photo 9), an attractive face is created in a practical and economical manner by dozing selectively handled large rocks to the downstream face, guided by batter boards or laser beams, such that the upper edge of each large rock is between the design line and plus 15 cm (38). The surface voids should not be filled in. Some stockpiling may be required to have large rocks available for the area near the crest, when the quarry production is low.

In addition to providing an aesthetically pleasing surface, this practice provides a stable surface slope, with minimum excess rockfill, and no need for later dressing of the slope. The practice permits slopes locally steeper than 1.3H : 1V for downstream face haul roads (10,38). Slopes of 1H : 1V–1.2H : 1V have been used.

The practice of dressing the face by backhoe is sometimes used. It can give a face of satisfactory appearance. It results in loose rocks rolling down the slope and does not permit steep face-road slopes.

Temporary Construction Slopes and Ramps. Different embankment sections within Zones 3B and 3C can be built to different levels at different times without restriction, using transverse and/or longitudinal internal temporary slopes of 1.3H : 1V between the different sections. Experience shows that building portions of the rockfill embankment in advance of others has negligible influence on the postconstruction movements of the crest or the face slab deflection under water load. The reason is that the compacted rock is relatively incompressible, and the highly granular and frictional nature acts strongly by arching and shear to redistribute stress and thus prevent localized areas of different compressibility. A term sometimes used for this is load-shedding. This property of the CFRD gives important flexibility in construction, permitting the maximum use of rock from required excavations to be placed directly in the dam without stockpiling, with important cost and schedule advantages. Ramping within the rockfill permits a minimum number of approach haul roads, which has cost and environmental benefits (11,12,19). The internal shear resistance on temporary construction slopes is of the same high value as for the rest of the rockfill.

A permanent road on the downstream face can be useful during construction, can help in optimum planning of internal ramps, and is the most reliable and shortest route between top and bottom of dam. This can be important for access between a powerhouse and power intake and spillway gates.

Compaction Control Tests. For the semipervious Zone 2 (26) tests of compacted density can be made as for earth dams. A requirement that the material should be compacted to 98% of the maximum density of the Standard Laboratory Compaction Test, based on the fraction finer than 3/4 in., is a satisfactory and readily achieved result. Experience has shown that a method specification for placing the material is satisfactory, and that density tests have their principal value in detecting any need for modified layer thickness or compactive effort.

For Zone 3 rockfill, no density control is possible or necessary. It has been routine practice to excavate pits to the depth of a layer thickness in the rockfill surface of Zone 3, with a volume of several cubic meters, and to determine the density and particle size distribution. This is only done several times to provide a record of the construction for comparison with other projects, not to control the adequacy of compaction.

Instrumentation

Instrumentation on CFRD dams has been important in gaining knowledge that has led to design improvements in joint and face-slab design, evaluation of rock and rockfill, and zoning in the rockfill. The results have given confidence in proceeding with higher dams. Instrumentation is not a requirement for safety monitoring. However, a minimum amount is used even on low dams, since some measure of performance and condition of any dam is necessary. For CFRD dams under 60 m in height, there is much data that indicates very small movements. Crest settlement movements and a leakage weir are adequate. During construction some water-level settlement devices are useful to determine the approximate vertical compressibility modulus of the rockfill. For CFRD dams in the range of 60–130 m, the scope of the instrumentation program is in the area of designer and owner preference. The height within this range, the magnitude of the project, the quality of rockfill, and other factors are involved. Instrumentation can be increased to include more hydraulic settlement devices, some surface monuments, joint meters, and face-slab stress and strain instrumentation. The owner is interested in the comparison of his dam with others, and with contributing to dam engineering.

Areia Dam would not be a CFRD without the pioneering engineering of the Hydro-Electric Commission of Tasmania (24,25), and the publication of instrumentation results. The Owners of the Areia (160 m) (11,12) and Salvajina (148 m) (16,17) Dams have extended the practice. The design of the CFRD is essentially empirical: It is based on experience and judgment (38). Instrumentation results are a major factor in the words "empirical," "experience," and "judgment." Instrumentation data on existing and new dams are important to continuing progress.

Conclusions

The following conclusions have been reached in this paper:

1. All grouting should be done through the toe slab acting as a grout cap. It should never be specified that the grouting must be done before the adjacent rockfill is placed.

2. There is a desirable trend toward running the horizontal steel through most of the vertical construction joints in the face slab. No waterstops or anti-spalling steel is needed in construction joints where the steel passes through the joint.

3. The selection of the width of the individual face slabs should be left to the contractor.

4. The current trend toward using thinner face slabs is reasonable and desirable.

5. The trend toward using less face-slab reinforcing in the central part of the dam where the slab is in compression is reasonable.

6. Parapet walls of 3–5 m in height are usually economical and desirable.

7. The top of the parapet wall should be considered to be the top of the dam, for freeboard purposes. Camber to compensate for postconstruction crest settlement is most conveniently provided by increasing the parapet-wall height.

8. It is not necessary or desirable to require that the entire embankment be completed before the concrete face slab is placed.

9. A perimeter joint with two waterstops (the copper sheet at the bottom and mastic covered with a rubber or plastic sheet on top) is generally adequate.

10. The membrane on top of the mastic perimeter waterstop should be bolted down to make a watertight seal against the concrete. This requires a stiff steel member between anchors.

11. The upstream embankment slope is moderately overbuilt to compensate for the compaction settlement. The face slab is placed with design thickness on the existing slope at the time of concrete placement and not to the design plane.

12. The pay thickness for the concrete face slab should be about 7.5 cm more than the design thickness.

13. Essentially, the entire reservoir water load is transmitted through the upstream shell of the dam into the upstream half of the foundation.

14. Except for about 15% of the total foundation width located just downstream from the toe slab, high rock cliffs and overhangs are not removed.

15. There is a desirable trend toward making the Zone 2, located directly under the face slab, a well-graded, minus-2-in. semipervious mixture.

16. The downstream portion of the rockfill (Zone 3C) should be placed in thicker layers than the upstream Zone 3B.

17. The maximum-size rock in a layer should be that which is incorporated in a layer providing a relatively smooth surface for compaction.

18. A good general specification for maximum fine content for the typical hard rock used for CFRDs (Zone 3) is that the average material should not have more than 20% smaller than the No. 4 sieve or more than 10% finer than the No. 200 sieve. Alternatively, a maximum of 50% minus-1-in. mesh generally defines a material having rockfill properties, the source of rockfill being known.

19. Visual observation of the stability of a wetted construction layer, under construction equipment operation, is the best method of judging whether the rock contains an excessive quantity of fines.

20. For CFRDs of low to moderate height built of mostly hard rock, the addition of water to rockfill during construction has negligible influence on the dam behavior.

21. If water is used, high-pressure monitors are not needed for washing fines into the rock voids.

22. Specifications for good general practice should require that large, selectively handled rocks be placed to a close tolerance on the downstream face.

23. Sections within the main rockfill embankment can be built to different levels at different times without restriction. Haul-road ramps, closure sections, and a partial cofferdam section have no influence on later smooth face deflection and rockfill settlement.

24. Generally hard rock with an unconfined compressive strength of as low as 300 kg/cm^2 is adequate for CFRD embankments. Rockfills from rocks with high strength have no technical advantage over those of low to moderate strength.

25. With suitable foundation treatment, CFRDs can be constructed at essentially all sites with rock foundations, even those with significant zones of erodible material.

Appendix. References

The following references (1–37) appeared in *Concrete face rockfill dams—Design, construction, and performance*, edited by J. Barry Cooke and James L. Sherard, Proceedings of a symposium held at the ASCE Convention, at Detroit, Mich., on October 21, 1985, ASCE, New York, N.Y., 1985:

1. Arrau, Luis, Ibarra, Ismael, and Noguera, Guillermo, "Performance of Cogoti Dam under seismic loading," pp. 1–14.
2. Leps, Thomas M., Cashatt, C. A., and Janopaul, R. N., "New Exchequer Dam, California," pp. 15–26.
3. Millet, J. C., Louis, F., and Robert, F., "Design, construction & performance of Fades Dam," pp. 27–39.
4. Good, R. J., Bain, D. L. W., and Parsons, A. M., "Weak rock in two rockfill dams," pp. 40–72.
5. Materon, Bayardo, "Alto Anchicaya Dam—Ten years performance," pp. 73–87.
6. Salembier, M., "Membrane sealing of Rouchain Dam," pp. 88–97.
7. Amaya, Fabio, and Marulanda, Alberto, "Golillas Dam—Design, construction and performance," pp. 98–120.
8. Dascal, O., "The Outardes 2 concrete-faced rockfill dam," pp. 121–139.
9. Casinader, R., and Watt, R. E., "Concrete face rockfill dams of the Winneke Project," pp. 140–162.
10. Beene, Ralph R. W., and Pritchett, Edward C., "The R. D. Bailey Dam—A concrete-faced, earth-rockfill," pp. 163–172.
11. Pinto, Nelson L. de S., Marques Filho, Pedro L., and Maurer, Edilberto, "Foz do Areia Dam—Design, construction, and behaviour," pp. 173–191.
12. Materon, Bayardo, "Construction of Foz do Areia Dam," pp. 192–207.
13. Mackenzie, P. R., and McDonald, L. A., "Mangrove Creek Dam: Use of soft rock for rockfill," pp. 208–230.
14. Bodtman, William L., and Wyatt, John D., "Design and performance of Shiroro rockfill dam," pp. 231–251.

15. Perdomo, D. A., Vasilescu, M. S., and Ferro, E. A., "Fortuna concrete face rockfill dam," pp. 252–265.
16. Sierra, Jesus M., Ramírez, Carlos A., and Hacelas, Jorge E., "Design features of Salvajina Dam," pp. 266–285.
17. Hacelas, Jorge E., Ramírez, Carlos A., and Regalado, Guillermo, "Construction and performance of Salvajina Dam," pp. 286–315.
18. Rogers, R. L., "Boondooma Dam," pp. 316–335.
19. Watakeekul, Somkuan, Roberts, Gordon J., and Coles, Andrew J., "Khao Laem—A concrete face rockfill dam on karst," pp. 336–361.
20. Morris, Michael, "Design and construction of Terror Lake Dam," pp. 362–378.
21. Gosschalk, Edward M., and Kulasinghe, A. N. S., "Kotmale Dam and observations on CFRD," pp. 379–395.
22. Phillips, P. R., "Batang Ai—Transition zone," pp. 396–409.
23. Pinkerton, Ivor L., Siswowidjono, Soetomo, and Matsui, Yutaka, "Design of Cirata concrete face rockfill dam," pp. 642–656.
24. Fitzpatrick, Mike D., Cole, Bruce A., Kinstler, Frank L., and Knoop, Bram P., "Design of concrete-faced rockfill dams," pp. 410–434.
25. Varty, Alan, Boyle, Rick J., Pritchard, Edward D., and Gill, Roger E., "Construction of concrete face rockfill dams," pp. 435–458.
26. Sherard, James L., "The upstream zone in concrete-face rockfill dams," pp. 618–641.
27. Seed, H. Bolton, Seed, Raymond B., Lai, S. S., and Khamenehpour, B., "Seismic design of concrete faced rockfill dams," pp. 459–478.
28. Bureau, Gilles, Volpe, Richard L., Roth, Wolfgang H., and Udaka, Takekazu, "Seismic analysis of concrete face rockfill dams," pp. 479–508.
29. Tsai, Chan-Feng, Yen, Bing C., Tucker, Keith D., and Askari, Shahen, "Design and stability evaluation of Balsam Meadow Dam," pp. 509–527.
30. Imaizumi, Hideo, and Sardinha, Álvaro E., "A study of deformations in concrete faced rockfill dams," pp. 528–540.
31. Hollingsworth, Harold, Conner, Timothy R., and Victor E. Anderson, "Design of Deer Creek Dam," pp. 541–558.
32. Vasconcelos, Aurélio A., and Eigenheer, Luiz Paulo, "The Xingó rockfill dam," pp. 559–565.
33. Pineros, Juan E., "Rockfill compression tests and other aspects of the design of Miel I Dam," pp. 566–586.
34. Pinto, Nelson L. de S., Marques Filho, Pedro L., and Maurer, Edilberto, "Segredo Dam—Basic design aspects," pp. 587–593.
35. Corrêa, Regis Danton, and Thielen, Fernando, "Machadinho Dam in Brazil," pp. 594–600.
36. Antunes Sobrinho, José, and da Cunha, Joaquim C., "Itá Dam—Selection of type," pp. 601–607.
37. Prusza, Z., De Fries, K., and Luque, F., "The design of Macagua concrete face rockfill dam," pp. 608–617.

Other References

38. Cooke, J. Barry. (1984). "Progress in rockfill dams," 18th Terzaghi Lecture. *J. Geotech. Engrg.*, 110(10), 1381–1414.
39. Discussion of "Progress in rockfill dams." (1986). Discussions by Casinader, R.; Chadwick, W. L.; Fetzer, C. A.; Fitzpatrick, M. D.; Fucik, E. M.; Hacelas, J. E., and Ramírez, C. A.; Houlsby, A. C.; Marulanda, A., and Ospina, C. S.; Materon, B.; Merritt, A. H.; Murti, N. G. K.; Pinkerton, I. L.; De Porcellinis, P.; Ripley, C. F.; Sherard, J. L.; Strassburger, A. G.; Swiger, W. F.; and Taylor, H. Closure by Cooke, J. Barry. *J. Geotech. Engrg.*, 112(2), 217–253.
40. Hacelas, J. E., and Ramírez, C. A. (1986). "Salvajina: A concrete face dam on a difficult foundation." *Water Power and Dam Constr.*, 38, Jun.
41. *Symposium on rockfill dams*. (1960). *Trans.*, ASCE, 125(II).

Introduction to Paper 21

Critical Filters for Impervious Soils

by J.L. Sherard and L.P. Dunnigan
Journal of the Geotechnical Engineering Division,
Vol 115, No. GT7, July 1989, pp 927-947

Prepared by Lorn P. Dunnigan[11], M. ASCE

Starting in June 1981, a series of concentrated laboratory testing efforts was initiated at the Soil Conservation Service (SCS) Soil Mechanics Laboratory in Lincoln, nebraska. These efforts were devoted to testing filters in a joint research project between Dr. Sherard and the SCS. There was no special funding and no fees were charged by either party. Jim believed that this project was the most interesting and potentially useful thing he could do with the time he had available for development work.

Each of our joint sessions at the SCS Soil Mechanics Laboratory lasted about ten days and we met at six to twelve month intervals. At the end of each period the results were analyzed and Jim prepared a summary memorandum. Between sessions these memoranda were thoroughly discussed to plan the next series of testing.

It was Jim's nature to share information. As the work progressed, he distributed each memorandum to colleagues he corresponded with regarding filter research.

In all, eight memoranda were prepared. Two papers were published in 1984 that covered the first five memoranda, and the following paper covered the last three memoranda. Following the completion of the eight memoranda, Jim prepared a rough draft manuscript of the following paper.

During the next year or so, as time permitted in his busy schedule, Jim would revist the draft manuscript. He would record his thoughts whenever he was working on this project. In December 1986, and early in January 1987, just before he found out he had cancer, Jim was debating whether or not we should write two papers. He planned to make this decision following his operation in January 1987.

Unfortunately he was not able to get back to this project. Following his death, his wife Elizabeth sent me his notes along with other information he had collected during our work together. Using his notes, many of which were dated December 1986 and January 1987, and the rough draft manuscript, I completed the paper for publication. The following paper was awarded the ASCE Thomas A. Middlebrooks Award in 1991.

---

[11]Former Head, Soil Mechanics Laboratory, Soil Conservation Service, U.S. Department of Agriculture, Lincoln, NE.

Critical Filters for Impervious Soils

By J. L. Sherard[1] and L. P. Dunnigan[2], Member, ASCE

Abstract: Final results of an extensive laboratory research program on "critical" downstream filters in embankment dams with impervious sections are presented. The investigation was carried out at the Soil Mechanics Laboratory, Midwest National Technical Center, Soil Conservation Service, United States Department of Agriculture, Lincoln, Nebraska. A new laboratory test, called the "no erosion filter" (NEF) test, was found to have substantial advantages over the slot and slurry tests used in the earlier part of the research program for silts and clays. It gave essentially the same results for silty and clayey sands with low fines content as was obtained with conventional filter tests on uniform sands. A wide range of different fine silts and clays and clayey and silty sands of different geologic origins were tested. The results of the investigation confirm conclusively that sand filters containing appropriate quantities of fine sand will reliably control and seal concentrated leaks through the impervious sections of embankment dams. The investigations show that for most fine silts and clays, a downstream sand filter with D_{15} ≤ 0.5 mm is conservative and that broadly graded soils, such as those from glacial moraines, need a relatively fine filter.

Introduction

This paper describes laboratory research directed at studying downstream filters that have been exposed to concentrated leaks developing in the protected impervious embankment material. We have called these "critical filters." The research was carried out during 1981–85 at the Soil Mechanics Laboratory, Soil Conservation Service, U.S. Dept. of Agriculture, Lincoln, Nebraska. Results of the first two years of this research were published in 1984 (Sherard et al. 1984a, 1984b), and a brief summary of the final results was given in 1985 (Sherard and Dunnigan 1985).

In this research, several different types of laboratory tests were employed using the same (base) soils. It was found that one test procedure is superior for evaluating critical downstream filters.

The research results confirm the conclusion that downstream filters with properties not greatly different from those used in current common practice will reliably prevent erosion of concentrated leaks, even when gradients and flow velocities are much greater than could possibly exist in a dam.

In this paper we are considering only the role of the filter in preventing erosion, i.e., we are not discussing design practice or the design problem of determining the needed permeability of a single-layer filter that also functions as a drain.

No Erosion Filter Test

The earlier paper on filters for clays and silts presented two different laboratory tests, the slot test and the slurry test, which gave the same results

[1]Consulting Engr. (deceased).

[2]Head, Soil Mech. Lab., Nat. Tech. Ctr., Soil Conservation Service, USDA, Lincoln, NE 68508-2919.

Note. Discussion open until December 1, 1989. To extend the closing date one month, a written request must be filed with the ASCE Manager of Journals. The manuscript for this paper was submitted for review and possible publication on June 21, 1988. This paper is part of the ***Journal of Geotechnical Engineering***, Vol. 115, No. 7, July, 1989. ©ASCE, ISSN 0733-9410/89/0007-0927/$1.00 + $.15 per page. Paper No. 23622.

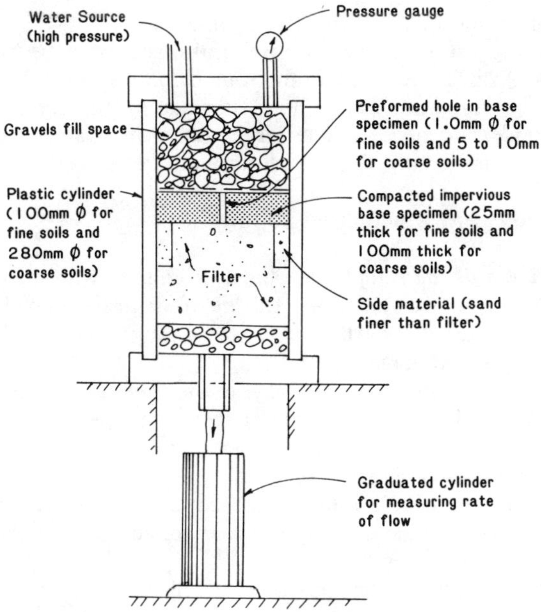

FIG. 1. No Erosion Filter Test Details (No Scale—Schematic)

(Sherard et al. 1984b). For any given silt or clay, these tests could be used to define a filter boundary size, D_{15B}, separating successful and unsuccessful tests.

For both the slot and the slurry tests, there was a small amount of erosion of the base specimen into the successful filter during the test. A filter was judged to be successful when the flow rate rapidly decreased and stabilized with a small constant flow of clear water. Later in the research program, another test was adopted in which it was possible to define a filter boundary size, D_{15b}, at which "no visible erosion" of the walls of the preformed leakage channel took place during the test. Tests with filters slightly coarser than the boundary had visible erosion. This test, the no erosion filter (NEF) test, was also found to work very well for coarse-grained impervious soils, whereas the slot and slurry tests were found to be not satisfactory for impervious soils with d_{85} size much greater than about 0.1 mm.

The details of the laboratory test setup are shown in Fig. 1. The filters were made as described for the slot and slurry test (Sherard et al. 1984b) by blending together measured weights of clean sand of various sizes to precisely control gradation and the D_{15} size of the filter.

The soil was compacted in place on top of the filter (Karpoff 1955; Lund 1949; Sherard et al. 1984b) at a water content near standard Proctor optimum. The water content and density of compacted base specimens were not measured routinely. The research showed that the test results were not influenced significantly by moderate differences in water content or compacted density. The fine-grained soils and the coarser impervious soils with plastic fines were cured at a water content satisfactory for compaction for at least

24 hr before compacting them in the test cylinder.

Base soils with d_{85} size up to two or three millimeters can be tested using the 1.0-in. (25-mm) thick specimen in the four-inch (100-mm) diameter cylinder. The initial 1.0-mm "preformed hole" is made by punching with a stiff steel wire or hypodermic needle. The larger holes in coarser samples are generally made by compacting the sample around a steel rod which is then pulled out.

The test procedure was generally as follows:

1. After preparation of the sample, as shown in Fig. 1, with the cylinder in the vertical position for downward flow, fill the voids in the gravel above the base specimen with water, letting the air escape through the open hole in the top plate where the pressure gage attaches.
2. Attach pressure gage and open the valve to the water source, applying full pressure about 413 kPa (4.2 kg/cm^2) to top of specimen rapidly. For most tests, water from municipal supply was used. For tests on dispersive clays, distilled water pushed by compressed air was used.
3. Observe behavior until equilibrium is reached, usually after five to ten minutes; this includes measuring the quantity of water coming through the filter and observing the color of the water.
4. Shut off water flow, dismantle apparatus, and observe erosion of base specimen.

The 413-kPa (4.2-kg/cm^2) pressure was used for convenience because this was the pressure in the local water supply system. The test results are independent of the water pressure used in the test as long as the pressure is high enough to cause the initial erosion on the bottom of the base specimen.

For each base soil, several of these tests were performed using filters with gradually changing coarseness, ranging from filters which are sufficiently fine to prevent all visible erosion of the base specimen to coarser filters that permit considerable erosion.

Earlier research (Sherard et al. 1984a) has shown conclusively that the D_{15} size of filter is a good quantitative measure of the pore sizes that prevents soil particles from passing through. Thus, the ability of normal (not strongly gap-graded) sand and gravel to act as a filter is well defined by a single parameter, D_{15}.

When a series of tests of this kind is made on a given base soil, a "boundary" filter D_{15b} can be defined. For tests using filters with D_{15} size smaller than D_{15b}, there is no visible increase in the diameter of the initial hole through the base specimen. For tests with filters having D_{15} sizes larger than D_{15b}, the initial performed hole is eroded; the eroded diameter increases as the D_{15} size increases.

A main result of the research was the discovery that there is a unique filter boundary, D_{15b}, for each base soil and that it can be reliably defined by these tests with a precision of the order of 0.1 mm.

In the research program, these NEF tests were made to determine D_{15b} for base soils of many kinds, generally covering the entire range of soil types used for the impervious sections of earth dams. In all cases it was possible to readily determine a filter boundary D_{15b}, but the test behavior varied somewhat for different soil types. From the standpoint of behavior in the NEF test and from test results, the entire range of soils used for impervious sections of embankment dams fell naturally into four broad soil groups:

1. Soil group 1—fine silts and clays with more than 85% passing the No. 200 sieve.
2. Soil group 2—silty and clayey sands and sandy silts and clays with 40–85% passing the No. 200 sieve.
3. Soil group 3—silty and clayey sands and gravelly sands with 15% or less passing the No. 200 sieve.
4. Soil group 4—soils intermediate between groups 2 and 3.

When determining the soil group for gravelly soils, the gravel fraction should be ignored. For example, if a given soil contains 20% gravel, the gradation curve of the finer 80% should be calculated and used for determining the classification and soil group.

A main conclusion of the four-year research effort is that the NEF test is the best test for evaluating the "critical" filter, both for general filter research and for routine laboratory investigations of filters in design practice. The test is suitable for use on all the different types of soil used for the impervious sections of embankment dams. The filter boundary D_{15b} determined in the NEF test for each given impervious soil is a unique value, which can be considered a fundamental property of the soil in the same sense that the Atterberg limits or the effective shear strength envelope parameters are fundamental properties of impervious soil.

The conditions in the NEF test simulate the most severe conditions conceivable that could develop in a dam from a concentrated leak through the core. In any other conceivable type of laboratory test with filter D_{15} smaller than D_{15b}, there can be significant erosion of the base material. The test is relatively simple and economical to make and gives reliably reproducible and easily interpreted results.

Fine Silts and Clays (Soil Group 1)

The NEF test for fine base soils is not greatly different from the previously used slot test (Sherard et al. 1984b). The main differences are that the base specimen is shorter, one inch (25 mm) instead of seven inch (175 mm), the initial hole is smaller, and the filter boundary is defined differently (no erosion). The shorter base specimen and smaller initial leakage hole allow a reliable judgment at the end of the test of whether or not erosion of the hole has taken place.

In previous successful slot tests, it was observed that that seal of the filter at the end of the test did not take place only at the small area at the end of the preformed leakage hole. Commonly, the water flowing in the leakage hole turned 90° at the filter face and eroded out radially from the hole, finally sealing all, or most, of the entire area of the filter upstream face in the testing device.

This same section takes place in the NEF tests. At the first instant after the water pressure is applied in a successful filter test, the velocity of the surge of water traveling through the initial leakage hole (1.0 mm) for the first instant is high and erosive. In this first instant, a small amount of eroded base material is carried to the filter face under the 1.0-mm hole, partially

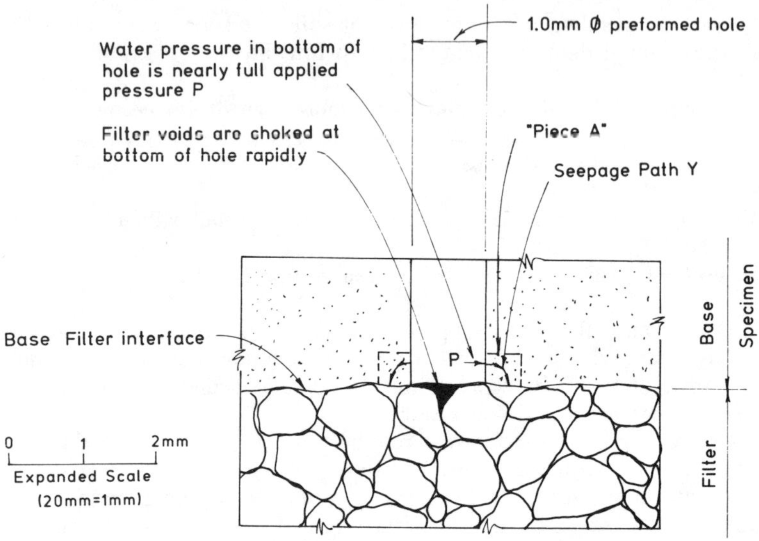

FIG. 2. Condition Existing at Bottom of 1.0-mm Hole at First Instant of Time after Water Value Is Opened

sealing the filter at the bottom of the hole. At this initial instant, the water pressure in the hole at the bottom is essentially the full applied test pressure 413 kPa (4.2 kg/cm^2). This creates a very high gradient on the small piece of base soil directly adjacent (piece *A*) to seepage path *Y*, as shown in sketches with expanded scale in Fig. 2.

Under this high gradient, piece *A* is moved by erosion or is simply pushed into the filter voids directly underlying it. This action creates a small space (formerly occupied by piece *A*) which then fills with water. The newly exposed portion of the filter face under piece *A* becomes partially sealed, in turn, by the material eroded from piece *A*. The water pressure then rises again to essentially full applied pressure, and the high gradient (new path *Y*) is created on "new pieces *A*," which are progressively moved downward into underlying filter voids as shown schematically in the expended scale in Fig. 3.

This action usually continues progressively, until within one to four minutes, a thin slot exists over the entire bottom of the base specimen (Fig. 4). After this time there is a thin, water-filled space (slot) between the bottom of the base specimen and the upstream face of the filter, which is often seen through the wall of the plastic cylinder that is used in the test. In addition, at the end of the test there is a thin zone (skin) over the entire upstream face of the filter in which the voids are partially choked with eroded soil particles from the base specimen (Fig. 5).

This thin skin is relatively impervious compared to the main body of the filter and becomes the "hydraulic control" that limits the quantity of water flowing through the apparatus. At the end of the test, the water pressure in the initial leakage hole and the thin eroded slot at the bottom of the base specimen is approximately equal to the full applied pressure 413 kPa (4.2

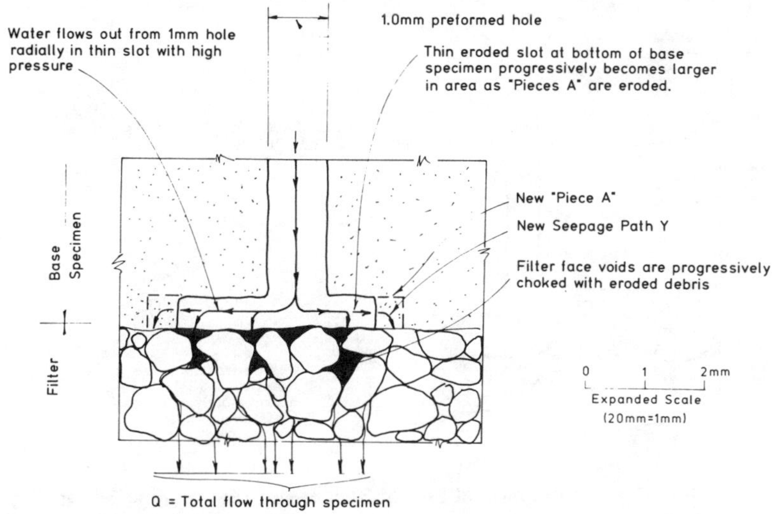

FIG. 3. Action Occurring at Bottom of Base Specimen during Test before Equilibrium Is Reached

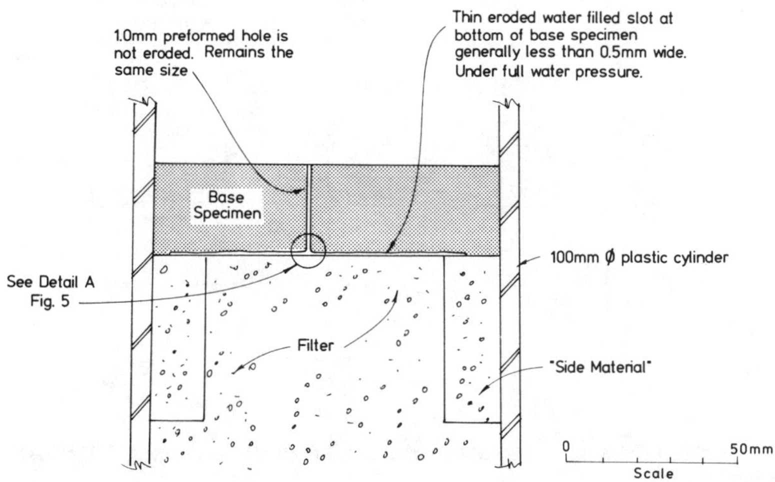

FIG. 4. Condition Existing at End of Successful No Erosion Filter Test on Fine Silt or Clay

kg/cm^2), and the pressure in the filter just below the filter surface is atmospheric pressure (Fig. 5). At the end of the test, therefore, the entire pressure head loss essentially takes place as the water seeps through the skin of choked filter at the upstream face. The water continues through the filter below the upper choked skin at atmospheric pressure by gravity flow.

The thin slot on the bottom of the base specimen develops by progressive

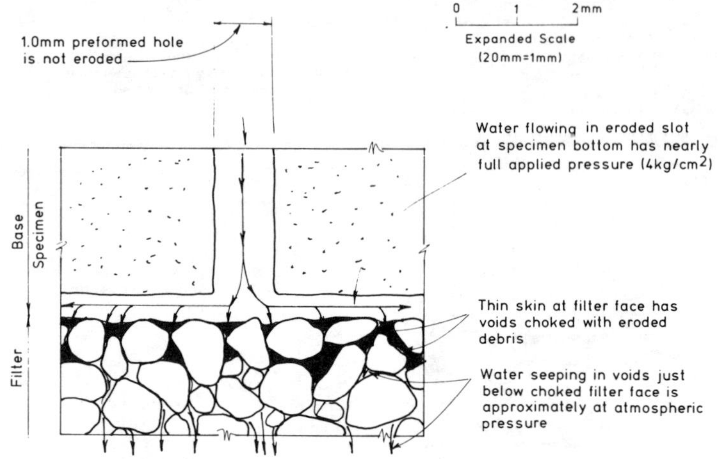

FIG. 5. Detail *A* (Fig. 4) Condition at End of Successful Test on Fine Silt or Clay (Enlarged Scale)

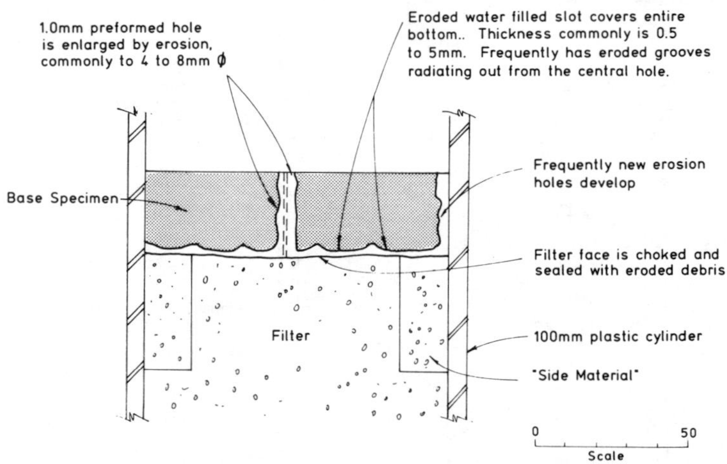

FIG. 6. Condition Existing at End of Unsuccessful (Failed) No Erosion Filter Test on Fine Silt or Clay

movement of pieces *A* into the underlying filter voids (Fig. 4). In tests judged to be successful, equilibrium is reached without visible erosion of the initial performed (1.0-mm diameter) leakage hole. The seal of the filter face is created wholly by the material from the thin slot on the bottom of the base specimen. In tests with the filter D_{15} size larger than D_{15b}, the smaller size soil particles pass through the filter so that the eroded material from the thin slot on the bottom of the base specimen is not adequate in volume to reduce the velocity through the preformed hole to a nonerosive velocity. The walls of the hole erode, and the hole diameter becomes larger (Fig. 6).

For tests with relatively coarse filters (compared to D_{15b}), new holes frequently are eroded through the base specimen by leaks through hydraulic fracturing cracks. This supplies an additional volume of eroded material to seal the filter face. These new holes frequently develop along the wall of the four-inch (100-mm) diameter cylinder, as shown in Fig. 6, but they also develop occasionally in the interior of the specimen.

At the end of the unsuccessful tests, the water pressure in the thin eroded slot along the bottom of the base specimen is practically equal to the applied pressure acting on top of the base specimen. Thus, at the end of the tests, there is essentially no seepage gradient through the base specimen because the water pressure is nearly the same at its top and bottom. All the applied pressure head is dissipated by seepage through the relatively impervious thin skin of choked voids at the filter face (Fig. 5). The thin skin of filter with choked voids has a thickness of about 1.0 mm or less and with a head of 40 m of water (413 kPa), the gradient across the thin skin is about 40/0.001 = 40,000. The research shows conclusively, however, that these choked filter faces are permanently stable, and gradients of this general magnitude are real and common in the tests. The eroded particles from the base specimen that are coarser become wedged securely in the voids at the filter face and cannot be made to move subsequently. The water pressure and gradient that can be put on this thin, choked filter face is only limited by the bursting strength of the laboratory equipment.

For successful tests, with filters having D_{15} size equal to or smaller than D_{15b}, the final rate of flow through the apparatus as limited by the choked filter face was commonly in the range from 0.02–0.3 ml/s/cm^2. Assuming an effective thickness of choked filter of 1.0 mm and head loss across it of 413 kPa (40 m of water head), the calculated coefficients of permeability of the 1.0-mm thick skin of choked filter are about 7×10^{-6} to 5×10^{-7} cm/s.

In successful tests on fine silts and clays, the eroded slot at the bottom of the base sample is usually less than about 0.5 mm wide (Fig. 4). Often this eroded bottom slot covers the entire area. For some base soils, however, the easily identified bottom erosion apparently extends out over only part of the sample area. For a few very fine silt and clay soils, there was very little visible bottom erosion, although there was always a question of whether there may have been an erosion slot that was too thin to be seen.

In unsuccessful tests, the initial 1.0-mm preformed hole is enlarged, and the eroded slot at the bottom of the base specimen is wider than for successful tests. Nevertheless, for filters with D_{15} sizes up to two to five times larger than D_{15b}, the filter face is also finally sealed.

In summary, the behavior of a typical test with filter having D_{15} near D_{15b} is as follows (Figs. 4 and 6):

1. When the initial erosive leak traveling in the preformed hole encounters the filter, it turns 90°. The water then flows radially out from the end of the initial hole in all directions, eroding a thin slot of impervious embankment just opposite the filter face and progressively sealing the filter face to the full diameter of the test apparatus (up to 280 mm in diameter).
2. For tests with filters having D_{15} size smaller than D_{15b}, there is no significant internal erosion of the base specimen. The walls of the initial leakage channel are not eroded during this action.

3. For tests with filters having D_{15} larger than D_{15b}, the filter cannot be sealed adequately with the material in the thin, eroded slot opposite the filter face. An additional volume of eroded material is needed from inside the base specimen (from the walls of the initial leakage channel).

This behavior of filter tests with erosive, concentrated leaks passing through a base specimen of fine-grained soil has not previously been reported in published descriptions of filter research as far as we know. A discovery that at first seemed to be only of academic interest, now appears to have the substantial, practical significance that follows:

1. For tests with filters having D_{15} size equal to or smaller than D_{15b}, the results are independent of the dimensions of the specimens.
2. Conversely, for tests with filters having D_{15} size larger than D_{15b}, the volume of erosion of the initial, preformed leakage hole in the base specimen that is needed to seal the filter and reach equilibrium is dependent on the diameter and thickness of the base specimen tested.

These statements can be illustrated by a comparison of Figs. 4 and 6. In the test of Fig. 4, the same result would be obtained if the sample of compacted base material had been made much thinner than 1.0 in. (25 mm) or if the base specimen had been of much greater diameter; this is true because no internal erosion (of the initial, preformed leakage hole) was needed to seal the filter face. In contrast, if the test shown in Fig. 6 had a larger diameter, say 24 in. (60 cm) instead of 4 in. (10 cm), the area of the filter face to be sealed would be $(24/4)^2 = 36$ times greater, and much more erosion of the preformed hole in the base specimen would be required to seal the filter face.

The area of filter face that would have to be sealed by a concentrated leak through a dam core is very much larger than the small areas of specimens used in the laboratory tests. It, therefore, would be difficult to be completely confident about extrapolating the results of laboratory tests to the prototype if the acceptable filter in the laboratory test required any significant amount of internal erosion of the walls of the initial erosive concentrated leakage channel. Thus, the discovery that no internal erosion of the walls of the concentrated leakage channel through the impervious specimen is required to seal filters with D_{15} smaller than D_{15b} provides support for the conclusion that the laboratory test results can be used to establish reliable design criteria for dams.

Tests were made on 28 different fine silts and clays. Twenty of these were soils which had previously been tested with slot and slurry tests (Sherard et al. 1984b). Usually two to four tests were made to bracket and determine the NEF filter boundary D_{15b} for each soil. In all these tests, the results were consistent. There was no erosion of the initial 1.0-mm diameter hole where the filter had D_{15} size smaller than the boundary D_{15b}. Conversely, there were no tests using filters with D_{15} larger than D_{15b} in which the initial 1.0-mm diameter hole did not erode.

The boundary D_{15b} was usually determined with a precision of about 0.1 mm or less. As a representative example, the following results were obtained from three tests on a typical, very fine clay of medium plasticity (sample *A*): for 0.42 mm, the initial 1.0-mm hole eroded to about 12-mm diameter;

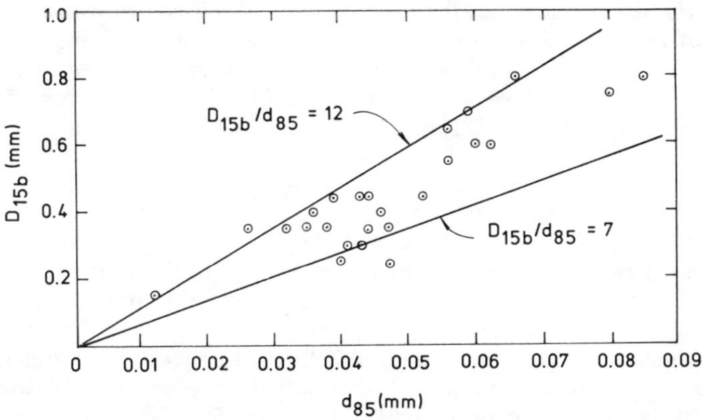

FIG. 7. Relationship between d_{85} and Experimentally Determined Filter Boundary D_{15b} for Fine Silts and Clays

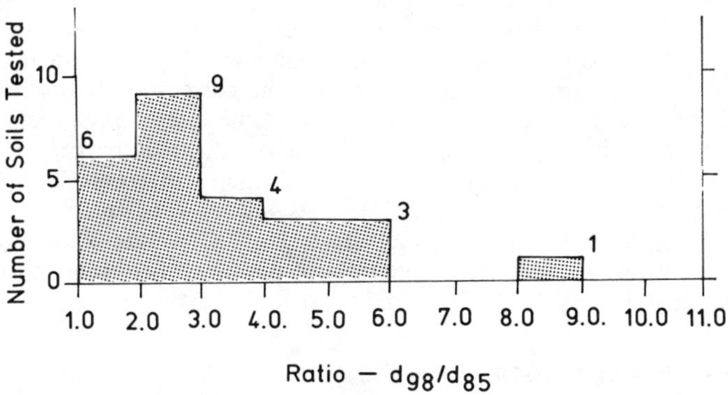

FIG. 8. Frequency Distribution of d_{98}/d_{85} for Fine Clays and Silts Tested

for 0.35 mm, the initial 1.0-mm hole eroded to about 6-mm diameter; and for 0.25 mm, there was no visible erosion of the initial 1.0-mm hole. For this soil, D_{15b} was chosen as 0.3 mm.

Sample *A* had been tested previously using the slurry test, and a filter boundary D_{15B} of 1.1 mm was established. Thus, the filter boundary established by NEF and slurry tests on this clay was $D_{15b}/D_{15B} = 0.3/1.1 = 0.27$, which is about the average relationship found for 20 different clays and silts investigated with both tests.

Fig. 7 shows the relationship between the d_{85} (of the base soil) and the experimentally determined boundary D_{15b} for 26 soils tested with d_{85} of 0.1 mm or less. This shows that there is a fairly strong correlation for these soils between D_{15b} and d_{85}, with 24 of 26 soils having D_{15b}/d_{85} greater than seven (generally ranging from eight to twelve). For the other two soils, both clays of low plasticity (CL-ML), the D_{15b}/d_{85} ratios were 5.7–6.2.

Some special gradation tests were made with fine clays and silts to determine the d_{98} size of each and the ratio d_{98}/d_{85}, which is a measure of the uniformity of the gradation at the coarse end. It was found that the d_{98}/d_{85} ratio ranged from 1.1 to more than 20, but most were in the range from 1.1–5.8 (Fig. 8). Of the five soils that had D_{15b}/d_{85} less than 8 (from 5.7–7.5), four of them had d_{98}/d_{85} less than 2 (very uniform gradation at the coarse end).

From these results, we concluded that generally fine-grained silts and clays had D_{15b}/d_{85} greater than 8 (usually greater than 9). For silts and clays which are very uniform at the coarse end, with d_{98}/d_{85} less than 2, the experimentally determined ratio, D_{15b}/d_{85}, is smaller and often in the range from 6–7.5. This smaller ratio is easily explained for these uniform soils because the d_{98} size is not much larger than the d_{85} size.

As in our previous paper (Sherard et al. 1984b), there was no correlation between the experimentally determined filter boundary D_{15b} and the plasticity as defined by the Atterberg limits or the tendency for dispersive erosion. Although the different soils tested obviously had greatly different resistance to the erosion of concentrated leaks, in the NEF test the initial flow velocity is sufficiently high that all the samples eroded at first in the test. Thus, the filter boundary D_{15b} determined experimentally in these tests is independent of the erosion resistance of the base soil.

From our 1981–82 research on fine silts and clays, we concluded that the slot and slurry tests defined the same filter boundary D_{15B}, which ranged roughly from 0.7–3.0 mm. For most fine clays, the correlation with d_{85} was very poor. Later, we determined that the slot and slurry tests gave the same result only for those base soils with a D_{15B} of about 2.0 mm or smaller.

Because of these problems and the many advantages of the NEF test as presented in this paper, the slot and slurry tests were subsequently abandoned. For 20 fine silts and clays for which both filter boundaries were determined, the ratio D_{15b}/D_{15B} was generally in the range between 0.2–0.5, averaging about 0.3.

Sandy Silts and Clays (Soil Group 2)

NEF tests were performed on a large number of different sandy silts and clays having 40–85% passing the No. 200 sieve. Soils of this type are common in nature from many different geologic sources and are widely used in embankment dams. The behavior of the material in these tests was essentially the same as described for the fine silts and clays of soil group 1.

The filter boundary D_{15b} was always in the range from about 0.6–1.5 mm, with no apparent correlation with the particle size distribution. For the large majority of soils tested, the filter boundary D_{15b} was in the range between 0.7–1.0 mm. For these sandy silts and clays, with less than 40% sand sizes, the behavior in the NEF test apparently is dominated by the fines. The tests behaved as though the sand grains were floating in a matrix of fines and could not be mobilized to assist in sealing the filter. This result was somewhat surprising at first, but the result was confirmed by many tests on different soils.

The NEF test results for this group of soils are different from those obtained with the previously used slurry test (Sherard et al. 1984b). Using the slurry test, the criteria $D_{15}/d_{85} \leq 5$ appeared to be conservative for soils in the range of d_{85} from 0.1–0.5 mm. Extensive research carried out later showed

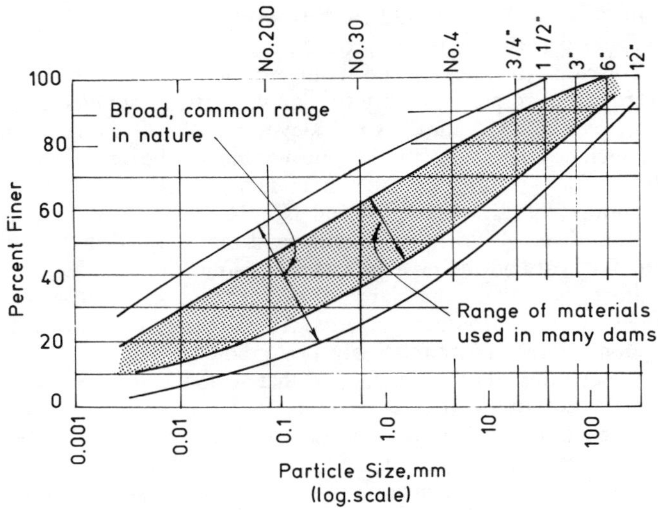

FIG. 9. Coarse, Broadly Graded, Glacial Moraine Soils Given Special Attention in Filter Research

that the difference between the results obtained from the two types of tests increased greatly for sandy soils with increasing d_{85} size and that the slurry test defined a filter boundary size D_{15B} at which large internal erosion would take place in the NEF test.

One of the typical soils in this group is from a broadly graded glacial moraine that contains all particle sizes from cobbles down to clay sizes. Soils like this have been widely used for impervious sections in embankment dams. In this section, we are discussing soils of the type graded generally as shown in Fig. 9. In many places, soils like this are cohesionless, although they contain some clayey fines. For soils from glacial moraines, there has been some question in the industry on the appropriate way to apply current filter criteria. There has been considerable trouble with internal erosion of impervious cores from these soils in many recently constructed dams where coarse filters were used. In this investigation, a great deal of testing was performed on typical broadly graded glacial soils of this type with cohesionless fines, using the NEF test with specimens of various sizes. As a result, it was definitely concluded that it is necessary to use a filter with D_{15} of the order 0.7 mm, or finer, to prevent all internal erosion. Because some of the erosion-damaged dams with soils of this type (Sherard 1979) had sand-gravel filters with D_{15} of two millimeters, it is apparent that the laboratory filter boundary D_{15b} of 0.7 mm is not excessively conservative if compared to prototype experience.

There have been many problems with erosion damage and piping in dams with impervious cores of coarse, broadly graded glacial soils, graded generally as shown in Fig. 9. Since publication of Sherard (1979), there have been a number of other similarly bad experiences, including one of the Soil Conservation Service (SCS) (Talbot and Ralston 1985). This research was directed at these particular glacial soils. A typical glacial till, from an SCS project site in New York, was used for this work.

For this glacial till, different test procedures were used to make tests on specimens of various thicknesses and diameter and on fractions of the material that were screened progressively to remove particles smaller than three inch (75 mm), one inch (25 mm), No. 4 sieve, and No. 10 sieve. In nature this material includes 10–25% cobbles and boulders. The principal conclusion from this effort was that the filter boundary D_{15b} measured in tests on the finer fractions (passing the No. 4 and No. 10 sieves) was the same as for tests on the −3 in. (−75 mm) material and that D_{15b} is of the order of 0.7–1.0 mm. For tests on the coarser material using filters with D_{15} of two or three millimeters and no initial preformed leakage channel, very large erosion tunnels developed on erratic paths. The larger gravels formed roofs over the erosion tunnels under relatively low gradients.

This experience supports the conclusion that for soils in this group containing gravels and cobbles, it is satisfactory to make laboratory filter tests on the fine fraction, such as the fraction finer than the No. 4 sieve. Because of the high fine content, the gravels are wholly surrounded by a matrix of the finer soil and have no influence on the action of the material in the NEF test.

Silty and Clayey Sands with low Fine Contents (Soil Group 3)

Silty and clayey sands with low contents of fines (in the general range from 8–15% finer than the No. 200 sieve) make excellent impervious sections for embankment dams. Soils of this group include decomposed granites and residual soils from other rock types, such as some sandstones and soils of various types of alluvial origin. Some of the highest and largest dams in the world have been constructed with impervious sections of these group 3 soils.

Although the behavior of the group 3 soils in the NEF test differs from the clays and silts of soil groups 1 and 2 described, the test is equally satisfactory for defining the filter boundary.

In tests with filters having D_{15} smaller than D_{15b} (successful tests), there is essentially no visibl erosion of the initial preformed hole and no erosion of the specimen bottom. Generally, the initial hole stays open, with the grains arching around the hole. At times, however, the hole collapses. Whether the hole collapses or not, the test result is the same, i.e., if the filter D_{15} is smaller than D_{15b}, there will be no visible erosion or penetration of base material into the filter, and if D_{15} is larger than D_{15b}, the sample will erode in the same way.

Fig. 10 is a sketch at expanded scale showing the common condition that exists in a successful test at the bottom of the preformed hole. Because the initial hole commonly remains open, there is no doubt that nearly the full applied presure 413 kPa (4.2 kg/cm^2) acts in the water in the bottom of the hole and that the seepage path that radiates out from the bottom of the hole through the adjacent base specimen to the nearby filter voids is very short (of the order of a few millimeters, giving a maximum gradient of about 40/0.002 = 20,000).

Although there are obviously soil fines in the base specimen along the seepage paths radiating out of the bottom of the preformed hole (Fig. 10),

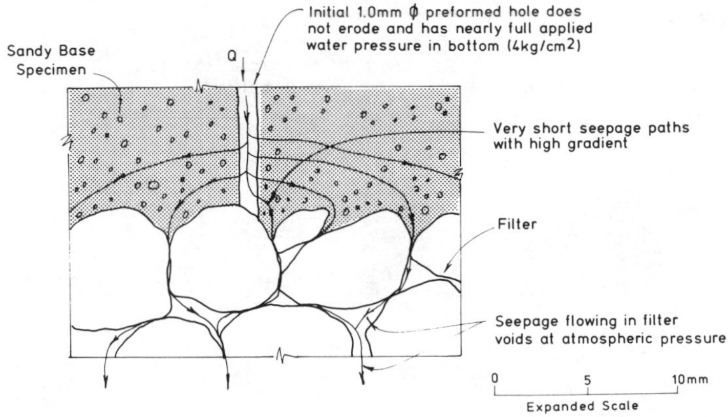

FIG. 10. Condition Existing at End of Successful Test on Sandy Soil of Group 3

these are not carried into the filter voids. For successful tests, the initial water flowing through the specimen is usually slightly colored for a very short time. Then the emerging water becomes completely clear and remains so indefinitely. Only a very small amount of the soil fines located just at the base filter interface are washed through the filter. The rest are trapped inside the pore spaces between the sand grains of the base soil.

In the successful tests on clays and silts of soil groups 1 and 2 described, the water pressure acting at the bottom of the hole created a thin water-filled slot on the bottom of the base specimen by progressive removal of the "pieces *A*" (Figs. 2–5). This progressive action probably is caused by a combination of erosion and shearing displacement. After observing many tests, we were not able to form a strong opinion whether the bottom slot in tests on fine-grained soils was created primarily by erosion or by the progressive action of the water pressure pushing the pieces *A* into the underlying filter voids. As seen in Fig. 2, the 413-kPa (4-kg/cm^2) water pressure acting horizontally on the vertical face of piece *A* could simply overcome the shearing resistance holding it in place, thereby causing the intact piece *A* to be forced downward into the adjacent open filter void. We believe that this lack of shearing resistance may be the dominant mechanism that creates the thin water-filled slot on the bottom of fine-grained specimens.

In the case of the sandy base soil with 15% fines or less (finer than the No. 200 sieve), the volume of the silty or clayey fines is small enough to be confined inside the voids of the coarser sandy material, i.e., the sand skeleton is in grain-to-grain contact carrying the compressive stresses in the embankment and capable of developing high frictional resistance against movement. Thus, when the high water pressure is exerted on the base specimen at the bottom of the preformed hole (such as on the vertical face of piece *A*, Fig. 2), the water pressure also creates high compressive stresses locally in the sand skeleton. We believe it is probable that these high compressive stresses generate high frictional resistance which immobilizes the particles of the sand base material at the bottom of the preformed hole in successful tests (Fig. 10). Only some action of this kind can explain the

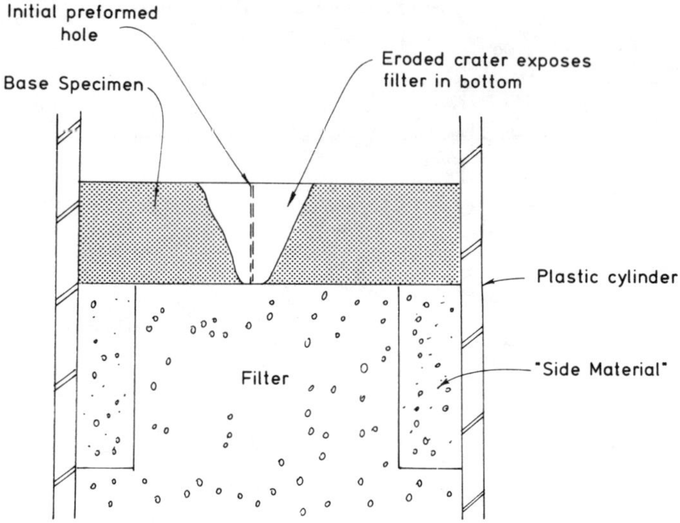

FIG. 11. Condition Existing at End of Unsuccessful (Failed) No Erosion Filter Test on Sandy Soil of Group 3

radically different results in successful tests on the various soil types (Figs. 5 and 10).

As the water pressure applied in the test is increased, there is an increased tendency for the base soil particles at the bottom of the preformed hole to be carried into the immediately adjacent filter voids (Fig. 10) by erosion or shearing action. As the water pressure is increased, however, the effective compressive stresses acting on the base soil particles at the bottom of the hole is also increased (by the seepage drag forces), thus increasing the frictional resistance and preventing particle movement in a compensatory amount. Therefore, the test results (no specimen erosion) are independent of the water pressure applied in the test. This phenomenon is considered one of the more interesting experimental results of the research program.

For unsuccessful tests, in which the filter D_{15} size is larger than D_{15b}, these sandy materials with 15% or less passing the No. 200 sieve also act quite differently from clays and silts. In tests on these sands when the filter has D_{15} only slightly coarser than D_{15b}, a large hole (crater) immediately develops through the base specimen (Fig. 11). This crater is usually, but not always, centered on the initial preformed leakage channel. Occasionally, two or more craters develop simultaneously.

Tests commonly define a very narrow boundary between successful and unsuccessful tests for soils of this group. Fig. 12 shows gradation curves for four materials tested with the NEF test. These four samples are all the same decomposed granite, screened progressively to remove coarser particles. These decomposed granites have angular particles and consequently higher shear strength than similarly graded alluvial sands with rounded grains.

Fig. 13 shows the results of a series of tests on the soils of Fig. 12, showing that there is a narrow boundary between successful and unsuccessful tests and that the experimentally determined D_{15b} is about $9d_{85}$–$10d_{85}$.

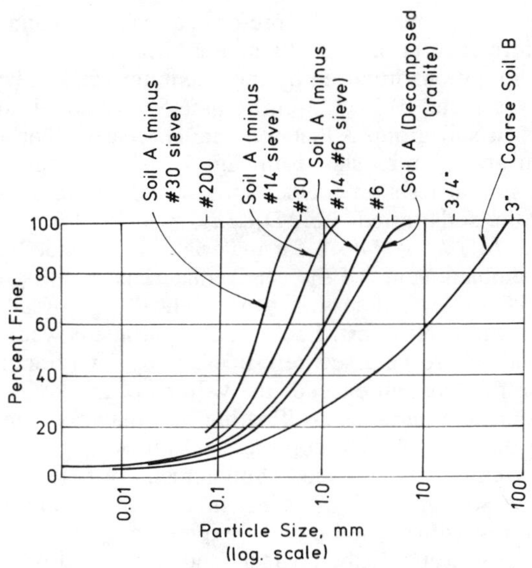

FIG. 12. Group 3 Sandy Soils Tested in No Erosion Filter Test

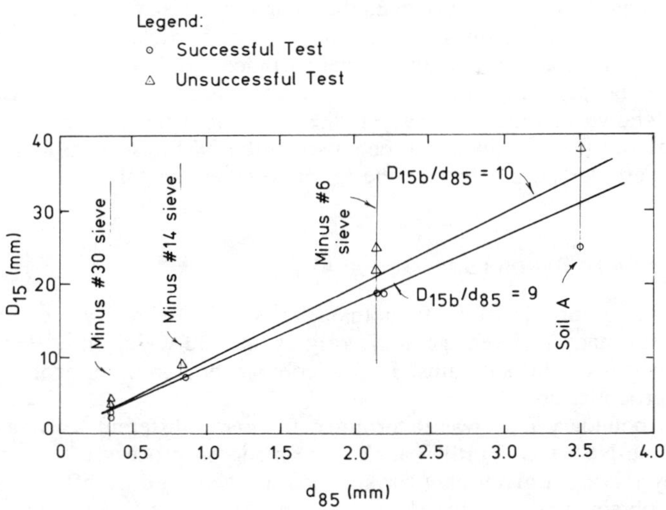

FIG. 13. Results of No Erosion Filter Test on Family of Sandy Soils of Group 3 (Decomposed Granite, Soil *A*, Fig. 12)

Similar tests on other soils in this group that had rounded and subrounded alluvial sand particles gave somewhat lower boundaries, generally in the range of D_{15b}/d_{85} between seven and eight. Thus, the filter boundary D_{15b} is slightly lower for sands with rounded grains. These results additionally support the conclusion that the frictional resistance developed by the grains of the base specimen at the bottom of the preformed leakage channel (Fig. 10) has a significant influence on the results of the NEF test.

Because of the limitations imposed by the maximum-sized laboratory cylinder used in our research, 10-in. (25-cm) diameter, we could not carry out tests on base soils in soil group 3 that were much coarser than the decomposed granite with gradation as shown in Fig. 13 (d_{85} 3.5 mm and D_{15b} 35 mm). We believe similar tests on coarser soils would confirm that the same rules apply as for the soils tested, i.e., D_{15b}/d_{85} = 9 or 10 for material with angular grains and d_{15b}/d_{85} = 7 to 8 for material with rounded grains.

For compacted embankments of the sandy materials of soil group 3 with fine content of 15% or less, unless they are radically gap-graded, friction dominates behavior in the NEF test. Each smaller particle is wedged between larger particles, and all are held securely in place by intergranular friction. It is only when the fine content exceeds the volume of the voids in the sand skeleton, so that the sand particles are floating in a matrix of fines, and the fines are carrying the embankment compressive stresses, that the results of the NEF test are significantly reduced. Supplementary research testing on coarser materials in soil group 3 (with d_{85} in the range from 10–75 mm) would be valuable to confirm the opinions expressed.

For materials of soil group 3, the coarser sand and gravel particles are all in contact and form a granular skeleton. This skeleton of sand and gravel particles carries the internal compressive stresses in the embankment that generate high intergranular friction forces to resist movement of particles by erosion. In materials of soil group 2, the coarse sand and gravel particles are floating in a matrix of fines and cannot generate particle-to-particle friction; therefore, the coarser sand and gravel particles should be ignored when evaluating the needed filter. In contrast, for materials of group 3, the results of the tests shown in Fig. 13 clearly indicate that it is the coarser sizes of the base material that dominate the behavior in the NEF test; consequently, the coarser particles do not have to be ignored when evaluating needed filters.

Clayey and Silty Sands (Soil Group 4)

Soils in this group (15–40% finer than the No. 200 sieve) are common in nature from many different geologic origins and are widely used for impervious sections of earthen dams. In fine content they are intermediate between soil groups 2 and 3.

The filter boundary D_{15b} was determined for many different soils in this group with the NEF test. Further, tests were made on processed soils manufactured by mixing a given clay (or silt) and a given sand in different proportions to obtain materials that had the same basic constituents and fine content varying in the range from 15–40%.

The results showed that the test behavior and the D_{15b} boundary determined experimentally vary greatly with the fine content as follows:

1. Tests on soils with high content of fines (such as 35–40%) have results similar to those of soil group 2. The erosion patterns for successful and unsuccessful tests are generally of the types of Figs. 4 and 6, and the filter boundary D_{15b} is of the order of 0.8–1.5 mm, independent of the d_{85} size.
2. Tests on soils with low fine content (such as 15–20%) have results simliar

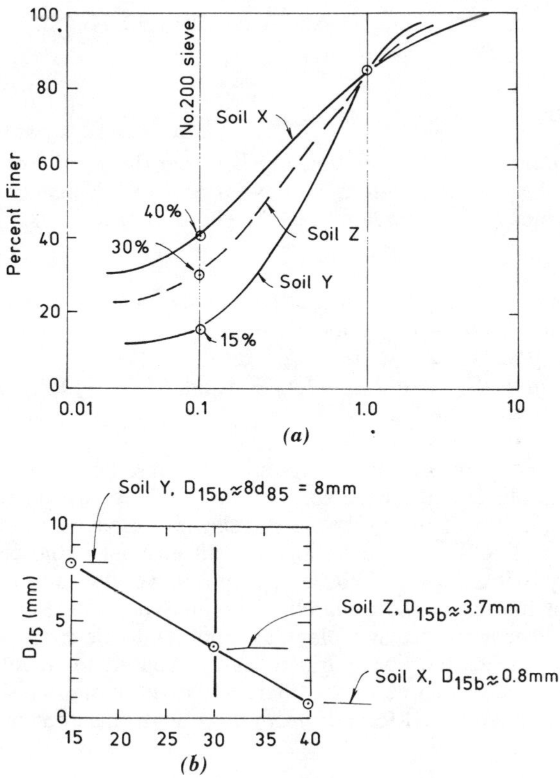

FIG. 14. Example of Relationship of D_{15b} for Soil Group 4 and for Soil Groups 2 and 3: (*a*) Particle Size, mm (Log Scale); (*b*) Percent Finer than No. 200 Sieve

to those in soil group 3. The erosion of unsuccessful tests causes craters (Fig. 11), and there is no erosion of the bottom of the base specimen in successful tests (Fig. 10). The filter boundary D_{15b} is roughly seven to eight times the d_{85}.

3. For soils with intermediate fine content over the range from 15–40%, the test behavior and filter boundary D_{15b} determined vary approximately linearly with the fine content between those of soil groups 2 and 3.

As an example, Fig. 14(*a*) shows grain size distributions for three different silty (or clayey) sands, all having the same d_{85} = 1.0 mm but with different fine contents. Soil *X* (40% fines) and Soil Y (15% fines) are in soil groups 2 and 3, respectively. As previously shown, it would be expected that the filter boundaries D_{15b} for soils *X* and *Y* would be about 0.8 and 8 mm. For the intermediate soil *Z*, the D_{15b} measured would be about 3.7 mm, as shown in Fig. 14(*b*).

The strong influence of the fine content on the test results for soils in group 4 initially surprised us. The results could not have been predicted from the results of previous filter researches. It is probable that the decrease in D_{15b} with increasing fine content is caused by the decrease in the shear strength available to resist the high water pressure exerted in the bottom of the preformed hole, as described earlier.

General Points

Our opinions concerning the relationship between the results of the laboratory tests for various impervious soil types and the design criteria are presented in Sherard and Dunnigam (1985). These criteria are summarized as follows:

1. For soil group 1, $D_{15} \leq 9 \times d_{85}$, but not smaller than 0.2 mm.
2. For soil group 2, $D_{15} = 0.7$ mm.
3. For soil group 3, $D_{15} \leq 4 \times d_{85}$,.
4. For soil group 4, $D_{15} \leq (40 - A/40 - 15)(4 \times d_{85} - 0.7 \text{ mm}) + 0.7$ mm.

The d_{85} values except the one for soil group 3 are based on a mathematical adjustment or regrading of the base soil so that 100% is passing the No. 4 sieve (4.75 mm). For soil group 3, the d_{85} can be based on the total soil with no regrading. The A in the equation for soil group 4 is the percentage of the base soil passing the No. 200 (0.075-mm) sieve after any regrading.

One significant finding of the research, previously not understood, is that the base material shearing strength plays an important role in the behavior of tests in which an erosive concentrated leak is caused to travel through the base specimen and discharge into a filter. Soils with more than 40% silty or clayey fines behave quite differently from those with fine contents of 15% or less.

Summary and Conclusions

1. The NEF test is the best available test for evaluating critical filters located downstream of impervious cores in embankment dams. This is considered the most valuable single conclusion from the four-year long research effort. The conditions in the test duplicate the most severe conditions that can develop inside a dam from a concentrated erosive leak through the core discharging into a filter. For tests with filters finer than the filter boundary (D_{15} smaller than D_{15b}), there is no visible erosion of the walls of the initial preformed leakage hole passing through the base specimen.

2. The NEF test is a simple test that can be made in any soil mechanics laboratory. It gives reliably reproducible and easily interpreted results, and it is well adapted for testing the entire range of impervious soils used for dam cores.

3. The filter boundary D_{15b} separating successful and unsuccessful tests for a given impervious soil, as determined by the NEF test, is unique. The boundary D_{15b} is independent of the dimensions of the laboratory apparatus and is dependent only on the properties of the protected impervious soil (base). The filter boundary D_{15b} can be considered a property of the base soil in the same sense that results of tests to determine the Atterberg limits and effective shear strength parameters are considered properties of the impervious soil.

4. Based on the results of NEF tests, soils used for the impervious sections of embankment dams fall into the four general categories shown in Table 1 depending only on fine content.

TABLE 1. Filter Boundaries (D_{15b}) as Determined by Fine Content for Four Soil Groups

Soil group (1)	Fine content by No. 200 sieve (%) (2)	Filter boundary (D_{15b}) determined by test (3)
1	85–100	D_{15b} = $7d_{85}$–$12d_{85}$ (average D_{15b} = $9d_{85}$)
2	40–80	D_{15b} = 0.7–1.5 mm
3	0–15	D_{15b} = $7d_{85}$–$10d_{85}$
4	15–40	Intermediate between groups 2 and 3, depending on fine content

The test results are not influenced by the relative erosion resistance of the base soils, the plasticity as measured by the Atterberg limits, or the potential for dispersive erosion.

5. For soil group 2, the materials of which contain an appreciable quantity of sand sizes and, frequently, gravels, the sand and gravel fraction has no influence on the test behavior. The volume of the fine matrix is so great for these soils that the sand and gravel particles can be considered to be simply floating in the fines. The test results are the same as if the soil contained no sand or gravel.

6. Because of the many problems in the industry with serious erosion damage to dams with impervious sections of coarse, broadly graded soils (glacial moraines and similar materials) (Fig. 9), many tests were made with this type of soil, which falls generally in soil group 2. The results confirmed that this material needs a relatively fine filter, typical of soil group 2.

7. The behavior of successful NEF tests is fundamentally different for fine impervious soils than for sandy impervious soils with low content of fines. For successful tests performed on the finer soils (groups 1 and 2), the initial concentrated leak passing through the preformed hole in the base specimen turns 90° at the base-filter interface. It then works radially out in all directions and creates a thin, eroded, water-filled slot on the bottom of the base specimen. At the end of the successful test, the entire area of the filter face is choked and partially sealed with the base material debris removed from the thin, eroded slot. For successful tests, on the sandy soils of group 3, there is no eroded slot on the bottom of the base specimen.

8. For the sandy soils of Group 3, the filter boundary apparently varies as a function of the angularity of the grains: for soils with angular grains, D_{15b}/d_{85} = 9–10; for soils with subrounded grains, D_{15b}/d_{85} = 7–8.

Appendix I. References

Karpoff, K. P. (1955). "The use of laboratory tests to develop design criteria for protective filters." *Proc., Am. Soc. Testing Materials,* 55, 1183–1198.

Lund, A. (1949). "An experimental study of graded filters," thesis presented to the University of London, at London, England, in partial fulfillment of the requirements for the degree of Master of Science.

Sherard, J. L. (1973). "Embankment dam cracking." *Embankment dam engineering,* S. Poulos and R. Hirschfeld, eds., John Wiley and Sons, New York, N.Y., 273–353.

Sherard, J. L. (1979) "Sinkholes in dams of coarse broadly graded soils." *Proc., 13th ICOLD Congress,* New Deli, India, 11, 25–35.

Sherard, J. L. (1985). "The upstream zone in concrete-face rockfill dams." *Proc.,*

Symp. on Concrete Face Rockfill Dams—Design, Construction, and Performance, ASCE, New York, N.Y., 618–641.
Sherard, J. L. (1986). "Hydraulic fracturing in embankment dams." *J. Geotech. Engrg.,* ASCE, 112(10), 905–927.
Sherard, J. L., Decker, R. S., and Ryker, N. L. (1972). "Piping in earth dams of dispersive clay." *Proc., Specialty Conf. on Performances of Earth and Earth Supported Structures,* ASCE, New York, N.Y., 589–620.
Sherard, J. L., and Dunnigan, L. P. (1985). "Filters and leakage control in embankment dams." *Seepage and leakage from dams and impoundments,* ASCE, New York, N.Y., 1–29.
Sherard, J. L., Dunnigan, L. P., and Decker, R. S. (1970). "Identification and nature of dispersive soils." *J. Geotech. Engrg. Div.,* ASCE, 96(4), 287–301.
Sherard, J. L., Dunnigan, L. P., and Talbot, J. R. (1984a). "Basic properties of sand and gravel filters." *J. Geotech. Engrg.,* ASCE, 110(6), 684–700.
Sherard, J. L., Dunnigan, L. P., and Talbot, J. R. (1984b). "Filters for silts and clays." *J. Geotech. Engrg.,* ASCE, 110(6), 701–718.
Talbot, J. R., and Ralston, D. C. (1985). "Earth dam seepage control, SCS experience." *Seepage and leakage from dams and impoundments,* ASCE, New York, N.Y., 44–64.
Vaughn, P. R., et al. (1970). "Cracking and erosion of the rolled clay core at the Balderhead Dam." *Proc., 10th ICOLD Congress,* Montreal, Canada, 3, 73–93.

Appendix II. Notation

The following symbols are used in this paper:

D_{15} = particle size in filter for which 15% by weight of particles are smaller;
D_{15B} = D_{15} size of filter found from slot and slurry testing to be boundary between successful and unsuccessful tests for any given base soil;
D_{15b} = D_{15} size of filter found from no erosion filter test to be the boundary between successful and unsuccessful tests for any given base soil, where "successful" is defined as no visible erosion of the walls of the preformed channel; and
d_{85} = particle size in base soil for which 85% by weight of soil particles are smaller (similarly d_{98}).

List of Publications

by James L. Sherard

1. Sherard, J.L., Influence of Soil Properties and Construction Methods on the Performance of Homogeneous Earth Dams, Sc.D. Dissertation, March 1952, Harvard University, Cambridge, Massachusetts (also published as Bureau of Reclamation Technical Memorandum 645, January 1953).

2. Sherard, J.L., "Influence of Soil Properties and Construction Methods on the Performance of Homogeneous Earth Dams," U.S. Department of the Interior, Bureau of Reclamation, Technical Memorandum 645, January 1953, Denver, Colorado.

3. Sherard, J.L., "A Report on the Damage to Hebgen Dam in the West Yellowstone Earthquake," Engineering News Record, Vol 163, 10 September 1959, pp 26-28.

4. Sherard, J.L., "A Steel-Faced Rockfill Dam," Civil Engineering, American Society of Civil Engineers, Vol 29, No. 10, October 1959, pp 698-701 (also published in Transactions, American Society of Civil Engineers, 1960).

5. Sherard, J.L., "Engineered Earth Fills: Foundation Soils Made to Order," Architectural Record, Vol 126, No. 4, October 1959, pp 257-260.

6. Sherard, J.L., Discussion of "Tractive Resistance of Cohesive Soils," by Dunn, I.S., Journal of the Soil Mechanics and Foundations Division, American Society of Civil Engineers, Vol 85, No. SM6, December 1959, pp 157-162.

7. Sherard, J.L., Discussion of "Rockfill Dams: Montgomery Dam," by Scheidenhelm, F.W., Snethlage, J.B., and Vanderlip, A.N., Transactions, American Society of Civil Engineers, Vol 125, Pt. 2, 1960, pp 461-462.

8. Sherard, J.L., "Rockfill Dams: Steel-Faced Dam," Transactions, American Society of Civil Engineers, Vol 125, Pt. 2, 1960, pp 465-472.

9. Sherard, J.L., "A Review of Experience with Asphaltic Concrete Impervious Membranes on the Upstream Slope of Earth and Rockfill Dams," Proceedings of Conference on the Use of Asphalt in Hydraulic Structures, Asphaltic Pacific Coast Division, Asphaltic Institute, May 1961, Bakersfield, California, pp 1-20.

10. Sherard, J.L., Woodward, R.J., Gizienski, S.F., and Clevenger, W.A., Earth and Earth-Rock Dams, 1963, John Wiley and Sons, Inc, New York.

11. Sherard, J.L., A Study of the Influence of the Earthquake Hazard on the Design of Embankment Dams, 1966, Report for the State of California Department of Water Resources, Woodward, Clyde, Sherard and Associates, Oakland, California.

12. Sherard, J.L., "Earthquake Considerations in Earth Dam Design," Journal of the Soil Mechanics and Foundations Division, American Society of Civil Engineers, Vol 93, No. SM4, July 1967, pp 377-401.

13. Sherard, J.L., "Some Considerations Concerning Underseepage Control for Earth Dams," Proceedings, Symposium on Earth and Rockfill Dams, Indian National Society of Soil Mechanics and Foundation Engineering, Vol 1, November 1968, pp 204-214.

14. Waggoner, E.B., Sherard, J.L., and Clevenger, W.A., "Geologic Conditions and Construction Claims on Earth- and Rockfill Dams and Related Structures," Legal Aspects of Geology in Engineering Practice, G.A. Kiersch and A.B. Cleaves, eds., Engineering Geology Case Histories No. 7, Geological Society of American, Boulder, Colorado, 1969, pp 33-43.

15. Sherard, J.L., "Mixing-in-Place Soil and Portland Cement," Journal of the Soil Mechanics and Foundation Engineering Division, American Society of Civil Engineers, Vol 96, No SM6, November 1969, pp 1357-1363 (presented at the ASCE Specialty Conference on Placement and Improvement of Soil to Support Structures, held at MIT, 26-28 August 1968).

16. Sherard, J.L., Discussion on Question No. 36 on Recent Developments in the Design and Construction of Earth and Rockfill Dams, Proceedings of the Tenth International Congress on Large Dams, Montreal, Canada, 1970, Vol 6, pp 377-381.

17. Sherard, J.L., "Study of Piping Failures and Erosion Damage from Rain in Clay Dams in Oklahoma and Mississippi," March 1972, U.S. Department of Agriculture Soil Conservation Service, Washington, D.C.

18. Sherard, J.L., Decker, R.S., and Ryker, N.L., "Piping in Earth Dams of Dispersive Clay," Proceeding, Specialty Conference on Performance of Earth and Earth-Supported Structures, American Society of Civil Engineers, Vol 1, Part 1, 1972, pp 589-626.

19. Sherard, J.L., Decker, R.S., and Ryker, N.L., "Hydraulic Fracturing in Low Dams of Dispersive Clay," Proceedings, Specialty Conference on Performance of Earth and Earth-Supported Structures, American Society of Civil Engineers, Vol 1, Part 1, 1972, pp 653-689.

20. Sherard, J.L., "Embankment Dam Cracking," Embankment-Dam Engineering, Hirschfeld, R.C., and Poulos, S.J. eds., John Wiley & Sons, New York, 1973, pp 271-353.

21. Sherard, J.L., "Potentially Active Faults in Dam Foundations," (Lecture Summary) Proceedings of Engineering Foundation Conference on Inspection, Maintenance, and Rehabilitation of Old Dams, American Society of Civil Engineers, September 1973, pp 768-770.

22. Sherard, J.L., Cluff, L.S., and Allen, C.R., "Potentially Active Faults in Dam Foundations," Geotechnique, Vol 24, No. 3, September 1974, pp 367-428.

23. Sherard, J.L, "Arcillas Dispersivas" (Dispersive Clay), Proceedings of the Fifth PanAmerican Conference on Soil Mechanics and Foundation Engineering, Buenos Aires, Argentina, 1975, Vol 4, pp 181-183.

24. Sherard, J.L., "Dams with Upstream Membranes of Reinforced Concrete and Asphaltic Concrete," Proceedings of the Fifth PanAmerican Conference on Soil Mechanics and Foundation Engineering, Buenos Aires, Argentina, 1975, Vol 4, pp 335-337.

25. Sherard, J.L., "Improved Understanding of Dispersive Clay," Proceedings of the Fifth PanAmerican Conference on Soil Mechanics and Foundation Engineering, Buenos Aires, Argentina, 1975, Vol 5, pp 441-442.

26. Sherard, J.L., Discussion of "Pore and Eroding Fluid Influences on Surface Erosion of Soil," by Arulanandan, K., Loganathan, P., and Krone, R.B., Journal of the Geotechnical Engineering Division, American Society of Civil Engineers, Vol 101, No. GT12, December 1975, pp 1299-1300.

27. Sherard, J.L., Dunnigan, L.P., Decker, R.S., and Steel, E.F., "Pinhole Test for Identifying Dispersive Soils," Journal of the Geotechnical Engineering Division, American Society of Civil Engineers, Vol 102, No GT1, January 1976, pp 69-85.

28. Sherard, J.L., Dunnigan, L.P., and Decker, R.S., "Identification and Nature of Dispersive Soils," Journal of the Geotechnical Engineering Division, American Society of Civil Engineers, Vol 102, No. GT4, April 1976, pp 287-301.

29. Sherard, J.L. and Decker, R.S. eds., Dispersive Clays, Related Piping, and Erosion in Geotechnical Projects, American Society for Testing and Materials, Special Technical Publication 623, 1977, American Society for Testing and Materials, Philadelphia, Pennsylvania.

30. Sherard, J.L. and Decker, R.S., "Some Engineering Problems with Dispersive Clays," Proceedings, Symposium on Dispersive Clays, Related Piping, and Erosion in Geotechnical Projects, American Society for Testing and Materials, Special Technical Publication 623, 1977, American Society for Testing and Materials, Philadelphia, Pennsylvania, pp 3-12.

31. Sherard, J.L. and Decker, R.S., "Summary-Evaluation of Symposium on Dispersive Clays," Proceedings Symposium on Dispersive Clays, Related Piping, and Erosion in Geotechnical Projects, American Society for Testing and Materials, Special Technical Publication 623, 1977, American Society for Testing and Materials, Philadelphia, Pennsylvania, pp 467-479.

32. Sherard, J.L., Discussion of "Load Transfer and Hydraulic Fracturing in Zoned Dams," by Kulhawy, F.H. and Gurtowsky, T.M., Journal of the Geotechnical Engineering Division, American Society of Civil Engineers, Vol 103, No. GT7, July 1977, pp 831-833.

33. Sherard, J.L., "Sinkholes in Dams of Coarse, Broadly Graded Soils," Transactions, Thirteenth International Congress on Large Dams, Vol 2, 1979, pp 25-34.

34. Sherard, J.L., "Piezometers in Earth Dam Impervious Sections," Recent Development in Geotechnical Engineering for Hydro Projects, Kulhawy, F.H., ed., American Society of Civil Engineers, 1981, pp 125-165.

35. Sherard, J.L., Discussion of "Design of Filters for Clay Cores for Dams," by Vaughn, P.R. and Soares, H.F., Journal of the Geotechnical Engineering Division, American Society of Civil Engineers, Vol 109, No. GT9, September 1983, pp 1195-1197.

36. Sherard, J.L, Dunnigan, L.P., and Talbot, J.R., "Basic Properties of Sand and Gravel Filters," Journal of the Geotechnical Engineering Division, American Society of Civil Engineers, Vol 110, No. GT6, June 1984, pp 684-700.

37. Sherard, J.L., Dunnigan, L.P., and Talbot, J.R., "Filters for Silts and Clays," Journal of the Geotechnical Engineering Division, American Society of Civil Engineers, Vol 110, No. GT6, June 1984, pp 701-718.

38. Sherard, J.L., Discussion of "Erosion in Relation to Filter Design Criteria in Earth Dams," by Arulanandan, K. and Perry, E.B., Journal of the Geotechnical Engineering Division, American Society of Civil Engineers, Vol 110, No. 7, July 1984, pp 1000-1002.

39. Sherard, J.L., "Trends and Debatable Aspects in Embankment Dam Engineering," Water Power and Dam Construction, Vol 36, No. 12, December 1984, pp 26-32.

40. Sherard, J.L. and Dunnigan, L.P., "Filters and Leakage Control in Embankment Dams," Proceedings, Symposium on Seepage and Leakage from Dams and Impoundments, Volpe, R.L. and Kelly, W.E., eds., American Society of Civil Engineers, 1985, pp 1-30.

41. Sherard, J.L., "Hydraulic Fracturing in Embankment Dams," Proceedings, Symposium on Seepage and Leakage from Dams and Impoundments, Volpe, R.L. and Kelly, W.E., eds., American Society of Civil Engineers, 1985, pp 115-141 (also published in Journal of the Geotechnical Engineering Division, Vol 112, No. GT10, October 1986, pp 905-927).

42. Cooke, J.B. and Sherard, J.L., eds., Concrete Face Rockfill Dams - Design, Construction, and Performance, American Society of Civil Engineers, 1985.

43. Sherard, J.L., "The Upstream Zone in Concrete-Face Rockfill Dams," Concrete Face Rockfill Dams - Design, Construction, and Performance, Cooke, J.B. and Sherard, J.L., eds., American Society of Civil Engineers, 1985, pp 618-641.

44. Sherard, J.L. and Dunnigan, L.P., Discussion of "Internal Stability of Granular Filters," by Kenney, T.C. and Lau, D., Canadian Geotechnical Journal, Vol 23, No. 3, August 1986, pp 418-420.

45. Sherard, J.L. and Cooke, J.B., "Concrete-Face Rockfill Dam: I. Assessment," Journal of the Geotechnical Engineering Division, American Society of Civil Engineers, Vol 113, No. GT10, October 1987, pp 1096-1112.

46. Cooke, J.B. and Sherard, J.L., "Concrete-Face Rockfill Dam: II. Design," Journal of the Geotechnical Engineering Division, American Society of Civil Engineers, Vol 113, No. GT10, October 1987, pp 1113-1132.

47. Sherard, J.L, "Lessons from the Teton Dam Failure," Proceedings of the International Workshop on Dam Failures, held at Purdue University, August 1985, Engineering Geology, vol 24, Nos. 1-4, December 1987, pp 239-256.

48. Sherard, J.L., "Retrospective Comments, Teton Dam Failure," Proceedings of the International Workshop on Dam Failures, held at Purdue University, August 1985, Engineering Geology, Vol 24, Nos. 1-4, December 1987, p 283-293.

49. Sherard, J.L. and Dunnigan, L.P., "Critical Filters for Impervious Soils," Journal of the Geotechnical Engineering Division, American Society of Civil Engineers, Vol 115, No. GT7, July 1989, pp 927-947.

SUBJECT INDEX

Page number refers to first page of paper.

AUTHOR INDEX

Page number refers to first page of paper.